Television Broadcasting:
Equipment, Systems,
and Operating Fundamentals

Television Broadcasting: Equipment, Systems, and Operating Fundamentals

by

Harold E. Ennes

HOWARD W. SAMS & CO., INC.
THE BOBBS-MERRILL CO., INC.
INDIANAPOLIS · KANSAS CITY · NEW YORK

FIRST EDITION

FIRST PRINTING—1971

International Standard Book Number: 0-672-20786-9
Library of Congress Catalog Card Number: 76-135991

Preface

The purpose of this book is twofold: to serve as a basic and practical course for new and prospective television broadcast technicians and operators, and to serve as a source of reference information for practicing technical personnel. It is assumed that the reader has had basic electronics training, and that he possesses solid-state knowledge at least equivalent to that contained in this writer's *Workshop in Solid State*.[1]

Over the years, the technical aspect of telecasting has grown into a highly advanced and very broad field that involves ultrasophisticated equipment and operating techniques. Yet, organized training programs have been limited largely to a few schools and manufacturers' seminars on specific new equipment. The practical use of broadcast equipment is a highly specialized field. Hence, any treatment of this subject must be applied specifically to broadcast-system installations and techniques.

This book covers the fundamentals of the entire television broadcast system. Consequently, it cannot be expected to meet the complete needs of more advanced personnel in every specific department of telecasting. But the foundation for more specific and advanced study is firmly established. For example, the basics of NTSC color are presented in an exhaustive treatment in Chapter 2. This material is arranged and designed to serve as a practical foundation for the study of all advanced applications of color equipment. Because of the extreme importance of this material, Chapter 2 is followed by a longer and more detailed exercise section than is the case for the average chapter.

Similarly, the coverage of camera chains, sync generators, television recording systems, and transmitters is more introductory than detailed or advanced. Sufficient coverage of cameras, switching systems, and transmitters is given to assure competent *operation* of the equipment. More highly specialized operations and maintenance of cameras and recording systems, interpretation of vertical-interval test (VIT) signals, and transmitter proof of performance must, of necessity, be assigned to more advanced volumes.

---

[1]Harold E. Ennes, *Workshop in Solid State* (Indianapolis: Howard W. Sams & Co., Inc., 1970).

PREFACE

The author is indebted to the following manufacturers and stations for their cooperation in supplying information and photographs vital to this book: American Telephone and Telegraph Company; Ampex Corporation; Cohu Electronics, Inc.; D. B. Milliken; Electro-Voice, Inc.; International Video Corporation; Kliegl Bros. Lighting; Mincom Division 3M Company; Philips Broadcast Equipment Corp.; RCA; RHG Electronics Laboratory; Shibaden Corp. of America; Shure Brothers, Inc.; Tektronix, Inc.; TeleMation, Inc.; Telesync Corp.; Visual Electronics Corporation; Westel Company; WBAL-TV; WBBM-TV; and WTAE-TV.

HAROLD E. ENNES

Contents

CHAPTER 1

CHAPTER 2

CHAPTER 3

CHAPTER 4

CHAPTER 5

CHAPTER 6

CHAPTER 7

CHAPTER 8

CHAPTER 9

CHAPTER 10

CHAPTER 11

CHAPTER 12

CHAPTER 13

CHAPTER 14

APPENDIX A

APPENDIX B

APPENDIX C

Introduction to Television
Broadcast Systems

This chapter is an elementary examination of the function performed by each basic piece of apparatus in the television system. This material should be studied by every reader who has not received basic training in television-broadcast theory. The more advanced reader will find it an excellent review to help clarify the overall picture of television broadcasting.

1-1. INTRODUCTION

We are about to study the major components that act on the video signal at a TV broadcast station. In practice, these various units are so interdependent that it is difficult to explain clearly the exact operation of any one unit without some mention of another unit. With this thought in mind, it is the purpose of this introductory section to give an overall view of the problems encountered, so that the content of the following sections may be understood more easily.

Fig. 1-1 is a simplified block diagram of all studio and transmitter units discussed in this chapter. It would be well for the reader to refer to this diagram often during the rest of the chapter so that the orientation of equipment may be seen clearly.

The pickup head and viewfinder constitute the television camera. The camera lens focuses the scene to be televised upon the photosensitive surface of the pickup tube, and the image imparts upon the surface a charge pattern that corresponds point-by-point with the light content of the picture.

Before we can "pick off" this charge pattern element by element, some means must be found to establish precisely the time at which this action is started. The synchronizing generator performs this function. Every electrical action that takes place in the scanning process is controlled by

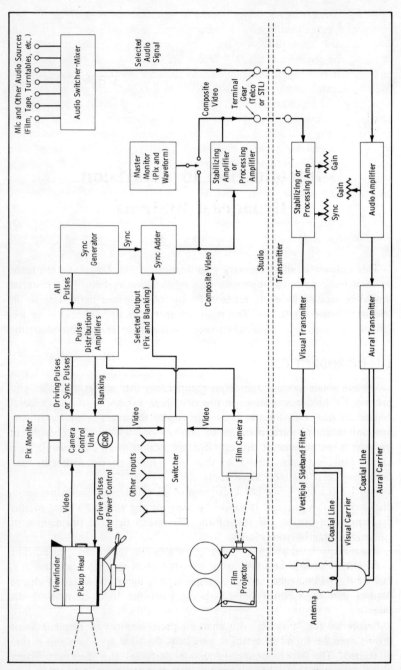

Fig. 1-1. Basic block diagram of TV broadcast installation.

this unit. Thus, when a driving pulse is supplied to the camera for the purpose of exact timing of a certain function, a so-called "sync pulse" is simultaneously transmitted on the video carrier to "trigger" the receiver action at the same time. Fig. 1-2 represents the scanning of one line of a picture that consists of black, white, and shades of gray. (The "aperture effect" indicated in Fig. 1-2 results from the fact that the scanning beam, or aperture, is not infinitely small, and cannot reproduce an abrupt change between gray levels. This effect will be discussed further in Chapter 3.)

Refer to Fig. 1-2 during this description. Within the pickup head are circuits that generate a sweep current to scan the image. This means that a beam of electrons similar to that in a cathode-ray tube is caused to sweep across the image and become modulated by the charges at the various points on the image surface. At the end of the time required for the beam to sweep exactly one line of the whole picture, the driving pulse is received. This pulse triggers the circuit and causes a rapid reversal of the scanning beam so that it is in a position to start another line of image scanning. At this same time, the horizontal (or line) sync pulse is transmitted on the video carrier; this pulse causes the receiver sync circuit to be similarly triggered. Also at approximately this same time, a "blanking" pulse is transmitted. This pulse drives the receiver picture tube into cutoff, extinguishing the beam so that the retrace line will not be visible to the viewer.

This action is repeated line by line until the camera scanning beam has reached the bottom of the image. Now the beam must be returned to the top of the picture to start another complete scanning sequence. At this time, a *vertical* driving pulse is received from the precisely timed sync generator, and the scanning beam is caused to return to the top of the picture. Also, a *vertical* sync pulse is transmitted on the video carrier so that the receiver is triggered in the same way, and a blanking pulse is transmitted simultaneously to cause the retrace line to be invisible.

NOTE: The difference between a "horizontal" pulse and a "vertical" pulse is primarily one of *frequency*. Blanking and sync pulses as delivered by the synchronizing generator are "composite"; that is, both horizontal pulses (at a rate of 15,750 Hz) and vertical pulses (at a nominal rate of 60 Hz) are present. Frequency-selective circuits then separate the horizontal-rate information from the vertical-rate information.

As shown in Fig. 1-1, the pulses from the sync generator are fed to a pulse-distribution panel. Camera-driving pulses are fed from this panel to the camera-control unit, and through this unit (via coaxial cables) to the camera in the studio. The camera-control unit also receives the video output signals from the camera, amplifies and controls them in amplitude and quality, and passes them on to the switching panel. It should be noted that in Fig. 1-1 sync pulses are fed to a sync-adder amplifier that feeds the line to the transmitter. This practice is optional and varies with individual sta-

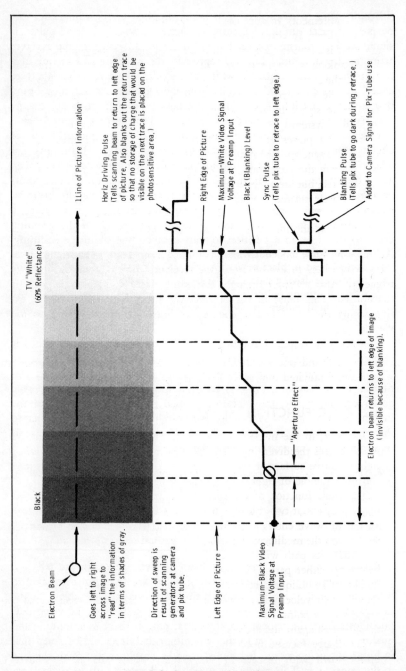

1 Line of Picture Information

TV "White"
(60% Reflectance)

Horiz Driving Pulse
(Tells scanning beam to return to left edge of picture. Also blanks out the return trace so that no storage of charge that would be visible on the return trace is placed on the photosensitive area.)

Right Edge of Picture

Maximum-White Video Signal Voltage at Preamp Input

Black (Blanking) Level

Sync Pulse
(Tells pix tube to retrace to left edge.)

Blanking Pulse
(Tells pix tube to go dark during retrace.)

Added to Camera Signal for Pix-Tube use

Electron beam returns to left edge of image (invisible because of blanking).

"Aperture Effect"

Black

Electron Beam

Goes left to right across image to "read" the information in terms of shades of gray.

Direction of sweep is result of scanning generators at camera and pix tube.

Left Edge of Picture

Maximum-Black Video Signal Voltage at Preamp Input

Fig. 1-2. Scanning of one line.

tion requirements. In some cases, the sync pulses are inserted in the stabilizing amplifier. In other stations, *composite switching* is used; this means the sync is added to each individual camera source before switching. (Video and blanking without sync are termed a *noncomposite* signal. After sync is added, the signal is termed a *composite* signal.) Modern switching systems can handle a composite signal, but older switching systems required noncomposite signal inputs, so the sync pulses were inserted *following* the switching systems.

Blanking pulses usually are inserted at the camera control unit. The composite sync and blanking are therefore transmitted along with the video signal from the same output amplifier. The camera driving pulses are not transmitted, but are supplied only to the camera and film equipment. They are precisely related in time to the sync by the functioning of the sync generator. Thus, the transmitter and receiver are exactly coordinated in operation at any instant.

NOTE: Some modern camera chains feed the composite sync signal to the camera head in place of driving pulses. The sync is then used to generate camera-driving pulses that control the camera sweep the same as individual driving pulses from the sync generator would do. In either case, camera-head pulses are termed "driving pulses" as distinguished from "sync pulses."

As shown in Fig. 1-1, the TV transmitter is actually two transmitters, one visual and one aural. The signals from these transmitters are combined and radiated into space from the antenna.

1-2. BASIC FUNCTION OF THE LENS

There is, in the human eye, a device known as the "crystalline lens," which causes the diverging waves of light to *converge* upon the *retina,* or light-sensitive surface at the back of the eye. Without such a lens, only a confused jumble of light would strike the retina.

The basic function of the lens is illustrated in Fig. 1-3A. Light waves travel only about two-thirds as fast through glass as through air. When an advancing wave enters a medium in which it is retarded in velocity, and if it strikes the medium obliquely (as the two outer rays in Fig. 1-3A strike the lens), its path will be bent because one part of the wave is slowed before the other part. The rays through the center of the lens, since they do not strike obliquely, continue on through the lens in a straight line (although their velocity is reduced in the glass). Thus, the light waves converge upon a point, known as the *focal* point, behind the lens, after which they diverge again and are no longer "in focus."

The image focused upon the light-sensitive surface behind the lens is upside-down with respect to the original. This fact is illustrated in Fig. 1-3B. An arrow is the original object. The rays from the top of the

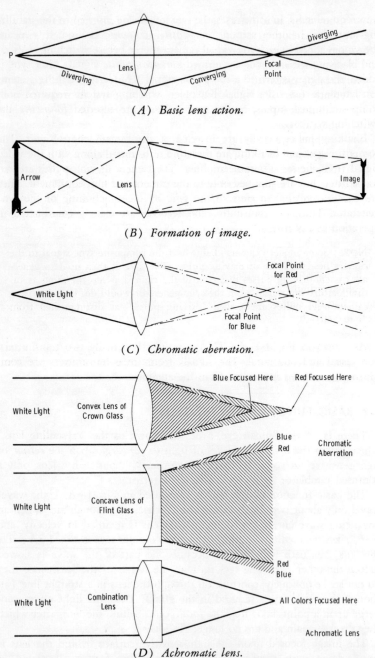

(A) Basic lens action.

(B) Formation of image.

(C) Chromatic aberration.

(D) Achromatic lens.

Fig. 1-3. Development of achromatic lens.

arrow (shown by solid lines) are bent in passing through the lens and converge at the bottom of the image being formed. Conversely, rays of light from the bottom of the object (shown by dash lines) will be converged upon a point above the tip in the image.

Colors, being essentially different wavelengths of the visible spectrum, act differently from one another upon passing through an ordinary lens. Fig. 1-3C illustrates the effect of an ordinary "uncompensated" lens when white light (all colors) is being passed. Two colors, red and blue, are shown. The blue rays are bent a greater amount than the longer-wavelength red rays, and therefore are focused at a point nearer the lens.

In actual practice, TV-camera lenses are not the simple type of biconvex lens shown thus far. As described above, an ordinary lens will not focus the various colors of the spectrum at the same point behind the lens. Therefore, the TV camera lens may be composed of a number of lenses of varying curvatures to correct certain defects. Fig. 1-3D shows the basic idea of a color-corrected lens. The original fault is known as *chromatic aberration*. A lens corrected as shown in Fig. 1-3D is known as an *achromatic* lens, meaning that chromatic aberration has been corrected. Another fault with an ordinary single lens is that the edges of the image are apt to be less sharp than the center. Such a defect is known as *astigmatic aberration*. A lens corrected for this fault is an *anastigmatic* lens, sometimes called an *anastigmat*.

Many TV-camera lenses are coated with a layer of magnesium fluoride approximately 4×10^{-6} inch thick. This coating reduces the amount of light *reflected* from the surface of the lens, thereby increasing the efficiency

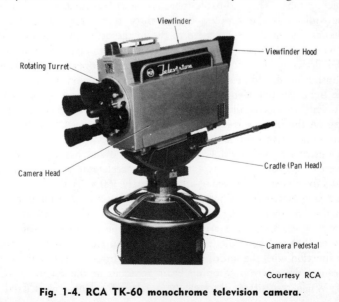

Courtesy RCA

Fig. 1-4. RCA TK-60 monochrome television camera.

of light passage through the lens. Such lenses may be recognized by their pale tint.

Monochrome TV-camera lenses are mounted on a *lens turret* accommodating four different types as shown in Fig. 1-4. The turret is rotated by the operator from the rear of the camera.

NOTE: Color cameras use one variable-focus (zoom) lens, as discussed further in Chapter 4.

1-3. THE PICKUP TUBE

A microphone is an audio transducer; that is, it converts varying air pressures constituting sound waves into corresponding electrical waves. The TV camera is an optical-video transducer, which converts the reflected light patterns reaching the lens to corresponding electrical impulses.

The entire camera (not including the viewfinder) is termed the *pickup head*. It contains the lens and iris system, pickup tube, horizontal and vertical deflection coils, alignment coil, focusing coil, horizontal and vertical deflection amplifiers, blanking amplifier, and video-signal preamplifier.

There are three basic types of pickup tubes used in television cameras: the *image orthicon,* the *vidicon,* and the *lead oxide.* The image orthicon is used primarily in monochrome studio and field cameras for live pickups. It also is used in some color cameras, again for live pickup. The other types will be found in both studio and film cameras. Fig. 1-5 illustrates on a comparative basis the two basic sizes each of the image-orthicon and vidicon pickup tubes. The lead-oxide tubes are described further in Chapter 4.

The vidicon is the simplest of the three types from an operational standpoint. Its small size (relative to the image orthicon) allows it to be used with very compact component arrangements. Fig. 1-6 is an exaggerated drawing of this pickup tube to illustrate operational functions; this illustration should be referred to during the discussion that follows.

The light-sensitive element may be visualized as consisting of two elements that are electrically separate: (1) a transparent, conducting film coating on the inner surface of the glass faceplate, and (2) a thin layer of photoconductive material on the scanning side.

Five grids are used. Grid 1 is the control grid and has a picture-cutoff voltage of −55 to −110 volts. Grid 2 is an accelerator grid usually operated at a fixed positive voltage of approximately 300 volts. Grids 3, 4, and 5 are focusing electrodes. When dynamic focusing is used, parabolic waveforms (to improve shading and corner resolution) are applied to grid 3; otherwise, grids 3, 4, and 5 are tied together. The potential of these grids (between 200 and 300 volts positive) creates an electrostatic field which, in conjunction with the uniform magnetic field from the external focusing coil, causes the electron scanning beam to focus at the photoconductive target. When the current through the external focusing coil is fixed, the

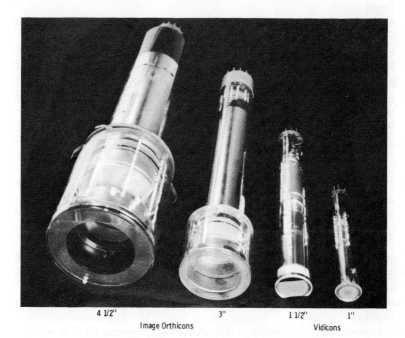

Fig. 1-5. Television-camera pickup tubes.

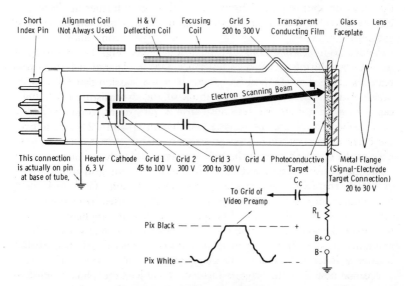

Fig. 1-6. Operation of vidicon tube.

focusing-grid voltage is made variable to allow optimum electrical focus of the beam.

The concentrated electromagnetic field provides a "stiff" beam which is held on a course down the tube axially toward the target coating on the inner faceplate. Thus, electrons in the beam are prevented from being captured by the highly positive attraction of grids 3 and 4. If current through the focus coil were decreased from the normal value, the beam of electrons would be less "stiff," and a given deflection current through the deflection yoke would cause a larger scan across the target. This feature is often used to provide a fixed amount of overscan for test purposes.

Grid 5 is actually a fine-mesh screen adjacent to the photoconductive layer. This grid is connected to grid 4 and is therefore at the same positive potential. Its purpose is to provide a uniform field on the beam side of the target so that the scanning beam impinges on the photoconductive layer perpendicularly, regardless of the angle from which it approaches. The target is maintained at a much lower positive potential (20 to 60 volts), and the arrangement may be seen to provide deceleration of the scanning beam. Hence, grid 5 is termed the *decelerator grid*.

The electron gun is conventional. A 6.3-volt heater is used to heat a thermionic cathode, which may be placed at ground potential as shown in Fig. 1-6.

Consider first the signal path under dark conditions (no light being transmitted by the camera lens). The metal ring around the front of the tube is the signal-lead connection, and the load resistor is connected in series between this electrode and the B-plus supply. The complete circuit then may be seen to extend from ground through the scanning beam, the light-sensitive surface, load resistor R_L, and the power supply back to ground.

Under no-light conditions, the photoconductive layer is essentially an insulator exhibiting a very high resistance. One of the electrically separate plates is charged to the positive potential of the signal electrode, and the other plate is "floating." We may think of the two plates as forming a capacitor with a dielectric resistance that is variable under conditions to be described. In the present analysis (no illumination reaching the target), the beam of electrons being swept across the target area, under the influence of the scanning current through the deflection yoke, will be deposited upon the positive target surface until this surface is reduced to cathode potential. Thereafter, the remaining beam electrons are turned back under the influence of the positive grids to form a return beam. This beam is not used in vidicon application. Although a considerable potential difference now exists between the opposite plates of the light-sensitive element, the resistance is so high that there is very little current. What little current does exist is termed the *dark current* of the tube.

Assume now that an image is focused by the lens on the light-sensitive element. The transparent conducting film at a particular point on the inner surface of the glass faceplate will now conduct a slight amount depending on the intensity of the light at that point. The lowered resistance adjacent to this particular conducting element robs a few electrons from the plate on the gun side, causing it to rise slightly toward the positive potential of the target supply. In this way, a pattern of positive potential is caused to exist on the gun side of the target in accordance with the light distribution in the focused image. Thus, more electrons are extracted from the total beam current to satisfy this deficiency of electrons on the target. The increased current in the signal path causes a greater voltage drop across resistor R_L, and the junction of R_L and coupling capacitor C_c is caused to swing in the negative direction. Since no-light conditions result in minimum current and high lights cause maximum current, the grid of the first preamplifier tube is caused to swing in the negative direction for high lights in the scene, and in the positive direction for dark portions of the scene.

The polarity of a television signal is always given in terms of *picture black,* since the blanking signal amplitude is always near picture black and is held at a given reference level for any particular system. Therefore, the polarity of the signal at the output of the vidicon is said to be black positive, or simply of positive polarity. This output polarity is just opposite to that of the image-orthicon tube, described below.

Fig. 1-7 shows a simplified diagram of an image orthicon. This device is similar to the vidicon in only one respect: It operates on the "storage" principle. The electron beam from the gun is of a "low-velocity" type. The camera lens focuses the reflected light from the scene onto a translucent photocathode, and corresponding photoelectrons are caused to be emitted from the back surface of the element. By "electron-lens" action, these emit-

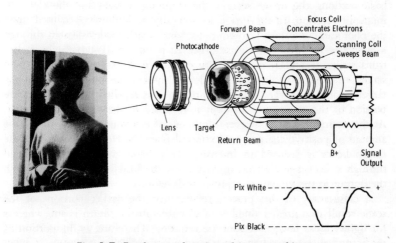

Fig. 1-7. Fundamental action of image orthicon.

ted photoelectrons cause an electron image to appear on the face of the target. The target consists of a very thin plate glass with a face (toward the front of the tube) of screen wire with an extremely fine wire mesh. An ordinary pinhead would cover some 7000 of the tiny openings of this screen mesh.

The number of electrons that strike the front of the glass target depends on the amount of light that strikes the photocathode. Thus, at each light portion of a scene, electrons strike the glass target and cause the emission of secondary electrons that are collected by the screen mesh and passed to ground. An area of electron deficiency (a positively charged area) is thus produced. Electrons on the back face of the glass leak through to the front at this point, and the back becomes positively charged at the corresponding point.

The scanning beam from the electron gun is aimed at the back of the target glass. As the beam sweeps across the target, points that still possess a negative charge (black portions of the scene) repel the electron beam and cause all of the emitted electrons from the gun to return in the form of a *return beam* (Fig. 1-7). However, when the beam is in the vicinity of a positive charge on the target (light portion of the scene), some of the electrons are "robbed" from the beam to neutralize the positive charge at that point. Therefore, the return beam varies in a way that corresponds to the light content of the scanned image on the target. This is the video signal; this signal then is passed through a five-stage electron multiplier for amplification.

Obviously, only the very basic characteristics of the image orthicon have been discussed thus far. They may be summarized as follows: The image orthicon is a video transducer that approaches the human eye in sensitivity. It is used primarily in studio and field pickup heads. It consists of three basic sections, the image section, the scanning section, and the electron-multiplier section. In the image section, any optical image focused upon the photocathode causes it to emit electrons, which are accelerated through a mesh screen in front of a glass target. At points on the target where electrons strike, secondary electrons are emitted to be collected by the wire screen and passed to ground. This leaves the corresponding point on the target positively charged. The scanning beam is repelled from the negative portion of the target, and all of it is returned to the electron multiplier. At the positive points, however, a number of electrons (depending on the amount of positive charge) are extracted from the electron beam, and the return beam is reduced in intensity. This "modulated" beam is passed through a five-stage electron multiplier to the load resistor, across which the output to the video preamplifier is developed.

A close study of this process reveals that the darker portions of the scene result in a greater number of electrons in the return beam, whereas for lighter areas fewer electrons are returned. Therefore, as the portion of the scene being scanned swings toward white, less output current is available. This results in less voltage drop across the load resistor, and the output signal swings positive. For dark areas, the signal swings negative relative to high lights, so the video output polarity of the image orthicon is negative. This is just opposite to the output polarity of the vidicon.

Bear in mind that a phase reversal occurs through each common-cathode or common-emitter stage of a video amplifier. Therefore, any picture polarity required can be obtained by proper circuit design.

Refinements of the vidicon and image orthicon, as well as details of other pickup tubes, are presented in Chapter 4.

1-4. THE SCANNING PROCESS

From the content of Section 1-3, it is evident that a means must be incorporated in the pickup head to "pick off" the image in the tube point-by-point, and to relay it to the video amplifiers in the form of electrical impulses corresponding to the brightness of the image at various points.

Electrons from the electron gun in the tube first are focused into a very narrow beam. Then this beam is caused to sweep back and forth across the image on the target with a definite time interval and sequence. Such movement of the electron beam is called the *scanning process*.

The beam is focused by a coil that creates a magnetic cross field that forces the emitted electrons into a beam of constant diameter. This cross section of the electron beam is termed the *scanning aperture*. (This term probably is a carry-over from the days of revolving mechanical discs with small holes that traversed the projected area of the scene.)

The beam is caused to scan the image by horizontal (H) and vertical (V) deflection coils constituting a *yoke* around the neck of the pickup tube. The H and V deflection coils are provided with sawtooth current waves that cause the electronic beam to be deflected electromagnetically.

Before the fundamentals of the scanning action are described, it may be well here to orient the directions of scan to prevent any confusion in future discussions. It should be recalled that the lens system of the camera inverts the image on the photosensitive surface of the pickup tube; that is, the picture is upside down. Refer now to Fig. 1-8. If the pickup tube is in a studio camera, the "top" of the image itself is on the bottom of the target. Therefore, if we consider the scanning process to sweep the target from the top left of the picture to the lower right (just as you are reading this page of print), the scanning beam, as shown, must start at the *lower left of the target* (as viewed from the lens side). The beam moves to the right until it reaches the edge of the target, then retraces to the left to start the next line scan. How this sequence compares to the receiver kinescope scan also is shown in Fig. 1-8. Throughout this book, when the scanning is said to be from top left to lower right, *we are referring to the picture itself*. There should be no confusion then as to why the picture is not transmitted upside down because of the lens action.

As will be discussed in Chapter 5, in televising motion picture film the image is projected directly onto the target by the lens of the film projector. In this case, the image is right-side up on the target. Direction of scan is easily changed at the sweep coils by reversing the leads from the deflection source, rotation of the yoke, or adjustment of direction of sweep currents. The point to remember is this: The image is said to be scanned from upper left to lower right, which means that *if you were looking directly at the scene in person, you would visualize the scene as being scanned from upper left to lower right.*

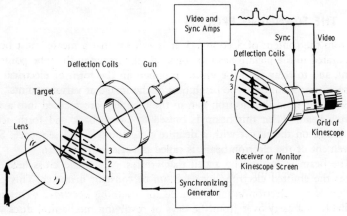

Fig. 1-8. Direction of scan.

The H and V *driving pulses* (synchronizing or timing pulses) are generated in the control room in a part of the main synchronizing generator. They are then fed into individual camera-control units in the video control room; these control units in turn are connected to their respective studio cameras via coaxial cables. Fig. 1-9 illustrates the scanning process. The generator of the sawtooth scanning waveform is generally in the studio-camera pickup head, and is "triggered" in operation by the driving pulses from the studio sync generator.

The question now arises as to why sawtooth waveforms instead of sine waves must be used for the deflecting coils. It should be recalled that a sine wave does not change linearly in amplitude with respect to time; i.e., the slope of the curve at any point depends on the angle corresponding to that point. This would cause the scanning spot (aperture) to move across the image with varying velocity, causing varying brightness across the reproduced picture.

A sawtooth wave is illustrated in Fig. 1-10. Such a wave increases linearly with respect to time (slope of curve constant), returns quickly to the "x" axis, and then repeats the cycle. Such a current waveform through the H and V deflection coils in the yoke about the camera pickup tube

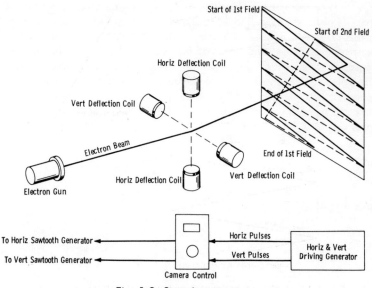

Fig. 1-9. Scanning process.

will cause the scanning spot to move at constant velocity across the scanned surface.

Fig. 1-9 shows the basic principles of the *odd-line, interlaced* scanning system that is standard for modern TV broadcasting. Odd-line scanning means that the total number of scanned lines is an odd number. It may be seen from Fig. 1-9 that this method of scanning allows the spot to return to the top of the surface to start scanning the second field at exactly the same *height* that it had when it started scanning the first field. Fig. 1-11 illustrates the action of the horizontal and vertical sawtooth currents on the scanning process. Current in the horizontal deflection coil causes the electron beam to deflect from left to right, then rapidly retrace as shown. The start of the second line is farther from the top of the raster because the sawtooth current in the vertical coil is lower in value, as shown.

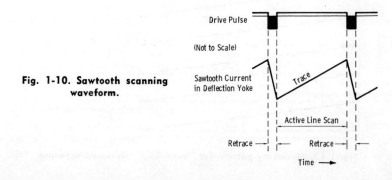

Fig. 1-10. Sawtooth scanning waveform.

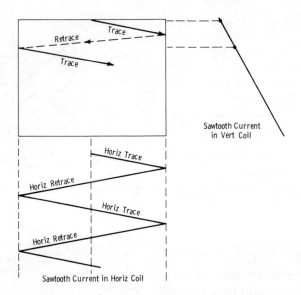

Sawtooth Current
in Vert Coil

Horiz Trace

Horiz Retrace

Horiz Trace

Horiz Retrace

Sawtooth Current in Horiz Coil

Fig. 1-11. Action of horizontal and vertical deflection currents.

Consider now what would happen if an even number of lines were scanned. Fig. 1-12 shows six lines of scanning. Here, alternate fields must be displaced *vertically* by one-half line with respect to each other. Thus, the perfectly uniform vertical scanning period of the odd-line system could not be used, and more complicated scanning and synchronizing circuits would be required.

The question now arises as to why "interlaced" scanning is used instead of allowing the spot to scan each successive line in turn. In the interlaced scan system, the scanning spot moves horizontally across alternate lines of

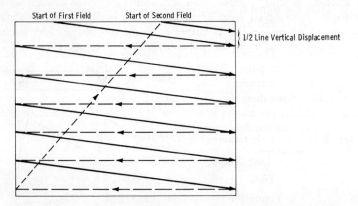

Fig. 1-12. Necessary pattern for "even-line" interlaced scanning.

the entire frame during one downward sweep, then returns to the top and scans the remaining lines during the next vertical sweep. Each one of these processes is termed a *field,* and two fields constitute one complete *frame.* The reason for the choice of interlacing lies mainly in the fact that such a process conserves bandwidth without sacrificing freedom from flicker of the reproduced image.

This will be made clearer if we consider the fundamentals of the motion picture. When still pictures are flashed in rapid succession on the screen, a sense of movement is imparted when each successive picture is slightly displaced in position from the preceding one. The process depends on the ability of the eye to retain for a split second the visual impression of a scene after the original stimulus is removed. This effect is termed *persistence of vision.* In early movies (often referred to as "flickers"), 24 frames of the film were flashed upon the screen each second. The flicker was very noticeable in these movies. It was then discovered that if each frame was flashed on the screen twice instead of once, the flicker disappeared. Thus, although the frame frequency is still actually only 24 per second, the *picture rate* is 48 per second.

This effect results from a basic law relating to the properties of flickering: The sensation of flicker of a reproduced image in motion is related to the *frequency of illumination* of the entire scene. A close study reveals that interlaced scanning is used to reduce the sensation of flicker for a given bandwidth of transmitted signal. If, for example, sequential scanning were used, "field" and "frame" would be one and the same. Frame repetition rates of 30 per second, if the scene is scanned (illuminated) only once in the frame time, produce noticeable flicker. The picture rate must be 60 per second, and if sequential scanning were used, all of the scanning lines would be traversed in 1/60 of a second. In the interlaced system, only half of the scanning lines are traversed in the same period, and the horizontal velocity of the scanning aperture is one-half that in the sequential system. Thus, the signal frequencies making up the radiated TV composite waveform are reduced by the same factor.

Interlaced scanning therefore allows 30 frames per second to be scanned at *twice* this rate. The field frequency of 60 per second is easily synchronized by the 60-Hz power lines standard in the United States. The total number of lines constituting a frame is 525; thus, 262.5 lines are scanned for each field. Since there are 30 complete frames per second, the number of horizontal lines per second is 525 × 30 = 15,750 lines per second. Thus, we may set up our standards as follows:

Line Frequency = 15,750 per second.

Field Frequency = 60 per second.

Frame Frequency = 30 per second.

Time of One Complete Line = 1/15,750 = 63.5 microseconds.

At the end of each line, the aperture moves back to the left to start the scan of the next alternate line. This time is called the *retrace,* or *fly-back,* period. The retrace time is very short, being equivalent to the steep slope of the curve of Fig. 1-10 where it drops toward the time axis.

Within the camera pickup head designed for studio operation are the H and V sawtooth generators. Such waveform generators are of the "driven" type, which means that their operation is controlled by synchronizing or triggering impulses. By this means, the scanning is controlled by the main sync generator, which must coordinate the entire TV system in relation to time.

The fundamental requirements of the scanning system may be outlined as follows: The electron in motion carries with it a minute magnetic field existing at right angles to the direction of electron motion. Therefore, the electron can be influenced by an external magnetic field through the linking (either aiding or opposing) of the two magnetic fields of force. Thus, the entire beam of electrons may be caused to deflect under the influence of coils carrying a current.

Examine Fig. 1-9 again. The beam of electrons from the electron gun must be caused to travel from left to right across the area to be scanned (looking toward the picture area). Also, a sufficient downward slope is provided so that each line is spaced the width of one line below the previous one. (Also see Fig. 1-11). This process of alternate line scanning is carried on at a linear rate to the bottom of the field (262.5 lines); then the beam is returned to the top of the area to start the scan of the next field. The "next field" means that the aperture is caused to continue its horizontal and downward trace, this time falling in the spaces left blank (unscanned) on the preceding field scan. At the end of this second field (525 lines, one frame), the aperture is returned to the position at the top of the scanned area where the first field started.

It may now be pointed out that the driving pulses supplied to the sawtooth-generator circuits are also "blanking" pulses. This term *should not be confused* with the blanking (pedestal) signal transmitted with the composite video signal to operate the receiver kinescope. For this reason, it is better for the beginner to term the camera signals received from the sync generator the *driving pulses,* although they are often termed *camera blanking* in technical literature. When the sawtooth generator is "blocked" (no driving pulse being received), the capacitor across which the sawtooth wave is formed receives its long charge interval from the B-plus supply. This is the "trace" portion of the sawtooth. When the negative driving pulse is received, this capacitor is discharged rapidly, causing retrace (which occupies considerably less time than the full line scan). The same H and V pulses are applied to a blanking-pulse generator in most cameras; the blanking pulses drive the target into the "black" region so that the retrace is not visible. The mechanics of scanning and blanking are presented in greater detail in Chapter 4.

1-5. THE CAMERA-CONTROL AND MONITOR UNIT

The picture output of the preamplifier in the camera pickup head is connected via coaxial cable (part of the complete camera cable) to the individual control unit in the control room. Television camera chains (systems), whether monochrome or color, have two basic configurations, as illustrated in Fig. 1-13.

The arrangement of Fig. 1-13A has all camera electrical adjustments as well as operational adjustments located in the control unit. In the arrangement of Fig. 1-13B, all camera electrical setup controls are located in the camera itself; the control panel may have only CONTRAST, BRIGHTNESS, and IRIS (exposure) controls that need occasional adjustment under operating conditions.

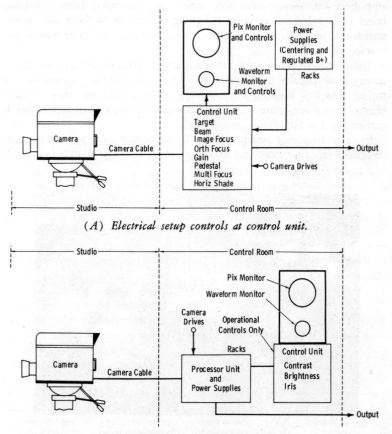

(A) Electrical setup controls at control unit.

(B) Electrical setup controls at camera.

Fig. 1-13. Basic camera systems.

The picture and waveform monitor shows the picture and waveform output of the individual camera system feeding the switcher. On the monitor panel are located adjustments for the kinescope focus and brightness (this applies to the monitoring kinescope, or picture monitor), oscilloscope (waveform monitor) focus and brightness, and a switch to accommodate monitoring either the line waveform (horizontal rate) or field waveform (vertical rate) on the oscilloscope tube. Constant brightness level is maintained by the operator through observing and correcting the video waveform on the oscilloscope in conjunction with observations of the picture monitor.

1-6. THE MIXING (SWITCHING) AND MONITOR UNIT

The outputs of the camera chains are fed to rack-mounted switching amplifiers and mixer amplifiers, which are controlled from a mixing panel (Fig. 1-14) where video switching and fading facilities are concentrated. It is here that the particular camera output desired to be transmitted is selected.

The switching panel is used to select any signal from as many as 24 (or more) input circuits, to switch or fade any signal into the program line, or to "lap-dissolve" between any two signals. It is here also that "special-effects" panels are located for use in matting or transitional techniques as outlined in Chapter 7.

In early switching systems, the switcher inputs were necessarily non-composite signals (video and blanking only), with sync pulses inserted at

Part of Monitor Deck   Switching Operator's Position   Alternate Program Director's Position

Program Director's Intercom   Preview Panel   Special-Effects Pattern Selector   Remote Controls for Film and Slide Projectors   Switching and Lap-Dissolve Panel

Fig. 1-14. One of two main switching panels at WTAE-TV.

a point in the circuit beyond the switches. This method was necessary because of the disturbance in the sync region that would occur because of the switching operation, allowing the receiver to "lose sync" momentarily when sources were changed. Modern switchers, either those that switch during the vertical blanking interval or other "high-speed" types, can accommodate composite-signal switching without causing such disturbances. These switches are covered in Chapter 7.

The "monitor deck" shown in Fig. 1-14 enables the operator to observe the camera outputs (or remotes, network, etc.) before they are switched to the program line. The program line itself is monitored as well.

Remote controls for film projectors and video tape systems also are normally a part of this operating position. The required preroll time for these units then may be initiated before actual switching occurs.

1-7. TV TRANSMITTERS AND ANTENNAS

Fig. 1-15 illustrates the aural and visual panel controls and meters for the RCA TT-50AH transmitter, as installed at WISH-TV. In the foreground is the transmitter-control operating position.

The basic functions of a television transmitter are (1) to provide a video carrier, *amplitude* modulated with the composite TV waveform,

Fig. 1-15. Transmitter and control console at WISH-TV.

and (2) to provide an aural carrier, *frequency* modulated with the audio program signal.

One hundred percent modulation of a regular fm broadcast transmitter in the 88-108 MHz region is defined as ±75 kHz, or a total frequency swing of 150 kHz. In television, however, 100-percent modulation of the fm aural transmitter is defined as ±25 kHz, or a total frequency swing of 50 kHz.

In practice, one complete sideband of the visual carrier is transmitted together with only a small part (vestige) of the other sideband (Fig. 1-16). The total width of the television channel is 6 MHz. The visual carrier is located 4.5 MHz lower in frequency than the aural center frequency. The aural center frequency is 0.25 MHz below the upper frequency limit of the channel.

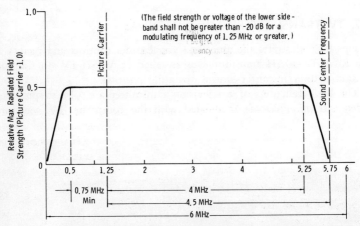

Fig. 1-16. Signal distribution within TV channel.

As a practical example, consider TV Channel 2 (54-60 MHz). Since the aural carrier frequency is 0.25 MHz below the upper limit, it is 59.75 MHz. The visual carrier frequency is 4.5 MHz lower than the aural carrier frequency, or 55.25 MHz. The frequencies for TV Channels 2 through 83 are shown in Table A-1 in Appendix A.

The upper sideband is allotted approximately 4.5 MHz of the total channel of 6 MHz. Thus, the modulator section must be "broadband" (good response to about 4.5 MHz) for proper transmission. Studio camera chains and other equipment generally are flat to 8 MHz for minimum accumulative picture distortion.

The number of broadband radio-frequency circuits in the visual transmitter depends on the method of modulation. When the final stage is modulated, the final power amplifier is the only broadband rf stage necessary. When a lower-level rf stage is modulated, all following rf amplifiers must respond to a bandwidth of 4.5 MHz. The tuned rf stages from the oscillator up to the modulated stage are ordinary high-frequency transmitter circuits that may be "meter-tuned" as in other transmitters.

There are advantages and disadvantages to both high-level and low-level modulation. In practice, the major manufacturers differ in their ideas of the best method to use, and transmitters, therefore, differ in this respect. When the final amplifier stage is modulated (high-level modulation), grid modulation is used rather than plate modulation, because of several design factors. When high-level modulation is used, the lower sideband is cut off by means of a device known as a *vestigial-sideband filter* (which will be discussed later).

When low-level modulation is used, the remaining rf carrier amplifiers must be broadband and tuned in such a way that the lower sideband is cut off by amplifier-circuit characteristics. To make tuned circuits broadband, low-impedance output circuits must be used at a sacrifice of gain per stage. Tuning must be done with special equipment using "marker" dots and oscilloscopes to obtain the proper bandwidth and suppression of the lower sideband. Lower-powered modulator sections may be used, however, and the need for an expensive vestigial-sideband filter is eliminated.

The possible input and output load impedances of a tube in a TV transmitter are determined largely by the tube capacitances, since at vhf and uhf these capacitances are the predominant impedances. The *figure of merit* of a vacuum tube is a ratio used to express the relative ability to amplify high video frequencies. Thus, at high frequencies, gain is proportional to transconductance and inversely proportional to shunt capacitance, and the figure of merit may be expressed as:

$$\text{Figure of Merit} = \frac{g_m}{C_t}$$

where,

g_m is the transconductance in micromhos,
C_t is the total shunt capacitance in picofarads.

It may be seen now why *plate modulation* of high-power stages is not practical for a TV transmitter. For such modulation of a 5-kW stage, the output capacitance of a modulator tube working into an ordinary shunt-compensated circuit would be on the order of 200 picofarads. Since transformers cannot be used for the modulated stage due to dc reinsertion (discussed below), the plate voltage for the modulated tube would have to be supplied directly by the modulator tube across the plate load. In the instance cited here, the modulator tube would "see" a capacitance of 200 pF, or a load of 199 ohms at 4 MHz ($X_c = 1/2\pi fC$). The required video signal of 3 to 4 kW peak-to-peak into 199 ohms would necessitate a power of approximately 100 kW from the modulator. Therefore, when high-level modulation is used, the modulating signal is applied to the grids, rather than the plate, of the final stage. When plate modulation is used, low-level modulation takes place at a power level of a few watts, and both the modulated-stage and modulator tubes are small, with correspondingly low capacitances.

Another function peculiar to TV transmitters is the matter of *dc reinsertion.* To understand the necessity for this function, it is advisable to review briefly the composite TV waveform. Fig. 1-17A shows a graphical representation of two lines of video information, with the horizontal blanking pedestals (at the end of each active line) upon which the horizontal-sync pulses are inserted. Maximum signal (at the transmitter out-

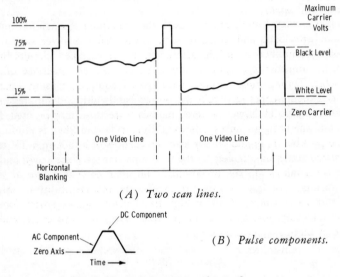

(A) *Two scan lines.*

(B) *Pulse components.*

Fig. 1-17. Dc component of waveforms.

put) exists for black portions of a scene. In television, as in any system of transmission, certain maximum values related to 100-percent modulation of the transmitter must exist. As shown in Fig. 1-17A, the "black level" of the picture equals or nearly equals the blanking signal level. TV standards set this value at 75 percent of 100-percent modulation. The sync-pulse peaks, then, occupy the remaining 25 percent of carrier amplitude, and the transmitter reaches 100-percent modulation only on sync-pulse peaks. It is also shown that maximum white level is held to 15 percent (actually 10 to 15 percent, for a nominal value of 12.5 percent) of the total carrier amplitude, never reaching zero.

Fig. 1-17B points out the fact that a pulse such as a blanking pedestal or synchronizing pulse has two components. The ac component forms the "sides" of the pulse where the value is changing with respect to time. The "flat top" of the pulse is a dc component, since it maintains a steady value of amplitude over a period of time. Obviously, the dc component is lost in ordinary transformer or RC coupled circuits, which act only on changing values of voltage or current. Therefore, in TV systems, dc reinsertion is provided at certain important points in the system.

Some reference must be used for establishing the dc level of the picture signal. It is recalled that the peaks of the sync pulses are maintained at a constant level above the pedestal (blanking) signal. At the transmitter, this relationship may serve as the reference voltage for establishing the dc level of the carrier amplitude. By means of the dc-restorer circuit, which is "keyed" by the horizontal sync pulses, the modulator-tube bias is automatically returned to the same predetermined value for each blanking pulse. This action, in effect, restores the dc component. The modulator plates are directly coupled to the grid or plate of the rf stage so that the dc value is maintained.

This type of circuit is known as a *clamping circuit*. Such a circuit *maintains a constant voltage-level output for an input waveform that contains amplitudes in the positive or negative sides above a predetermined value.* Clamping circuits are used in other points of the complete TV system and will be described where applicable in the following chapters. The basic function should be memorized at this time.

A *sync-pulse expander* is also common at transmitters and elsewhere in the TV chain of amplifiers. This type of circuit is used to compensate for any loss in the amplitude of the sync pulse. In a typical circuit, a Class-A amplifier that acts on all the composite signal is used to excite a Class-C amplifier that draws plate current only on peaks (the sync pulse) of the signal. This added current, combined with the output of the Class-A amplifier, serves to expand the signal at the time of the sync pulse.

The TV transmitter radiates a composite modulated rf carrier made up as follows:

1. The rf carrier wave, generated by the crystal oscillator and then amplified
2. The picture-signal content from the studio cameras, which is used to amplitude modulate the rf carrier
3. The sync pulses as follows:
   (A) Horizontal (H) blanking pulse (horizontal pedestal)
   (B) Horizontal sync pulse (constructed on the H pedestal)
   (C) Vertical (V) blanking pulse (vertical pedestal)
   (D) Vertical sync pulse (constructed on the V pedestal)
   (E) Equalizing pulses (also constructed on the V pedestal) preceding and following the serrated vertical sync pulse

Fig. 1-18 shows the appearance of the carrier envelope at the modulated rf stage (prior to removal of the lower sideband). Fig. 1-18A shows the character of the envelope for two lines of almost black content; the carrier wave is at maximum level. Fig. 1-18B represents two lines of highest brightness, and the carrier amplitude is at a minimum (approximately 15 percent). Fig. 1-18C represents two lines of "average" brightness. Note that the blanking pedestals and sync pulses remain of constant amplitude under video modulation (dc restoration).

The system of modulation illustrated in Fig. 1-18 is termed *negative transmission*—*more* light content in the picture causes the carrier ampli-

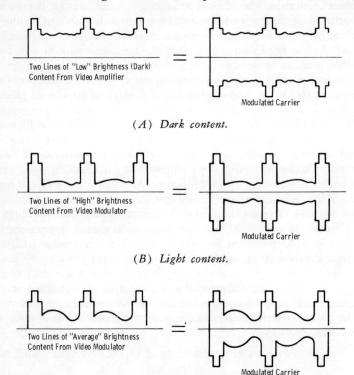

(A) Dark content.

(B) Light content.

(C) Average content.

Fig. 1-18. Waveforms at visual modulator and modulated stage.

tude to be *less*. This is the standard method for the United States. It has a number of advantages over positive transmission, as follows:

1. Since the black level is maintained constant, and the sync peaks (which have a very short duration) represent the maximum radiated power, a considerable reduction in average power results. Since most of the signal represents varying degrees of lightness, the average carrier power is relatively low.
2. Electrical "noise" impulses are more readily compressed.
3. Negative transmission allows use of the pedestal level (black level) and sync peaks to operate a comparatively simple type of automatic gain control (agc) at the receiver.

As was mentioned earlier, when high-level video modulation is used, a band-elimination filter must be used to attenuate the lower sideband. This filter is known as a *vestigial-sideband* filter and is placed between the final rf amplifier and the antenna as shown in Fig. 1-19.

The FCC standards state that the lower sideband shall not be greater than minus 20 dB for a modulating frequency of 1.25 MHz or higher. The vestigial-sideband filter accomplishes this attenuation. It is actually composed of coaxial elements that are equivalent to the circuit of Fig. 1-19. The low-pass filter accepts the lower sideband and dissipates its power in load resistor R. The high-pass filter allows the upper sideband to pass, and the *notch filter* is an arrangement so tuned that it produces a "notch" (that is, it dissipates energy) at a frequency 0.25 MHz below the lower limit of the channel. This "notch" provides protection to the aural carrier of the next lower TV channel.

The *diplex unit* is necessary because two transmitters, visual and aural, are feeding a common antenna system. The equivalent schematic of a diplex unit is shown in Fig. 1-20. The principle is that of a balanced bridge that prevents interaction between the two transmitters. The resistive elements labeled N-S (north-south) and E-W (east-west) radiators represent the elements of a superturnstile antenna. The balancing impedances are adjusted to balance the bridge circuit. Thus, each transmitter feeds between two points of equivalent potential with respect to the other, and interaction is negligible when the circuit is properly balanced. In order to effect a substantially circular radiation pattern from the elements of the antenna, which are at right angles to each other, a phasing loop is inserted in one transmission line. This loop is actually an extra quarter-wave line section that delays the signal 90 degrees for the associated radiator.

There are several forms of TV transmitting antennas, the most popular (particularly for low-band vhf) being the *superturnstile,* or "bat-wing,"

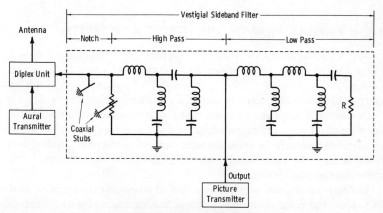

Fig. 1-19. Vestigial-sideband filter.

antenna mentioned above. Regardless of the type, these antennas are characteristically broad-band for operation over a six-MHz range. They are also capable of power gain, since radiation is concentrated near the horizontal plane.

The *antenna field gain* is a rating commonly encountered in literature on TV antennas. The reference is an ordinary dipole with a 1-kw input power, corresponding to a field intensity of 137.6 millivolts per meter at one mile. Thus, antenna field gain is the *ratio* to 137.6 millivolts per meter of the effective free-space field intensity at one mile produced by a given antenna with 1-kw power input. *Antenna power gain* is the square of the field gain. A greater number of antenna elements vertically "stacked" results in greater gain.

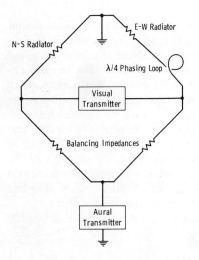

Fig. 1-20. Principle of diplexer.

1-8. A CAPSULE VIEW OF THE NTSC COLOR SYSTEM

The NTSC color system is the FCC-approved type of color transmission in the United States. NTSC stands for "National Television Systems Committee," which formed a panel (or panels) of experts to organize and carry out research and development of a *compatible* color system.

Definition: Compatibility means (A) monochrome receivers must receive color signals in monochrome; (B) color receivers must receive monochrome signals in monochrome; and (C) the color transmission must occupy the same bandwidth (Fig. 1-16) as that occupied by a monochrome transmission.

The above definition simply states that all receivers existing prior to the FCC-approved color transmission would not be made obsolete. Color information "interleaved" with the regular monochrome information would not detract from the ability of a monochrome receiver to reproduce (in black and white) the color signal. Color receivers could utilize the additional information for full-color reproduction, while also being capable of reproducing a monochrome signal in black and white.

The system is based on the utilization of three color primaries: red, green, and blue.

Definition: A primary combination must be such that the addition of any two primaries will not result in the third primary color.

See Fig. 1-21, which represents the three primaries (red, green, and blue) as spotlights overlapping each other. Note that where all three overlap, white results. Yellow results from red and green, purple results from red and blue, and cyan (greenish blue) results from blue and green.

The "brightness" component of a color signal is termed luminance and is designated by the symbol Y. Color information consists of two basic components:

1. Hue (actual color, such as blue, yellow, etc.).
2. Saturation (degree of "purity" of the color. This means the difference between "pale yellow" and "deep yellow," etc., and is a measure of contamination of a "pure" color of single wavelength with "white," or all other colors).

Let us start at the "end of the line," the color picture tube, and see what this device requires. See Fig. 1-22. A phosphor screen of some one million dots is arranged in *triads,* or red-green-blue elements, as shown. Each triad is *one picture element;* that is, the dots are spaced so closely that at normal viewing distance a triad looks like a single spot of picture information. Therefore, if all three elements of a triad are excited equally, a "white dot" results. If only red is excited, the picture element is red. If red and green are excited but blue is zero, the picture element is yellow, etc.

Three separate electron guns are spaced 120° apart around the tube axis, as shown in Fig. 1-22. The "blue beam" is so termed because it re-

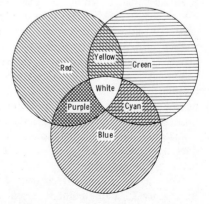

Fig. 1-21. Primary colors for television.

ceives only "blue" information from the composite color signal. Similarly, the red and green beams each receive only information for the corresponding colors.

Each beam is "aligned" to its respective phosphor dot by the voltage on the aperture mask so that the triad forming a single picture element is properly *converged* to eliminate color *fringing* that would occur if a portion of the red beam should strike the blue phosphor, etc. Since we have pointed out that there are some one million dots on the screen, it can be seen that there are some one-third million aperture-mask holes in the tube (one hole for each three dots in a triad).

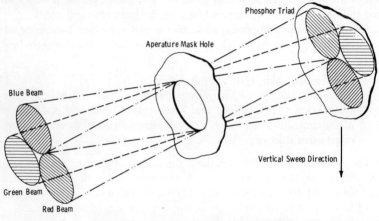

Fig. 1-22. Principle of color picture tube.

Fig. 1-23 illustrates one example of how guns are controlled in the color picture tube. The luminance (Y) amplifier drives all three cathodes in parallel. The intensity of each beam is controlled by the individual red, green, and blue signals (R, G, and B) on the grids. Note that if R, G, and B are all equal, a "white" or "gray" results. If any one hue predominates, the respective grid will swing further positive, and the corresponding phosphor will be excited more than the others.

Each grid is actually driven by a *color-difference* amplifier. The term "R − Y" means "red-minus-brightness"; "G − Y" means "green-minus-brightness"; and "B − Y" means "blue-minus-brightness." Thus, it may be inferred that the system has transmitted brightness (luminance) information separately from the chrominance (hue and saturation) information. The chroma portion of the signal carries *only* the hue and saturation information *minus* the brightness component. Thus, a color transmission does not seriously affect the image being reproduced on a conventional monochrome receiver.

Note that G − Y is *derived* from R − Y and B − Y. Details are covered in Chapter 2; at this point it is sufficient to understand that this method

enables a two-phase chroma subcarrier to carry information for the reproduction of three colors. Since the color-transmission primaries carry information from all three primaries, green is conveniently extracted at the receiver by matrixing of $R - Y$ and $B - Y$. (A matrix is simply a cross-
connected voltage divider). The reader will discover in Chapter 2 that
the NTSC color system is a fascinating "game of numbers"!

The color-difference amplifiers receive their information from synchronous demodulators (a form of product demodulation). As was stated
above, the color information is transmitted by means of a two-phase subcarrier with components 90° apart. Color information is carried only in

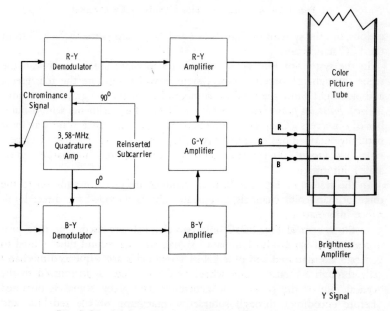

Fig. 1-23. Application of signals to color picture tube.

the sidebands, so the original subcarrier is suppressed to avoid needless
beat interference. The subcarrier is reinserted in the receiver from a reference subcarrier *burst* transmitted on the back porch (interval following
horizontal sync pulse) of the composite color signal.

Fig. 1-24 shows the relative location of the color-subcarrier sidebands
in the composite color signal. The color sidebands are in the upper range
of the channel, where monochrome information is weak, to aid in minimizing interference in black-and-white receivers. The "interleaving"
process in which the color subcarrier is working at an odd harmonic of
one-half the line scanning frequency is explained in Chapter 2.

Fig. 1-25 illustrates the three basic color systems in use. Strictly speaking, the system of Fig. 1-25A is the method as defined by NTSC specifi-

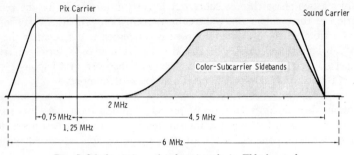

Fig. 1-24. Location of color signals in TV channel.

cations, but the systems of Figs. 1-25B and 1-25C are perfectly "legal" from the FCC standpoint.

In the system of Fig. 1-25A, three pickup tubes are used to scan the scene through colorimetric optical systems which separate the red, green, and blue light. Since the need for compatibility exists, the gray scale (luminance) *must be preserved* as in nature. Detailed experiments with a number of observers (Chapter 2) revealed that the luminance scale was most naturally obtained by deriving 59 percent green, 30 percent red, and 11 percent blue (total 100 percent) to obtain the required luminance (grayscale) information.

In the system of Fig. 1-25B, four channels are used. In addition to the three primary-color channels, a separate tube is provided for direct luminance information.

In the system of Fig. 1-25C only two channels are used. One tube produces the full-bandwidth luminance signal, and the second tube is used to generate alternate red and blue fields. These fields are sequenced mechanically through a rotating filter wheel (red and blue) synchronized to the vertical rate of the studio sync generator. The green signal is obtained (before encoding) through subtractive matrixing of the red/blue and luminance signals.

Details of the NTSC color system are given in Chapter 2, and most existing color-signal sources and their variations are covered in Chapters 4 and 5.

1-9. TV LINKING FACILITIES

Wideband linking facilities are necessary for relaying TV signals from remote points to studio, from studio to studio (network), and from studio to transmitter. Such relaying is accomplished by either coaxial cable or microwave transmitter-receiver equipment.

Three types of TV auxiliary stations are defined as follows:

Television Pickup Station: A mobile station for TV broadcast licensed for transmission (audio and video) of temporary types of programs (such

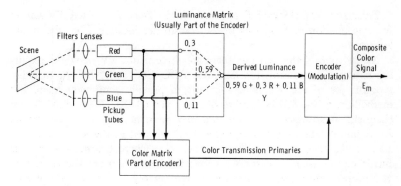

(A) Three channels.

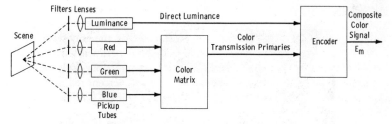

(B) Four channels.

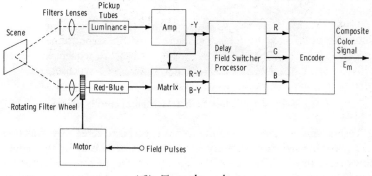

(C) Two channels.

Fig. 1-25. Basic color-camera systems.

as football games, spot news events, etc.) where coaxial-cable service is not practical.

Television Intercity Relay Station: A fixed TV station, owned either by TV broadcast-station licensees or by a communications common carrier, used to relay TV programs from one station to others, as for network broadcasting.

Television STL (Studio-Transmitter Link) Station: A fixed TV station for transmission of programs to the transmitter location from the studio location.

At the time of this writing, there are three major TV networks requiring nation-wide transmission facilities: ABC, CBS, and NBC. As in radio, it is the job of the telephone company to provide equipment, men, and methods to fit the desired pattern of any given telecast. Many transmission channels, so arranged as to provide maximum flexibility of application, are required. The necessary equipment is provided in the long lines division of the telephone company in each city where broadcast service is furnished to stations. For the TV networks, these offices are known as television operating centers. Centers vary considerably in size and complexity, depending on the amount of activity to be accommodated. The TV operating center in New York (Fig. 1-26) is the control office for the entire television network service.

Courtesy American Telephone and Telegraph Company

Fig. 1-26. AT&T New York television operating center.

Technicians at these operating centers perform three key functions: (1) make the switches necessary to route TV programs along the required network channels, (2) check the quality of the picture and sound on suitable monitoring equipment, and (3) test and maintain the transmission performance of the network channels.

Switching schedules, known as the *daily operations orders,* are made up as far in advance as possible from information received from the stations being served. In some cases, these orders are transmitted via teleprinter directly from the station to the control center. From that point, they may be retransmitted to the various operating centers about the networks, since overall switching may involve several different operating centers. Direct telephone order circuits also are used where a last-minute change in the daily operations schedule may be made anywhere across the country.

It should be pointed out here that the audio portions of TV programs are seldom transmitted in the same network channel as the video portions. In practice, a TV broadcaster located directly on a video route may actually receive the audio service from another city. This is so because the telephone company has found it both impractical and uneconomical always to provide a fixed route for any particular service. This use of separate routings for video and audio emphasizes another problem, the difference in transmission speeds of signals in different media. Video signals, via radio relay, have propagation speeds closely approaching the speed of light. On the other hand, audio-channel transmission speed is relatively slow when the usual loaded cable facilities are used. Although this delay is entirely acceptable for ordinary standard broadcast service, it is not tolerable for TV service since the sound portion would lag behind the video signal. Therefore, *carrier-current* systems are used to carry the sound portion of television programs. As a result, the audio signal is propagated along the wires at the same speed as the picture signals.

Fig. 1-27. Microwave relay station at Green Pond, N.J.

Courtesy American Telephone
and Telegraph Company

The relay system now in use across the country for TV-network interconnection uses microwave links for most of the service. Such relay stations are spaced an average of 30 miles from each other, depending on the terrain. At microwave frequencies, line-of-sight propagation must be maintained for satisfactory service. The energy is focused into a beam in much the same way that light is focused into a beam in a searchlight. Because of the tremendous gain of microwave transmitting and receiving antennas, "flea power" on the order of $\frac{1}{2}$ watt is all that is necessary for transmitter output power.

One microwave installation, as used by the American Telephone and Telegraph Company (AT&T), is shown in Fig. 1-27. One radio relay of the type shown may transmit or receive up to six broad-band channels.

Thus, only four antennas at each point suffice to handle six broad-band channels in two directions. One transmitting and one receiving antenna are placed side by side facing in one direction, and a similar pair faces in the opposite direction. The receiving antenna feeds the relatively weak microwave signal into an amplifier to compensate for the original transmission loss. The output of this amplifier then feeds the opposite transmitting antenna, which sends the signal on to the next relay station. A TV program signal transmitted over such a network from New York to San Francisco passes through 105 microwave relay stations.

Of special consideration in the efficient maintenance of such a highly important service is the method of quickly locating any fault in the complex system. The repeater stations are normally unattended and automatic in operation. A highly developed control system is therefore used to relay information concerning the operating condition of each individual station to special maintenance or alarm centers. When trouble occurs, both a visual and aural signal are produced at the alarm center. There are 42 different alarm conditions that quickly convey to the maintenance center information about such exact conditions as a rectifier failure, an open door, or failure of an aircraft-warning light on a tower. Appropriate corrective measures then can be undertaken with a minimum of delay.

In the hope of reducing costs, studies are being made with regard to the use of a domestic satellite system for television-program distribution. A plan for such a system was explained in considerable detail to the FCC in connection with a proposal by the Bell System in Docket 16495 (Domestic Satellite). It was proposed to use satellites for provision of telephone circuits over great distances, as well as TV distribution, since studies show that satellites will be most economical when used jointly for both services. Under this plan, some terrestrial television facilities would continue to be used in conjunction with the satellite channels.

The Bell System proposal also recommended the eventual use of broad frequency bands above 18 GHz for satellite transmissions. While the use of these frequencies entails certain difficulties, such as rain absorption, it is believed that diversity measures could overcome them. Among the advantages of using these higher frequencies would be an expanded system capacity and the ability to locate receiving and transmitting antennas on top of telephone buildings. The latter factor would result in lower operating expense and minimize the use of connecting facilities.

Fig. 1-28 shows how a typical TV service (pickup to home viewer) might be provided by a satellite system. The television signal of the football game is transmitted from a mobile unit to a television operating center (TOC). The signal is then transmitted from the ground station on the West Coast to the satellite, which in turn retransmits the signal to the ground station on the East Coast. The signal is then fed through another TOC to the customer's network studio. It is returned to the satellite and transmitted to numerous ground stations as depicted in the center portion of Fig. 1-28. The signal is then fed to a TOC, the local TV studio, and the TV station transmitter. The only difference from the present service is that some long terrestrial paths would be replaced with satellite channels. In certain high-density areas, it appears that terrestrial intercity facilities will be more economical. In either case, existing local distribution systems would still be needed. Any economies resulting in the use of satellites for television depend on employing the system for both message and television services.

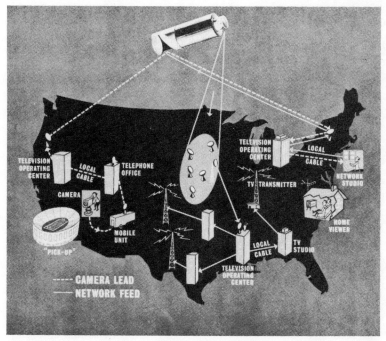

Courtesy American Telephone and Telegraph Company

Fig. 1-28. Television relay by satellite.

Work continues on land transmission systems that also appear to offer future savings. The most promising today is the use of pulse code modulation on coaxial cable and waveguide systems. Digital techniques offer unique advantages since signal distortion is no longer a function of circuit length. Also, with the signals confined to coaxial conductors or waveguides, valuable radio-frequency spectrum space would be conserved. One transmission system now under development at Bell Telephone Laboratories will have a capacity of three TV video signals per coaxial cable (three times the capacity of present cable systems); the waveguide system appears to have a potential capacity of 150 TV channels.

EXERCISES

Q1-1.   What is the basic function of a TV-camera pickup tube?

Q1-2.   What is the purpose of "camera-driving pulses"?

Q1-3.   Why is the retrace in a camera pickup tube blanked out if blanking at the picture tube makes retrace invisible?

Q1-4.   Why is the H sync pulse delayed from the start of the H blanking pulse?

Q1-5.   Define interlaced scanning.

Q1-6.   Give the ratio of field frequency to frame frequency for interlaced scanning in standard TV broadcasting.

Q1-7.   What is the significance of the ratio of the horizontal scan frequency to the frame frequency of the picture?

Q1-8.   What time is consumed in scanning one picture line?

Q1-9.   What time is consumed in scanning one field of a picture?

Q1-10. What time is consumed in scanning one frame of a picture?

CHAPTER 2

NTSC Color Fundamentals

This color training is slanted to the television broadcast engineer, who is concerned with the generating and transmission equipment involved in studio and film color telecasts. Consequently, receivers and monitors are covered only to the extent necessary to ensure adequate comprehension of the entire color system. The beginner should review Section 1-8 before proceeding.

2-1. METHOD OF ATTACK

This Chapter is concerned with "Colormath," or the practical mathematics required for a fundamental understanding of FCC-approved color. The mathematics involved is not advanced and will be reviewed only for a few isolated and unusual applications.

Regardless of how much effort is expended in some "practical color courses" to bypass the subject of colorimetry, the broadcast engineer who wants to *understand* his subject cannot afford this bypass. Therefore, we will treat this subject in a way that is used later in practical applications. It is pertinent to note that a review of luminance principles is incorporated, because of the importance of this subject in colorimetry.

One of the primary advantages of understanding basic colorimetry is to be able to transform from any given set of primaries to another primary combination that is linear to the first. For example: RGB (red-green-blue) primaries in color TV systems are transformed to transmission primaries Y, I, and Q. Most modern receivers (and some color monitors) demodulate on still another set of primaries termed the "color-difference" primaries, $R - Y$, $G - Y$, and $B - Y$, where "Y" represents luminance.

NOTE: Some courses and books on color TV, in an attempt to "simplify" terminology and symbols, have used the letter "M" (for monochrome) to signify luminance. This practice can be confusing, since in FCC terminology E_M (read "E sub M") is used to specify the composite color signal, which includes the color component. Therefore, in this book, the letter "Y" will be used to signify luminance only.

2-2. WHAT IS COLORIMETRY?

Colorimetry is a basic tool in the modern color-TV system. The present FCC-authorized system would not be possible without an understanding of the science of colorimetry.

Any work involving color must specify a definite basis upon which a given color is classified. Imagine the difficulty of buying blue thread to match "blue," or green paint to match "green," without some basis of exact classification. Colorimetry provides a basis for specifying a color in either relative or absolute numerical values.

The study of colorimetry is a necessary prerequisite to the study of color television. Analogies are the study of the nature of radio waves before the study of methods of transmission and reception, and the study of the nature of sound waves prior to a study of audio systems.

Colorimetry concerns the limits of human vision (intensity threshold, contrast range, visual angle, and time threshold) and the entire nature of human vision. Naturally, we will be concerned only with those elements important to the handling of color television systems.

2-3. INTRODUCTION TO THE NATURE OF HUMAN VISION

The nature of human vision is psychophysical. Observers watching the same event (which means interpreting reflection of light rays from objects) may describe that event quite differently. According to psychologists, we "see" by any one of three attitudes:

1. Attention to physical objects perceived by reflected light
2. Attention to the light itself
3. Attention toward a mental image (partly by previous experience)

Any one of the above attitudes can lead to any one of three results, or a combination of more than one result:

1. Direct physical knowledge
2. Illusion
3. Hallucination

In reality, very few of us obtain results in terms of direct physical knowledge. This result would involve evaluation of light in terms of surface chromaticities and texture, color in shadows, and all other "real" attributes of light and objects. Hallucination cannot be discounted even in sane people. However, most of us "see" by illusion, or largely in terms of past experience. In nature, if we observe a man with a white shirt standing in daylight and watch him move into a room with incandescent lighting, we still "see" a white shirt on that person. This same event observed under simulated types of lighting on a color-TV screen, compressed into the relatively small area of the picture tube under different psychological conditions, will be interpreted quite differently. We can sense the actual shift in "white" of the shirt between the two different spectrum distributions of daylight and incandescent light. Also, if the "shadows" in the picture should be reflecting some light from a red drape not in the picture, the shadows will be noticed to have a reddish cast. This effect ordinarily is not noticed in nature.

Experts tell us that the left halves of both retinas are connected to the left half of the portion of the brain devoted to vision, and the right halves of both retinas are connected to the right section of this portion of the brain. Central areas of the retina are connected to both halves in the brain. Specific areas of the brain are associated with specific areas of the retina.

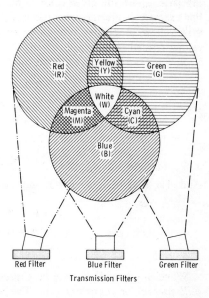

Fig. 2-1. Combinations of primary colors (red, green, and blue).

It is possible that an optical image on the retinal surface is duplicated as a distorted electrical image in the brain. The degree of distortion differs among individuals.

Since the above conditions prevail, the nebulous subjects of luminance (or, more correctly, brightness sensation) and color must be considered in terms of a "Mr. X," or "standard observer." The standard observer was derived from an average of tests with a large number of observers in colorimetric experiments conducted by the CIE. (The American term is ICI, or International Commission on Illumination. It should be understood that CIE and ICI are one and the same. The French designation, CIE, has been recommended as standard by the Standards Committee of the Institute of Electrical and Electronic Engineers, or IEEE.)

First of all, we start with three *primary colors*. The requirement for a "primary" is that the addition of any two primaries will not produce the third color.

Fig. 2-1 symbolizes three projectors illuminating a white screen with colors of light corresponding to the filters that select the wavelength (frequency) to be passed. The three primary colors are red, green, and blue; they are assumed to be "pure" hues, highly saturated, undiluted with white light. Red plus green produces yellow. Red plus blue produces a bluish red termed *magenta*. Blue plus green produces a bluish green termed *cyan*. A combination of red, blue, and green produces white. Also, a combination of cyan, magenta, and yellow contains all three primaries, and will produce white (all colors). Black denotes lack of any color. The three primaries allow any given color to be matched closely by proper mixtures. For example, a greater proportion of red to green produces orange rather than yellow. A greater proportion of green to red results in a greenish yellow termed "lemon yellow," etc.

Now we can begin to establish color relationships (refer to Fig. 2-1):

$$R + B = \text{Magenta}$$
$$R + G = \text{Yellow}$$
$$G + B = \text{Cyan}$$
$$R + G + B = \text{White}$$
$$C + M + Y = \text{White}$$

And also, if color A = color B and color C = color D, then:

$$A + C = B + D$$
$$A = B + D - C,$$

etc.

Wavelengths of light may be expressed in terms of *nanometers* (nm), *millimicrons,* or *angstroms* (A). The relationships of these and other units of measurement are as follows:

1 nanometer = 1 millimicron = 10^{-9} meter
1 angstrom = 10^{-1} nanometer = 10^{-10} meter
1 micron = 10^{-6} meter
1 nanometer = 10 angstroms = 10^{-7} centimeter

Wavelengths to which the eye is a natural receiver lie between approximately 400 and 700 nm. This range is the frequency (wavelength) limitation of vision and is important only in that a given color can (and must) be specified in terms of a single wavelength. For example, "red" (in CIE specifications) lies between roughly 630 and 700 nm. When measurements are made, the exact single wavelength of "red" must be specified.

A device known as a *colorimeter* was used in CIE tests in which mono-chromatic primaries of specified wavelength were adjusted by logarithmic attenuators until the observer felt he had a match with a given spectral hue. The basic principle is shown by Fig. 2-2. The CIE primaries were specified as follows:

Red: 700 nm
Green: 546.1 nm
Blue: 435.8 nm

This procedure is known as color matching on a power basis, in which the relative amounts of light flux (lumens) are adjusted to match the

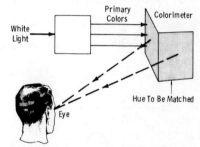

Fig. 2-2. Principle of colorimeter tests.

unknown. The logarithmic attenuators used are calibrated by unit amounts so that 1 unit of red, 1 unit of green, and 1 unit of blue produce white at a specified luminance level. We will see later that "white" also must be specified as a certain value.

The results of the averaged tests described above may be presented as shown in basic form by Fig. 2-3A. It was found that in the region (cyan) between 450 and 550 nm, red must be subtracted to obtain a match. This would mean that the spectral hue to be matched must be added to the amount of red light shown below the X axis. In practice, a small amount of red was added to the color to be matched. This technique results in a new set of primaries ($\bar{x}$, $\bar{y}$, and $\bar{z}$), linear to the first and shown by Fig. 2-3B. (At the moment, take our word for this. You will develop this technique as we go along.) This form is more easily handled in camera filter (taking) characteristics.

The luminosity response of the human eye as a function of wavelength is shown by Fig. 2-4. This curve shows that the maximum contribution to brightness is made by the yellow-green region of approximately 555 nm.

2-4. THE STANDARDIZATION OF TELEVISION COLOR

Our study of color thus far has been concerned with single-wavelength hues (hues of maximum saturation). To completely specify color, not only the hue is involved; the purity or saturation must be considered also.

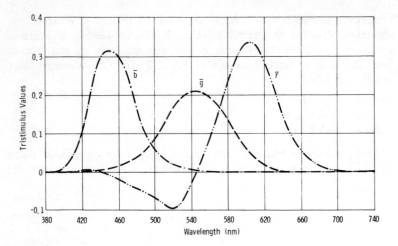

(A) Basic test results.

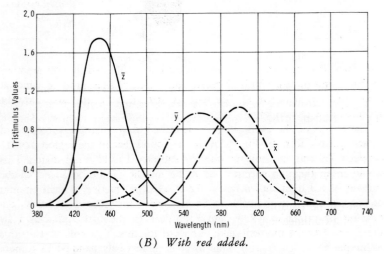

(B) With red added.

Fig. 2-3. Color-primary mixture curves.

For example, the difference between red and pink is the amount of white contained in the red.

One basic method of graphically illustrating hue and saturation is the circle method of Fig. 2-5. The R, G, and B primaries are vectors 120° apart. Degree of saturation (degree of mixture with white) is measured along a vector from center to circumference (zero saturation to maximum saturation). Some point on the circumference could arbitrarily be given a reference phase, and all vectors could be specified relative to this

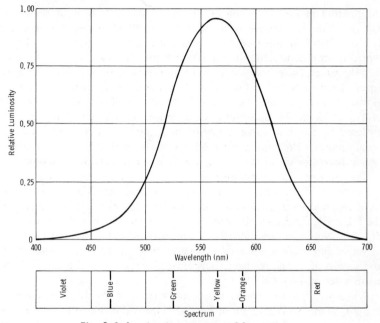

Fig. 2-4. Luminosity response of human eye.

phase. Hue would then be specified by phase, and saturation by amplitude of the vector. It may be seen that for equal amplitude of the primary vectors, the resultant of R, G, and B at 120° would be zero (white). Therefore:

$$R + G + B = White$$

Also, $R + G =$ Yellow, $Y + B =$ White, etc. In this circle diagram, white would be *equal-energy* white at the center of the circle.

Color can be visualized best in terms of a three-dimensional representation, but two-dimensional diagrams can be used also. For example, the

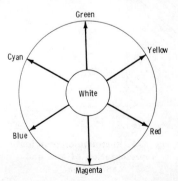

Fig. 2-5. Circle representation of hue and saturation.

information in Fig. 2-3A can be plotted in its RGB form as shown by
Fig. 2-6. This is a simple rectangular-coordinate RGB diagram in which
$R + G + B = 1$. In this case, B has zero value in the x-y plane and is represented in amplitude along the z axis. We must visualize the z axis as going
into the paper, away from the point of observation. Since $R + G + B$ must
equal 1, then $B = 1 - (R + G)$. For equal-energy white, $R = 0.33$,
$G = 0.33$, and (by definition) $B = 0.33$. Also in this case, the $-R$ area
of Fig. 2-3A in the blue-green region must be represented by the area
enclosed by the dash curve. Study of this simple diagram shows that an
equal mixture of red and green produces yellow, with the degree of saturation indicated by the distance along the dash line from white to yellow.

The cumbersome relationship of the $-R$ area can be eliminated by
using an *XYZ diagram.* This diagram is a linear transformation of the
RGB diagram so that all values are positive. Transformation is based on
the following fundamentals: Assume we have two vectors, R and G, as in
Fig. 2-7. Note that the addition of r units of R and g units of G results
in vector sum C. We also may express color C in terms of x units of X
and y units of Y, and we have transformed from one set of axes (RG) to
another set of axes (XY).

We now apply this theory to the conversion of the RGB diagram of
Fig. 2-6 to the XYZ diagram of Fig. 2-8. Line YZ is drawn tangent to the
spectrum locus at 505 nm, and line YX is drawn tangent to the spectrum
locus in such a way that the value of Z is negligible in the red region. The
luminance of X and the luminance of Z are made essentially zero, and the
luminance of Y is given the value of unity. Then:

$$L_X = L_Z = 0$$
$$L_Y = 1$$

where L stands for luminance, or brightness contribution.

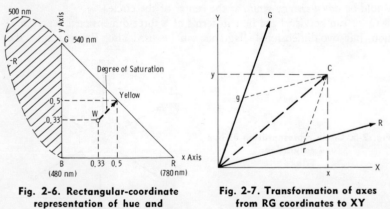

**Fig. 2-6. Rectangular-coordinate
representation of hue and
saturation.**

**Fig. 2-7. Transformation of axes
from RG coordinates to XY
coordinates.**

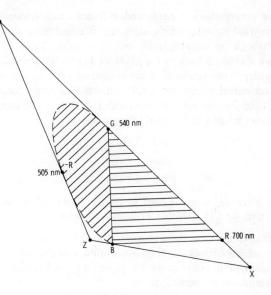

Fig. 2-8. XYZ color diagram.

We can now consider "color space." If we use rectangular coordinates X, Y, and Z and view in three dimensions as in Fig. 2-9, we can observe the relationships of RGB to XYZ, the "neutral" (or "white") axis, and the "constant-luminance" planes. Remember that the X-Z plane (indicated by dotted lines) has zero luminance, and luminance units are expressed along the Y axis as shown. Note that the blue primary, having a small luminance coefficient (see standard luminosity curve, Fig. 2-4), projects

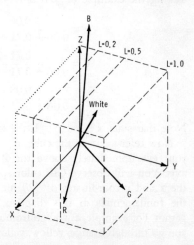

Fig. 2-9. Cube to show color space.

upward at a very shallow angle and will not reach constant L for some distance beyond the edge of the drawing. We will learn more about "constant luminance" as we progress.

Now we should go back to the graph of Fig. 2-3B. This plot of tristimulus values shows the amounts of the primaries (red, green, and blue) that must be combined to achieve a color match at a given frequency (wavelength). Take for example the wavelength 520 nm. The amounts of spectral primaries needed to match this color are:

$$Red\ (\bar{x}\ curve) = 0.06$$
$$Green\ (\bar{y}\ curve) = 0.71$$
$$Blue\ (\bar{z}\ curve) = 0.08$$

NOTE: Since the curve is hard to read to exact values, we have used values from the CIE Table of Tristimulus Values.

(The bars over the x, y, and z markings on the curves in the diagram simply designate instantaneous values, or x units, y units, and z units.)

The chromaticity coordinates (x, y, and z) are found by solving a set of three simultaneous equations:

$$x = \frac{\bar{x}}{\bar{x} + \bar{y} + \bar{z}}$$

$$y = \frac{\bar{y}}{\bar{x} + \bar{y} + \bar{z}}$$

$$z = \frac{\bar{z}}{\bar{x} + \bar{y} + \bar{z}}$$

Now we can see how the three-dimensional color diagram is plotted. Taking the example of 520 nm:

$$x = \frac{0.06}{0.06 + 0.71 + 0.08} = \frac{0.06}{0.85} = 0.07$$

$$y = \frac{0.71}{0.06 + 0.71 + 0.08} = \frac{0.71}{0.85} = 0.84$$

$$z = \frac{0.08}{0.06 + 0.71 + 0.08} = \frac{0.08}{0.85} = 0.09$$

Note that since $x + y + z = 1$, the total of the above results is equal to 1.

Now refer again to the cube of "color space" in Fig. 2-9. Mentally place your eye immediately above the Z axis and look downward along this axis. You will observe CIE color space as plotted in Fig. 2-10. Note that the x and y coordinates of 520 nm are as we derived above. Remember the fundamentals of this diagram and you can always recall the color gamut. For example, as you go from red (700 nm) toward green (520 nm) you go through orange, yellow, yellow-green, and green. The outer perime-

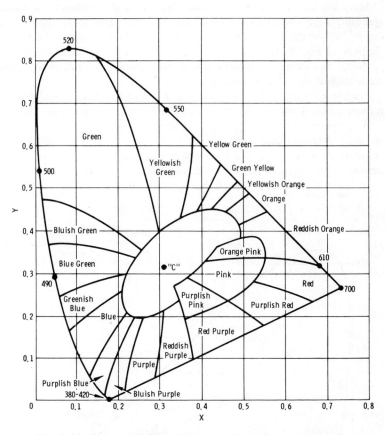

Fig. 2-10. Color perception of the standard observer enclosed in CIE color space.

ter of the horseshoe curve designates hues of maximum saturation (pure, or not mixed with white). As you leave red toward white, you encounter desaturated reds, or pinks. As you travel from green toward blue, you pass through the various blue-greens termed "cyan." The straight line from blue to red designates purples, which actually cannot be described by any single wavelength; therefore, these are known as *nonspectral hues.* They are expressed as complements of the actual spectral colors. Note also that white is here designated as "C," for "illuminant C."

2-5. TELEVISION WHITE AND COLOR STANDARDS

The "color of white" to which given primaries are balanced must be specified and located on the chromaticity diagram. See Fig. 2-11 and note the following definitions:

Spectrum Locus: The locus of points representing the chromaticities of spectrally pure stimuli in a chromaticity diagram. This locus defines the curved part of the outline (outer perimeter) of the rectangular-coordinate diagram of Fig. 2-11.

Purple Boundary: The straight line drawn between the ends of the spectrum locus to complete the outline.

Planckian Locus: The locus of chromaticities of Planckian (black-body) radiators having various temperatures. This concept is expanded below.

A temperature rating is given to various light sources. In this context, the unit of temperature is the *Kelvin* (abbreviated K). The Kelvin scale is based on the concept of a *black-body,* which is a body that absorbs all incident light rays and reflects none; therefore, it is a theoretical concept only. If the black-body is heated, it begins to emit visible light rays start-

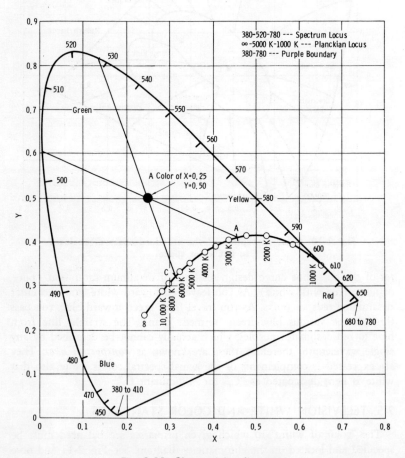

Fig. 2-11. Chromaticity diagram.

ing with dull red, then red through orange to "white heat." We are familiar with this concept through the heating of metals. A low Kelvin rating is given to a light source with most of the visible spectrum in the red region. The higher the Kelvin rating, the closer the light becomes to daylight.

The three standard *illuminants* are designated simply as A, B, and C. Be careful in your concept of this "illuminant." It is a "point" in the three-dimensional color space. It obviously is not a single wavelength; the energy distribution is among all colors. This is shown roughly by Fig. 2-12. Note that illuminant C has a slight peak in the blue region. Television-receiver "white" is very slightly bluish with proper white balance.

A word of caution here: The color-temperature scale (the Planckian locus) shown on Fig. 2-11 is one-dimensional. If you have a given light source of (for example) 6500 K color temperature, the *chromaticity* might fall anywhere on a line drawn across the Planckian locus at 6500 K.

We can also see here why "white" must be specified. Suppose we have a color that is plotted as $x = 0.25$ and $y = 0.5$, as shown on Fig. 2-11. The dominant wavelength (spectrally pure color) is determined by placing a straightedge between the color and the reference-white point, and then extending the line to intersect the spectrum locus. Note that if illuminant A is the reference, the dominant wavelength (therefore the resultant color) is about 503 nm (a color in the green sector approaching bluish green—see Fig. 2-10). If illuminant C is the reference, the color is a "pure green" point on the spectrum locus, or about 530 nm.

Remember this: As the reference white (to which all the colors of a given system are balanced) shifts toward the yellow region (illuminant A), the resultant color shifts counterclockwise around the spectrum locus toward the cyan and blue region.

Now the final steps in standardizing primary chromaticities for color television are in order. This standardization is fixed by the types of phosphors available for the end of the system, the color picture tube.

The luminance contributions of the CIE red and blue primaries, since these primaries are close to the outer ranges of visibility, are very low.

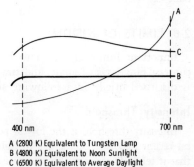

Fig. 2-12. Energy distribution of illuminants A, B, and C.

400 nm　　　　　　　　700 nm

A (2800 K) Equivalent to Tungsten Lamp
B (4800 K) Equivalent to Noon Sunlight
C (6500 K) Equivalent to Average Daylight

Thus, red and blue would require excessively large energies to obtain sufficient brightness. Tables drawn from the standard luminosity curve show the following luminances for CIE primaries.

$$L_R = 0.004$$
$$L_G = 0.98$$
$$L_B = 0.01$$

If the primaries are changed to the FCC values shown in Table 2-1 (which lists both the CIE and FCC primaries), the luminance of the red and blue primaries is increased to a value that is practical for color-phosphor manufacturing.

Table 2-1. Primary Colors

CIE			FCC		
Wavelength (nm)	Chromaticity Coordinates		Wavelength (nm)	Chromaticity Coordinates	
	x	y		x	y
700 (Red)	0.735	0.265	610 (Red)	0.670	0.330
564 (Green)	0.274	0.717	535 (Green)	0.210	0.710
435 (Blue)	0.167	0.0089	460 Blue)	0.140	0.080

Also implicit in the FCC primary specification is that white balance (one unit each of R, G, and B yields white) be for illuminant C. The chromaticity coordinates for illuminant C are:

$$x = 0.310$$
$$y = 0.316$$
$$z = 0.374$$

All of these values are plotted on the diagram of Fig. 2-13, which fits the television color gamut within the CIE color space as balanced to illuminant C.

2-6. LIMITS OF VISION

Important limits of vision to be considered are: intensity threshold, contrast range, visual angle, and time threshold. Each of these limitations is discussed briefly in a following subsection.

Intensity Threshold

Intensity threshold is the lowest perceptible brightness level; it depends on recent exposure of the eye to light. The eye is an ac-coupled device in that it tends to average the range of luminances when interpreting bright-

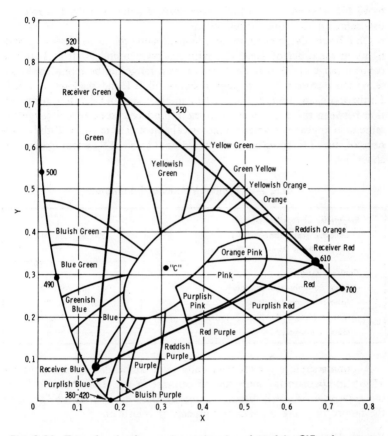

Fig. 2-13. Triangle of color-receiver primaries plotted in CIE color space.

ness sensation. For example, one foot-lambert (ft-L) can be either "white" or "black" depending on the luminance of surrounding areas. The foot-lambert is an absolute measurement of luminance and cannot be construed as resulting in a given brightness sensation. (The foot-lambert is defined in Chapter 9.) Due to the added burden placed on the luminance function in color television, properly designed dc-restorer circuits are important.

Contrast Range

The contrast limit is the least brightness difference that can be perceived. The eye is sensitive to percentage change, *not* to absolute change. If 10 foot-lamberts are increased to 15 ft-L, the absolute change is 5 ft-L, and the percentage change is 50 percent. If an increase from 20 ft-L must produce the same sensation of brightness change, the increase must be to 30 ft-L (50 percent), which is an absolute value of 10 ft-L. Table 2-2 lists some typical values of luminance.

On a bright day, a spectator at a football game might be observing high lights in the order of 1000 ft-L with shadows around 20 ft-L. This is a contrast range of 50 to 1. The same contrast range can be reproduced on a television picture tube with high lights of 50 ft-L and shadows of 1 ft-L. Furthermore, an observer stationed in an enclosure and dividing his attention between the actual scene and the reproduced scene would judge the same total appearance, after psychological adjustment to the 1/20th brightness of the TV receiver, and providing overall TV-system gamma is about 1.4.

Table 2-2. Typical Values of Luminance

Brightness of Familiar Objects	
Outdoor Scene (Bright Day)	600 to 1000 ft-L
Lighted Page (Minimum Recommended)	10 ft-L
Contrast Range	
Clear Sunlight and Shadow	100/1 up to 1000/1
Interior (Normal Artificial Light)	30/1 to 50/1
Motion Pictures for Theater Projection	50/1 to 100/1
Good Photographs	25/1 to 50/1

A gamma slightly greater than unity is recommended for monochrome TV to compensate for some loss of contrast in the rendition of color to the gray scale. We will discuss gamma in more detail after we "get over the hump" on more fundamental aspects of colormath.

Visual Angle

As an object decreases in size, the angle formed by imaginary straight lines from the extremities of the object to the eye (visual angle) becomes smaller. An image must occupy a certain minimum area on the retina to be visible. This minimum area of visibility depends on whether we are concerned with brightness (monochrome) only, or the eye must be able to interpret hue. For color, the minimum size required to distinguish correct hue depends on the particular color. A much smaller area can be defined by the eye for monochrome than is true for any color. In this case (monochrome), the smallest area of visibility depends on brightness and *detail contrast*.

As the visual angle becomes smaller, blues and blue-greens cannot be distinguished as different hues, and gradually merge into grays (neutral). At the point where this just occurs, *orange* can *still be distinguished*. Still further reduction causes orange to become indistinguishable from brown, then finally grays.

Thus, the major share of fine detail in the transmitted picture should be carried by a wideband signal that depends on the luminance information. Then, by properly proportioning R, G, and B, two "transmission primaries" are formed. These primaries are transmitted as signals of less bandwidth than the luminance signal (but greater for the orange-cyan axis than for the yellow-green–purple axis). Thus, chrominance information is conveyed on a principle of band sharing with the luminance signal. Since the chrominance channels have different bandwidths, simple filtering can be used to minimize cross talk in a practical system.

Time Threshold

Time threshold is the minimum time a stimulus must act in order for an image to be perceived. This time depends on size, brightness, and color.

2-7. TRANSMISSION PRIMARIES

In the practical color broadcast system, these specifications apply:

1. The Y signal must be proportional to luminance.
2. Subcarrier signals (two in number) do not affect luminance (not considering type of gamma correction).
3. The subcarrier signals become zero for white (illuminant C).

Under the second requirement above, it is further specified that the subcarrier signals control zero-luminance primaries similar to X and Z, and that:

Primary 1:
Y signal = luminance (Illuminant C)
Chrominance Primaries:
   Primary 2 can be matched by mixture of R and G only. This primary has nothing to do with the amount of blue and is termed the zero-blue primary, or $R - Y$ primary.
   Primary 3 can be matched by mixture of G and B only. This primary has nothing to do with the amount of red, and is termed the zero-red primary, or $B - Y$ primary.

First, we will consider the math function for a three-channel color-camera system. Some color cameras use a total of four channels, one for the luminance function only and three relatively narrow-band primary-color channels; these systems will be covered later.

2-8. SIGNAL PROPORTIONS FOR ILLUMINANT C

The three primary channels of the color-camera output are combined to form the luminance signal. The individual primary-color signals also are

fed to a separate system to derive corresponding electrical signals relative to chrominance only. See Fig. 2-14.

Since the eye is most sensitive to detail in the green region, less in the
red region, and still less in the blue region, the Y signal (luminance channel) is proportioned accordingly. As shown in Fig. 2-14, we are consider-

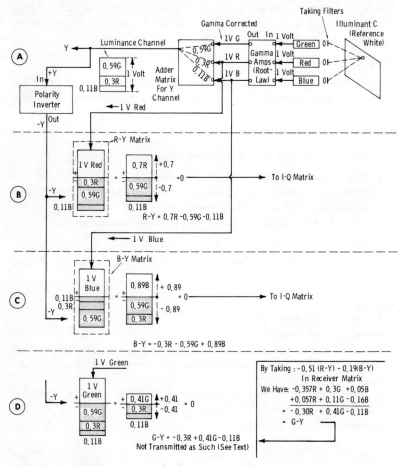

Fig. 2-14. Standard three-camera signal proportionment for illuminant C.

ing the proportions for transmission of a "white" signal. In this case,
$R = G = B = 1$, the unity value of 1 volt being used arbitrarily to show
signal proportions. It should be understood that channel outputs may
vary anywhere between zero and unity, depending upon hue and intensity.
For white (illuminant C), the voltage outputs from the initial amplifiers
of the cameras are equal.

After the signals pass through the adder matrix for the Y channel, one luminance volt is made up of 0.3R, 0.59G, and 0.11B, where R, G, and B represent 1-volt signals from the R, G, and B channels, respectively. This division of energy, made in accordance with the proportionate detail needed, conserves bandwidth and allows narrow-band chrominance information to be interleaved conveniently within the luminance channel; the interleaving process will be described later. (The adder matrix referred to is simply a cross-connected voltage divider.)

Now each of the primaries must contribute to the color subcarrier only information relative to the chrominance, i.e., hue and saturation minus the brightness information contained in the luminance (Y) channel. This method is desirable for a number of reasons, chief of which is the fact that any interference occurring around the color subcarrier frequency will cause a change only in hue, which is far less noticeable to the observer than a change in brightness would be.

The chrominance-system takeoff is located after the gamma amplifiers and before the Y adder matrix (Fig. 2-14A). Fig. 2-14B illustrates how the red-difference signal ($R - Y$) is obtained. The Y signal is fed through a polarity inverter to obtain a $-Y$ signal. Note that the output of the $R - Y$ matrix contains plus 0.7 units and minus 0.7 units; the luminance (Y) therefore is zero. The $R - Y$ color-difference signal is then:

$$R - Y = 0.7R - 0.59G - 0.11B$$

The same function is performed to obtain the $B - Y$ signal (Fig. 2-14C).

$$B - Y = -0.3R - 0.59G + 0.89B$$

Note carefully that each color-difference signal (signal containing only chrominance information) includes all three color primaries. This fact permits the use of only a two-phase color system instead of a three-phase system for conveying the information of three color primaries. The $R - Y$ and $B - Y$ signals modulate the color carrier in quadrature; hue information is contained in the phase, and saturation information is contained in the amplitude. The receiver may then combine the color-difference signals to obtain the $G - Y$ component (Fig. 2-14D).

NOTE: Although we have shown "color-difference" matrices in Fig. 2-14, the latest systems derive I and Q directly from the Y matrix. We want to emphasize in your thinking at this point that I and Q contain not only R, G, and B, but also the "color-difference" primaries.

$R - Y$ and $B - Y$ are easily found simply by subtracting the expression for Y from R and B:

$$R - Y = R - (0.3R + 0.59G + 0.11B)$$
$$B - Y = B - (0.3R + 0.59G + 0.11B)$$

Therefore:
$$R - Y = 0.70R - 0.59G - 0.11B$$
$$B - Y = -0.30R - 0.59G + 0.89B$$

The results of this derivation are indicated on the $R - Y$ and $B - Y$ matrices in Figs. 2-14B and 2-14C.

Now remember we are presently studying the signal proportions for transmission of reference white. This fact sets two conditions:

1. The color-carrier sidebands disappear.
2. The luminance channel contributes "white" (maximum depth of modulation of the video carrier), referred to in our study as unity, or 1 volt.

Condition 2 is obvious from Fig. 2-14A. The luminance signal, composed of $0.3R + 0.59G + 0.11B$ (in which R, G, and B are each 1 volt) has a magnitude of:

$$
\begin{array}{l}
0.3 \\
0.59 \\
\underline{0.11} \\
1.00 \text{ Luminance Volt} = \text{Reference White}
\end{array}
$$

Now consider the chrominance signal, which actually consists of three color-difference signals:

$$
\begin{array}{l}
R - Y = 0.7 - 0.59 - 0.11 \\
B - Y = -0.3 - 0.59 + 0.89 \\
G - Y = \underline{-0.3 + 0.41 - 0.11} \\
R + G + B = 0.1 - 0.77 + 0.67
\end{array}
$$

and:

$$0.1 + 0.67 - 0.77 = 0.77 - 0.77 = \text{zero}$$

Zero modulation of the subcarrier occurs, and the subcarrier sidebands disappear.

2-9. I AND Q TRANSMISSION PRIMARIES

Note from Figs. 2-14B and 2-14C that the $B - Y$ and $R - Y$ signals are fed to a following "I-Q" matrix. This matrix places the color information in such form that zero modulation of the chrominance channel occurs on illuminant C, yet for color information, the $B - Y$ and $R - Y$ components still may be used in the receiver to extract the $G - Y$ signal for the green information. Note that "I" designates the "in-phase" subcarrier and "Q" designates the "quadrature" subcarrier, as we will see later.

Fig. 2-15 shows the action of the I-Q matrix. The I channel combines
−0.27 of B − Y with 0.74 of R − Y to obtain:

$$I = 0.60R - 0.28G - 0.32B$$

(Therefore I can be obtained directly from R, G, and B.)
The Q channel combines 0.41 of B − Y with 0.48 of R − Y:

$$Q = 0.21R - 0.52G + 0.31B$$

(Therefore Q also can be obtained directly from R, G, and B.)
Since we are still considering conditions for reference white:

$$I = 0.60 - 0.28 - 0.32 = 0.6 - 0.6 = 0$$
$$Q = 0.21 - 0.52 + 0.31 = 0.52 - 0.52 = 0$$

Thus, modulation of the color subcarrier is zero. The luminance transmission primary is still $0.3 + 0.59 + 0.11 = 1$ volt.

Before modulation of the color subcarrier, the I and Q signals are fed to separate filters that have characteristics as shown in Fig. 2-15. The I

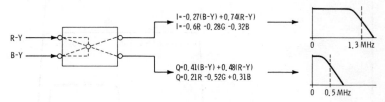

Fig. 2-15. Action of I-Q matrix.

channel is broad-banded, but has vestigial-sideband characteristics because of the upper-frequency cutoff of the transmitter. This channel is single-sideband for frequencies higher than 500 kHz. Transmission of frequencies up to 500 kHz is double-sideband on both the I and Q channels. Such operation allows two types of receiver action: (1) The receiver may utilize the extra color information in the wideband I channel, or (2) receivers may ignore this extra information and reproduce only chrominance detail supplied up to 500 kHz. We will go into more detail on this later, as well as the reason for the difference in bandwidths for I and Q.

Now we want to consider what happens in transmitting a monochrome picture at points other than "reference white." Continue to consider that white corresponds to maximum depth of modulation, 1 volt in our example. Consider now that the scanned point in question is a gray between maximum black and maximum white. Since no specific color exists, the outputs of the camera pickup tubes still will be equal. However, since the luminance is not as great for gray as for white, the amplitudes are reduced accordingly. For example, assume the amplitudes are 0.5 volt each.

In the Y channel, the 0.5 volt will be made up of 30 percent from the red, 59 percent from the green, and 11 percent from the blue channel. The resulting 0.5 volt will reproduce gray on the receiver picture tube, since the depth of carrier modulation is one-half that for white. This is conventional monochrome action.

As long as the initial camera outputs are equal, the combined chrominance proportions will add up to zero as in the case for reference white. The color subcarrier therefore is *not modulated* for *any* condition of monochrome transmission, and no sidebands occur. The receiver picture tube then reacts only to the brightness information in the Y channel.

Fig. 2-16 indicates the band-sharing proportionment of the composite color signal (E_m) as far as we have gone. The exact color subcarrier frequency is 3.579545 MHz, normally referred to as 3.58 MHz.

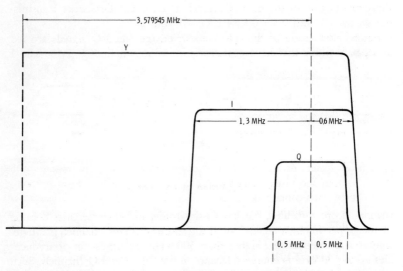

Fig. 2-16. Frequency distribution of color signal.

2-10. LUMINANCE AND CHROMINANCE MODULATION LEVELS

The word "compatibility" implies that conventional monochrome receivers be able to reproduce a color telecast in black and white. This requirement immediately fixes the bandwidth for colorcasting as no more than the established 6 MHz per television channel. Therefore, the addition of the color information must be by a method that will not add to the required bandwidth.

The pickup tube in a monochrome camera is scanned at the line frequency of 15,750 Hz. When the scanning beam sweeps across the target at this frequency, the beam is changed in amplitude in accordance with the charge pattern that corresponds to the focused image. The resultant rate of voltage changes (frequency of voltage change) will always be some multiple of the initial scanning rate of 15,750 Hz (Fig. 2-17A). Therefore, the major signal components lie at integral multiples of the line-scanning rate.

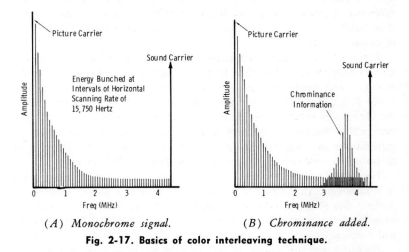

(A) Monochrome signal.   (B) Chrominance added.

Fig. 2-17. Basics of color interleaving technique.

These clusters of signal information decrease in amplitude as they get farther from the visual carrier frequency. Gaps formed between the signal components contain no great amount of information at all. These *gaps* occur at *odd multiples of one-half the line frequency*. Thus, if we generate a subcarrier at some odd multiple of one-half the line frequency, subcarrier sidebands will lie in the gaps formed by harmonics of the line frequency (Fig. 2-17B). This process is termed "interleaving" in color-TV systems. The color subcarrier is placed high in the band, and its maximum-amplitude sidebands occur where the monochrome sidebands are small. As the color sidebands get farther from the subcarrier and nearer the monochrome (main visual) carrier, they decrease in amplitude and produce minimum effect in that region, where the monochrome sidebands are larger.

Since there are an odd number of lines (525), an odd multiple of one-half the line frequency is also an odd multiple of one-half the frame rate. A frame is composed of two (interlaced) fields. For each field, a point on one line that is made brighter by the color-subcarrier sidebands lies directly above a point on the succeeding line, which is made darker. When the viewer is far enough from the screen that the lines are not distinguishable, this "space-integration" effect cancels any brightness variations caused by the presence of the color sidebands in the composite signal. Also, a "time integration" occurs since brightness variations in corresponding lines of successive frames are 180 degrees out of phase, provided the color sidebands fall exactly at odd multiples of one-half the frame frequency. (Since during a frame the color frequency passes through a whole number of cycles plus one-half cycle, the patterns produced by the color subcarrier in corresponding lines of successive frames are one-half cycle, or 180 degrees out of phase.) Since interference between the two carriers cancels, it is only ncessary for the receiver to demodulate each signal with respect to its own carrier.

The frequency of the chrominance subcarrier is 3.579545 MHz. This is an odd multiple of half the line frequency for proper multiplexing (interleaving) with the luminance signal, as pointed out above. This frequency is not so high as to cause interference to the sound carrier or restriction of the chrominance bandwidth. At the same time, it is not so low that conventional monochrome receivers will be visibly affected by chrominance information in nonlinear circuits. At frequencies above 3.5 MHz, most monochrome receivers have attenuation of 10 to 25 dB or more.

The chrominance subcarrier frequency also was chosen so as to result in minimum beat-frequency interference with the sound carrier. To achieve this result, it is necessary that the frequency offset between the sound and chrominance carriers also be an odd multiple of one-half the line frequency.

Thus, the necessary choice of the chrominance subcarrier frequency has resulted in a slight change in actual line and frame frequencies from those of previous monochrome standards. To make the chrominance carrier frequency an odd multiple of half the line rate, the *new* line rate becomes 15,734.26 Hz. This is a reduction of only about 0.1 percent and is well within the range of existing monochrome-receiver hold controls. The new field rate then is:

$$\frac{15,734.26}{525/2} = 59.94 \text{ Hz}$$

Thus, stations no longer use "line-lock" circuits to lock the field rate (and, hence, all synchronizing signals) to the power-line frequency (60 Hz). The color subcarrier frequency is generated by a crystal oscillator, and counters are used to obtain a driving signal for the conventional sync generator, as shown later.

The color-sync burst consists of 8 to 10 cycles at a frequency of 3.579545 MHz. Its average value coincides with blanking level, and its peaks extend above and below this level by about one-half the sync-pulse amplitude. The actual peak-to-peak excursion is the same as the sync amplitude. Extension below blanking level has a tendency to brighten retrace lines. This effect is minimized by proper reference-black *setup* at the studio and by the retrace-blanking circuits in modern receivers. (Setup refers to the difference between black level and blanking level.)

Review the signal proportionments for Y, I, and Q, and then study Fig. 2-18. The chrominance-sideband amplitudes are added to the luminance (Y) levels as shown in Fig. 2-18A. The solid lines represent the respective Y levels as observed on an oscilloscope connected at the encoder (RCA terminology is "colorplexer") output when the I and Q channels in the encoder are turned *off* (or as observed through a low-pass filter that eliminates the higher-frequency signals). The amplitudes are for an encoded composite color signal (E_m) for *fully saturated* color bars. Note that the setup level is 7 percent, and that picture black, not the blanking level, is used as zero reference level.

Important: Modern color-bar generators produce color bars in descending order of luminance as follows: yellow, cyan, green, magenta, red, and blue. Color-bar generators are covered further later in this book. The bars of Fig. 2-18 are given to allow detailed analysis by the student of all color video possibilities.

Note that the luminance of yellow (red plus green) is 0.89, which is the sum of red luminance (0.3) and green luminance (0.59). The chrominance amplitude for yellow (Fig. 2-18B) is 0.447, which added to 0.89 is 1.337, resulting in a 33 percent overshoot for a *fully saturated* yellow signal. You will see later how the chrominance modulation must actually be limited to avoid greater than a 33 percent excessive amplitude for fully saturated signals. Of course, a color-bar signal is artificial, and such conditions are seldom encountered in normal pictures, as will be obvious in future discussions.

Note also (for practice) that cyan is green plus blue, and 0.59 plus 0.11 equals 0.70. Magenta is red plus blue, or 0.41.

Note from Fig. 2-18C the effect on the I and Q additions. Since yellow is R and G minus blue, yellow is the *complement* of blue. Note that yellow and blue have identical amplitudes in I and Q but are of inverse polarity. The complements (cyan, magenta, and yellow) of the primary colors (red, green, and blue) have the same I and Q amplitudes as their respective primaries, but are opposite in phase (see Fig. 2-18B).

The next section starts a study of the encoding system. In it, you will learn how the chrominance amplitudes and phases result.

2-11. FUNDAMENTALS OF CHROMINANCE MODULATION

Let us review briefly some fundamental ac theory important to the handling of colormath. See Fig. 2-19. This diagram shows point A moving about point O in a circle of radius R. Point A is moving at a constant angular velocity, which is $2\pi f$ (hereafter designated ω). Point B indicates the instantaneous projection of A on the x axis. The angular displacement measured counterclockwise from the x axis is indicated by θ

and is equal to ωt. (2π times the frequency times the time from reference point).

Note that as point A moves about its circular path, projection B moves back and forth on the x axis so that line OB varies in length between zero and R, which is to say between zero and E_{max}. The length for any angle is:

$$x = R \cos \theta = R \cos \omega t$$

And since $R = E_{max}$:

$$x = E_{max} \cos \omega t$$

The projection on the y axis is:

$$y = E_{max} \sin \omega t$$

Note also that x is really OB and that in the above example it is assumed that the reference starting point is zero degrees. Now if we assume a refer-

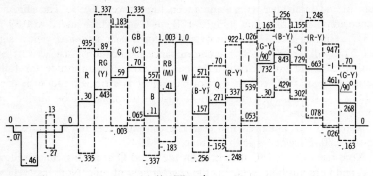

(A) Waveform.

Color	Y	I	Q
Red	0.30	+0.60	+0.21
Yellow (R + G)	0.89	+0.32	−0.31
Green	0.59	−0.28	−0.525
Cyan (G + B)	0.70	−0.60	−0.21
Blue	0.11	−0.32	+0.31
Magenta (R + B)	0.41	+0.28	+0.525
White	1.0	0	0

(C) Y, I, and Q amplitudes.

Fig. 2-18. Signal levels

ence starting point at some other angle, at any time later the projection is given by:

$$OB = E_{max} \cos (\omega t + a)$$

where a is the additional angle.

Signal	R	G	B	Y	Subcarrier	
					Amplitude	Phase
R	1	0	0	0.30	0.635	103.4°
RG (Yellow)	1	1	0	0.89	0.447	167.1°
G	0	1	0	0.59	0.593	240.8°
GB (Cyan)	0	1	1	0.70	0.635	283.4°
B	0	0	1	0.11	0.447	347.1°
BR (Magenta)	1	0	1	0.41	0.593	60.83°
RGB (White)	1	1	1	1	0	—
B − Y (R − Y = 0)	0.1571	0	1	0.1571	0.4135	−0.367°
Q (I = 0)	0.5371	0	1	0.2711	0.4265	33°
R − Y (B − Y = 0)	1	0	0.3371	0.3371	0.5848	90.03°
I (Q = 0)	1	0.4056	0	0.5393	0.4865	123°
(G − Y)/90° (G − Y = 0)	1	0.7317	0	0.7317	0.4313	146.38°
Color Burst	—	—	—	—	0.20	180°
−(B − Y) (R − Y = 0)	0.8429	1	0	0.8429	0.4135	179.6°
− Q (I = 0)	0.4629	1	0	0.7289	0.4265	213°
−(R − Y) (B − Y = 0)	0	1	0.6629	0.6629	0.5848	270.03°
− I (Q = 0)	0	0.5944	1	0.4607	0.4865	303°
−(G − Y)/90° (G − Y = 0)	0	0.2683	1	0.2683	0.4313	326.4°

(B) *Color-bar signal table.*

corresponding to color bars.

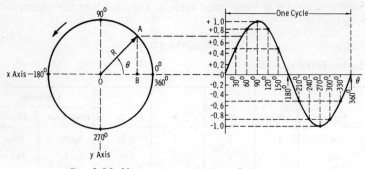

Fig. 2-19. Vector representation of sine wave.

Amplitude Modulation of Carriers in Quadrature

Now assume we are going to modulate two carrier waves that are phased in time by 90 degrees. Since this phase relationship must be maintained, the carriers must have exactly the same frequency.

See Fig. 2-20A. The vector sum of two equal-amplitude quadrature carriers forms an angle (θ) of 45 degrees. When carrier Q is reduced in amplitude (I remaining the same), the resultant angle is increased as

(A) I and Q equal.

(B) I larger than Q.

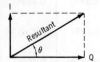

(C) Q larger than I.

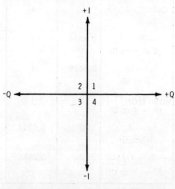

(D) Quadrants for resultant.

Fig. 2-20. Vectors in quadrature.

measured from the x axis, and reduced as measured from the y axis (Fig. 2-20B). When carrier I is reduced in amplitude (Q remaining the same), the resultant angle is decreased as measured from the x axis, and increased as measured from the y axis (Fig. 2-20C). Therefore, amplitude modulation of carriers in quadrature produces phase modulation as well as sidebands resulting from the amplitude-modulation components.

From Fig. 2-20D we can see that $+I$ and $+Q$ result in an angle (θ) in the first quadrant; $+I$ and $-Q$ result in θ in the second quadrant; $-I$ and $-Q$ result in θ in the third quadrant; and $-I$ and $+Q$ result in θ in the fourth quadrant.

Basic Chrominance Modulation Principles

Sync and blanking pulses are generated in the conventional sync generator at the studios. The difference from conventional monochrome transmission lies in the practice of driving the sync generator from circuits that derive a countdown from the color-subcarrier generator. In this way, the chrominance and luminance control pulses are locked into an integral relationship.

Fig. 2-21 illustrates the basic action in block form. The subcarrier generator feeds the color modulators with signals delayed 57° for the I modulator and 57° + 90° = 147° for the Q modulator. The subcarrier is therefore split into quadrature components. The hue of the color is represented by the phase of the vector sum relative to burst, while the saturation of the particular hue is represented by the amplitude of the vector sum. The modulators are double-balanced, eliminating both the subcarrier and the original I and Q video signals. Only the sidebands exist in transmission. The net result is that the color sync burst *leads* the I and Q signals, which are in quadrature. The synchronous demodulator in the receiver uses the sync burst as a reference from which to decode separately the information in the I and Q channels.

We are now in a position to see the formation of the actual transmission primaries. We know that the $B-Y$ and $R-Y$ signals were modified in amplitude in the I-Q matrix to form:

$$\left.\begin{array}{l} I = -0.27(B-Y) + 0.74(R-Y) \\ Q = 0.41(B-Y) + 0.48(R-Y) \end{array}\right\} \begin{array}{l} \text{Each contains} \\ \text{both } R-Y \text{ and } B-Y \end{array}$$

Now from Fig. 2-21 we see that the I and Q video signals will be 90 degrees apart because the I-Q subcarrier drives are phased in quadrature. The I components lead the Q components by 90 degrees.

Now we will look at what happens to the color-difference signals that form the I and Q channels. First consider the $R-Y$ component. Observe Fig. 2-22A (also note that Fig. 2-22B shows that the $B-Y$ signal is arbitrarily assigned the zero reference axis). The I channel, which contains 0.74 of $R-Y$, leads the Q channel, which has 0.48 of $R-Y$, by 90 degrees. (See I and Q equations above for amplitude values.) Note that the vector sum is $0.877(R-Y)$. The I and Q vectors are drawn with the respective phase angles shown by Fig. 2-21.

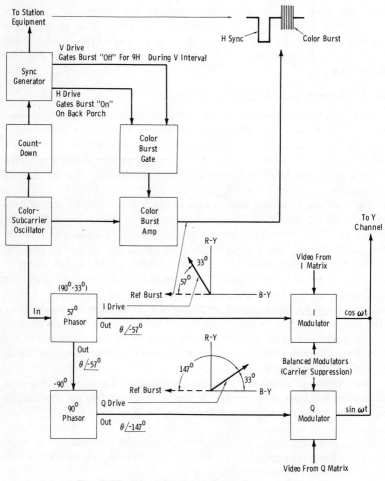

Fig. 2-21. Principle of color-signal generation.

For the $B-Y$ component (Fig. 2-22B), the I vector goes negative by $-0.27(B-Y)$, and the Q vector goes to $0.41(B-Y)$. The vector sum is $0.439(B-Y)$.

We will see later that this reduction in amplitude is necessary to prevent more than a 33 percent overshoot on certain maximum saturated colors. Respective amplitudes are restored in the receiver decoding system.

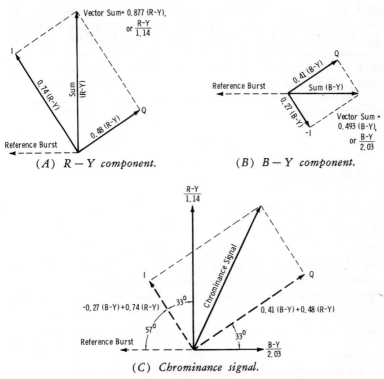

(A) R — Y component.

(B) B — Y component.

(C) Chrominance signal.

Fig. 2-22. Relationships of R — Y, B — Y, I, and Q vectors.

Fig. 2-22C shows the total as given by FCC standards. Note that I and Q are in quadrature, and that the color-difference components also are in quadrature. The R — Y and B — Y components have a simple 90- and 180-degree phase relationship to the reference color-sync burst. Narrow-band color receivers utilize this arrangement in direct demodulation of the color-difference components with comparatively simple circuits. Wide-band receivers utilizing full I-channel bandwidth are more complicated, not so much because of the phase relationships as because of the more elaborate matrixes necessary and the facilities used to modify the effect of cross talk caused by the vestigial sideband.

The vectors shown in Fig. 2-22 indicate the relationships of the chrominance information (only) in terms of I, Q, R — Y, and B — Y. This is not a fixed relationship, as witness the fact pointed out previously that on "television white" the sum of I and Q, and of R — Y and B — Y, is zero. (The vectors then "collapse," and the chrominance signal is zero.) Also, the quadrature system may express any phase angle from 0 to 360 degrees.

The phase angles of the I and Q transmission primaries were chosen so that the I (wideband) axis lies along the orange-cyan region of the color gamut. The reason for this already has been pointed out: orange can be distinguished in areas so small that other hues are indistinguishable from grays. A wider "color bandwidth" therefore is employed for I than for Q. Remember, however, that all color receivers and some color monitors use only $R - Y$ and $B - Y$; only the most costly color monitors demodulate on the I and Q axes.

At this point, you should be able to visualize the equation of the complete color signal as follows:

$$\underbrace{E_M =}_{\substack{\text{Composite} \\ \text{Color} \\ \text{Signal}}} \underbrace{E_Y'}_{\substack{\text{Luminance} \\ \text{Infor-} \\ \text{mation}}} \underbrace{+ \ [E_Q' \sin(\omega t + 33°) + E_I' \cos(\omega t + 33°)]}_{\text{Chrominance Information}}$$

$$E_Y' = 0.30E_R' + 0.59E_G' + 0.11E_B'$$

$$E_Q' = 0.41 \ (E_B' - E_Y') + 0.48 \ (E_R' - E_Y')$$

$$E_I' = -0.27 \ (E_B' - E_Y') + 0.74 \ (E_R' - E_Y')$$

E_M is the total video voltage, corresponding to the scanning of a particular picture element, applied to the picture-transmitter modulator.

E_Y' is the gamma-corrected voltage of the monochrome portion of the color picture signal, corresponding to the given picture element.

E_R', E_G', and E_B' are the gamma-corrected voltages corresponding to red, green, and blue signals during the scanning of the given picture element.

The gamma-corrected voltages are suitable for a color picture tube having primary colors with the following chromaticities in the CIE specification:

	x	y
Red (R)	0.67	0.33
Green (G)	0.21	0.71
Blue (B)	0.14	0.08

2-12. DERIVATION OF CHROMINANCE PHASE

Assume the camera to be scanning a highly saturated red surface, at maximum brightness (Fig. 2-23A). We now may trace the action as we did previously for white. The output of the gamma amplifiers will have 1 volt for red, zero for blue, and zero for green (see Table 2-3). The luminance channel then has just 0.3 volt, the luminance level of the red signal. As shown by Fig. 2-23A, these voltages result in an $R - Y$ chrominance primary of 0.7, and a $B - Y$ chrominance primary of -0.3. When put into transmission primaries as discussed above, $R - Y$ becomes 0.614 at 90 degrees, and $B - Y$ becomes -0.148 (Fig. 2-23B). Note that the

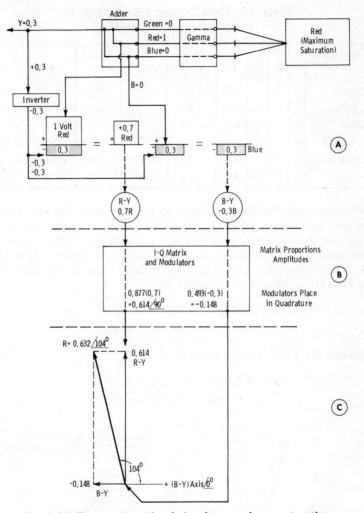

Fig. 2-23. Transmission of red signal at maximum saturation.

positive $B - Y$ axis is taken as the reference of zero degrees; the color sync burst is designated as plus 180 degrees. The vector sum for the red primary is therefore 0.632 at 104 degrees (Fig. 2-23C). This amplitude is for maximum saturation; mixture with white would simply reduce the amplitude. Contamination with any other *hue* would change the phase.

Fig. 2-24 illustrates in detailed form the derivation of the vector for maximum saturated red in terms of I and Q.

Now examine Fig. 2-18B again. You will see here that the red phase is spelled out as 103.4°. Note also that the $B - Y$ axis (on the right-hand,

Table 2-3. Color System Relationships for Primaries and Complements

Transmitted Color	E_G	E_R	E_B	E_Y	G — Y	R — Y	B — Y	Q	I
Green	1	0	0	0.59	0.41	—0.59	—0.59	—0.525	—0.28
Yellow	1	1	0	0.89	0.11	0.11	—0.89	—0.31	+0.32
Red	0	1	0	0.3	—0.3	0.7	—0.3	+0.21	+0.60
Magenta	0	1	1	0.41	—0.41	0.59	0.59	+0.525	+0.28
Blue	0	0	1	0.11	—0.11	—0.11	0.89	+0.31	—0.32
Cyan	1	0	1	0.7	0.3	—0.7	0.3	—0.21	—0.60

or positive, side) is designated in terms of full counterclockwise rotation of the vector and is not *exactly* 0° or 360°, but is —0.367°; the R — Y axis (on the upper, or positive, axis) is specified as 90.03°. These are the absolute values as derived to a large number of decimal places from the NTSC color formulas. For example, we have taken the practical values of I and Q (in terms of R, G, and B) as:

$$I = 0.6R - 0.28G - 0.32B$$
$$Q = 0.21R - 0.52G + 0.31B$$

whereas, the actual absolute values are:

$$I = 0.599R - 0.2773G - 0.3217B$$
$$Q = 0.213R - 0.5251G + 0.3121B$$

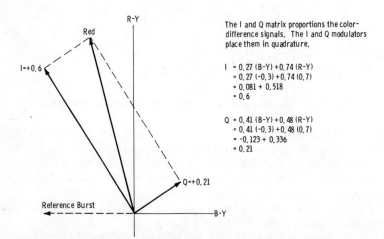

The I and Q matrix proportions the color-difference signals. The I and Q modulators place them in quadrature.

$$I = 0.27 (B-Y) + 0.74 (R-Y)$$
$$= 0.27 (-0.3) + 0.74 (0.7)$$
$$= 0.081 + 0.518$$
$$= 0.6$$

$$Q = 0.41 (B-Y) + 0.48 (R-Y)$$
$$= 0.41 (-0.3) + 0.48 (0.7)$$
$$= -0.123 + 0.336$$
$$= 0.21$$

Fig. 2-24. I and Q vectors for red of maximum saturation.

Therefore, do not be confused by small differences in values of color specifications you may encounter in your studies of this system. Fig. 2-18B was given for convenience in finding the absolute values of the mathematics of color. For all practical purposes, in our measurements we may consider the positive B − Y axis as 0°, with the reference burst as 180° and the six "major" colors (three primaries and their complements) as follows:

Color	Phase (θ)
B − Y	0°
Magenta	61°
R − Y	90°
Red	104°
Yellow	167°
Reference Burst	180°
Green	241°
Cyan	284°
Blue	347°

These relationships for maximum-saturation conditions (maximum values of vector amplitudes) are illustrated in polar form in Fig. 2-25. Note that these amplitudes explain why the red and blue chrominance

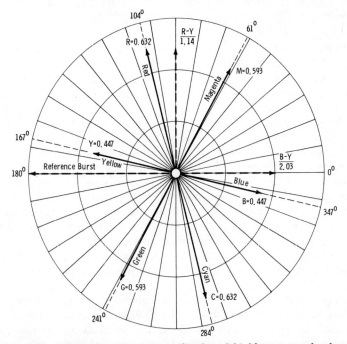

Fig. 2-25. Phase and maximum amplitudes of highly saturated colors.

signals exceed luminance levels into the sync region, and why yellow extends beyond carrier cutoff (Fig. 2-18A). Also note that cyan is in quadrant 4 directly opposite red in quadrant 2; yellow is directly opposite blue; magenta is directly opposite green; and that these "complimentary colors" have the same amplitude as their primaries. The amplitudes must be held within ±20 percent for color fidelity. Further deviation results in both luminance and chrominance errors, since the ratio of color carrier to luminance carrier affects saturation information in colored areas.

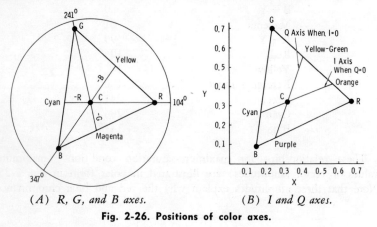

(A) R, G, and B axes.   (B) I and Q axes.

Fig. 2-26. Positions of color axes.

We may place the respective color axes on the chromaticity diagram of Fig. 2-26A, assuming an overall linear system (gamma = 1). Departure from linear gamma would be shown by curved lines. Note first the red-cyan axis. As the red vector collapses into illuminant C (white), the minus red axis continues to cyan. Now observe from Fig. 2-25 that cyan is actually the vector sum of the green and blue components, as could be proved by completing the parallelogram as in Fig. 2-24. Thus, when the beam for the red phosphor is biased off, and the blue and green phosphors are excited by the amplitudes of Fig. 2-25, cyan results. Now follow the green-magenta axis of Fig. 2-26, and note that as green goes through white, the minus green axis projects to magenta along the red-blue axis. See also from Fig. 2-25 that magenta is actually the vector sum of red and blue. In this case, the green phosphor is not excited, and excitation of the red and blue dots produces magenta. Yellow is the vector sum of red and green. If the red, green, and blue vectors are added, the vectors collapse (vector sum = 0), and white (illuminant C) is contributed entirely by the luminance channel.

In Fig. 2-26B, I and Q are placed on their respective axes on the color triangle. The I axis (with Q = 0) is able to define colors along the orange-cyan line; the Q axis (with I = 0) defines colors along the line from yellow-green to purple. The wideband information (I signal) was chosen along this axis after careful field testing revealed the advantage in defining smaller color elements in this region. The color areas from yellow-green to purple require less definition for satisfactory reproduction. Frequencies up to 500 kHz allow a definition of approximately 50 lines, and frequencies up to 1.3 MHz (the I channel) allow resolution of approximately 140 lines.

For convenience and further reference to follow, Table 2-3 lists color-system relationships for given hues, at *maximum saturation*.

Thus, we have a three-primary (full-color) system capable of four general sets of conditions:

1. The luminance voltage (E_Y) carries the fine detail of the smallest picture elements in monochrome.
2. If only E_Y and E_I are active, we have a two-primary orange-cyan system.
3. If only E_Y and E_Q are active, we have a two-primary yellow-green to purple system.
4. If E_Y, E_I, and E_Q are active, we have the three-primary full-color system.

2-13. CHROMINANCE MODULATION

The heart of the encoder is the modulation process by which the chrominance information is translated into sidebands of the subcarrier frequency. The modulation process (for each of the two subcarriers, I and Q) consists of applying both carrier and video in push-pull at the inputs, with the modulator outputs connected in parallel (Fig. 2-27A). Note that the subcarrier is injected into the suppressor grids of V3 and V4 180° out of phase by means of phase-splitter stage V1 (or a transformer may be used). Since the space currents of V3 and V4 are out of phase, the carrier-frequency voltage produced at the common plate connection is zero. In a similar way, the video is fed to the control grids of V3 and V4, and again we have cancellation of the signal at the common plate output. Whenever the input is applied push-pull and the outputs are paralleled (or vice-versa), the fundamental frequency of the input signal is cancelled at the output.

Thus, if the two tubes are operating identically, and if the input signals are perfectly balanced, there will be zero output in the common plate load. In practice, because of asymmetries of circuit components or tubes, or unbalance of subcarrier signals, some subcarrier output would occur in the absence of video. So a carrier-balance control must be provided, as shown, to result in zero subcarrier output when no video is applied. Similarly, if an unbalance results so that some video signal appears in the output, a video-balance control must be provided so that only carrier sidebands exist in the modulator output.

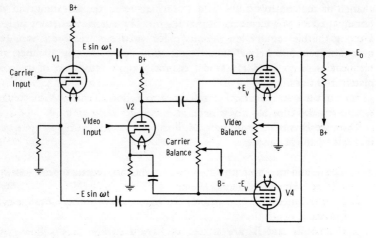

(A) Double-balanced modulator.

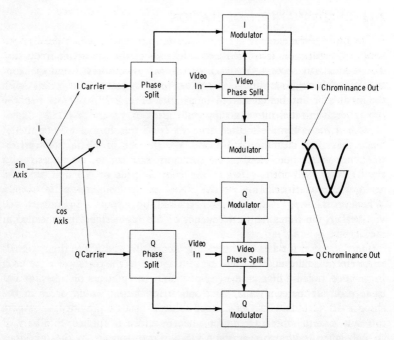

(B) I and Q modulation.

Fig. 2-27. Subcarrier modulation.

Each of the controls interacts on the other to a certain extent, but the principal effect of each is marked enough to make the distinction useful. Varying the bias on the control grids sharply affects the total space current. The suppressor grids control the *division* of this current between plate and screen. For a given voltage swing on the suppressor grid, the plate-current swing will be greater if the total space current is greater, and smaller if the space current is smaller. Hence, the carrier-balance control, which sets the control-grid biases, may be used to control the subcarrier output of each side. Naturally, changing the bias on Grid 1 also changes the video gain of the tube, and this change appears as a secondary effect.

Adjusting the video-balance control varies the amount of cathode degeneration (and thus the video gain) of each side and therefore acts as a video balance. However, it also affects the net control-grid bias and therefore has a secondary effect on the subcarrier balance. In all encoders, you will find these two adjustments must be repeated a number of times to obtain complete cancellation of both the subcarrier and video frequencies.

The function is illustrated in block form by Fig. 2-27B. Note that the I and Q carriers are of the same *frequency* as the subcarrier, but are in quadrature, or 90 degrees apart in phase. Note also that the carriers, since they are introduced by the method indicated by Fig. 2-27A, cancel in the output. When no video (chrominance) modulation is introduced, no sidebands result. With chrominance modulation, sidebands occur in the output, with the fundamental carrier frequency suppressed. Actually, we have amplitude-modulated sidebands that are $90°$ apart in phase so that the receiver may separate the two pieces of information as outlined later. We know that the amplitudes depend on the degree of saturation of the color at the point being scanned, and that the output contains a phase angle, relative to the color-sync burst, that depends on the hue.

Since white is composed of all primary colors, any particular hue that is not at maximum saturation will have some red plus green plus blue added to the dominant hue. Note particularly that addition of *all* colors by an *equal amount* will not change the dominant hue, but will change its shade, or color characteristic. This particular characteristic is termed the *degree of saturation*.

Fig. 2-28 illustrates the development of the I-Q vector for transmission of a red that is *not* at maximum saturation (red plus white). Compare this to the action for a red of maximum saturation (Fig. 2-24).

Assume that the degree of white mixture is such that 0.5 unit of each primary is added (Fig. 2-28A). Since Y is 30 percent red, 59 percent green, and 11 percent blue, the luminance channel carries (in monochrome) 0.3R plus 0.295G plus 0.055B for a total of 0.65 volt. Compare this to a Y of 0.3 volt for red of maximum saturation.

Now note the formation of the color-difference signals (Fig. 2-28B):

$$R - Y = +0.35$$
$$B - Y = -0.15$$

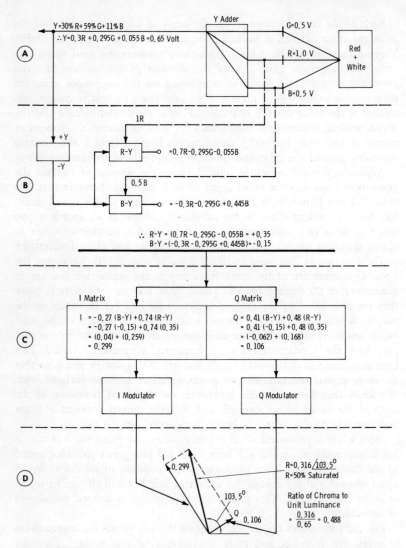

Fig. 2-28. Chrominance-luminance ratio for red of 50-percent saturation.

Then the I-Q matrix (Fig. 2-28C) results in:

$$I = +0.299$$
$$Q = +0.106$$

Compare this result to the condition of maximum saturation where I equals +0.60 and Q equals +0.21.

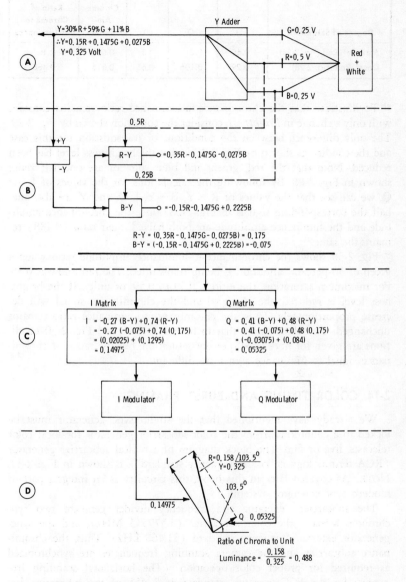

Fig. 2-29. Chrominance-luminance ratio, red of 50-percent saturation, low brightness.

Interpretation of saturation occurs in the ratio of chrominance signal to luminance signal. The difference between Fig. 2-24 and the example given by Fig. 2-28 may be tabulated as follows:

Degree of Saturation	I	Q	Y	Chroma Amplitude	Ratio of Chroma to Unit Luminance
Max. Sat. Red (Fig. 2-24)	0.60	0.21	0.3	0.632	2.1
Desat. Red (Fig. 2-28)	0.299	0.106	0.65	0.316	0.488

To illustrate that the chrominance-to-luminance ratio remains constant with only a change in *brightness,* consider the condition shown by Fig. 2-29. The only difference between the conditions of transmission for this case and the conditions shown by Fig. 2-28 is that the brightness level has been reduced. Note that the red, green, and blue voltages are one-half those shown in Fig. 2-28. By following the calculations for the values of I and Q, we can see that the values of $R - Y$, $B - Y$, I, Q, and Y are also one-half the corresponding signals in Fig. 2-28. Since the chrominance amplitude and the luminance amplitude are both halved, their *ratio* (0.488) remains the same.

Fig. 2-30 shows the chrominance-to-luminance amplitude ratio when a green of 50-percent saturation is being transmitted. The ratio now is 0.38. For maximum saturation, the ratio is 0.59 to 0.59, or unity. If the brightness level is reduced, the Y signal and the chrominance signal will decrease proportionately. Thus, the chrominance-to-luminance ratio remains unchanged. You should note that in both Fig. 2-28 and Fig. 2-30 conditions are given for hues of 50-percent saturation, but the ratios of chrominance signal to luminance signal are different.

2-14. COLOR TIMING AND BURST PHASING

We already have mentioned that the studio sync generator must be locked to a countdown from the color-subcarrier generator for local color telecasts, live or film. The block diagram of a typical subcarrier generator (RCA terminology is "color frequency standard") is shown in Fig. 2-31. NOTE: As covered later in this book, this circuitry is an integral part of modern sync-generator systems.

The subcarrier generator and frequency divider generate two synchronous signals: the color subcarrier (3.579545 MHz), and the sync-generator external frequency standard (31.468 kHz). Thus, the chrominance subcarrier and the camera scanning frequencies are synchronized as required for proper color operation. (The horizontal scanning frequency for the FCC-approved standard is 2/455 of the subcarrier frequency, or 15,734.26 Hz, and the sync-generator external standard is twice this frequency.) The reference frequency for the sync generator is obtained by first generating the subcarrier in a crystal-controlled oscillator and then performing the following frequency divisions and multiplications on the subcarrier:

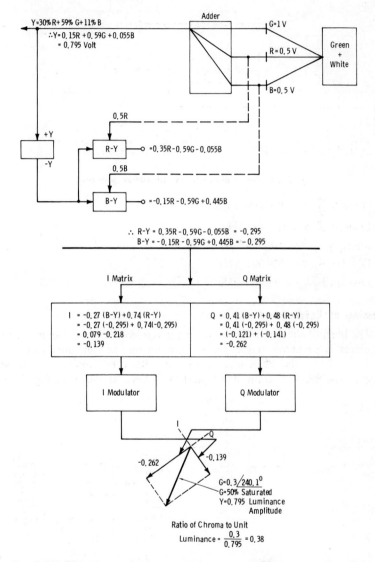

Fig. 2-30. Chrominance-luminance ratio, green of 50-percent saturation.

Divided by 5 = 715.909 kHz
Divided this by 7 = 102.273 kHz
Multiply this by 4 = 409.091 kHz
Divide this by 13 = 31.468 kHz

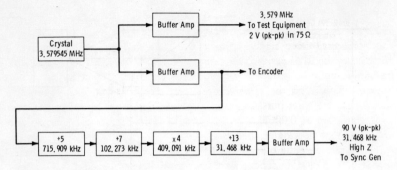

Fig. 2-31. Block diagram of subcarrier generator.

Another sequence commonly used is:

Divided by 5 = 715.909 kHz

Multiply by 4 = 2.86364 kHz

Divided by 7 = 409.091 kHz

Divided by 13 = 31.468 kHz

Timing of Reference Burst

The burst of subcarrier that serves as the reference in color monitors and receivers is gated on at the proper position on the horizontal-blanking back porch by the burst-keyer generator (RCA terminology is "burst flag"). The block diagram of a typical burst keyer is shown in Fig. 2-32.

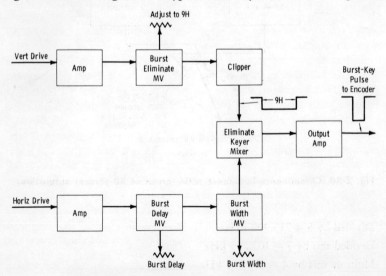

Fig. 2-32. Block diagram of burst-key generator.

The keying signal may be generated from a composite-sync input, or, as in the example of the block diagram, it may be derived from vertical and horizontal drive signals. Horizontal drive is amplified to trigger a burst-delay multivibrator. This multivibrator is adjusted to delay the keying-on pulse so that the leading edge of this pulse occurs approximately 0.5 μs after the trailing edge of horizontal sync. FCC standards specify that the color burst must follow the trailing edge of horizontal sync by a minimum of 0.006H, or 0.38 μs. The burst therefore occurs on the back porch of horizontal blanking.

The duration of the burst must be sufficient to encompass 8 to 10 cycles of the 3.579545-MHz subcarrier. One cycle of this frequency has a duration of 0.28 μs; therefore:

$$8 \text{ cycles} = 2.24 \ \mu\text{s} \ (\text{minimum})$$
$$9 \text{ cycles} = 2.52 \ \mu\text{s} \ (\text{nominal})$$

The burst-width multivibrator, which receives the delayed pulse from the delay multivibrator is adjusted so that the proper number of 3.58-MHz sine waves occurs.

Further, the burst must be eliminated during the 9-line (9H) interval of vertical sync in each field. Vertical drive pulses are amplified to trigger the burst-eliminate multivibrator. This multivibrator is adjusted for a width equivalent to 9H. The waveform is then clipped for a clean, flat-topped pulse, which is used to gate off the burst-key pulse for the vertical-sync interval.

Burst Phasing

Although most color-equipment instruction books contain a fairly complete description of how to adjust the encoder for proper burst phase, they are particularly lacking in explanations of the circuits associated with this adjustment. Most books make no mention at all of this theory of operation that is so vital to maintenance personnel. Yet nearly all color encoders at the time of this writing employ essentially the same method of obtaining the reference burst phase.

Color specifications require that on a vector diagram showing the subcarrier phase for a pure Q signal as 33°, and for an I signal as 123°, the burst phase must be 180°. We already have explained that the subcarrier is (in effect) delayed (from the reference subcarrier providing the burst signal) and then split into quadrature components for the I and Q modulators. Therefore, we know that I and Q lag the subcarrier, and that the initial subcarrier leads I and Q. But the "reference phase" of the gated subcarrier burst is obtained from the I and Q modulators themselves (in most modern encoders) by the following action:

See Fig. 2-33A. At tube V1, red is grid-injected, and green and blue are cathode-injected. Therefore, in the plate circuit green and blue add in

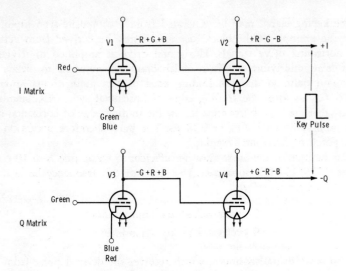

(A) I and Q matrixes.

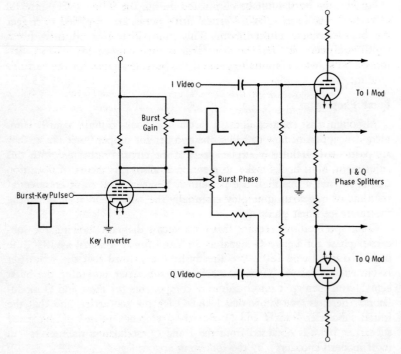

(B) Keying of burst.

Fig. 2-33. Circuits for phasing of color burst.

the same polarity (+G and +B), but red is reversed in polarity (−R). These signals are inverted in tube V2 so that +R, −G, and −B result. Now from previous studies, and disregarding the amplitudes, we know that:

$$+I = +R -G -B$$
$$-I = -R +G +B$$

and also that:

$$+Q = +R -G +B$$
$$-Q = -R +G -B$$

So at the plate of V2 we have a +I signal. The gating-on pulse of the burst keyer, if inserted at this point, will affect the modulators along the +I axis.

In tube V3, blue and red are cathode-injected, and green is grid-injected. Therefore, the signal at the plate of V3 contains −G +R +B. Tube V4 inverts this signal to obtain +G −R −B, which we know is a −Q signal. Therefore, a pulse at this point affects the modulators along the −Q axis.

Now see the simplified schematic of Fig. 2-33B. The amplitude of the key pulse is set by the burst-gain potentiometer. The ratio of pulse amplitude applied to I and Q is set by the burst-phase potentiometer. See Fig. 2-34 for the resultant effect on the modulators. If the *amplitude* of the pulse is increased while the *ratio* of amplitudes remains the same, the resultant vector sum (burst amplitude) increases. If the amplitude remains fixed while the ratio is varied, the phase of the vector sum (burst phase) will change in accordance with the ratio change. When the ratio of burst I pulse to burst Q pulse is −tan 33°, the resultant burst phase is correctly placed along the −x axis. Since the keying pulse is timed on the horizontal back porch, no video is present, and only the burst of subcarrier appears at the modulator outputs.

Subcarrier Quadrature Network

Subcarrier division into quadrature components to feed the I and Q modulators is based on the action of a network similar to that of Fig. 2-35.

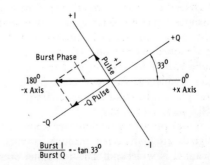

Fig. 2-34. Phase relationships for burst generation.

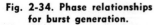

A signal, E sin ωt, is converted to a pair of push-pull signals by means of a bifilar-wound transformer, L1-L1'. To obtain the push-pull quadrature (90°-phased) signals, the original E sin ωt is applied to series-resonant circuit C-L2. The voltage across L2 leads the input signal by 90° and has an amplitude equal to the circuit Q times the input voltage. To make the output equal to the input, Q is made equal to unity, or 1 ($R = X_L$). The actual design calculations for R, L, and C are complicated by the presence of stray capacitance and the reflected impedance of L' into L. The coefficient of coupling is very nearly unity because of the bifilar winding.

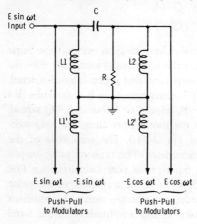

Fig. 2-35. Circuit for producing quadrature signals.

2-15. FUNDAMENTALS OF THE COLOR RECEIVER AND MONITOR

Previous sections have been concerned primarily with the mathematics of the sending (encoding) end of the color signal. It is time now to consider the basics of the receiving (decoding) end of the signal path.

Bandpass and Delay Sections

The FCC standards specify that E_Y, E_Q, E_I, and the components of these signals must match each other in time to 0.05 microsecond. These standards assume a picture element to be one-tenth microsecond in duration. Thus, the relative timing of luminance and chrominance sideband frequencies is maintained within one-half picture element, or 0.05 microsecond. This difference is for the range of frequencies close to the chrominance-subcarrier frequency. At frequencies in the spectrum approximately 0.6 MHz away from the color subcarrier frequency, a tolerance of 1 picture element (0.1 μs) is allowed.

From preceding sections, we know that the luminance and chrominance signals are transferred through circuits of differing bandwidths. The Y channel is wideband. The I channel occupies a much narrower band and is vestigial sideband, as is the Y channel from the transmitter. The Q channel is of still narrower bandwidth. Without compensation at both transmitter and receiver, the narrow-bandwidth signal components would lag behind the wider-band components, since filters delay the signal envelope. Such incorrect phasing results in poor transient response, causing color fringing on the edges of colored objects. The effect is more apparent to observers of a color picture than is the case for monochrome.

This subject is illustrated by Fig. 2-36. Since delay is a function of the associated bandpass filters at the sending and receiving ends, the I channel is delayed from Y, and the Q channel suffers the most delay. Delay lines are used in both the encoder and the receiver to bring all channels into time coincidence, as shown in Fig. 2-36. The locations of delay functions are shown in the receiver block diagram of Fig. 2-38.

Delay lines consist either of lumped inductances and capacitances, or of helix cable. The number of sections determines the amount of signal delay. The longest line is therefore found in the luminance (Y) channel. The amounts of delay shown in Fig. 2-36 are approximate, being a function of the particular receiver design. A delay line must effectively delay the entire signal envelope; hence it must not offset the frequency response of the channel concerned.

The transmitter attenuation of the lower sideband (vestigial-sideband transmission) and the receiver low-frequency cutoff (amplitude response 50 percent of maximum at visual carrier frequency) result in envelope delay. Phase-correction circuits are incorporated at the transmitter to compensate for this source of phasing error (Chapter 13).

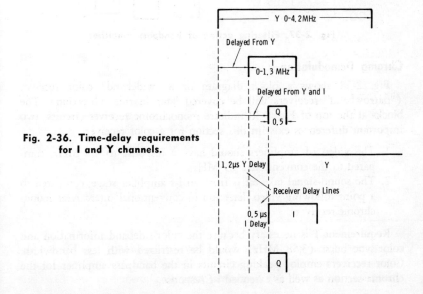

Fig. 2-36. Time-delay requirements for I and Y channels.

The output of the video detector feeds the luminance amplifiers and a chroma bandpass amplifier as shown in Fig. 2-37. Detection is illustrated here for green, red, and blue color bars. The bandpass amplifier has a "passing" characteristic of approximately 2.3 to 4.2 MHz. This response eliminates the lower-frequency Y components (including sync pulses), and passes the region around the chrominance subcarrier frequency, including the higher frequencies in the Y channel. These higher frequencies passed from the luminance channel are filtered out in the I and Q demodulator load networks.

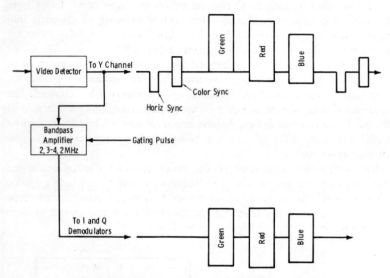

Fig. 2-37. Filtering action of bandpass amplifier.

Chroma Demodulation

Fig. 2-38 shows a block diagram of a "wideband" color receiver ("narrowband" receivers will be covered later in this subsection). The blocks at the top of Fig. 2-38 indicate monochrome receiver circuits; two important differences exist in this section for a color receiver:

1. The video i-f amplifiers should have a response to 4.2 MHz, compared to the conventional 3.5 MHz.
2. The sound-signal take-off is from an i-f amplifier stage, compared to a point following video detection in conventional intercarrier monochrome receivers.

Requirement 1 is necessary because the color-sideband information and color-sync burst (3.58 MHz) would be restricted with less bandwidth. Color receivers employ peaking circuits in the bandpass amplifier for the chroma section as well as extended i-f response.

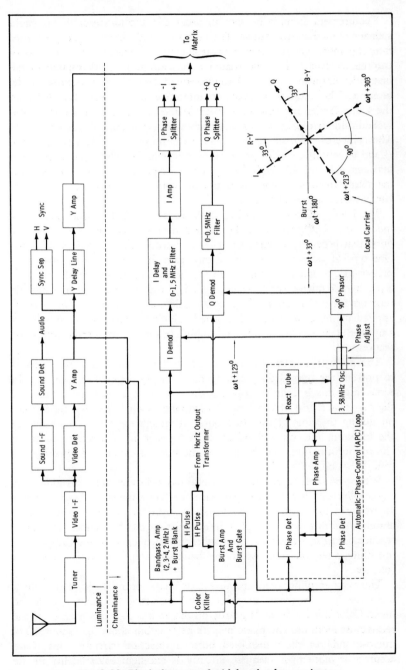

Fig. 2-38. Block diagram of wideband color receiver.

Requirement 2 must be met in order for a color receiver to obtain sufficient sound signal voltage. The ratio of sound carrier to picture carrier must be more greatly attenuated at the video detector than for monochrome receivers. (In some receivers, this attenuation takes place in the first video amplifier stage.) The video-sound carrier ratio should be 40 dB or more. This ratio is necessary to minimize the 920-kHz beat between the sound-carrier and color-subcarrier frequencies. It is a precaution taken in addition to the choice of color-subcarrier frequency discussed earlier. The choice of color-subcarrier frequency minimizes the beat through the rf and i-f amplifiers, with proper tuning of circuits. The rapid attenuation between 4.2 and 4.5 MHz in the video amplifiers in the receiver then maintains freedom from this beat that would cause shimmering soundbar patterns on the picture tube.

The bandpass amplifier in Fig. 2-38 feeds the I and Q synchronous demodulators. The purpose of the burst gate is to pass *only* the color-sync burst that occurs approximately 0.4 microsecond after each H sync pulse. One type of gate circuit holds the tube gain to zero with a negative suppressor-grid voltage. A pulse from the horizontal-output transformer, delayed by 0.4 microsecond, when applied to the suppressor then gates the tube on for the duration of the 8-10 cycle color-sync burst. The burst is applied to the control grid. The gated color-sync burst is fed to a burst-amplifier stage to obtain sufficient amplitude for controlling the color-sync section.

The pulse that gates the color burst on also is used in some receivers to prevent the burst from reaching the I and Q demodulator circuits. In some cases this is done by applying the gate pulse in negative polarity to the suppressor grid of the bandpass amplifier itself. Thus, the tube is cut off during the time of the color burst. Although the burst occurs during the blanking interval and might seem to have no theoretical effect on chroma information, the presence of the burst is capable of affecting dc-restorer action.

When no color-sync burst exists (monochrome transmission), the color killer disables the bandpass amplifier by applying a high negative control-grid bias. Although no I and Q signal is present during monochrome transmission, the higher frequencies in the Y channel, and interference occuring around the color subcarrier frequency, would cause spurious chroma-section response if the bandpass amplifier were allowed to function.

The burst amplifier feeds the automatic-phase-control (apc) loop, which locks the local oscillator in phase (and therefore frequency) with the burst. Color fidelity depends greatly on the accuracy of this portion of the receiver or monitor. The phase may be held within about 5° (about 0.004 microsecond) of the transmitted burst in practical circuits. This phase accuracy is more than adequate to provide excellent color fidelity in the reproduced picture.

The 3.579545-MHz local oscillator, controlled as described above, feeds the I demodulator directly and the Q demodulator through a 90° phasing circuit. One typical design is to excite the suppressor grids of the demodulators with the cw signal from the local carrier, while the control grids receive the chrominance signals from the bandpass amplifier.

Now we can see how the I and Q chrominance information is separated and used by the synchronous demodulators. The local carrier supplied to the I demodulator may be represented as $\cos(\omega t + 33°)$. Since the local carrier supplied to the Q demodulator is fed through a 90-degree phasor, the Q-demodulator excitation is $\sin(\omega t + 33°)$, which is in quadrature with that of the I demodulator.

As a specific example, examine the recovery of the red vector. This recovery is shown in Fig. 2-39A. The chrominance signal along the I axis

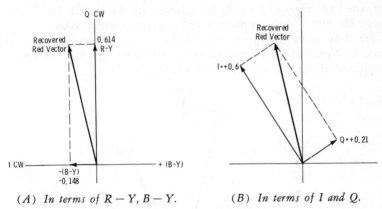

(A) *In terms of* $R - Y, B - Y$.   (B) *In terms of I and Q.*

Fig. 2-39. Recovery of red vector.

(for this specific example) is the $B - Y$ component of the red signal. That along the Q axis is the $R - Y$ component. The outputs of the I and Q demodulators then contain a vector corresponding to that of the transmitted red signal. Synchronous demodulators provide an output amplitude depending upon instantaneous voltage and phase with respect to the reference color-sync burst phase. Instantaneous output is a product of signals applied to the demodulator grids, and these circuits are sometimes termed "product demodulators."

To become thoroughly familiar with the chrominance system, you may watch this action in terms of the actual I and Q vectors for a red signal. Although it is useful practice to figure the values of color-difference signals and the I and Q components for each definite color (from the equations given earlier), Table 2-3 may be used.

Note that for a red transmission, $I = 0.6$ and $Q = 0.21$. Placing these vectors along their respective axes as in Fig. 2-39B, you again see that

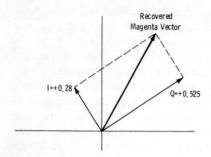

Fig. 2-40. Recovery of magenta vector in terms of I and Q.

these respective amplitudes and polarities of I and Q express the amplitude of the red vector.

For another example, consider magenta (red plus blue). From Table 2-3, we note that values of I and Q for magenta are +0.28 and +0.525, respectively. Fig. 2-40 shows recovery of the magenta transmission.

Fig. 2-41 shows all primary colors and their complements in terms of I and Q. Note that the quadrature system rotates through 360° by simple amplitude relations of positive or negative values.

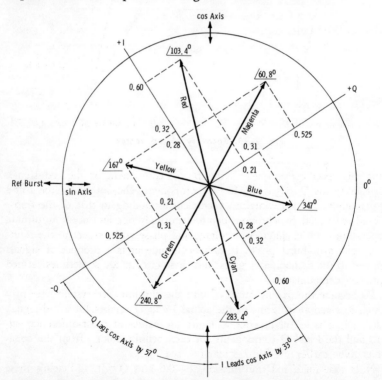

Fig. 2-41. Primary colors and complements in terms of I and Q.

For frequencies up to 500 kHz, double-sideband chrominance signals prevail. For this condition, the double sidebands of the I signal add together in the I demodulator, and the double sidebands of the quadrature (Q) signals cancel and produce no output in the I channel. Similarly, the Q sidebands add in the Q demodulator, and the I signals cancel and produce no output in the Q channel.

Now it should be recalled that the I channel is wideband; that is, for frequencies up to 500 kHz, double sidebands prevail, but above this frequency to 1.3 MHz, the I signal is single sideband. This type of modulation results in two *sets* of equal sidebands, one set in phase with the carrier and the other set *in quadrature* with the carrier. The quadrature components of the I channel obviously cause cross talk in the Q channel, since the Q channel is actually a quadrature component of the I channel. This is the reason for the Q filter in the block diagram of Fig. 2-38. Cross talk from the I channel into the Q channel occurs at frequencies above 500 kHz. By limiting the response of this channel to its useful range of 500 kHz, this interference is prevented. Since the Q sidebands are double, no cross talk occurs from the Q channel to the I channel.

The I-channel delay network also filters above 1.5 MHz. Thus, the I-channel delay and filter, and the Q-channel filter, eliminate the higher-frequency Y-channel components passed by the chroma bandpass amplifier. The I and Q outputs are fed to phase splitters so that positive and negative values of each are fed to the matrix along with Y signal.

Narrow-Band (Color-Difference) Demodulation

While the wideband receiver just described gives 140-line resolution in the orange-cyan color areas, the "narrow-band" receiver of Fig. 2-42 is more

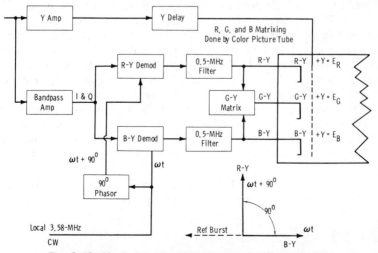

Fig. 2-42. Block diagram of narrow-band color receiver.

economical to manufacture and somewhat simpler in circuitry. Note that the locally injected carrier corresponds to the $R - Y$ and $B - Y$ axes instead of the I and Q axes. The color-difference signals are then directly demodulated, and a single matrix is used to extract the $G - Y$ component.

This receiver must limit both color-difference channels to 500 kHz to prevent spurious response. Since the bandwidths are equal, the only delay line necessary is that for the Y channel.

The wideband receiver must use a matrix to obtain the color-difference components. The receiver of Fig. 2-42, since it obtains the color difference signals directly from demodulation, needs no additional matrix (other than $G - Y$) for this purpose. With $R - Y$ voltage applied to the red cathode, and Y voltage applied to the common grids, the grid-cathode voltage is E_R. A similar condition exists for the other two sections, for their respective colors.

Receiver Matrix

The matrix extracts the color-difference signals, $R - Y$, $G - Y$, and $B - Y$, from the filtered output of the I-Q demodulators. At each instant, these signals, along with the luminance (Y) signal, excite the color picture tube so that the overall function "matches" the corresponding scanned point at the studio camera.

It was shown earlier that the I-Q matrix at the sending end reduced the $R - Y$ component to $0.877(R - Y)$, or $(R - Y)/1.14$. Similarly, the $B - Y$ component was reduced to $0.493(B - Y)$, or $(B - Y)/2.03$. In this manner, both the I and Q channels contain some of both color-difference components so that only a two-phase color signal is required for three chrominance primaries. Now, it is the purpose of the matrix in a monitor containing an I-Q demodulator to recover $(R - Y)$ (1.14) and $(B - Y)$ (2.03). That is, the color-difference components will be recovered in their original forms before I-Q matrixing at the transmission end; therefore, $R - Y$ gain will be 1.14 and $B - Y$ gain will be 2.03. In narrow-band color receivers (Fig. 2-42), the gain of the $R - Y$ demodulator is 1.14, and the $B - Y$ demodulator has a gain of 2.03. In the wideband (I-Q) monitor, this signal relationship is achieved in the matrix operation.

The foregoing is emphasized by Fig. 2-43. We have simply multiplied $(R - Y)/1.14$ by 1.14 and $(B - Y)/2.03$ by 2.03. You may then note the values of I and Q necessary to extract the color-difference components existing before modulation. The necessary amplitudes are found to be:

$$R - Y = 0.95I + 0.63Q \qquad \text{(Eq. 2-1.)}$$
$$B - Y = -1.11I + 1.71Q \qquad \text{(Eq. 2-2.)}$$

We know that Y is:

$$Y = 0.30R + 0.59G + 0.11B \qquad \text{(Eq. 2-3.)}$$

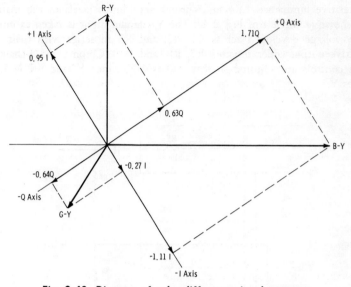

Fig. 2-43. Diagram of color-difference-signal recovery.

Rearranging Equation 2-3:

$$G - Y = -0.51(R - Y) - 0.19(B - Y) \qquad \text{(Eq. 2-4.)}$$

This is the action performed in receivers, such as in Fig. 2-42, in which the color-difference signals are directly demodulated in the matrix to extract the $G - Y$ component. Now, in terms of I and Q for the wideband color receiver, substituting Equations 2-1 and 2-2 into Equation 2-4:

$$G - Y = -0.27I - 0.64Q \qquad \text{(Eq. 2-5.)}$$

Equation 2-5 is also shown in Fig. 2-43 for matrix operation necessary to extract the $G - Y$ color difference signal.

Fig. 2-44 shows a typical monitor matrix. We know that the color-difference signals (Equations 2-1, 2-2, and 2-5) plus the luminance (Y) signal produce the primary colors:

$$R = 0.95I + 0.63Q + Y \qquad \text{(Eq. 2-6.)}$$
$$G = -0.27I - 0.64Q + Y \qquad \text{(Eq. 2-7.)}$$
$$B = -1.11I + 1.71Q + Y \qquad \text{(Eq. 2-8.)}$$

The matrix takes the plus and minus gain-adjusted signals from the I-Q phase splitters and proportions the signals according to Equations 2-6, 2-7, and 2-8. That is, the Y signal with unity gain is mixed with properly proportioned I and Q signals.

Relative luminance (Y) and chrominance input levels to the matrix are shown at the left of Fig. 2-44. The Y channel input is taken as unity, with voltage developed across R1, R2, and R3 in series with their respective output adder resistors R7, R8, and R10. Chroma and I-channel gain controls are adjusted so that +Q is 1.71 times Y, and −I is 1.11 times Y.

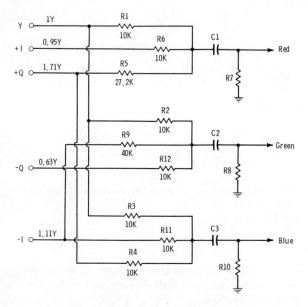

Fig. 2-44. Typical monitor matrix.

Note (for example) that the +Q voltage is applied in parallel to blue-matrix resistor R4 and red-matrix resistor R5. Resistor R5 provides the 1.71/2.72, or 0.63, of Q required for the red matrix.

Normally, the phase splitters feeding the matrix become a part of the required voltage division. Assume for example that +Q appears at the plate of a phase splitter, and the plate load resistor is 33k. This resistance is in parallel with matrix resistor R5 (27.2k) and in parallel with R4 (10k) for an effective plate load of about 5.8k. The −Q signal appears at the cathode of the phase splitter. Assume the cathode load to be 3k in parallel with matrix resistor R12 (10k), or 2.3k effective resistance. Thus, the plate provides 5.8/2.3, or 2.52, times the gain at the cathode. With the chroma gain adjusted so that +Q is 1.71 times Y, then −Q is 1.71/2.52, or 0.68, times Y, which appears at one input of the green matrix, resistor R12.

For the I channel (prior to the matrix), −I is available at the plate and +I is available at the cathode of a phase splitter. The gain of the I channel is adjusted so that −I at the plate is 1.11Y. The plate load is 10k in parallel with R11 (10k) in parallel with R9 (40k); the effective plate load is 4.4k. The cathode load is 6.2k in parallel with R6 (10k), or 3.8k. Thus, the plate provides 4.4/3.8, or 1.17, times the cathode gain. With the I gain adjusted so that −I is 1.11Y, then +I is 1.11/1.17, or 0.95, times Y for the +I input at the red matrix.

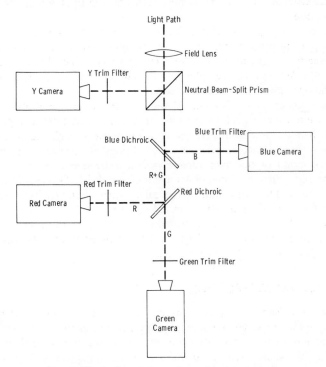

Fig. 2-45. Optics of four-channel color camera.

Output resistors R7, R8, and R10 are the adders, across each of which appears the algebraic sum of the Y, I, and Q signals as determined by the three preceding resistors. Only those signals containing red appear across R7, since the green and blue components are cancelled by the polarity and amplitude relationships of Y, I, and Q used to form the red output. The same principle holds for the green and blue outputs.

2-16. THE YRGB CAMERA SYSTEM

Thus far, we have considered only the three-camera (RGB) system in which luminance is proportional to the sum of the three camera outputs. Another method utilizes four cameras as shown in Fig. 2-45; Y (luminance) is handled by a single wideband camera.

The optical arrangement shown in Fig. 2-45 is identical to the three-camera system with the exception of the added camera, which does not have color-discriminating optical components (except that given by the associated Y trim filter). The Y-trim filter is such as to give the spectral response of the standard luminosity curve shown by Fig. 2-4, when the spectral characteristics of the camera tube and the light source are taken into account.

The color image from the projector is focused onto the field lens. The beam-splitting prism transmits part of the light toward the Y camera, and the remaining light to the color-separating system. The blue dichroic reflects blue and transmits red plus green (yellow). The red dichroic reflects red and transmits the remaining green.

The main practical advantage of the four-camera system is that it provides a luminance signal (wideband) that is obtained from a single tube rather than all three, and therefore is independent of registration. The result is a sharper picture even under conditions of "normal" registration. Misregistration of the color camera is practically unnoticed on a monochrome reproduction, although color fringing will be observed on a color reproduction. The bandwidth of the color camera is much less than that of the Y camera, simplifying associated circuitry.

It should be noted in passing that the optical lengths from the projector gate to the pickup-tube faceplates must be identical so that the same size image occurs at focus for each camera.

2-17. GAMMA CORRECTION

The term "gamma" as used in television differs somewhat from the photographic usage. For photography, gamma defines the *maximum slope* of a curve of density versus exposure, with exposure on a logarithmic scale (Fig. 2-46). When used in television application, gamma is the *exponent* of a number defining the slope of the transfer characteristic at a specified voltage level. We are concerned with light input to the camera tube

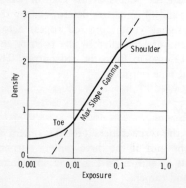

Fig. 2-46. Typical photographic transfer characteristic.

versus light output of the picture tube. If the plot of light output versus light input is a straight line (slope A of Fig. 2-47A), the device is linear. (In Fig. 2-47A, the plot is on linear coordinates. In Fig. 2-47B, the plot is on log-log coordinates.) If the line is curved on linear coordinates (slope B or C of Fig. 2-47A), the device is nonlinear. Also, the slope of the line on log-log coordinates defines the attenuation or gain of the device at various light (voltage) levels.

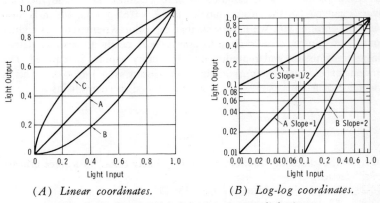

(A) *Linear coordinates.*　　　　　(B) *Log-log coordinates.*

Fig. 2-47. Graphs of light output versus light input.

Assume for the moment that a color voltage is picked up by a linear device and the picture tube also is linear. Assume further that this voltage is such that the green kinescope gun receives 1 volt and the red gun receives 0.5 volt. This condition may be expressed as 1G plus 0.5R, or a greenish yellow. But if the picture tube has a transfer characteristic such that its exponent is 2, then:

$$(1)^2 G + (0.5)^2 R = 1G + 0.25R$$

and there is an excess of green in the "greenish yellow."

Now let us go the other way and assume the picture tube is linear but the pickup tube has a transfer exponent of 0.5:

$$(1)^{1/2} G + (0.5)^{1/2} R = \sqrt{1}\, G + \sqrt{0.5}\, R$$
$$= 1G + 0.71R$$

Now we have almost a pure yellow instead of "greenish yellow."

In practice, the camera pickup tube does have a gamma of less than unity, and the picture tube does have a gamma greater than unity. From the above example, it may be seen that a system exponent exceeding unity will shift all hues (made up of two or more primaries) toward the larger of the primaries. Likewise, a gamma of less than unity will shift the hue toward the lesser of the primaries.

See Fig. 2-47B. If slope B is representative of the picture tube (exponent 2), then the system prior to the picture tube must have an exponent (gamma) of 0.5 (slope C) to result in the linear transfer of slope A. The entire gamma is the product of the individual gammas, or $0.5 \times 2 = 1$, or unity.

For example, the gamma of the vidicon tube (used in color cameras in film chains) is around 0.65 under normal operating conditions. The average color picture tube is assumed to have a gamma of 2.2, so the total overall gamma (of the two terminals) is $2.2 \times 0.65 = 1.43$, which is ideal for monochrome. But we have seen that the color system requires tighter control of transfer characteristic to approach as nearly as possible strictly linear transmission and reproduction; thus unity gamma is required. The amount of additional gamma correction necessary is $1/1.43 = 0.7$. This is the value of gamma correction most often found in color chains.

EXERCISES

Q2-1.   Give the basic requirement for a "primary color" in a three-primary system.

Q2-2.   What is the fundamental basis for CIE data?

Q2-3.   What is a wavelength of 520 nm in angstroms? Approximately what color is represented by this wavelength?

Q2-4.   What region of the visible spectrum appears brightest to the human eye? Why?

Q2-5.   Name the three primaries used for color television. What are the three secondary colors, and how do they come about?

Q2-6.   If you lose the blue output from a color system, what happens to "whites" in the scene?

Q2-7.   What is meant by "degree of saturation"?

Q2-8.   From Fig. 2-6, if red is 0.4 and green is 0.3, what is the value of blue?

Q2-9.   What are the relationships of all colors drawn along the line from W to the outer triangle limit of Fig. 2-6?

Q2-10. See Fig. 2-10. What is the approximate value of z at the dot on the horseshoe curve designated as 550 nm?

Q2-11. On Fig. 2-10, how would you interpret a line drawn along a straight-edge from "C" to 500 nm?

Q2-12. What three attributes are used to describe any one color, or to differentiate among several colors?

Q2-13. Can you draw an analogy between a color as specified under question 2-12 above and a radiated radio wave?

Q2-14. Name the one and only receiver primary that is fully saturated according to CIE specifications.

Q2-15. Note from Fig. 2-13 that the hue "brown" is not shown. Can you figure out how a brown would be reproduced?

Q2-16. When a color camera looks at a scene that is colorless (black to white only), what can you say about the outputs from the pickup tubes?

Q2-17. When a monochrome receiver is tuned to a color transmission, what primary color reproduces as the darkest shade of gray?

Q2-18. Give the polarities of the I and Q signals for each of the primary colors.

Q2-19. Name the bandwidth limitations for the I, Q, and Y signals.

Q2-20. Why does each color-difference signal include all three color primaries?

Q2-21. What were the three basic considerations involved in the exact choice of the color-subcarrier frequency?

Q2-22. How would you describe the "cancellation effect" of the chrominance signal?

Q2-23. What colors produce the maximum overshoot in the white direction?

Q2-24. What colors produce the maximum overshoot in the blacker-than-black region?

Q2-25. From Figs. 2-18A and 2-18B, you see that the luminance value of blue is 0.11 and the chroma amplitude is 0.447. This results in a maximum amplitude of $0.11 + 0.447 = 0.557$ as shown. How do you explain the value of -0.337 in the blacker-than-black region?

Q2-26. See Fig. 2-19. What does line AB represent?

Q2-27. In Fig. 2-19, what does line OA represent?

Q2-28. The angular velocity times the time from reference point really indicates what?

Q2-29. In Fig. 2-19, if line OB (plus-x axis) has a value of 7.07 volts, and the plus-y voltage (line AB) is 7.07 volts, what is the value of the maximum voltage (E_{max}), and what is the angle with the reference axis?

Q2-30. Solve Fig. 2-22A on paper.

Q2-31. Prove on paper the amplitude and phase of the red vector, in terms of $R - Y$ and $B - Y$ (Fig. 2-22C).

Q2-32. Prove on paper (in terms of I and Q) the magenta vector specification of Fig. 2-25.

Q2-33. Why are the chrominance amplitudes of $R - Y$ and $B - Y$ reduced in the transmission primaries by 0.877 and 0.493, respectively? This is what you found from the development of Fig. 2-22.

Q2-34. How can you arrive at the necessary chrominance reduction factors?

Q2-35. When the polarity of the input signal to a double balanced modulator is reversed, what can you say happens to the output signal?

Q2-36. In the color system, you have two double balanced modulators whose output signals are added. How would you describe the nature of the resulting signal when the modulator output signals have different amplitudes?

Q2-37. What color characteristic is interpreted by chroma-to-luminance ratio?

Q2-38. For a given degree of saturation, do all colors have the same chroma-to-luminance ratio?

Q2-39. Does degree of saturation change when brightness level changes?

Q2-40. What is the function of the burst-key generator?

Q2-41. What adjustment in the burst-key generator affects the placement of the reference burst in the composite color signal?

Q2-42. What adjustment in the burst-key generator affects the number of cycles of subcarrier appearing in the reference burst?

Q2-43. What is the usual method of obtaining quadrature components from the subcarrier signal?

Q2-44. How is the reference burst phase made to lead the I and Q subcarrier components?

Q2-45. Why is it necessary to use delay compensation in a color monitor or receiver?

Q2-46. What does the bandpass amplifier do?

Q2-47. How would you describe the nature of the signal at the output of the bandpass amplifier?

Q2-48. What are the basic requirements for the tuner and i-f amplifiers in a color receiver?

Q2-49. In a color receiver, why is the sound i-f carrier severely attenuated in the video i-f amplifier?

Q2-50. In an $R-Y$, $B-Y$ color receiver, why is only luminance-channel delay compensation required?

Q2-51. Name the color-system primaries and their respective complementary colors.

Q2-52. Describe the chrominance signal for a primary color as contrasted to the chrominance signal for its complementary color.

Q2-53. Describe the relationship of all colors that lie on a line drawn on the chromaticity diagram from point "C" to the spectrum locus.

Q2-54. What is the purpose of the color burst?

Q2-55. How is color-information "interleaving" achieved in the transmission process?

Q2-56. How would you derive the color-standard line-scanning frequency?

Q2-57. Derive the color-standard field frequency and subcarrier frequency.

Q2-58. Why is the output of the subcarrier divider chain $4/455$ of the subcarrier frequency?

Q2-59. What is the purpose of the "color-killer" circuit in a color receiver?

Q2-60. (A) Give the polarity of the Y signal when the composite signal shows sync to be in the positive direction. (B) Give the polarity of the Y signal when a positive voltage excursion causes an increase in brightness of the picture tube.

Q2-61. How many cw reference signals are developed in the color-sync circuits of a receiver, and what is their phase relationship?

Q2-62. What is the basic function of a synchronous demodulator?

Q2-63. Describe the difference between the cw reference signals used for I-Q demodulation and color-difference $(R-Y$, $B-Y)$ demodulation.

Q2-64. What would you say would be the effect of carrier unbalance in a color-encoding system on white and gray areas of the picture?

Q2-65. In case of carrier unbalance in the encoder, will dominant hues be shifted?

Q2-66. What would you say would be the effect of video unbalance in a color encoder?

Q2-67. What does a receiver matrix circuit do?

Q2-68. Assuming a fully saturated color-bar pattern, can you give the amplitude of the green signal at the color picture tube for the following bar sequence: white, yellow, cyan, green, purple, red, blue?

Q2-69. From the contents of Chapters 1 and 2 only, do you feel you are now in a position to set up and service a color-encoding system?

CHAPTER 3

Basic Television Waveforms
and System Fundamentals

This chapter will deal with elementary waveforms and the circuits involved with producing them. Technicians and engineers will find this section a helpful refresher for a clear understanding of operations covered later in the book. The subjects will be approached from a more practical standpoint than ordinarily is used in textbooks, and this study will be applied specifically to television-broadcast circuitry.

3-1. BASIC DEFINITIONS

The operator and maintenance engineer is continuously concerned with interpreting a variety of waveforms. Actually, there are just two basic pieces of information in a waveform, amplitude and time. The all-important "waveshape" is *not* a third basic piece of information, because waveshape = amplitude versus time (Fig. 3-1).

The video system is concerned with a pulse response as illustrated in Fig. 3-2. Ample high-frequency response is required to pass fast amplitude changes at the leading and trailing edges of the pulse. It also should be noted that a pulse requires good low-frequency response down to dc. The dc requirement is met by dc restorers and clamping circuits, which will be described later.

Rise time (or fall time) is the time during which the amplitude is changing between the 10- and 90-percent levels, as shown by Fig. 3-2 for both positive and negative pulses. (Note that a "positive pulse" might actually start at, for example, a −10-volt level and go to zero volts. A "negative pulse" might start at, for example, a +10-volt level and go to zero volts. A more precise term is "positive-going pulse" or "negative-going pulse," as the case may be.)

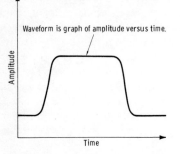

Fig. 3-1. Fundamentals of a waveform.

The base pulse width is the width at 10 percent of peak amplitude. Peak pulse width is the width at 90 percent of peak amplitude.

An important aspect of a video amplifier is the gain over a sufficiently broad bandwidth. This capability is expressed in the familiar *gain-bandwidth* product as follows:

$$\text{Gain} \times \text{Bandwidth} = \text{Upper Frequency Limit}$$

The upper frequency at which the gain of a tube is reduced to unity is:

$$f_u = \frac{g_m}{(2\pi)(C_t)}$$

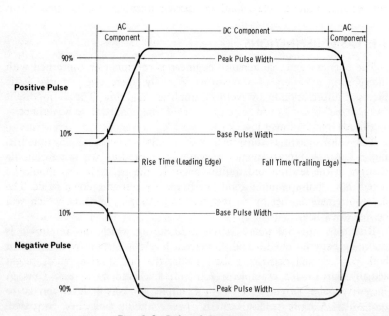

Fig. 3-2. Pulse definitions.

This expression shows that the frequency limit is proportional to the ratio of tube transconductance to total shunt capacitance. The upper frequency limit (f_u) is expressed in MHz when total tube and wiring capacitance (C_t) is given in picofarads (pF) and transconductance (g_m) is given in micromhos.

For example, the input capacitance (from tube data sheets) of a single triode section of the Type 12AT7 is 2.2 pF, and the output capacitance is 1.5 pF. To this it is necessary to add a typical value of stray circuit capacitance of 15 pF. The total, C_t, is therefore 18.7 pF, which we may round off to 18 pF for convenience. Then, since g_m of a 12AT7 is approximately 4000 micromhos:

$$f_u = \frac{4000}{6.28 \times 18} = 35 \text{ MHz (approx)}$$

This triode will have unity gain at approximately 35 MHz.

The bandwidth for gains other than unity (practical amplifier circuit) can be calculated easily. For instance, in this example:

$$\text{Gain} \times \text{Bandwidth} = 35 \text{ MHz}$$

The bandwidth for a gain of 10 is:

$$\text{Bandwidth} = \frac{35}{10} = 3.5 \text{ MHz}$$

We see that achieving a gain of 10 (in an uncompensated amplifier) will limit the frequency response to 3.5 MHz.

NOTE: Modern high-frequency transistors and integrated circuitry have much less capacitance than tubes; hence, they have higher gain-bandwidth products.

3-2. SIGNIFICANCE OF BANDWIDTH-CURVE SHAPE

Video amplifiers may be compensated by means of peaking circuits to achieve a broader bandwidth for a given gain. When peaking is used, the resultant phase shift across the passband makes it necessary to compromise between frequency response and ideal transient response.

The "ideal" amplitude-frequency response curve is usually drawn as a flat line across the desired passband (curve 1 in Fig. 3-3A). When a square wave is passed through such a system, overshoots result from the sharp cutoff. Since the gain-bandwidth-product limitation exists, frequency response must be obtained with a more ideal rolloff for good transient response. The *Gaussian* (normal-error) curve (curve 2 in Fig. 3-3A) gives ideal transient response, with overshoots (or undershoots) limited to less than 3 percent. In this case, the gain at the highest frequencies in the desired passband is 3 dB lower than at the middle frequency, and detail contrast becomes less than 100 percent in fine picture detail. The Gaussian

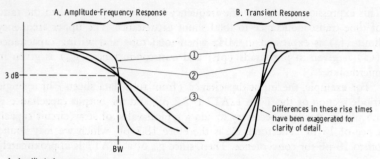

A. Amplitude-Frequency Response   B. Transient Response

3 dB

BW

Differences in these rise times
have been exaggerated for
clarity of detail.

1. Amplitude-frequency response curve falls off too rapidly, causes transient response to show overshoot.
2. Amplitude-frequency response curve falls off along a Gaussian curve and produces the optimum transient
   response: the sharpest corner free from overshoot and ringing.
3. Amplitude-frequency response curve falls off too slowly, causing undershoot.

Fig. 3-3. Relationship of response curve to pulse reproduction.

curve illustrated by Fig. 3-4 simply reveals a rolloff such that if (for example) the 3-dB point occurs at 4 MHz, the response at twice this frequency (8 MHz) is down 12 dB. Since a 2:1 frequency range is one octave, it is noted that the slope of the ideal curve is 9 dB/octave.

If the rolloff is too slow (and starts at a lower frequency) as in curve 3 of Fig. 3-3A, transient response again suffers and poor picture resolution results. The edges of sharp vertical transitions become smeared.

Note that the *shape* of the rolloff does not appreciably affect the rise time, only the transient characteristic of the leading and trailing edges. To gain a practical insight into the relationship between bandwidth and rise time, observe Fig. 3-5. The product of bandwidth and rise time is a factor (k) which is dependent upon the amount and type of high-frequency compensation. Its value lies between 0.33 and 0.5; 0.35 is typical. If we limit

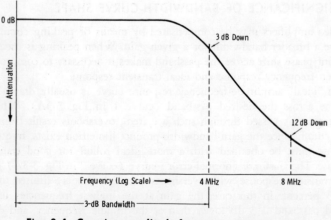

0 dB

3 dB Down

Attenuation

12 dB Down

Frequency (Log Scale) ──►   4 MHz   8 MHz

3-dB Bandwidth

Fig. 3-4. Gaussian amplitude-frequency response curve.

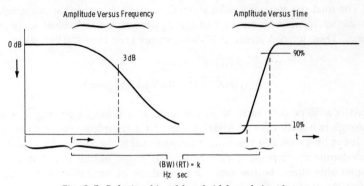

Fig. 3-5. Relationship of bandwidth and rise time.

leading-edge overshoot to less than 3 percent, factor k can be given a value of 0.35, and we can put down the following relationships:

$$(BW)(RT) = 0.35$$

$$(BW) = \frac{0.35}{(RT)}$$

$$(RT) = \frac{0.35}{(BW)}$$

where,

BW is bandwidth in Hz,
RT is rise time in seconds.

We observe that the greater the bandwidth is, the shorter the possible rise times are. Table 3-1 tabulates this relationship.

Table 3-1. Rise Time Versus Bandwidth

BW (MHz)	RT (μs)	H Resolution (TV Lines)*
1	0.35	80
2	0.18	160
3	0.117	240
4	0.087	320
5	0.07	400
6	0.058	480
7	0.05	560
8	0.044	640
9	0.039	720
10	0.035	800

*NOTE: For the horizontal resolving power in TV lines/MHz bandwidth, see derivation in Section 3-3.

The total rise time of a pulse through a series of cascaded stages is equal to the square root of the sum of the squares of the individual stage rise times. Thus, for two identical 10-MHz stages (rise time = 0.035 μs):

$$RT = \sqrt{0.035^2 + 0.035^2}$$
$$= \sqrt{0.002450} = 0.05 \ \mu s \ (\text{approx})$$

This is a 40-percent increase in rise time as a result of passing a signal through two identical (10-MHz) stages of the complete video system.

It can be shown that rise time is proportional to the *area* under an amplitude-frequency response curve; hence, the shape of the rolloff does not appreciably affect the rise time. But the shape of the rolloff *does* directly affect the *transient response* of the system.

Observe again the "ideal-rolloff" curve of Fig. 3-4. Then remember that the transmitter sound carrier is just 4.5 MHz above the picture carrier, and that the video response must be down at least 20 dB 4.5 MHz above the picture carrier. Note that this rolloff is extremely rapid in comparison to the "ideal" shape. The practical results are illustrated by Fig. 3-6, which shows the two separate effects occurring from the FCC spectrum limitations. In practice, these waveform defects are minimized by applying amplitude and phase "precorrection" to the video signal before transmitter modulation, as discussed in Chapter 13.

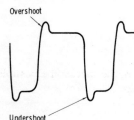

(A) *Effect of rolloff slope.*   (B) *Effect of bandwidth.*

Fig. 3-6. Effects of bandwidth and rolloff slope on pulse shape.

3-3. WHAT TO EXPECT IN PICTURE RESOLUTION

There are two resolution factors for a television picture: (1) vertical resolution, which is independent of system bandwidth, and (2) horizontal resolution, which is directly related to system bandwidth.

Vertical resolution determines how well horizontal lines in a picture are resolved. The maximum vertical resolution is fixed by the number of active scanning lines. The United States standards call for a total of 525 lines. Vertical blanking time is approximately 7.5 percent of the total frame time; therefore, $525 \times 0.075 = 39.375$, or about 40, lines are blanked out. This leaves a total of 485 active picture lines (Fig. 3-7A).

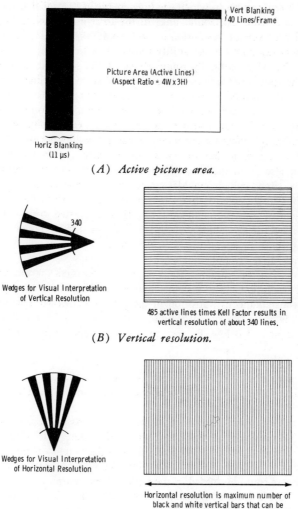

(A) *Active picture area.*

(B) *Vertical resolution.*

(C) *Horizontal resolution.*

Fig. 3-7. Resolution in television picture.

From the foregoing, it might appear that 485 horizontal lines in the image would be resolved. But in practice there is a slight spacing between the scanning lines, and the scanning spot straddles some of the lines. Both these conditions tend to reduce the utilization of the maximum number of active lines by a factor that can be taken as 0.7 times the total number of active lines; this is known as the *Kell factor.* Thus, the effective resolution is $485 \times 0.7 = 340$ lines, approximately. Fig. 3-7B shows the horizontal wedges of a test pattern in which the lines merge at a point that represents 340 black and white horizontal lines in the total image height. This is a typical value of vertical resolution at both studio and transmitter outputs.

Horizontal resolution is the ability to define vertical lines in the image. See Fig. 3-7C. The essentially round shape of the scanning spot and the fact that the spot is not infinitely small both place an immediate limitation on the ability to reproduce rapid picture transitions. When the beam suddenly encounters a sharp vertical line representing a transition from black to white, the signal is not a square wave but more nearly a sine wave.

The "rounding" effect on the video waveform results not only from the round shape and finite size of the scanning beam (often termed "aperture"), but also from the fact that a straight vertical line represents an infinite rise time—which would require infinite bandwidth. But infinite bandwidth is impossible to obtain in practice, and the rise time of the signal representing the instantaneous transition is limited by the practical system bandwidth available. When the total rise times and decay times equal the spacing between lines, the lines are not visible as separate picture elements, and they are not resolved.

Assuming that the scanning beam is properly focused, the limitation on horizontal resolution is the system bandwidth. The pulse rise time representing an instantaneous transition in the picture is directly related to the system bandwidth. As a rule of thumb, 80 TV lines require a 1-MHz bandwidth, as explained below. Be sure to study this derivation until you understand it.

1. The aspect ratio of the picture is 4 units of width to 3 units of height. This requires the horizontal resolution of a test chart to be related to the height.
2. If black and white lines with the same thickness as those indicated at the 340 position on the horizontal wedge were placed adjacent to one another, a total of 340 could fit into the height of the chart. For an aspect ratio of 4:3, $340 \times 4/3 = 452$ vertical lines of the same thickness could be placed in the width of the chart.
3. There are 2 lines of horizontal resolution per cycle of video signal: One cycle (positive and negative) represents two picture elements— 1 white, 1 black.
4. Then the horizontal resolution factor equals 2/1.33, or 1.5. (Note: $4/3 = 1.33$).
5. For a frequency of 1 MHz, there is one cycle per microsecond. Therefore, the number of lines of resolution is the horizontal resolution factor times the active line interval in microseconds.
6. The total line interval is 63.5 microseconds. Horizontal blanking is usually 11 microseconds so that the active line interval is $63.5 - 11$, or 52.5 microseconds.
7. $52.5 \times 1.5 = 80$ TV lines/MHz (approx).

The maximum video bandwidth of the transmitter is 4.18 MHz. Therefore, the maximum horizontal resolution to be expected from the transmitter is 4.18 times 80, or 334 lines, about equal to the vertical resolution. Thus, the transmitter operator will observe essentially the same resolution on the vertical and horizontal wedges of the test chart for an optimized tuning condition. The test chart transmitted to the home viewer normally is arranged to have a minimum wedge equivalent to horizontal resolution of 320 lines. Table 3-1 relates bandwidth to rise time and horizontal resolving power.

Although the transmitter is limited to essentially a 4-MHz video bandwidth, it is well known that because of accumulative factors of frequency-phase distortions, the better the picture going into the transmitter is, the better the picture received in the home will be. For this reason, studio facilities normally are maintained to approximately twice the bandwidth employed in the transmitter, or 8 MHz. Specifications of modern television studio equipment are well within 1 dB to 8 MHz, and within 3 dB to 10 MHz. Thus, the studio operator will normally observe about 640 lines of horizontal resolution $(8 \times 80 = 640)$, which approaches the upper limit of the resolving power of most picture monitors and is almost twice the vertical resolution possible.

Resolution as defined for television does not correspond directly to resolution as defined for the optical, film, and printing industries. Television resolution measurements refer to the maximum number of discernible black and white lines within a dimension equal to the height of the picture. Optical resolution is referred to in terms of cycles per millimeter, line pairs per millimeter, or lines per millimeter. Each of these terms actually refers to pairs of lines, and the resolution capability is generally accepted as the maximum number of line pairs per millimeter that can be visually resolved. Since television resolution is relative to picture height and optical resolution to the millimeter, no direct conversion factor exists.

The resolution of a television system is independent of the image or monitor-screen size. The definition refers to the number of resolvable lines within the picture height, whatever that height may be.

The aperture effect of the scanning beam was indicated in Fig. 1-2. Fig. 3-8 will make the actual result more apparent. Fig. 3-8A shows a "checkerboard" pattern consisting of alternate black and white squares. Fig. 3-8B shows the "ideal" camera output signal for such a pattern. In Fig. 3-8C is an illustration of the aperture as it sweeps across a single square of the pattern, when the diameter of the spot is just equal to the sides of the square. In Fig. 3-8D, it may be seen that the resultant waveform is distorted from the ideal, because of the finite size of the scanning aperture. In Fig. 3-8D, the resultant output waveform of the camera is compared with the ideal camera signal for the checkerboard pattern. Such deviation of signal content is known as *aperture distortion,* and it occurs for elements in the image on the tube screen that approach the size of the spot. In practice, a picture element is defined as being the smallest area of a scene that may be resolved by the pickup tube. Thus, such a picture element may be limited by the spot diameter.

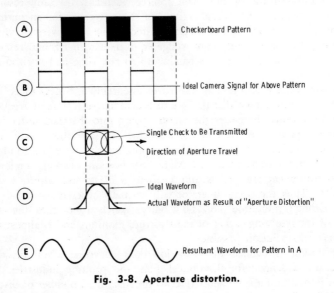

Fig. 3-8. Aperture distortion.

The phenomenon of aperture distortion should now be more apparent. In practice, such distortion is of minor consequence compared to other factors that limit resolution, *if the beam current is adjusted exactly for minimum spot size.* When such is not the case, either from faulty adjustment or from circuit troubles, the aperture distortion increases rapidly.

3-4. THE T PULSE AND DETAIL CONTRAST

In Fig. 3-9A, the scanning beam encounters a sharp transition; the resultant waveform is a sine wave superimposed on a ramp. Now consider the beam scanning across a thin white bar on a black background. Fig. 3-9B shows the resultant pulse output when the fine detail approaches the size of the scanning beam. The significant energy content of the pulse is measured by the half-amplitude duration (h.a.d.). Note that one picture element (black to white) occurred in one time T and another picture element (white to black) occurred in another time T.

The relationship between half-amplitude duration and cutoff frequency (f_c) is:

$$h.a.d. = \frac{1}{f_c}$$

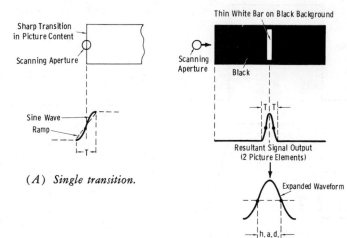

(A) Single transition.

(B) Thin white bar.

Fig. 3-9. Waveforms for sharp transitions.

Therefore:

$$f_c = \frac{1}{\text{h.a.d.}}$$

For example, if the h.a.d. of the pulse is 0.125 microsecond, the cutoff frequency is:

$$f_c = \frac{1}{\text{h.a.d.}} = \frac{1}{0.125(10^{-6})} = 8 \text{ MHz}$$

If the cutoff frequency is 4 MHz:

$$\text{h.a.d.} = \frac{1}{f_c} = \frac{1}{4(10^6)} = 0.250 \ \mu\text{s}$$

The most exact basis for measuring television-picture resolution is by means of stating the "detail contrast" of resolution. For example, the relative voltage amplitude of a 4-MHz transition might be only 50 percent of the amplitude of a 0.5-MHz transition. Such measurements are normally made relative to "flat field" response of 100-percent reference.

The pulse of Fig. 3-9B is very similar to a "sine-squared" pulse generated by test equipment for TV systems maintenance. We will learn more about the characteristics of test pulses as we progress.

3-5. THE TIME-CONSTANT CONCEPT IN VIDEO TRANSMISSION

The television system is a pulse system. Picture information is contained in the fundamental and harmonics of the 60-Hz field frequency and the 15,750-Hz line-scanning frequency.

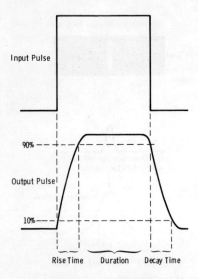

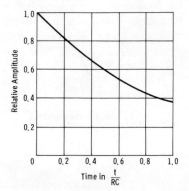

Fig. 3-10. Terminology used to describe a pulse.

Fig. 3-11. Voltage decrease as a function of time constant.

See Fig. 3-10. As we have seen, the rise and decay times of a pulse depend on high-frequency response and the shape of the frequency-response curve. We are concerned now with the duration response (t_d), which depends on t/RC, or time divided by the RC product. This, of course, is the low-frequency characteristic in practice.

The output voltage as a function of t/RC is shown in Fig. 3-11. At the instant the pulse is applied ($t = 0$), the output voltage is 1 times the input voltage. As t increases, t/RC increases, and the output voltage decreases until at $t/RC = 1$ the output voltage drops to 0.37 of the applied voltage. This theory, of course, can be found in any basic text that covers the operation of RC circuits.

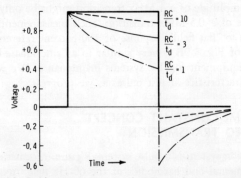

Fig. 3-12. Pulse response as a function of time constant.

Since the pulse durations required in a TV system are known, it is most convenient to use the reciprocal of the above relationship in thinking of practical RC-coupled circuits. Fig. 3-12 is a plot of the output voltage of a pulse in relation to the RC/t_d ratio. Note that it is necessary to have an RC product of 10 times the pulse duration (t_d) to avoid more than 10-percent tilt over the duration of the pulse. It is obvious that the time-constant problem becomes severe in any practical circuit when the duration of a field is 1/60 second, or 16,666 μs.

The time constant (TC) is the product of R times C; TC is in seconds when R is in ohms and C is in farads or when R is in megohms and C is in microfarads. It is in microseconds when R is in ohms and C is in microfarads or R is in megohms and C is in picofarads. The microsecond relationships are most useful for TV circuits.

For example, a 0.1-μF coupling capacitor connected to a 1-megohm grid resistor results in TC = 100,000 μs. We can see that this time constant is not 10 times the field duration. The TC value in practical circuits is limited by the stability factor (motorboating, large capacitances to ground, etc.), and this is the reason why either negative feedback to flatten the lows (as well as the highs) is used, or the "low-frequency boost" circuit is employed. In amplifiers incorporating clamping circuits, the low-frequency characteristics are almost entirely dependent upon proper operation of the clamp pulse former and clamping circuit. We will examine clamping circuits soon.

See Fig. 3-13. This cathode-follower circuit has a sine-wave frequency response that is only 3-dB down (relative to 1 MHz) at 5 Hz. But the time constant of this circuit is, for all practical purposes, $1000 \times 12 = 12,000$ μs. This is an RC/t_d ratio of less than 1 for the field duration in a television signal.

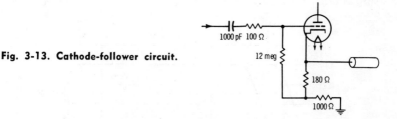

Fig. 3-13. Cathode-follower circuit.

Some readers may recognize this circuit as being from an oscilloscope probe. Since capacitive loading of the circuit by the probe must be minimized, a small coupling capacitor must be used. The value of this probe lies in making medium- and high-frequency circuit checks without a sacrifice of gain such as results from a 10-1 capacitance-divider probe. Such a probe is not intended to be used where low-frequency duration checks are important. This serves to provide a sharp differentiation in the reader's thinking between a "flat to 60 Hz" response to *sine waves,* and passing a 60-Hz *square wave* without tilt.

A video amplifier resembles the average audio amplifier only in the basic method of RC coupling the stages. The bandwidth of an audio amplifier is based on a concept that does not apply to the video amplifier. That is, the mid-frequency gain is given a value of unity, and the upper and lower limits of the bandwidth are taken as those points at which the gain falls to 0.707 of the mid-frequency gain.

Fig. 3-14 illustrates the comparison of a response curve of a good audio amplifier with that of a good video amplifier. Notice that the mid-frequency gain is given as unity, or 1. The points on the audio response curve that correspond to a gain of 0.707 of the mid-frequency gain are at frequencies of 60 Hz and 15 kHz, respectively. These points correspond to a power loss of 3 dB, or 50 percent, and they generally define the effective bandwidth of an audio amplifier.

Note, however, that the desired passband of the video amplifier is taken over that portion of the curve that is essentially flat (in practice, within 1 dB, or 10 percent on the voltage scale), denoting constant gain. Over this region of constant gain in an RC-coupled circuit, the angular phase shift should be proportional to frequency, resulting in an equal time delay for all input frequencies within this range. In the upper and lower end regions, where the gain changes rapidly, phase shifts cannot be proportional to frequency, and the time delay therefore will not be constant, resulting in phase distortion in the video amplifier. Obviously, the rolloff characteristic should not be as rapid as depicted in the curve of Fig. 3-14.

Since this type of distortion is detrimental to picture quality, its effects will be analyzed. A study of phase shift and the effect of circuit elements on this phase shift will serve to clarify the overall bandwidth and transient-response requirements for video amplifiers.

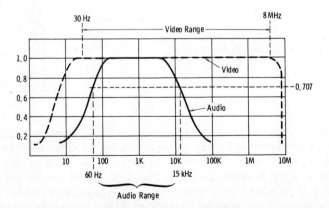

Fig. 3-14. Ideal response curves.

In the midfrequency range of an amplifier, the shunt capacitances and coupling capacitances may be considered to have negligible effect on the amplification, and the amplifier may be represented by an equivalent circuit such as the one in Fig. 3-15. In this range of frequencies, the gain of the amplifier may be assumed to be approximately the product of the transconductance of the tube and load resistance R_L.

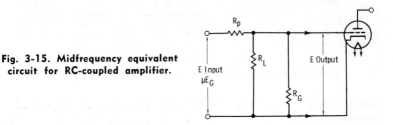

Fig. 3-15. Midfrequency equivalent circuit for RC-coupled amplifier.

At the higher frequencies, shunt capacitances across the load resistor become effective, and the amplifier response is reduced with an increase in frequency. An equivalent circuit for the higher frequencies is illustrated in Fig. 3-16. At the lower frequencies, the impedance of the coupling capacitor serves to attenuate the signal; the equivalent circuit for low frequencies is shown in Fig. 3-17. (Also, the screen and cathode bypass capacitors become less effective at low frequencies, and the resulting negative feedback contributes to reduced low-frequency gain.)

To increase the passband of the video amplifier so that frequency and phase distortion may be held to a minimum, low-frequency and high-frequency boosting circuits are used. In addition to these special circuits, a relatively low value of R_L is used, at a sacrifice of gain, to achieve a broader flat response than is possible with conventional values of load resistors. This effect is illustrated by the curves in Fig. 3-18 drawn for a typical high-transconductance pentode. In commercial equipment, coupling resistors generally range from 680 to 2000 ohms.

The internal plate resistance of pentode TV video amplifiers is much greater than R_L, and the grid resistor for the following stage also is much greater than R_L. With these conditions prevailing, the current change

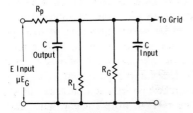

Fig. 3-16. High-frequency equivalent circuit for RC-coupled amplifier.

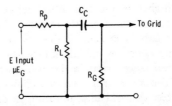

Fig. 3-17. Low-frequency equivalent circuit for RC-coupled amplifier.

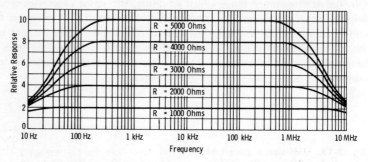

Fig. 3-18. Effect of plate resistor on pentode gain and bandwidth.

through R_L is in phase with the generator voltage, since the internal resistance (pure resistance) of the tube is actually the principal impedance in the load circuit (see Fig. 3-17). Also, the internal plate-current change is essentially equal in magnitude and phase to the current change through R_L, and the voltage drop across R_L is in phase with the plate-current change. However, since coupling capacitor C_C is reactive at low frequencies, the current through R_G will lead the voltage change applied and is therefore displaced in phase. The amount of leading phase shift is determined by the ratio of the capacitive reactance of C_C to the resistance of R_G, and it may therefore be seen to increase with a decrease in frequency (larger capacitive reactance with a decrease in frequency). Since the current change through resistor R_G is displaced in phase, the corresponding voltage drop across this resistor is likewise displaced in phase, and low-frequency phase distortion results.

To keep the voltage change across the grid resistor in phase with the plate-current change through R_L, the time constant of the coupling network (C_C-R_G) must be effectively increased so that the reactance of the coupling capacitor becomes negligible even at the lowest frequencies in the passband. The two most obvious means of accomplishing this result are *not* practical in design. First, the coupling capacitor could be made extremely large in value to provide negligible attenuation at the lowest frequencies. Larger capacitors, because of their physical size, however, increase the effective shunt capacitance to ground, severely attenuating the higher frequencies. Also, circuit instability in the form of motorboating may occur. The grid resistor (R_G) could be greatly increased so that the relative reactance of the coupling capacitor would be small. This cannot be done beyond the limits determined by the maximum allowable grid resistance given in the manufacturer's data for the particular tube used. Too much grid resistance allows gas current (positive-ion current) to affect the grid bias, resulting in excessive average plate current. The practical solution, therefore, is to shift the phase of the voltage changes across R_L so that the effects of the coupling capacitor are compensated, and the current changes through R_G are in phase with the current changes through load resistor R_L.

This is the function of the low-frequency boosting circuit shown in Fig. 3-19. The desired relative phase shift across R_L results when the two parallel branches of the load circuit have similar phase angles. This is accomplished when the product of the plate load resistance and the decoupling capacitance is equal to the product of the coupling capacitance and the grid resistance (equal time constants):

$$R_L C_D = C_C R_G$$

The current changes through the grid resistor will be in phase with the current changes through the plate load resistor. Decoupling resistor R_D must be used to provide a dc path to the plate of the tube, and it must be much larger than the plate load resistor. This value is limited in size by the available plate supply voltage (plate current must pass through this resistor as well as R_L), and if R_D cannot be made large in ratio to the reactance of C_D, then compensation is made by shunting the coupling capacitor with a suitable value of resistance. This value is such as to restore the similarity of the two parallel branches, and the maintenance engineer will find this method used in some cases. To avoid a dc path to the grid, an extra capacitor is used (dash lines in Fig. 3-19).

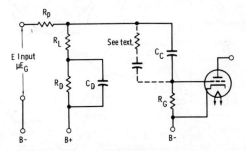

Fig. 3-19. Low-frequency decoupling (boost) circuit for video amplifier.

For a system to pass a pulse with the same rise time and shape as the input pulse, it must have a bandwidth of:

$$BW = \frac{1}{2\ RT}$$

This equation says the bandwidth (BW) must be equal to half the reciprocal of the rise time (RT). For a pulse with $RT = 0.02\ \mu s$:

$$BW = \frac{1}{2(0.02)} = \frac{1}{0.04} = 25\ \text{MHz}$$

Since the rise time is in microseconds, the result is in MHz.

Now immediately the reader should note that we have used a different relationship for conversion of rise time to bandwidth than in Section 3-2. The relationship in Section 3-2 is based on an assumption that overshoot will be "under 3 percent." This is where the "k factor" enters, and says that the bandwidth is 0.35 divided by the rise time. For RT = 0.02 μs, this calls for a bandwidth of 17.5 MHz. The rise time will still not be materially affected, but a slight amount of overshoot will occur, on a rise time of 0.02 μs.

The relationship that states that the bandwidth must be equal to half the reciprocal of the rise time is based upon Fourier analysis. The Fourier theorem says that any recurring nonsinusoidal waveform can be shown to be made up of a sum of sine waves, cosine waves, or both, of various amplitudes, phases, and frequencies.

Fig. 3-20A illustrates a fundamental sine wave combined with its third harmonic. Note that the resultant shows a tendency to start formation of a

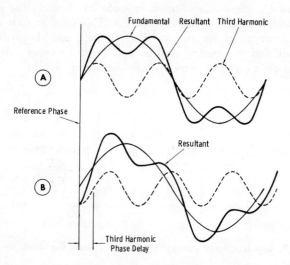

Fig. 3-20. Harmonic content of waveforms.

square wave, by the steepening of the sides. (The "dip" in the center is filled in by higher-frequency harmonics.) As odd-order harmonics are added, this effect increases, and a perfect square wave would be composed of an infinite number of odd harmonics. The amplitudes of the added harmonics decrease with increasing frequency; thus the higher the harmonic frequency, the less its amplitude is in relation to the fundamental. Now, provided the frequency-phase bandwidth is adequate, these harmonics are retained in the original amplitude and phase relationship, and the pulse is passed without distortion.

To maintain the harmonics of the square wave in the proper *time* relationship with the fundamental component, the *phase shift must be proportional to frequency*. We will expand on this in a moment.

Suppose the third harmonic of Fig. 3-20A undergoes a different time delay than its fundamental. Fig. 3-20B shows that the result is a "square-wave" response with a "tilt" across the top and an undershoot at the trailing edge.

Tube or transistor amplifying action in itself causes a normal inversion of the signal between the input and output of common-cathode or common-emitter circuits. This means that a uniform time delay occurs for all frequencies in the passband. If each frequency is considered separately, it may be seen that for a uniform time delay a different phase shift must occur at each frequency, so that the resulting phase displacement is proportional to frequency. At the horizontal scanning frequency, the time of scan of one line is 63.5 microseconds. Across a 10-inch monitor this corresponds to about 6.35 microseconds per inch. Should any part of the video signal be delayed only one microsecond above the normal phase inversion, that portion of the picture would be displaced approximately one-sixth of an inch. Obviously, if all frequencies in the passband were delayed one microscond, the entire picture simply would be displaced about one-sixth of an inch, but it would be a satisfactory picture. However, the resistance-capacitance elements of the coupling networks (when uncompensated) cause a shift in phase that differs both in direction and number of degrees for different frequencies. Phase distortion may be seen to be directly related to amplitude distortion, since minimum phase distortion is obtained only with a broad, flat-topped response over the desired passband and with *no sharp cutoff* above the passband.

The significance of amplifier response may be emphasized by considering the content of the video signal when the image to be scanned is the extreme case of a black bar on a white background (Fig. 3-21). Waveform A represents the ideal response in which an abrupt rise in the tube current of an image orthicon would occur as the scanning aperture encountered the leading edge of the black bar. (Fig. 3-21 shows waveforms for a stage in which positive-black video polarity exists.) At the trailing edge of this bar, the current should fall abruptly to the "white" level. Amplifier circuits that must faithfully reproduce such current changes will have exceptional amplitude and phase characteristics. If the amplifier has insufficient high-frequency gain, the leading edge of the amplified wave becomes a gradual slope instead of a sharp rise (waveform B, Fig. 3-21). The reproduction is that of a gradual shading from gray to black on the leading edge of the black bar, and a gray-to-white "smear" on the trailing edge. Waveform C shows the effect of overcompensation of the low-frequency response. The effect is similar to that of insufficient high-frequency response, but it is not as pronounced. Insufficient low-frequency response with attendant phase shift is illustrated by waveform D. Since loss of lows causes the flat top of the ideal square-wave response to become tilted as shown, black-to-gray shading occurs at the leading edge, and a white-to-gray smear results at the trailing edge.

One possible effect is illustrated in Fig. 3-22. Such an effect may be produced also in peaking circuits by transient oscillations that result from shock excitation by suddenly changing square-wave currents through the

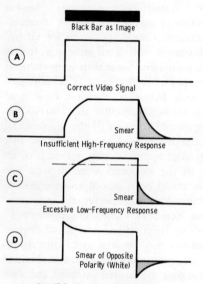

Black Bar as Image

(A)

Correct Video Signal

(B)

Smear

Insufficient High-Frequency Response

(C)

Smear

Excessive Low-Frequency Response

(D)

Smear of Opposite Polarity (White)

Insufficient Low-Frequency Response

Fig. 3-21. Effects of amplifier deficiencies on square wave.

coils. Damping resistors are used across series peaking coils, which are often adjustable in commercial circuits for proper peaking and damping characteristics.

The curves in Fig. 3-23 show the correct phase-shift characteristic (proportional to frequency—uniform time delay at all frequencies), in comparison to that of an average amplifier not compensated for flat high-

Fig. 3-22. Example of reversed-polarity smear.

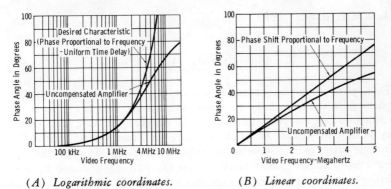

(A) Logarithmic coordinates.   (B) Linear coordinates.

Fig. 3-23. Phase-shift requirements for video amplifier.

frequency response. The relation of time delay, phase shift (θ), and frequency is the following:

$$\text{Time Delay} = \frac{\text{Phase Shift in Degrees}}{360° \times \text{Frequency in Hz}}$$

From examination of the desired characteristic curves at a frequency of two megahertz, the phase shift is 30°. Therefore:

$$\text{Time Delay} = \frac{30°}{360°(2 \times 10^6)} = \frac{30°}{720° \times 10^6} = 0.041 \text{ microsecond}$$

From the same curve, at 3 megahertz $\theta = 45°$. Therefore:

$$\text{Time Delay} = \frac{45°}{360°(3 \times 10^6)} = \frac{45°}{1080° \times 10^6} = 0.041 \text{ microsecond}$$

From the same curve, at 4 megahertz $\theta = 60°$. Therefore:

$$\text{Time Delay} = \frac{60°}{360°(4 \times 10^6)} = \frac{60°}{1440° \times 10^6} = 0.041 \text{ microsecond}$$

Thus, for phase shift proportional to frequency, a uniform time delay occurs throughout the video amplifier at any frequency. This results in a uniformly shaded picture, other factors being equal.

Let us see now the effect of the uncompensated-amplifier phase-shift curve on time delay at various frequencies. From the uncompensated curve, at 2 MHz $\theta = 27°$, and:

$$\text{Time Delay} = \frac{27°}{360°(2 \times 10^6)} = \frac{27°}{720° \times 10^6} = 0.037 \ \mu s$$

At 3 MHz, $\theta = 38°$, and:

$$\text{Time Delay} = \frac{38°}{360°(3 \times 10^6)} = \frac{38°}{1080° \times 10^6} = 0.035 \ \mu s$$

At 4 MHz, $\theta = 48°$, and:

$$\text{Time Delay} = \frac{48°}{360°(4 \times 10^6)} = \frac{48°}{1440° \times 10^6} = 0.033 \ \mu s$$

Thus, phase distortion occurs. It can be noted that for high-frequency phase shift, the time delay decreases as frequency increases. This is the effect of a lagging phase shift across the coupling network; it is caused by shunt capacitances. At low frequencies, the leading phase shift results in time delays that increase as the frequency decreases. Most of the ringing effect (smearing after black bars in the picture) is caused by low-frequency phase distortion. This distortion may also be noticed as a gradual shading in backgrounds from top to bottom of the reproduced picture. High-frequency phase distortion results in general deterioration of resolution.

When 75-ohm transmission lines are to be fed from an amplifier, the low-frequency time constant becomes a problem. This fact is obvious when it is considered that the internal output impedance of the device should provide an impedance that matches the transmission-line characteristic impedance. All modern video and pulse amplifiers provide this sending-end matching impedance for driving the low-impedance line. Earlier video and pulse amplifiers normally provided a rather high output impedance with the termination for the line appearing only at the receiving end.

The internal output impedance of a tube-type cathode follower is the reciprocal of the transconductance. Thus, for a Type 6AG7 tube ($g_m = 11,000$ micromhos), the internal output impedance (R_C) is:

$$R_C = \frac{1}{g_m} = \frac{1}{0.011} = 90 \ \text{ohms}$$

Since this impedance is in parallel with the cathode resistor (so far as the load is concerned), suitable choice of this resistor provides a good sending-end match, and the cathode follower is thus an ideal low-impedance coupling device. However, it is obvious that the coupling capacitor (to block dc from the line) would be of a very large value to provide a good low-frequency time constant.

Design engineers ascertain the figure of merit of any coupling circuit at low frequencies by simply multiplying the time constant of the circuit by the lowest frequency to be passed with good fidelity:

$$(\text{LFM}) = RCf$$

where,

    (LFM) is the low-frequency figure of merit,
    RC is the time constant,
    f is the lowest frequency to be passed.

In practice, the value of LFM should not be less than 20. So, if LFM = 20, then:

$$C = \frac{20}{Rf}$$

where,

    R is the coupling resistance of the time constant,
    f is the low frequency to be passed.

For the 6AG7 cathode follower mentioned (R = 90 ohms) and an f of 30 Hz:

$$C = \frac{20}{(90)(30)} = 0.0074 \text{ farad (approx)} = 7400 \ \mu F$$

The physical size of such a capacitor that could withstand even the relatively low voltage encountered would be prohibitive.

Fig. 3-24 shows one example of how this problem is solved in modern solid-state coax-line drivers. The output impedance of emitter-follower Q2

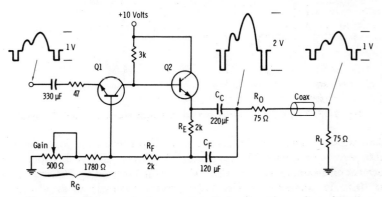

Fig. 3-24. Time-constant correction in low-impedance line driver.

is only a fraction of an ohm; hence, build-out resistor R_O is added to provide sending-end termination. Capacitor C_F feeds the output signal back to the input (inverse-polarity, or negative, feedback) after the dc-decoupling capacitor, C_C, providing correction for loss of low-frequency time constant caused by C_C. Thus, the tilt across a low-frequency pulse is completely corrected.

The voltage gain of the feedback pair in Fig. 3-24 is determined by R_F and R_G. The output voltage is given by:

$$E_{out} = E_{in} \frac{R_F + R_G}{R_G}$$

Thus, if R_G is 2000 ohms (nominal value) and the video input is 1 volt (peak-to-peak), then:

$$E_{out} = (1) \frac{4000}{2000}$$

$$= 2 \text{ volts (peak-to-peak) at Q2 emitter}$$

Note that R_O and R_L form a 2/1 voltage divider so that with a 2-volt signal at the emitter of Q2, a 1-volt signal will appear across R_L. This is a normal situation for a feedback pair driving a low impedance when the internal output impedance of the line driver is extremely low.

3-6. OPERATIONAL MEASUREMENTS OF TV WAVEFORMS

From the beginning, it is important that we understand the difference between the video-to-sync ratio of the composite video signal itself, and the translation of this ratio in terms of percentage of transmitted carrier. Fig. 3-25 represents the transmitter carrier wave modulated with a stan-

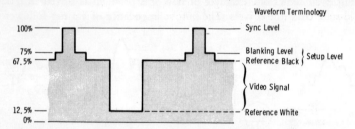

Fig. 3-25. Transmitter carrier modulated with standard window signal.

dard window signal (white bar on a black background). Sync tips are 100 percent of carrier, blanking level 75 percent, and peak white 12.5 percent of maximum carrier. This "white setup" is fixed by FCC standards at 10 to 15 percent of carrier (12.5 percent nominal) to avoid carrier cutoff. The latter results in audio buzz in intercarrier-type receivers as a result of loss of one of the carriers that beat to produce the i-f.

It is evident that the signal amplitude from the studio line is represented on the modulated carrier as that portion between 12.5 and 100 percent of maximum carrier value. Since full modulation of the transmitter represents an 87.5-percent change in the carrier, the necessary sync-to-video ratio of the input signal to result in 25-percent sync is $25/87.5 = 0.286$, or 28.6 percent of the total composite signal. Thus, disregarding special circuits such as sync-stretching stages of transmitters or stabilizing amplifiers, the input signal (to a linear transmitter) must be 71.4 percent video and 28.6 percent sync to obtain the FCC requirement of 25-percent sync in the radiated signal. Notice also that, on recovery of the video by detection of the transmitted carrier, *exactly the same ratio of video to sync,*

71.4 percent video and 28.6 percent sync, should exist for a modulated-carrier ratio of 75 percent video and 25 percent sync.

Before going ahead, it will be necessary to review briefly the history of video signal-level measurements. This is mandatory in order for the reader to "make sense" of the IEEE scale that is used for such measurements.

NOTE: *IEEE* is an abbreviation for the Institute of Electrical and Electronic Engineers. *IRE* is an abbreviation for the Institute of Radio Engineers. The scope graticule scale which we will discuss was formerly termed the IRE scale. Since the IRE has been absorbed into the IEEE, the latter term has been employed to designate the scale. The reader will find the two terms used interchangeably in practice.

The original standards for studio line output (transmitter input) set in 1946 established a 2-volt composite signal level consisting of 0.5 volt of sync and 1.5 volt of video, or 25 percent sync to 75 percent video. All early transmitters employed sync-stretching circuits to adjust the respective amplitudes and to compensate for the inherent sync compression.

In 1950, this standard was changed to a composite level of 1.4 volts, 1 volt of video to 0.4 volt of sync. This change was made largely because of difficulty in obtaining good amplitude linearity in existing equipment over the 2-volt range (particularly in a large number of cascaded amplifiers, as in network transmission). The new standard not only established better amplitude-linearity characteristics, but it also provided a compatible 28.6-percent sync to 71.4-percent video ratio. Some manufacturers deleted the usual sync-stretching circuits from transmitters and obtained proper sync/video ratios in external stabilizing amplifiers.

Since the advent of color television, stations (even those operating monochrome only) have established a 1.0-volt composite signal as a standard for the line output level. Because of the overshoot of the color subcarrier on color-bar transmissions, amplitude-linearity problems again manifested themselves with the 1.4-volt standard. Numerous tests indicated the desirability of reducing the transmission level to a 1.0-volt composite signal, and this is now standard practice for AT&T television operating centers and commercial broadcast stations.

The new 1.0-volt standard maintains the same sync/video ratio as the older 1.4-volt standard. The voltage ratios of the old and new standards, as correlated with the IEEE scale (adopted for use as an industry-wide standard), are shown in Fig. 3-26. Although the 1.0-volt signal is normally spoken of as consisting of 0.7 volt of video and 0.3 volt of sync, the actual values are 0.714 volt of video and 0.286 volt of sync. By expressing all values in IEEE units, a standard that eliminates confusion is established. It is only necessary to calibrate the scope gain so that 140 IEEE units correspond to 1 volt peak-to-peak.

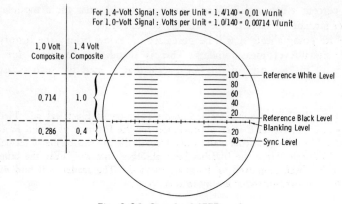

Fig. 3-26. Standard IEEE scale.

Fig. 3-27 illustrates the three scales recommended for the various points to be monitored. Operating scale No. 1 is for use at the camera control unit when sync *is not* added at this point. Reference white is at 100, reference black is at 10, and the blanking level is at 0. Note that the reference black level at 10 is a continuous line, as is the blanking level at 0. This *setup* level is very important. In a theoretically perfect transmission system, black level and blanking level could be maintained the same, thus utilizing to the fullest extent the video-amplifier gains. In practice, however, some amplitude distortion exists, resulting in at least slight amounts of overshoot in the black region. When this occurs, some retrace lines will be visible on the picture, unless the receiver controls are adjusted to clip black peaks. This results in *compression of blacks.* When the setup value is raised to between 5 percent and 10 percent (7.5 percent nominal value), optimum operating conditions are realized. The setup must be constant and of

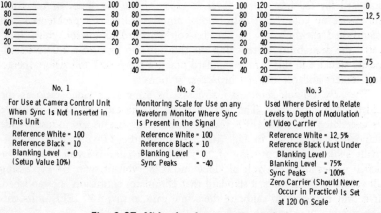

Fig. 3-27. Video-level measuring scales.

the same value between cameras (studio or film), studios, networks, etc., to maintain constant background brightness in the home receiver.

Operating scale No. 2 in Fig. 3-27 is recommended where the composite signal level (sync added) is to be measured. Reference white is at 100, reference black is at 10, the blanking level is at 0, and the sync peaks are at −40. This type of scale is common at the line monitoring position.

Operating scale No. 3 is recommended for use at transmitter locations, where it is desirable to relate the arbitrary IEEE units to depth of modulation of the video-transmitter carrier wave. Reference white of 100 represents 12.5-percent carrier, zero carrier being represented by 120 on the scale. Reference black is at 10, and blanking level at 0 represents 75-percent carrier. Sync peaks at −40 represent 100-percent carrier modulation. The relationship to the FCC specifications of carrier levels may now be observed. Zero carrier level (which should never occur in practice) is set opposite 120 on the IEEE scale, and maximum carrier is opposite −40. Blanking level (zero on scale) then occurs at 75 percent of maximum carrier, and reference white (100 on scale) occurs at 12.5 percent of maximum carrier.

All level checks not involving frequency response should be made with the scope response on the "IRE" position. If the scope is on wideband response, the small energy overshoots of the higher-frequency components will be apparent. If these overshoots are held below 100 IEEE units and the scene suddenly changes to one of much lower frequency content, the operator must adjust his gain to bring the overall level up. Since the luminance content is largely in the middle and lower frequency range, the operation results in a needless shift of apparent contrast in the home receiver. To avoid this result, the IRE response curve was standardized for the purpose of "riding gain" on the video and for checking levels of normal signals between the studio and the transmitter, or the studio and AT&T.

Since the waveform monitor of the master unit often is used to indicate the level of video as it leaves the studio for the transmitter, the operator will find a choice of two pattern characteristics associated with this circuit. For example, the master monitor incorporates a selectable vertical-amplifier response for the waveform oscilloscope. In the wideband position, the response is essentially flat to 5 MHz. When the oscilloscope is used for level measurement only, the response is rolled off at the high end, as illustrated in the IRE curves of Fig. 3-28. The solid line indicates the recommended rolloff, and the dash lines show allowable tolerances in the amplifier response for level measurements.

In Fig. 3-28, waveform 1 is the normal appearance of the waveform displayed on a wideband scope when the signal input is from a camera looking at a test pattern. Notice that this pattern can indicate the quality of a signal as well as the level, since it shows good shading characteristics and the distribution of lights and darks in the pattern. Waveform 2 is the same signal with artificially produced overshoots (transients) as viewed on the

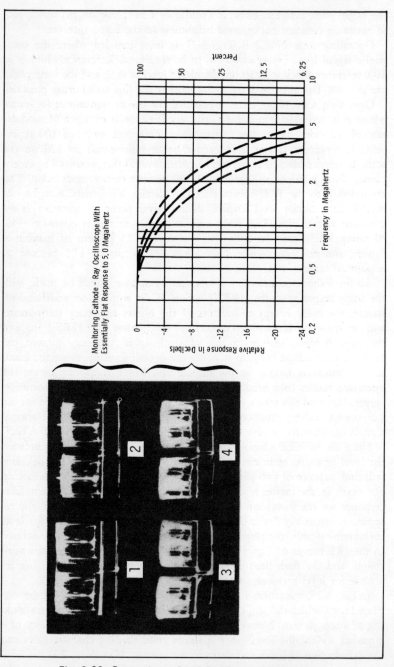

Fig. 3-28. Comparison of wideband and IRE response.

same wideband scope. These characteristics lead to major differences in handling overall gain by different operators. As a result of a study from which the IRE standard was compiled, it was found that appropriate rolloff characteristics reduced disagreements among operating personnel concerning levels, and still provided sufficient indication of overshoot so that an excessive amount would be apparent. Waveform 3 in Fig. 3-28 is the pattern produced by signal 1 on a narrow-bandwidth CRO. Note that, since highs are reduced in response, the entire pattern appears to be of the same intensity. With the same characteristics, the overshoots of waveform 2, when displayed as in waveform 4, are not as apparent (except in the "black" region), and a more uniform level indication results.

Scopes most commonly used in television broadcasting employ a time base calibrated in microseconds rather than frequency. Thus, the horizontal rate of 15,750 Hz is 63.5 microseconds per line. To display what is commonly termed "two lines" at the horizontal rate, the scope sweep is adjusted to 7875 Hz (one-half the line rate), or 127 microseconds.

On standard waveform monitors, either H- or V-rate pulses may be selected by means of a switch. For the "H" position, the sweep rate is set automatically to the 1/2H rate for a display such as is shown in Fig. 3-28. For the "V" position, the sweep rate is set automatically to the 1/2V rate (30 Hz).

NOTE: The rolloff characteristic is also termed the "IEEE rolloff response curve," but most scopes are simply labelled "IRE" when this response is selectable on a switch. In this book, video signal levels will be referred to in terms of "IEEE units," and the rolloff response as "IRE response."

3-7. AVERAGE PICTURE LEVEL (APL) AND DC RESTORATION

Fig. 3-29 provides a review of pulse repetition time, sometimes termed *pulse period*. This time is normally taken as the time interval between leading edges of consecutive pulses. It could be the time from 50 percent of the rise time of one pulse to 50 percent of the rise time of the next pulse; or it could be from 10 percent of the fall time of one pulse to 10 percent of the fall time of the next pulse. This is to say that the pulse period includes the rise time, the pulse duration, the fall time, and the interval be-

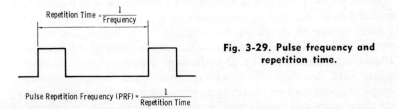

Fig. 3-29. Pulse frequency and repetition time.

tween pulses. The pulse repetition time is equal to the reciprocal of the pulse frequency, sometimes termed *pulse repetition rate* (PRR). Conversely, the pulse repetition *rate* is equal to the reciprocal of the pulse repetition *time*.

For example, the frequency (PRR) of horizontal-sync pulses is 15,750 Hz, so the pulse repetition time is $1/15,750 = 63.5$ microseconds. This is the time interval between adjacent pulses. Conversely, $1/63.5 \times 10^{-6}) = 15,750$ Hz. In this connection, bear in mind that you can have a chain of pulses at a certain PRR that may be "keyed on" at a much lower rate. Equalizing pulses are 31.5-kHz pulses "keyed on" at the rate of only 60 Hz (Chapter 6).

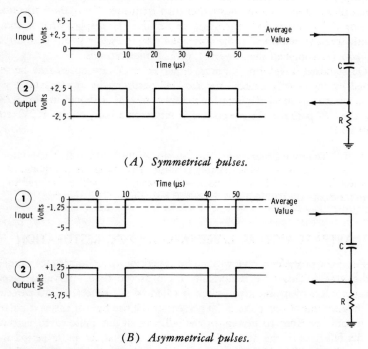

(*A*) *Symmetrical pulses.*

(*B*) *Asymmetrical pulses.*

Fig. 3-30. Effect of duty cycle on dc axis for an RC-coupled circuit.

Pulses are composed of a dc voltage plus a large number of harmonically related frequencies. The dc component of a pulse series is equal to the average voltage of the pulses. For example, in the input-pulse series in waveform 1 of Fig. 3-30A, each pulse has an amplitude of +5 volts for a 10-μs duration and zero for a 10-μs duration. This describes a symmetrical square wave. Assuming this waveform is transferred through a capacitive coupling circuit with an RC product sufficiently long to prevent distortion, the pulse series will arrange itself so that the ac axis is at the average dc value (waveform 2, Fig. 3-30A). Note that the same would be true for a sine wave or any symmetrical repetitive pulse.

In the pulse series of waveform 1, Fig. 3-30B, the input pulse has an amplitude of −5 volts for 10 μs, and the voltage then is zero for 30 μs. This describes a nonsymmetrical (asymmetrical) pulse series. On the other side of the coupling capacitor (waveform 2, Fig. 3-30B), the ac axis (zero axis) corresponds to the average dc value of −1.25 volts, so the base of the pulse is at +1.25 volts and the pulse tip is at −3.75 volts.

Power normally is measured as an average value over a relatively long period of time. But in nonsymmetrical pulse trains, average power delivered over one cycle of operation can be quite low compared to the peak power available during the pulse time. The basic relationship is:

$$\frac{\text{Average Power}}{\text{Peak Power}} = \frac{\text{Pulse Width}}{\text{Pulse Repetition Time}}$$

Either one of the above ratios is an expression of the fraction of total time that energy is supplied. This time relationship is termed the duty cycle:

$$\text{Duty Cycle} = \frac{\text{Average Power}}{\text{Peak Power}}$$

and:

$$\text{Duty Cycle} = \frac{\text{Pulse Width}}{\text{Pulse Repetition Time}}$$

Since a "resting time" occurs between pulses that are narrow in comparison to the repetition time, the average power can be quite low even though the peak power might be quite high. The horizontal sync pulse, with a width of 4.8 μs and a repetition time of 63.5 μs, has a duty cycle of $4.8/63.5 = 0.075 = 7.5$ percent. Suppose we arbitrarily assign a peak power of 1000 watts. Then for 4.8 μs, 1000 watts of power is dissipated. For the remaining 58.7 μs ($63.5 − 4.8 = 58.7$), zero power is available. Now from the basic relationship:

$$\text{Average Power} = (\text{Peak Power})(\text{Duty Cycle})$$
$$= (1000)(0.075) = 75 \text{ watts}$$

Of most importance at this time in our discussion is the effect of the varying dc axis in a video signal passed through RC-coupled circuits. The "duty cycle" when applied to a video signal is termed *average picture level* (APL). This can vary from a very dark scene to a very bright scene. For proper signal handling, "black" must hold a steady reference level from which the "shades-of-gray" video excursions can be referenced.

In Fig. 3-31A, a repetitive "stair-step" signal represents an APL of 50 percent, just as would a symmetrical square wave or a sine wave. The blanking level has been positioned on the CRO at zero IEEE units.

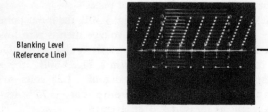

Blanking Level
(Reference Line)

(A) 50-percent APL.

Fig. 3-31. Shift of reference blank-

In Fig. 3-31B, the stair-step signal occurs only once every five lines, with zero setup on the four intervening lines. This represents an APL of 10 percent, or a very dark picture. The waveform has shifted upward by 35 IEEE units because dc restoration is not used.

In Fig. 3-31C, the stair-step signal again occurs only once every five lines, with a 100-percent ("white") pulse in each of the four intervening lines. This simulates an APL of 90 percent (bright picture), and the waveform has shifted downward about 20 IEEE units from reference level.

The peak-to-peak excursion for the three conditions of Fig. 3-31 is about 35 + 20 IEEE units, or 55 IEEE units. Since one IEEE unit = 0.00714 volt, then 55 units = (0.00714)(55) = 0.4 volt (approx). Thus, an RC-coupled stage without dc restoration must handle a 1.4-volt peak-to-peak signal for a 1-volt input signal, to handle the full range of APL.

At various points in video amplifiers, a *clamping circuit* is used. Such a circuit is feasible because of the wide bandwidth of amplification that is necessary for television signals. A clamping circuit accomplishes two things that at first appear to be contradictory: (1) it improves low-frequency video-signal response, and (2) it eliminates spurious low-frequency pickup, such as 60-Hz ac, from the amplifier response.

To understand how an electronic device can eliminate one low-frequency signal and improve the response to another low-frequency signal requires a study of the clamping circuit, which functions at high efficiency because of the nature of the TV waveform itself. Consider first Fig. 3-32, which shows an ac signal source with a series RC circuit and a switch. Waveform 1 represents the signal voltage that would appear at the output terminals with the switch open. Suppose now that the switch is closed (shorting the output terminals) for the duration of the shaded areas along the axis of the ac signal. Waveform 2 illustrates the severe attenuation of the signal appearing at the output terminals. It is understandable, then, that if the switch were opened and closed at a rate *much faster* than the frequency of the applied ac signal, the output voltage would be practically zero. A clamping circuit in a TV amplifier is actually an electronic switch, which does exactly what is described above. It opens and closes a switch

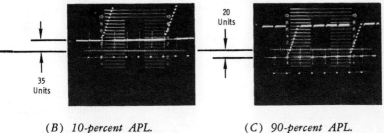

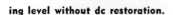

(B) 10-percent APL.   (C) 90-percent APL.

ing level without dc restoration.

at the horizontal frequency (15,750 times per second) so that any 60-Hz sine wave (such as would occur from a stray field) is greatly attenuated.

Consider now the action of the same circuit when the input waveform is not a pure sine wave, but is broken up by *pedestals* at a fixed level, such as a video signal from the pickup head, with inserted blanking pulses. Waveform 1 in Fig. 3-33 pictures the output waveform with the switch open, showing that the same waveform appears as is applied. Now assume that some circuit action occurs that results in poor low-frequency response. Waveform 2 shows the resulting waveform if the switch remains open as above. The low-frequency component is attenuated, but the pedestals (fixed levels) remain of the same amplitude. Thus, the tips of the pedestal peaks vary from constant level above the dc axis as shown. If the switch is closed for the Δt intervals shown, the output waveform will appear as in waveform 3, restoring somewhat the original waveform. Should the switch be operated electronically at a rate much faster than the applied waveform, *negligible attenuation* would result. In effect, the low-frequency video signals will be improved in response. It may be observed that the clamping action depends upon having a *fixed pedestal level for the duration of time in which the switch is closed.* In this way, low frequency sine waves are

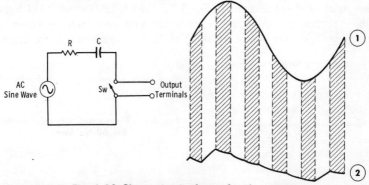

Fig. 3-32 Clamp-circuit theory for sine waves.

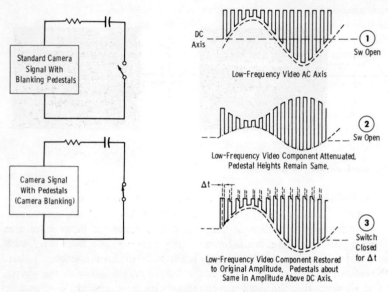

DC Axis

Low-Frequency Video AC Axis

① Sw Open

② Sw Open

Low-Frequency Video Component Attenuated.
Pedestal Heights Remain Same.

Δt

③ Switch Closed for Δt

Low-Frequency Video Component Restored
to Original Amplitude. Pedestals about
Same in Amplitude Above DC Axis.

Standard Camera Signal With Blanking Pedestals

Camera Signal With Pedestals (Camera Blanking)

Fig. 3-33. Clamp-circuit operation for TV signal.

severely attenuated, whereas video signals containing fixed pedestals are passed without attenuation, and the low-frequency response is improved without accentuation of stray-field response.

A clamper is known variously as a *keyed clamping* or *line-to-line clamping* circuit. It is used when it is desired to maintain all voltage amplitudes of either positive or negative parts of a waveform at some predetermined level. It is pulse operated, either during the horizontal-sync pulse or the back porch of the horizontal-blanking interval. The clamping process establishes a fixed level for repetitive pulses of a video signal.

The effectiveness of line-to-line clamping at 60 Hz may be evaluated by the following analysis: It takes 1/120 second for the hum signal to pass through its amplitude extremes (Fig. 3-34), and 1/120 second = 131 lines. The amplitude change at the peaks of the sine-wave hum waveform is negligible. Therefore, where the slope of the hum is maximum, the ampli-

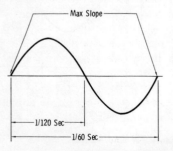

Max Slope

1/120 Sec

1/60 Sec

Fig. 3-34. Time-amplitude relationship for 60-Hz hum.

tude change during a single line may be considered to be $1/131 = 0.00764$, or 0.764 percent. Since the signal passes through maximum slope twice in this time, $2 \times 0.764 = 1.528$, or approximately 1.5 percent. A reduction to 1.5 percent is 38-dB attenuation of the 60-Hz hum component. Effectiveness at higher frequencies of sine-wave pickup decreases until complete lack of effect is noticed at approximately 2 kilohertz.

It may be seen that for 10 times the 60-Hz frequency (600 Hz) the wave passes through maximum slope at a rate 10 times greater, or $10 \times 1.5 = 15$ percent. This is only approximately 16.5-dB attenuation. Table 3-2 lists the effectiveness of the clamper at frequencies between 60 and 2400 Hz.

Table 3-2. Effectiveness of Clamper

Frequency (Hz)	dB Attenuation (Approx)
60	38
120	31
300	22.5
600	16.5
900	13
1200	10.5
1500	8.5
1800	6.9
2100	5.6
2400	4.4

Clampers are switch-operated devices that are timed to refer either sync-tip or back-porch-blanking level to a reference voltage, either ground or a regulated voltage displaced from ground. Thus, each scanned picture line starts from the same dc reference level regardless of APL.

Fig. 3-35 illustrates the pulse-transformer type of clamping circuit. The base of clamp driver Q1 receives a sharp negative pulse to drive it from cutoff to saturation. The resulting surge of collector current through the primary of the pulse transformer "rings" this circuit as a result of the inductive "kickback." But after the first positive alternation, diode X5 clamps the negative swing by shorting out the primary when the collector attempts to swing negative. Note that this polarity would be reversed if an npn transistor were used. Note also the polarity of the resulting pulses on the secondary and how this results in forward-biasing the quad diode circuit, closing the "switch" and applying −12 volts to the base of Q2. Coupling capacitor C_C is always small, since it must be charged or discharged quickly (during the approximate 1.5-μs duration of clamping) to the reference −12 volts. The time constant of R1C1 must be long compared to a line interval so that the charge on C1 will hold the switch open (nonconducting) between pulses during the active line (video) interval. As a rule of thumb, we will expect pulse amplitudes at each end of the quad at least three times the peak-to-peak video signal at the clamped base. This will be true for any type of driven clamping circuit.

Dc restoration normally should be found at the picture tube, at the wave-form-monitoring CRO, at the transmitter modulator stage, and in any circuit where a video reference black must be established. Examples of the latter are at blanking- or sync-insertion circuits, gamma-correction circuits (Section 3-8), sync-stretch circuits, etc.

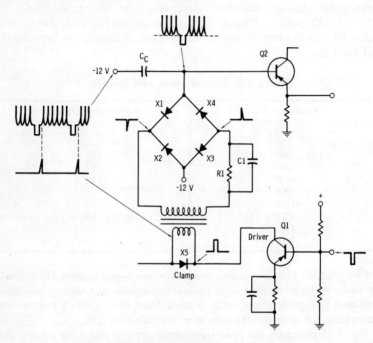

Fig. 3-35. Pulse-transformer clamp circuit.

3-8. THE "SYSTEM CONCEPT"

See Fig. 3-36. All signal sources must be corrected for frequency-phase deficiencies so that they have a "flat" response into the studio switcher and video distribution system. Precorrection must be employed at the transmitter for deficiencies in the vestigial-sideband transmission characteristic, as well as those of the transmitter system itself. In addition, precorrection must be made at camera sources for the nonlinear transfer characteristic of the picture tube.

The "ideal" situation is one in which the light output of the kinescope is directly proportional to the light input from the televised scene. But the kinescope is nonlinear in the direction of compressing blacks and stretching whites. The image-orthicon (I.O.) tube is nonlinear in the

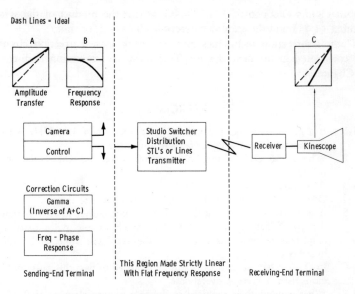

Fig. 3-36. System concept of TV transmission and reception.

direction of stretching blacks and compressing whites. Also, the transfer characteristic of the I.O. is dependent in a complex manner on whether the scene is "high-key" or "low-key," which is another way of saying that the dynamic transfer curve varies somewhat with APL.

The resultant overall characteristic of an uncompensated system is black compression, because the kinescope is more nonlinear than the I.O. tube. The same is true for the vidicon or *Plumbicon.*

Now let us be sure of our terminology. Amplitude linearity is a measure of the *shape* of the *transfer curve.* It is a function of output-vs-input luminance levels of the system. Gamma is the *exponent* of the transfer characteristic. This is the slope of the transfer characteristic plotted on log-log coordinates. As shown by the simple transfer blocks of Fig. 3-36, a gamma of unity (dash lines) is a strictly linear transfer slope. If the slope is greater than unity (kinescope), blacks are compressed and whites are stretched. If the slope is less than unity (pickup tube), blacks are stretched and whites are compressed. Overall system gamma is the product of the individual gammas.

For example, the vidicon has a relatively constant gamma of 0.65 over the normal beam-current operating range. The average kinescope gamma is around 2 (color standards assume an exponent of 2.2), which means that the picture-tube high-light brightness increases approximately as the square of the applied voltage above cutoff. Then assuming all other units of the system have unity gamma, the overall system gamma is $0.65 \times 2.2 = 1.43$, or greater than unity. The amount of gamma correction necessary to obtain a unity exponent is $1/1.43 = 0.7$ so that the product of the system gamma (1.43) and the gamma correction (0.7) is 1, or unity.

Gamma-correction techniques and frequency-phase correction circuitry are considered in the next chapter. Transmitter precorrection is discussed in Chapter 13.

EXERCISES

Q3-1. If a given stage has a gain-bandwidth product of 100 MHz, what is the bandwidth for a gain of 10 in this stage?

Q3-2. What is the term for the "ideal" rolloff curve, and what is the rate of rolloff?

Q3-3. What is an octave?

Q3-4. Does the rate of rolloff appreciably affect pulse rise time?

Q3-5. Assume the sync pulse at the sync-generator output has a rise time of 0.1 μs. What is the required bandwidth of the pulse-distribution amplifiers to pass this pulse with the same rise time, and with no overshoot?

Q3-6. In Q3-5, what bandwidth would be required to pass this pulse with no more than 3 percent overshoot?

Q3-7. What bandwidth is required to define a picture element of 0.05 μs? (Ideal transient response, no overshoot)

Q3-8. What is the finest picture element (in terms of TV lines) that can be resolved when the transmitter bandwidth is 4 MHz?

Q3-9. The FCC specifies that the radiated carrier must have a ratio of 25 percent sync to 75 percent video for a standard composite signal. What sync-to-video ratio is required at the transmitter input?

Q3-10. What is the actual peak-to-peak voltage excursion of a 1-volt (peak-to-peak) video signal in an RC-coupled circuit?

CHAPTER 4

The Television Studio
Camera Chain

We will start at the "front" of the camera and work backward. This sequence is: the camera lens, the optical system, the pickup tube, the camera-head amplifiers, the rack equipment, and the control panel.

4-1. THE BASIC MONOCHROME CAMERA LENS

Let us begin with a basic review of lens operating parameters. Observe the basics of lens angles illustrated in Fig. 4-1. Note that the half-angle (θ) is:

$$\theta = \tan^{-1} \frac{w}{2F}$$

where,
w is the diagonal of the film or pickup-tube picture area in inches,
F is the focal length of the lens in inches.

This formula simply states that angle θ is an angle whose tangent is the ratio of one-half the film (or pickup-tube) picture-area diagonal to focal length. For a film-projector lens, the diagonal concerns film dimensions. For a camera lens, the diagonal is that of the useful area of the pickup tube used (see Fig. 4-2).

Example 1: A 16-mm film projector has a lens with a 2-inch focal length. The diagonal of 16-mm film is roughly 0.475 inch. Then:

$$\theta = \tan^{-1} \frac{0.475}{2 \times 2}$$

$$= \tan^{-1} 0.118$$

$$= 6.7° \text{ (approx)}$$

153

The total angular field is then:

$$2\theta = 2 \times 6.7° = 13.4°$$

Example 2: An image-orthicon camera has a lens with a 2-inch focal length. The diagonal of the photocathode (for either the 3-inch or 4½-inch I.O.) is 1.6 inches. Therefore:

$$\theta = \tan^{-1} \frac{1.6}{4}$$

$$= \tan^{-1} 0.4$$

$$= 21.8° \text{ (approx)}$$

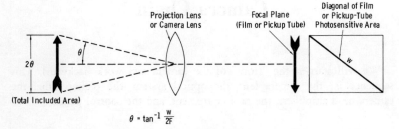

$$\theta = \tan^{-1} \frac{W}{2F}$$

Fig. 4-1. Lens angle.

The total angular field is:

$$2\theta = (2) \ (21.8) = 43.6°$$

But now we must go a step further. We know that TV scanning has a 4:3 aspect ratio, as shown in Fig. 4-2. Also bear in mind that the 16-mm film dimension is approximately 0.28 inch by 0.38 inch so that the same consideration applies to a projector lens as to a camera lens. Therefore, we have different effective angles for the vertical and horizontal dimensions; the vertical angle can be taken to be 3/4 of the horizontal angle. So, first find the horizontal angle:

$$\theta = \tan^{-1} \frac{1.28}{4}$$

$$= \tan^{-1} 0.32$$

$$= 17.8° \text{ (approx)}$$

The total horizontal angle is:

$$2\theta = (2) \ (17.8)$$

$$= 35.6° \text{ (in practice, about 34°)}$$

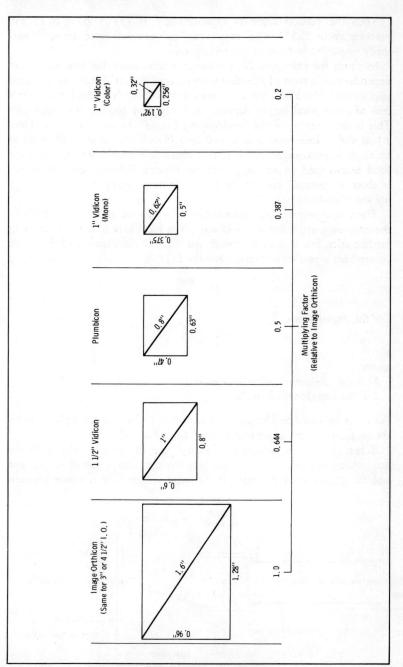

Fig. 4-2. Useful scanned area.

Then the vertical angle is approximately $(0.75)(35.6) = 26.7°$ (in practice, about $25.5°$). The "in practice" values result from the mask normally employed in front of the pickup tube.

So much for examples. Now remember this: Since the lens angles are dependent on a ratio of effective scanned area to focal length, all practical applications (field of view at a given distance, depth of field, etc.) for a lens of given focal length depend on the type of pickup tube employed. This is the meaning of the "multiplying factors" shown in Fig. 4-2. Thus, a lens with 2-inch focal length used on a *Plumbicon* camera will result in an angle approximately one-half that obtainable with a lens of the same focal length used on an image-orthicon camera. Whereas the 2-inch lens is about a "normal" eye for the I.O., it is more nearly a "telephoto" eye for the *Plumbicon*.

There is a very simple relationship by which we can quickly estimate the horizontal angle for a lens of any given focal length. It is not exact in mathematics, but it is close enough for a rapid visualization of the angle covered for a particular camera. For the I.O.:

$$H = \frac{68}{F}$$

For the *Plumbicon:*

$$H = \frac{34}{F}$$

where,

H is the horizontal angle in degrees,
F is the focal length in inches.

For a similar simplified formula for any other pickup tube, simply multiply the 68 figure by the appropriate factor from Fig. 4-2.

When the lens is focused at "infinity" and the pickup tube is all the way toward the lens, the distance between the optical center of the lens and the photosensitive surface is the focal length. The distance between

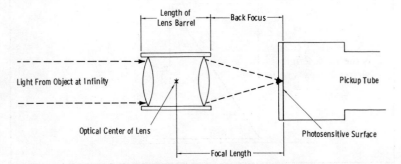

Fig. 4-3. Basic characteristics of a lens.

the rear of the lens and the photosensitive surface is the "back focus." The length of the lens barrel depends on the structure of the lens and its focal length; in general, the barrel is longer for longer focal lengths. These general characteristics are illustrated by Fig. 4-3.

In summary, focal length governs the angle of field covered and, therefore, the size of a given image at a given distance. The angle covered depends on the size of the pickup tube for a given focal length. Table 4-1 shows the horizontal field angle for various focal lengths when the lens is used with an image-orthicon camera. To find the angle for any other pickup tube, use the appropriate multiplying factor from Fig. 4-2.

Table 4-1. Horizontal Field Angles

Description	Typical Maximum f/ Number	Total Horizontal Field Angle (Width of Field)
Studio Camera Lens, 35 mm (1½")	f/3.3	51.5°
Studio Camera Lens, 50 mm (2")	f/1.9	34°
Studio Camera Lens, 90 mm (3½")	f/3.5	19°
Studio Camera Lens, 135 mm (5.3")	f/3.8	13°
Studio and Field Camera Lens, 8½"	f/3.9	8°
Studio and Field Camera Lens, 13"	f/3.5	5°
Field Camera Lens, 15"	f/5.0	4.5°
Field Camera Lens, 17"	f/5.0	4°
Field Camera Lens, 25"	f/5	2.75°

The *width of field* covered by a lens of given focal length is related to the active area of the scanned surface by the following ratio:

$$\frac{F}{\text{Width of Scanned Area}} = \frac{\text{Inches From Subject to Lens}}{\text{Width of Field in Inches}}$$

If the width of the scanned area is assumed to be 1.28 inches for an image orthicon and the above relationship is solved for the width of field in feet, the result is:

$$W = \frac{1.28D}{F}$$

where,

W is the width of field in feet,
F is the focal length of the lens in inches,
D is the distance focused upon in feet.

For example, to find the width of field when the lens on an I.O. camera has a focal length of 2 inches and is focused at 10 feet:

$$W = \frac{(1.28)(10)}{2} = \frac{12.8}{2} = 6.4 \text{ feet}$$

The the field height at 10 feet is $(0.75)(6.4) = 4.8$ feet.

Now how about the same lens when used with the *Plumbicon?* The computation is:

$$W = \frac{(0.63)(10)}{2} = \frac{6.3}{2} = 3.15 \text{ feet}$$

This result is approximately one-half the width of field for the image orthicon. Another way of looking at this is that a 1-inch lens on a *Plumbicon* is about equivalent to a 2-inch lens on an image orthicon.

Another important aspect of the TV lens is the rated *speed*. The speed of a lens determines the amount of light that must be used for satisfactory reproduction. Changing the speed of a lens affects the *depth of field*.

The relative amount of light that passes through a lens depends on the lens diameter. Obviously, a larger lens admits more light than a smaller one. Because of physical difficulties in grinding a high-quality lens and correcting it for certain deficiencies such as color aberrations, the maximum size of the lens is limited.

A lens is rated at its "widest stop," that is, at the maximum diameter of the *iris* opening. The lens iris performs the same function as the iris of the human eye. When we look at a relatively bright scene, the iris in the eye contracts and admits less light, allowing us to distinguish the object without "blinding" the sense of sight. Similarly, the TV-lens iris varies the size of the opening that allows light to pass through the lens.

The "stop opening" of the iris is designated by the small letter f, and is defined as follows:

$$f = \frac{F}{D}$$

where,

f is the stop value,
F is the focal length,
D is the lens diameter with iris fully open (diameter of iris opening).

Thus, it may be observed that the f number is rated *in proportion to the focal length*. If the widest stop of the iris is one inch and the lens has an 8½-inch focal length:

$$f = \frac{8.5}{1} = f/8.5$$

Notice that the result is written $f/8.5$, to indicate that 8.5 is the f number.

For the same 8½-inch focal length, a lens diameter of 2 inches would result in a "speed," or f number, of $f/4.25$; a diameter of 4 inches would give a very fast speed of about $f/2$.

A lens with a short focal length such as 50 mm (2 inches) gives a wider angle of field and a smaller image. A one-inch-diameter lens of this focal length would give a speed of $f/2$.

In practice, the iris of a TV camera is used to meet a variety of operating conditions. One of the important effects of the "stop value" is its influence on the *depth of field*. As is probably obvious, the camera rarely is called upon to focus on a single object. Usually, more than one object or a comparatively large area of a scene is distinguished by the eye as being "in focus." This area is determined by the depth of field.

Depth of field is the distance between the nearest object in focus and the farthest object in focus, when the lens is nominally focused on a given point. This is the distance between the far object and the near object to the left of the lens in Fig. 4-4A. *Depth of focus* is the distance between

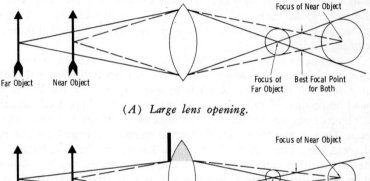

(A) *Large lens opening.*

(B) *Small lens opening.*

Fig. 4-4. Circles of confusion.

the nearest in-focus image behind the lens and the farthest in-focus image behind the lens, when the lens is nominally focused on a given point. This distance is represented to the right of (behind) the lens in Fig. 4-4A.

Study Fig. 4-4A. Remembering the basic principle of focusing, observe that the near object is focused at a certain point, and the far object is focused at another point. At a certain point between the two images is shown the point for "best" focus of both objects. Now study the point of focus for the *far* object; at this point, the rays of light from the *near* object form a small diffused circle of light (the drawing is exaggerated for this study). Now look at the point of focus for the *near* object; here, the rays of light from the *far* object (which are now diverging again) form a diffused circle of light. These circles are called *circles of confusion,* and in practice they must be kept *under* two scanned lines to be unnoticed by the viewer. Thus, it can be understood why objects outside the depth of field appear hazy or unfocused.

Fig. 4-4B shows the effect of *stopping down* the active lens area by means of the iris. The circles of confusion are reduced in size by an amount that depends on the f number, and the image is sharper, allowing a greater *depth of field*. In other words, when a smaller lens opening (higher f number) is used, the objects in good focus may be farther apart before the circles of confusion become great enough to cause a diffused (blurry) picture.

When the maximum allowable diameter of the circles of confusion is less than two scanned lines, what does this mean in terms of the I.O.? The useful photocathode area is 0.96-inch high; one inch is close enough for our computations. We have 490 active lines (525 minus 6- to 7-percent blanking per frame); therefore, one scanned line is 1/490 inch. This is essentially 1/500 inch, and since the maximum diameter of a circle of confusion must be *less than* two scanned lines, 1/500 inch (0.002 inch) normally is used.

To become familiar with depth-of-field characteristics, it is necessary to know the *hyperfocal distance* of a lens. The hyperfocal distance is that distance at which the nearest object is in sharp focus when the lens is focused on infinity. If, then, the lens is focused on this hyperfocal distance, all objects from *one-half* the focused distance to infinity will be in sharp focus. Thus, it may be observed that the shorter the hyperfocal distance of a lens, the greater the depth of field is. The shorter the focal lngth (F), the shorter the hyperfocal distance is and the greater the depth of field is.

To obtain the hyperfocal distance of a lens of given focal length at a given f stop, use the formula:

$$H = \frac{(1/d_c)F^2}{12f}$$

where,

H is the hyperfocal distance in feet,
F is the focal length in inches,
$1/d_c$ is the reciprocal of the maximum allowable diameter of the circle of confusion,
f is the stop number.

Example: To find the hyperfocal distance of a lens with a 3½-inch focal length when the lens is used with an I.O. camera and the f stop is $f/8$:

$$H = \frac{(500)(3.5)^2}{(8)(12)} = \frac{6125}{96} = 64 \text{ feet (approx)}$$

We know that if this lens is now focused at 64 feet, all objects from *one-half* this distance (32 feet) to infinity will be in focus.

The next point of interest is to visualize the nearest and farthest limits of depth of field when a given lens is focused at a certain distance from the lens. These limits are related mathematically as follows:

$$\text{Nearest Limit} = \frac{H \times D}{H + D}$$

$$\text{Farthest Limit} = \frac{H \times D}{H - D}$$

where,

H is the hyperfocal distance,
D is the distance from the lens to the point focused on.

For example, if the lens just discussed is focused at 10 feet:

$$\text{Nearest Limit} = \frac{(64)(10)}{64 + 10} = \frac{640}{74} = 8.7 \text{ ft (approx)}$$

$$\text{Farthest Limit} = \frac{(64)(10)}{64 - 10} = 11.8 \text{ ft (approx)}$$

The depth of field in this case is from 8.7 to 11.8 feet. This example illustrates that the depth of field is reduced radically for the same lens and same f stop when the lens is focused close up. Depth of field increases when the lens is focused farther away. This principle is illustrated further by the drawing for a 90-mm (3½-inch) lens in Fig. 4-5. This drawing shows the actual depth of field as judged in practice; these values are slightly greater than the theoretical values of Table 4-2. Fig. 4-5 simply illustrates that so long as the circle of confusion is *less* than two scanned lines, we will have good focus.

When a lens is stopped down to a higher f number (smaller iris opening), the image brightness is reduced as the square of the f number. The f number is a ratio of lengths ($f = F/D$) whereas light-passing properties depend on area. Since the area depends on the diameter (D), which in the foregoing equation is the denominator, a smaller value of f indicates a larger diameter and consequently a greater quantity of light gathered. A lens mount is usually marked in f numbers so that changing from one

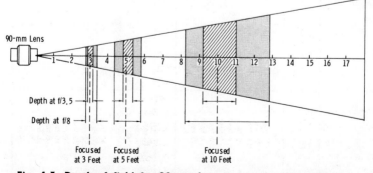

Fig. 4-5. Depth of field for 90-mm lens on image-orthicon camera.

Table 4-2. Depth of Field for Image-Orthicon Cameras

f/1.9, 50 mm — Circle of Confusion, 0.002 in

Distance* Focused On	f/1.9		f/2.8		f/4		f/5.6		f/11		f/22	
	ft in	to ft in	ft in	to ft in	ft in	to ft in	ft in	to ft in	ft in	to ft in	ft in	to ft in
Inf	85 —	inf	59 —	inf	41 —	inf	29 —	inf	15 —	inf	7 6	inf
50 feet	32 —	120 —	27 —	inf	23 —	inf	19 —	inf	11 9	inf	6 9	inf
25 feet	19 6	37 —	17 6	43 —	15 8	63 —	13 9	162 —	9 6	inf	6 —	inf
15 feet	12 9	17 9	12 —	20 —	11 —	23 —	10 —	32 —	7 9	inf	5 3	inf
10 feet	9 —	11 3	8 8	12 —	8 2	13 —	7 6	15 —	6 3	28 —	4 6	inf
8 feet	7 4	8 9	7 1	9 2	6 10	9 10	6 5	10 9	5 4	16 4	4 —	inf
6 feet	5 8	6 6	5 6	6 7	5 3	6 11	5 1	7 1	4 5	9 6	3 6	25 —
5 feet	4 9	5 3	4 8	5 4	4 6	5 7	4 4	5 10	3 11	6 6	3 2	13 —
4 feet	3 10¼	4 1¾	3 9¼	4 3	3 8⅜	4 4½	3 7	4 6½	3 3	5 3	2 9	7 9
3 feet	2 11⅛	3 1	2 10⅝	3 1½	2 10	3 2⅛	2 9	3 3	2 7	3 7	2 3	4 6
2 feet	1 11⅝	2 ⅜	1 11⅜	2 ⅝	1 11	2 1	1 10¾	2 1¼	1 9⅞	2 2½	1 8¼	2 5¾

f/3.5, 90 mm — Circle of Confusion, 0.002 in

Distance* Focused On	f/3.5		f/5.6		f/8		f/11		f/16		f/22	
	ft in	to ft in	ft in	to ft in	ft in	to ft in	ft in	to ft in	ft in	to ft in	ft in	to ft in
Inf	150 —	inf	94 —	inf	66 —	inf	48 —	inf	33 —	inf	24 —	inf
200 feet	88 —	inf	66 —	inf	50 —	inf	39 —	inf	29 —	inf	21 —	inf
100 feet	62 —	350 —	50 —	inf	41 —	inf	33 —	inf	25 —	inf	20 —	inf
50 feet	38 —	75 —	33 —	106 —	29 —	210 —	25 —	inf	20 —	inf	17 —	inf
25 feet	22 —	29 —	19 6	33 —	18 4	39 —	16 8	50 —	14 6	96 —	12 8	inf
15 feet	13 9	16 8	13 —	17 6	12 4	19 —	11 8	21 —	10 6	26 —	9 6	37 —
10 feet	9 6	10 8	9 —	11 —	8 9	11 8	8 6	12 4	7 10	13 9	7 4	16 3
8 feet	7 8	8 4	7 5⅝	8 8	7 3	9 —	7 —	9 4	6 8	10 3	6 3	11 4
6 feet	5 10	6 2½	5 8½	6 4	5 7	6 6	5 5	6 8	5 3	7 —	5 —	7 8
5 feet	4 10½	5 1½	4 9⅝	5 2½	4 8⅝	5 4	4 7½	5 5⅜	4 6	5 8	4 4	6 —
4 feet	3 11	4 1	3 10½	4 1½	3 9⅞	4 2⅛	3 9¼	4 3	3 8	4 4⅝	3 6¾	4 6¾

f/3.8, 135 mm — Circle of Confusion, 0.002 in

Distance* Focused On	f/3.8 ft in to ft in	f/5.6 ft in to ft in	f/8 ft in to ft in	f/11 ft in to ft in	f/16 ft in to ft in	f/22 ft in to ft in
Inf	310 — inf	250 — inf	150 — inf	130 — inf	74 — inf	54 — inf
200 feet	119 — inf	103 — inf	85 — inf	72 — inf	54 — inf	42 — inf
100 feet	75 — 148 —	69 — 176 —	59 — inf	52 — inf	42 — inf	35 — inf
50 feet	42 — 59 —	40 — 65 —	37 — 75 —	34 — 90 —	30 — 148 —	26 — inf
25 feet	23 — 27 —	22 6 28 —	22 — 29 —	21 — 32 —	19 — 36 —	17 6 44 —
15 feet	14 5 15 9	14 3 16 —	13 10 16 6	13 5 17 —	12 9 18 4	12 — 20 —
10 feet	9 9 10 4	9 8 10 5	9 6 10 7	9 4 10 10	9 — 11 3	8 9 11 9
8 feet	7 10 8 2	7 9 8 3	7 8 8 4	7 6 8 6	7 4 8 9	7 2 9 —
6 feet	5 10⅞ 6 1⅛	5 10½ 6 1½	5 9⅞ 6 2⅛	5 9 6 3	5 8 6 4	5 6 6 6
5 feet	4 11⅜ 5 ⅝	4 11 5 1	4 10¾ 5 1¼	4 10¼ 5 1¾	4 9½ 5 2⅝	4 9 5 5
4 feet	3 11⅝ 4 ⅜	3 11⅜ 4 ⅝	3 11⅛ 4 ⅞	3 10⅞ 4 1⅛	3 10½ 4 1½	3 10⅛ 4 2

*Distances are measured from the subject to the image plane of the camera.

Courtesy RCA

figure to the next higher one cuts the amount of light passed by one-half. Thus, if $f/2$ is the maximum aperture, the next marked value is usually $f/2.8$, for which the transmitted light is halved. The next value is usually $f/4$, for which the light is again halved, or reduced to one-fourth the amount transmitted by the $f/2$ opening. Another way of saying this is that an $f/2$ lens transmits four times as much light as an $f/4$ lens. Since the ratio of 4 to 2 is 2, it may be seen that image brightness is related to the *square* of the f-number ratio.

It can be seen that available lighting in the studio or in the field is a limiting factor with respect to how far a lens may be stopped down in practice. Lighting in the studio is usually ample and well controlled. Field events pose a greater problem in this respect, because of the variations that can occur after a camera is set up.

It should be mentioned that no lens has a 100-percent light-passing characteristic. An efficiency of 70 to 75 percent is average for the type of lens used in TV work. The barrel in which the lens is mounted must be dark, since any reflection here would result in severe distortions, destroying the ability of the lens to define the scene. Practically all TV lenses are fluoride coated to prevent internal reflections and thereby increase the efficiency of transmitting light through the lens structure.

A lens turret (Chapter 1) holds four lenses of various sizes; the turret may be rotated in a matter of seconds to change the operating position so that a different lens may be used. This action is accomplished by squeezing the large handle on the rear of the camera head, and rotating it to the desired position. Usually, an arrow engraved on the turret-handle indexing plate points to one of four lens identification spaces on the masking ring. These spaces provide a means for the operator to pencil in the type of lens mounted in each position.

Optical focusing (relative distance from the lens to the photocathode surface in the pickup tube) is accomplished by turning a knob on the side of the camera. (Some cameras have focusing knobs on both the right and left side for convenience. Others have the knob only on the right-hand side.) The focusing knob does not move the lens at all. The distance is varied by sliding the pickup tube and its associated yoke (deflection and alignment coils) back and forth on two tracks. This arrangement facilitates the use of the lens turret, since complicated mechanisms would be required with a turret to move the actual lens in and out. The end result is the same whether the lens or the pickup tube is moved.

Lenses of 13-inch and longer focal length are termed *telephoto* types. It is the purpose of such a lens to provide "close-ups" of objects that are relatively great distances from the camera. Telephoto lenses are extremely popular for field events and often are used indoors when distances of 20 to 50 feet or more are involved in the telecast. The image of the object upon which the lens is trained is made large by sacrificing vertical and horizontal fields covered, and depth of field.

4-2. LENS SYSTEMS WITH VARIABLE FOCAL LENGTH

In recent color cameras, the turret has been eliminated completely. This is made possible by using a lens that has a variable 5-to-1 or 10-to-1 range of focal length. Although such a lens contains many optical elements, the basic function can be visualized with the aid of the drawing of Fig. 4-6. When the focal length is being changed, the lens elements that remain stationary are the front element of the front lens (1), the "erector" (4), and the relay element (6). Those being moved to change the focal length are the rear element of the front lens (2), the "variator" (3), and the "compensator" (5).

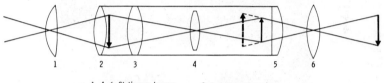

1, 4, 6 Stationary Lenses   2, 3, 5 Movable Lenses

Fig. 4-6. Basic elements of zoom lens.

With such a lens, and assuming a properly adjusted assembly, the focus will track throughout the range of focal-length adjustment, provided the object remains the same distance from the camera. If the camera is dollied or the object moves away from or toward the camera, the focus must be readjusted. The amount of readjustment, of course, depends on the degree of movement and the effective depth of field being employed.

Most recent *Plumbicon* color cameras have a built-in 10 to 1 focal-length range, for example 18 to 180 mm (0.7 to 7 inches). Remember the multiplying factor to relate this to an equivalent for the I.O. camera; the equivalent is 36 to 360 mm (1.4 to 14 inches).

Some color cameras employing image orthicons in combination with vidicons still use only a 5 to 1 focal-length range. With these cameras, a *range converter* is required for some applications. Table 4-3 summarizes this. The two "normal" ranges shown are usually selected by a switch on the rear of the camera. Note in particular that when this switch is on the 8- to 40-inch position (200 to 1000 mm) the maximum speed is $f/8$, and this does not change when the range converter is used to reduce the focal length to an effective 3.2- to 16-inch range. The range converter attaches to the front port of the camera as shown in Fig. 4-7.

Fig. 4-8 illustrates the optical system of the RCA TK-42 color camera. The optical path starts at the upper right and can be traced as follows:

1. *Range Converter*. This item was just discussed, and its effect is tabulated in Table 4-3.

2. *5 to 1 Variable-Focal-Length Lens.* This assembly contains the remotely operated iris that becomes the main operating control after the camera is set up and properly adjusted for the scene.

3. *Orbitor Wedge.* This device is driven by a 1.3-rpm motor, to rotate the image slowly to prevent "burn-in" on the image orthicon. It is not used with later image orthicons that do not "burn."

(A) Vidicon side.

(B) Image-orthicon side.

Courtesy RCA

Fig. 4-7. RCA TK-42 color camera.

4. *Beam-Splitter Prism.* This prism directs 20 percent of the light toward the left side (from rear) of the camera for the luminance tube, a 4½-inch image orthicon. The disc (shown partly by dash lines) on the side toward element 5 is a luminance trim filter (yellow-

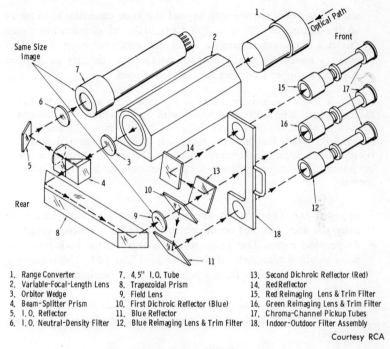

1. Range Converter
2. Variable-Focal-Length Lens
3. Orbitor Wedge
4. Beam-Splitter Prism
5. I.O. Reflector
6. I.O. Neutral-Density Filter

7. 4.5" I.O. Tube
8. Trapezoidal Prism
9. Field Lens
10. First Dichroic Reflector (Blue)
11. Blue Reflector
12. Blue Reimaging Lens & Trim Filter

13. Second Dichroic Reflector (Red)
14. Red Reflector
15. Red Reimaging Lens & Trim Filter
16. Green Reimaging Lens & Trim Filter
17. Chroma-Channel Pickup Tubes
18. Indoor-Outdoor Filter Assembly

Courtesy RCA

Fig. 4-8. Optics of TK-42 camera.

green) to bring the essentially blue response of the I.O. toward the standard luminosity curve. Without this filter, the brightness toward the red end of the spectrum is low, and reds are likely to be more orange or pink than a good red.

5. *I.O. Reflector.* A front-surface mirror is used to reflect the light into the luminance tube.

6. *I.O. Neutral-Density Filter.* Since the maximum f under certain operating conditions is $f/8$, this filter is selected by illuminating a chip chart with exactly 250 foot-candles and noting where the "knee" occurs. For a new I.O., the filter is normally about 0.3D. The value is

Table 4-3. Use of Range Converter

Normal			With Range Converter		
Focal Length (Inches)	Max f	Min Object Distance	Focal Length (Inches)	Max f	Min Object Distance
4-20	4	12 ft	1.6-8	4	20 inches
8-40	8	12 ft	3.2-16	8	20 inches

selected to permit a half-stop beyond the knee condition to occur at an iris setting of $f/8$, with 250 foot-candles of illumination. Since this is a maximum-sensitivity condition, somewhat more light normally is used for a studio program so that the iris may be stopped down to a higher f number for control of level.

NOTE: A loss of sensitivity occurs as the tube ages, and the value of neutral-density filter must be reduced with age. Eventually a point is reached at which no density can be used. But the filter element itself changes the optical length of the light path. When zero neutral density is called for, a "clear" (zero-value) filter must be used. If the filter is left out entirely, the camera will not focus at extreme ranges of the zoom control.

7. *4½-inch I.O. Tube.* When the tube is properly adjusted longitudinally, the image formed on the photocathode has a 1.6-inch diagonal.
8. *Trapezoidal Prism.* This prism serves to elongate the back focal distance to allow placement of the color field lens (9). The remaining 80 percent of the light from 4 is bent 180° to the right (from rear) for the color optics.
9. *Field Lens.* The optical path is such that the image formed here is the same size as that on the photocathode of the I.O. (1.6-inch diagonal).

The remainder of the system operation should be obvious from examination. The reimaging lenses (12, 15, and 16) serve to focus their respective vidicons on the field lens (9). Image size on the vidicon photosensitive surface is reduced to 0.32 inch (diagonal) to provide greater light density. This is normal for vidicon color applications.

The indoor-outdoor filter assembly (18) normally is left free of neutral-density filters except when outdoor applications are common. In this case, a neutral-density filter (about 0.2D) normally is inserted in one end (the center, green, channel is bypassed) so that in the "outdoor" position blue sensitivity is reduced because of the much higher response in this region outdoors. A clear filter normally is placed in the opposite port. Thus, when going from outdoors to indoors, it is only necessary to invert the indoor-outdoor filter without readjustment of the reimaging lenses.

4-3. COLOR LIGHT-SPLITTING OPTICS

The conventional color camera has a minimum of three channels. Fig. 4-9 illustrates the basic problem of forming three images that must come into focus (optical) at exactly the same time. These images must have exactly the same positioning (optical and electrical) with respect to the scanning rasters for good *registration*.

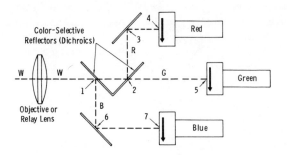

Fig. 4-9. Principle of light splitting.

The lens in Fig. 4-9 can be either an objective lens or a field relay lens. We can see that it would be possible to mount a number of these objective lenses of different focal lengths on a turret. But it is also obvious that the back focal distance of this lens must be rather large, so it would have to have a very long focal length and therefore would be unsuitable for average studio pickups. So now consider the lens as a field relay lens. This means that the objective lens is ahead of the relay lens, or a film-projector lens is throwing a real image on the relay lens. In this case, each pickup tube would have its own objective lens (normally termed a *reimaging* lens) focused on the real-image plane of the relay lens. This is the basic type of color-camera optics for both live and film applications.

Since the images on all three pickup-tube photocathodes must have identical size and must come into focus at one common adjustment, physical paths 1-6-7 (blue), 1-2-3-4 (red), and 1-2-5 (green) *must be identical*. This is the reason for the physical placement shown.

Dichroic mirrors divide the incoming light roughly into blue, red, and green components. The dichroic at point 1 reflects blue and passes red and green. The dichroic at point 2 reflects red and passes the remaining light, which is primarily green. At points 3 and 6 are front-surface mirrors.

Fig. 4-10 shows typical transmission and reflection characteristics of a pair of dichroic mirrors (red and blue). These curves show the reason why it was stated that dichroics split the light "roughly" into the three primary color components. It is obvious that additional *trim filters* in the light paths following the dichroic mirrors are necessary.

There are still two very important characteristics remaining to complete the story of how the color camera "sees" a color scene. These are:

1. Studio-lamp color temperature or, in the case of film, projector-lamp color temperature and film base (and processing) characteristics.
2. Pickup-tube spectral-response sensitivity curve.

The total story can be simplified by the chart of Fig. 4-11. The solid lines indicate typical spectral response of an image orthicon and the spectral distribution of an incandescent lamp with a *color temperature* of 3000 de-

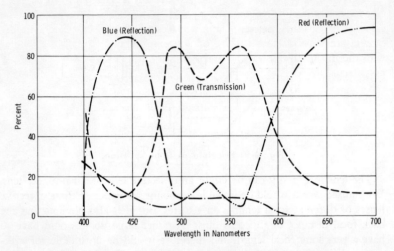

Fig. 4-10. Characteristics of a pair of dichroic mirrors.

grees Kelvin (3000 K). Although the Type 5820 or 1854 I.O. is indicated in Fig. 4-11, the curve shown is typical of most other present-day tubes used in color cameras (high peak in the blue region). To obtain the relative amplitudes and passbands of the three channels (shown by broken lines), color-correction and neutral-density filters must be used in the optical paths.

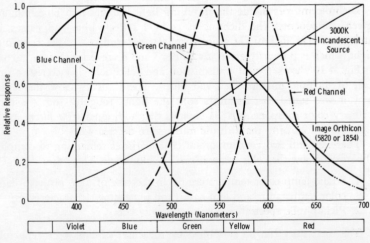

Fig. 4-11. Camera spectral-response characteristics.

In practice, light can become polarized when it is reflected from polished objects, specular surfaces, or even perspiring foreheads of persons in the scene. The red and blue reflecting surfaces of the color splitters in the camera are sensitive to the plane of polarization, since they are at angles with the incoming light. (The green path is straight through and hence is not affected by polarization.)

NOTE: Both the color discrimination and the difference-of-wavelength effects become worse with increasing angles of incidence.

All of these drawbacks are minimized by modern prism-type optics. The front part of the prism optical assembly contains a quarter-wavelength plate, in diagonal position, such that the characteristics of perpendicularly polarized and parallel-polarized light are averaged. Again, the effectiveness is dependent on minimum angles of incidence of the following reflective surfaces. For this reason, some cameras use pickup-tube yoke assemblies that are "fanned out" rather than parallel.

Without this correction for polarization, the red and blue outputs are reduced for certain critical angles or variations in scenic reflectance or light levels, leaving a predominantly green high light. This sometimes is noticed as a greenish flesh tone of a person (for example) on the left of the screen, while proper flesh tones are reproduced in the rest of the scene.

4-4. THE IMAGE ORTHICON TUBE

There are numerous types of image-orthicon tubes (often called "I.O.'s" or "image orths" for convenience). These types vary in sensitivity, spectral response, and operating characteristics. Fig. 4-12 illustrates the basic construction of all image orths, and will serve as a reference in descriptions to follow.

The three basic types of I.O.'s used in television are as follows:

1. The 3-inch non-field-mesh (monochrome cameras)
2. The 3-inch field-mesh (monochrome and color)
3. The 4½-inch field-mesh (monochrome and color)

Fig. 4-13 shows details of the Type 5820 image orthicon. (This is a 3-inch non-field-mesh tube.) It may be observed that such a tube has two pin bases, the keyed 7-pin base around the outer rim at the back of the front projection of the tube, and the small-shell 14-pin base at the end of the electron gun. The "keyed jumbo annular" 7-pin base fits into the socket that is part of the deflecting-coil assembly in the camera. The 14-pin base (end-base pins) fits into the diheptal socket at the center of the coil assembly to the rear of the deflection yoke.

The basic image-orthicon tube consists of a heater, a cathode, six grids, five dynodes, one anode, a photocathode, and a target. These components

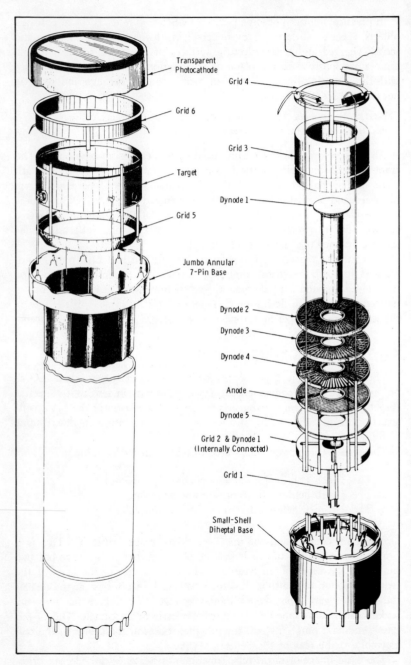

Fig. 4-12. Parts of an image orthicon.

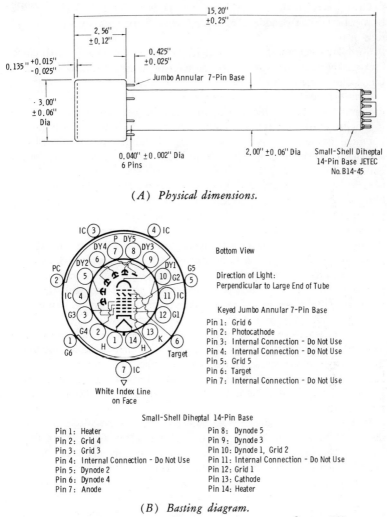

(A) Physical dimensions.

Bottom View

Direction of Light:
Perpendicular to Large End of Tube

Keyed Jumbo Annular 7-Pin Base
Pin 1: Grid 6
Pin 2: Photocathode
Pin 3: Internal Connection – Do Not Use
Pin 4: Internal Connection – Do Not Use
Pin 5: Grid 5
Pin 6: Target
Pin 7: Internal Connection – Do Not Use

White Index Line
on Face

Small-Shell Diheptal 14-Pin Base

Pin 1: Heater	Pin 8: Dynode 5
Pin 2: Grid 4	Pin 9: Dynode 3
Pin 3: Grid 3	Pin 10: Dynode 1, Grid 2
Pin 4: Internal Connection – Do Not Use	Pin 11: Internal Connection – Do Not Use
Pin 5: Dynode 2	Pin 12: Grid 1
Pin 6: Dynode 4	Pin 13: Cathode
Pin 7: Anode	Pin 14: Heater

(B) Basting diagram.

Courtesy RCA

Fig. 4-13. Type 5820 image orthicon.

will be defined immediately; then each section of the tube will be described as to function under an individual subheading.

Grid 1 is the control grid; it immediately surrounds the emitting cathode of the electron gun. The potential on grid 1 therefore serves to limit and control the beam current.

Grid 2 is next to grid 1, and is termed the accelerating grid. The potential on grid 2 sets the velocity at which electrons leave the cathode.

Grid 3 acts on the return beam. The potential of grid 3 controls electron collection of dynode 2, and is adjusted in practice for optimum picture quality.

Grid 4 creates an electrostatic field which, in conjunction with current through the focusing coil to form an electromagnetic focusing field, causes the electron beam to be focused on the rear of the target.

Grid 5 is termed the decelerator grid. The potential on this grid serves to adjust the shape of the decelerating field between the target and grid 4 to obtain uniform electron bombardment of the entire target area. Its potential is low (+25 volts) so that the forward beam is slowed down sufficiently to prevent the beam itself from producing secondary electron emission from the target.

Grid 6 is in the image-multiplier section at the front of the tube, between the photocathode and the target, and is termed the accelerator grid. The potential on this grid serves to accelerate the incident photoelectrons from the rear of the photocathode over to the face of the target.

The photocathode and target were basically described in Chapter 1. It now remains to define the basic functions of the dynodes. A dynode is capable of producing secondary electrons with high efficiency. It is recalled that the return beam, which is modulated in accordance with the charge pattern on the target, is fed to an electron-multiplier section before being impressed across the output resistor of the anode circuit. The partial schematic of Fig. 4-14 shows the connections of the dynode multiplier section. The return beam first impinges on dynode 1, which is the first dynode of the five-stage electrostatically focused multiplier. The multiplied electron beam then strikes dynode 2, and so on through the following stages to the anode. The modulated beam is thus amplified some 500 times before reaching the anode.

It should be remembered from Chapter 1 that less positive portions of the target correspond to dark portions of the scene. Thus, at the time of scanning the high lights of the scene, more electrons are taken from the beam to neutralize the charge on the target, and the return beam is *decreased* in amplitude. This action causes the signal output voltage across R_L in Fig. 4-14 to change in the *positive* direction (less voltage drop across the load resistor). Thus, the grid of the first video preamplifier stage swings in the positive direction for light portions of the scene and in the negative direction for dark portions of the scene. This is known as "black-negative" polarity.

In the following subsections, each principal part of the basic image-orthicon pickup tube is described in more detail.

The Image Section

Reflected light from the scene being televised is gathered by the lens, transmitted through the faceplate of the image orthicon, and optically focused on the photocathode (Fig. 4-15). The photocathode is a semi-

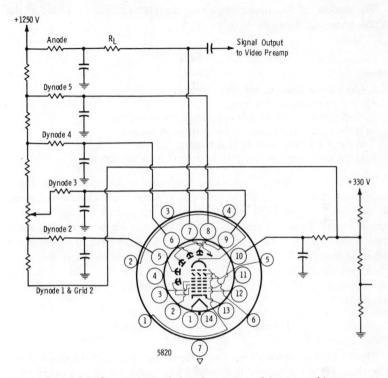

Fig. 4-14. Connections of dynode section of image orthicon.

transparent photosensitive surface. This means that when light rays strike the front surface, electrons, called *photoelectrons,* are caused to be released from the back of the surface. The number of these released electrons is in proportion to the intensity of the light.

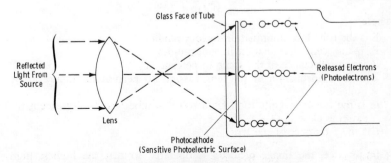

Fig. 4-15. Image section of image orthicon.

The amount of illumination reaching the photocathode is related to the actual scene illumination as follows:

$$I_s = \frac{4f^2 I_{pc} \ (m+1)^2}{TR}$$

where,

I_s is the scene illumination in foot-candles,
f is the f number of the lens,
I_{pc} is the photocathode illumination in foot-candles,
m is the linear magnification from scene to target,
T is the total transmission of the lens,
R is the reflectance of the principal subject in the scene.

This formula is not as complex as it appears at first glance. The linear magnification (m) may be neglected except for extreme close-ups. Assume we are using a Type 5820 tube, with a lens stopped to $f/8$ and having a transmission (T) of 70 percent. Assume also that we are televising a test card composed of blacks and whites, and we may take the reflectance value (R) to be 50 percent. Since we should allow some safety margin, a photocathode illumination of 0.02 foot-candle (instead of 0.01) is used in the calculations. Therefore, the theoreticeal intensity of scene illumination should be:

$$I_s = \frac{4 \times 8^2 \times 0.02}{0.70 \times 0.50} = \frac{5.12}{0.35} = 14.6 \text{ fc}$$

Thus, 14.6 foot-candles (fc) would be required for the test chart. Actually, there are other factors that should be considered in this calculation; these will be analyzed in Chapters 9 and 10. Also, the reader must realize that for scenes in which principal subjects may have reflectance values much lower than 50 percent, the intensity of illumination required for the pickup would be correspondingly higher.

For practical purposes, the illumination on the camera-tube face can be calculated from the following relationship:

$$E = \frac{E_s RT}{4f^2 (m+1)^2}$$

where,

E is the tube-face illumination in foot-candles,
E_s is the scene illumination in foot-candles,
R is the reflectance of the scene in percent,
T is the transmission of the lens (assign 70 percent for monochrome cameras),
m is the linear magnification from scene to tube face (unity except for extreme close-ups),
f is the f number of the lens.

In practice, the image orthicon is operated so that the highest high lights of the scene being televised bring the tube just above the *knee* of

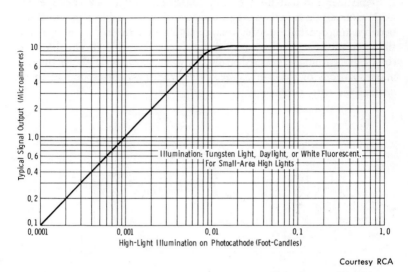

Illumination: Tungsten Light, Daylight, or White Fluorescent. For Small-Area High Lights

Fig. 4-16. Static transfer characteristic, type 5820 image orthicon.

the operating curve. Fig. 4-16 illustrates a typical curve (output signal in microamperes as a function of photocathode illumination), and shows the knee of the curve often referred to in manufacturer's instructions.

In general, the light-transfer characteristic of an I.O. changes for different illumination levels. The basic transfer characteristic shown by Fig. 4-16 is termed a *static* curve. This curve is representative only for small-area high lights. For larger-area high lights, the knee is not nearly as abrupt as that shown.

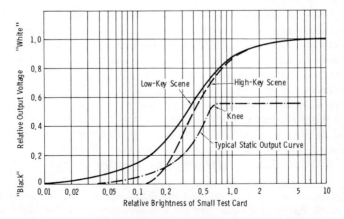

Fig. 4-17. Actual transfer characteristic of image orthicon.

Remember also that the eye has a logarithmic brightness response. Fig. 4-17[1] takes all these variables into consideration. To indicate the difference between the dynamic transfer characteristics and the static output curves shown in tube handbooks, a typical static curve is included. The difference comes about from electron redistribution when, as is nearly always the case, major portions of the scene exceed the brightness corresponding to the knee of the static curve.

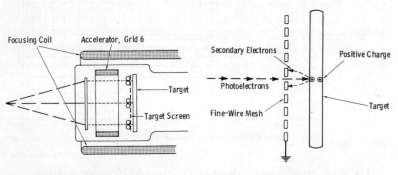

Fig. 4-18. Formation of charge pattern on I.O. target.

Fig. 4-19. Transfer of charge through target.

The Image-Multiplier Section

The image-multiplier section consists of the photocathode described in the previous section, an electron lens system (consisting of the electromagnetic field from the focusing coil and the electrostatic field of the accelerator grid), and a target (Fig. 4-18). The photocathode is approximately 1.6 inch across the diagonal (useful scanning area). The target is a very thin glass disc with an extremely fine wire-mesh screen closely spaced to it on the photocathode (face) side.

Under the influence of incident light rays from the lens, photoelectrons are released from the back of the photocathode and enter the electromagnetic and electrostatic fields of the electron lens system. Focusing of the electron beam (as distinguished from optical focusing of the lens) is accomplished by the focusing-coil current. The accelerator, grid 6, provides the accelerating field.

Typical operating potentials are: photocathode, −400 volts; accelerator grid, −320 volts; target voltage, near zero; and target screen, ground potential or slightly negative. Therefore, the photoelectrons are accelerated with about 400 volts by the time they pass through the fine wire screen to hit the target. The photoelectrons cause secondary electrons to be emit-

[1]The curves in Fig. 4-17 are from the Subcommittee on Lighting for Production of Television Programs of the Committee on Television Lighting of the Illuminating Engineering Society.

ted from the glass target, and these secondary electrons are collected by the adjacent wire screen. At points on the target where this action occurs, a deficiency of electrons results, and the point is positively charged.

Fig. 4-19 shows what happens when a beam of photoelectrons strikes a point on the glass target. The secondary electrons emitted from the point under consideration may be considered to be attracted to the screen and passed off to ground. This point of the target is left deficient in electrons, and therefore positively charged. Since electrons from the back of the target are attracted to the front to neutralize the electron deficiency there, the corresponding point on the back of the glass will become charged positively. Thus, it is evident that the charge at any point on the target depends on the intensity of the photoelectron beam striking that point, and the charge becomes more positive as the light intensity is increased.

The Scanning Section

The elements of the scanning section are illustrated in Fig. 4-20. The electron gun is conventional, of the type found in cathode-ray tubes. The

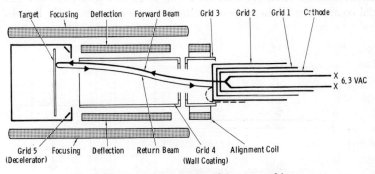

Fig. 4-20. Scanning section of image orthicon.

emitted electrons are formed into a beam by the combined action of the focusing coil and grids 2, 4, and 5. Grid 3 has little effect on the forward motion of the beam, but it acts on the return beam as described later.

As the beam of electrons is caused to focus on the rear of the target by the above action, control pulses (scanning sawtooth) applied to the horizontal and vertical deflection coils cause the beam to sweep across the target in accordance with the prescribed standards. If no positive charge exists at the target (lack of light, or black portion of the scene), the electron beam is repelled by the ground potential (or slightly negative potential) of the target. Therefore, all of the electrons are "reflected" in the form of a return beam that is equal to the forward beam. When a portion of the target is positive (corresponding to a light portion of the scene), some of the electrons are extracted from the beam to neutralize the positive charge at that point. Naturally, for a greater positive charge (brighter corresponding part of the scene), more electrons are extracted from the beam. Thus, the return beam from the rear of the target is amplitude modulated in accordance with the corresponding light variations of the original scene before the camera lens.

The coils through which the deflection currents pass are located within the long focusing coil that extends almost the full length of the image orthicon. The deflection coils extend only along the scanning portion of the tube, and are prevented from affecting the image section by a metal shield around the target section. The alignment coil shown at the end of the gun in Fig. 4-20 creates a transverse magnetic field that serves to impart an initial alignment to the electron beam, allowing exact control for any slight irregularities of alignment between gun and target and between different tubes.

The Electron-Multiplier Section

The sensitivity of the image orthicon as compared to earlier types of pick-up tubes is the result of both the image-multiplier section, described earlier, and the electron-multiplier action to be described now. The essential components of the electron-multiplier section are shown in Fig. 4-21.

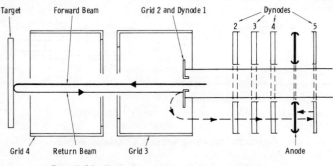

Fig. 4-21. Dynode section of image orthicon.

As the modulated electron beam starts its return journey from the rear of the target, it re-enters the field of grid 4, which is a wall coating along the tube and is held at a potential of approximately 120 volts. Therefore, by the time it enters the field of grid 3, the beam has a tendency to spread, which is called *fringing*. In so doing, the beam strikes the surface of grid 2, which is also the first dynode. A dynode is coated with material capable of high secondary-electron emission, and for each electron striking its surface it emits several additional electrons. Dynodes actually consist of flat vanes inclined at an angle of 40° to the axis of the structure, and covered by a fine screen. The screen is used to prevent distortion of the electrostatic field by the dynode vanes. These vanes radiate from the center of the tube much like the spokes of a wheel (Fig. 4-12).

The electrons released from dynode 1 enter the field of grid 3 and the accelerating field of dynode 2. Grid 3 provides a more complete collection by dynode 2 of the released electrons. The acceleration through the five-stage electrostatically focused multiplier may be more clearly understood by observing typical operating voltages of the Type 5280 image orthicon. Dynode voltages are as follows:

Dynode 1 (grid 2)	300 volts
Dynode 2	600 volts
Dynode 3	880 volts
Dynode 4	1160 volts
Dynode 5	1450 volts
Anode voltage	1500 volts

The much-amplified return-beam current from the final dynode is at-tracted to the anode, which serves as the signal plate for the pickup tube. Useful signal output current of several microamperes through the load resistor is obtained, and picture noise is dependent primarily on the actual shot effect of the electron beam itself.

Image-Orthicon Operating Temperatures

The electrical resistivity of the glass target is affected by temperature. The extreme limits of operating temperature are approximately 35 to 60° *Celsius.* (NOTE: Celsius is now the preferred term for the centigrade tem-perature scale. The name honors the inventor of this scale, Anders Celsius.) The optimum temperature is in the neighborhood of 45° C, which corre-sponds to 113° Fahrenheit. When the temperature is too low, the electrical resistivity is higher than it is at proper temperature, hindering the trans-fer of electrons to the screen. This results in a "sticking picture" of reversed polarity when the televised object (or the camera) moves. When the tem-perature is too high, the electrical resistivity is decreased to the point where loss of resolution occurs, since random electrons travel too easily from point to point. Permanent damage to the target may result from high temperature.

Temperature control of a television camera tube used in studios is pro-vided by a cooling blower, since heat from the coils and tubes is sufficient to bring the operating temperature high enough, and the cooling air then provides the necessary control. In cases where the camera is used outdoors in cold weather, such as in covering a football game, a target heater may be necessary to bring the temperature up to the normal value. This heater fits between the focusing coil and the glass envelope near the shoulder of the tube. The cooling blower directs air along the envelope surface.

In practice, a very close approximation of target temperature may be obtained by measuring the temperature of the glass envelope adjacent to the target. This temperature will be essentially the same as that of the target.

The Field-Mesh Image Orthicon

To understand why a "field-correcting mesh" has been added to some image orthicons, it is necessary first to examine some defects of the non-field-mesh I.O. that result in image impairments.

Refer to Fig. 4-22 during the following discussion. An "excessive" high light is represented by a direct source of light such as a candle flame. (In practice, reference white is a surface with a 60-percent reflectance value.) The focused image on the photocathode causes release of a large number of photoelectrons from the surface (point 1). The photoelectrons strike the glass target at point 3, and the resulting secondary electrons are attracted to the target mesh (point 2), since this screen is slightly positive

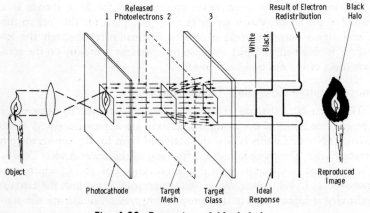

Fig. 4-22. Formation of black halo.

relative to picture cutoff. When an excess number of secondary electrons is released, the potential difference between target screen and glass is nullified, and the uncollected electrons rain back on the glass in the area surrounding point 3. Obviously, the surrounding area is driven to a negative potential, and the return beam for this area represents black. Shown at the right of Fig. 4-22 is the "ideal" waveform in which black reference is maintained relative to white reference. Also shown is the waveform resulting from the excess negative charge in areas surrounding the high light; this excess charge causes the black halo effect in the reproduced image.

The halo effect has been reduced greatly in more recent image orthicons by three factors:

1. Reduction of the spacing between target and target mesh, thereby increasing target-to-mesh capacitance. This increases the signal amplitude and decreases the area in which electrons rejected by the target mesh fall back on the glass target. This spacing cannot be re-

duced beyond the point at which *microphonics* (spurious signals caused by tube movement during camera handling) become troublesome.

2. Placement of a field-correcting mesh (Fig. 4-23) on the scanned side of the target between the decelerator (grid 5) and the target. This mesh "stiffens" the electrostatic field to obtain a more true beam landing on the scanned side of the target.

3. Increased physical size of the target, as in the 4½-inch image orthicon. The high light then occupies a greater area while the spread of rejected electrons remains the same. This approach, along with a high-capacitance target and field mesh, has resulted in the virtual elimination of halo.

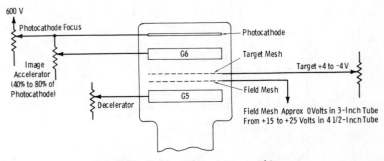

Fig. 4-23. Field mesh in image orthicon.

The basic action of the 4½-inch tube is shown in Fig. 4-24. Note the additional faceplate coil, which is a "pancake" focusing coil connecting in series-aiding with the main focusing coil (illustrated in Fig. 4-30). The resultant shape of the graded magnetic field is such that the optically focused photocathode image is electro-optically magnified between the photocathode and the target by about 1.7 times. (The useful area on the photocathode is exactly the same as for the 3-inch image orthicon.) The magnification of 1.7 times means that the *area* of the high light is increased three times, but because of target capacitance the spread of the excess electrons remains the same as before magnification. Thus, the ratio of halo to high light is reduced drastically. As shown, an excessive high light has only a closely spaced fringe of black in the reproduced image.

Details of the 4½-inch I.O. are presented in Fig. 4-25. Note from Fig. 4-25A that the general configuration is the same as that of the 3-inch tube, except the faceplate coil and the field mesh are added. Note also that, whereas the 3-inch tube employs pins on the shoulder (envelope), the 4½-inch tube employs wider spring-type contractors that slide into slots in the yoke.

We cannot leave the subject of picture impairment without discussing another type of I.O. problem known as *image-orthicon ghost*. This effect, which occurs in earlier tube types, is the opposite of the black halo effect. An I.O. ghost is an image displaced from the high light and of the same polarity. It is most evident when a marked contrast exists between the background (very dark) and the high light.

The I.O. ghost is the result of high-velocity secondary electrons released from the high-light area on the target. In this case, particularly when the background is dark, the photoelectrons from the photocathode are intercepted by the specularly reflected charges from the target. Thus, a spurious image of the high light appears at another area of the target. Depending

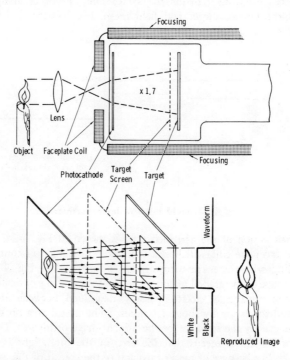

Fig. 4-24. Basic operation of 4½-inch image orthicon.

on the location of the high light on the target, the ghost may appear to the right, left, above, or below the original. If the original image is directly in the center of the target, no ghost can occur. The displacement of the ghost from the original image is proportional to its distance from the target center.

The ghost practically has been eliminated in more recent tubes by an "anti-ghost" image-section design. This design includes a modification of the grid-6 (image accelerator) shaping relative to the target area. There is no difference in schematic notation or use of the tube.

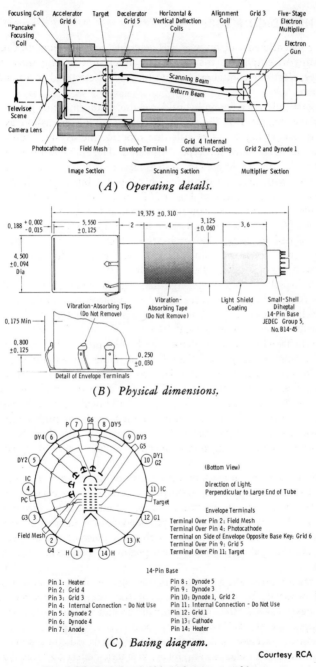

(A) Operating details.

(B) Physical dimensions.

(C) Basing diagram.

Courtesy RCA

Fig. 4-25. Details of 4½-inch image orthicon.

4-5. THE VIDICON TUBE

The conventional vidicon tube was described in basic terms in Chapter 1. It remains to emphasize the significant differences among these tubes—diameter, electrode configuration, and method of electrical focus and beam deflection.

The first difference we will consider is the arrangement of grids, as typified by the diagrams of Fig. 4-26. A review of the discussion associated with Fig. 1-6 (Chapter 1) reveals that tubes of the five-grid construction, such as the 6326, were the type considered. The major difference is that tubes of the 7038 variety (Fig. 4-26A) have grids 3 and 4 as a common connection, with grid 4 a continuation of grid 3 but in a different plane. The five-grid arrangement of Fig. 4-26B allows separate control of grid 3 for vernier adjustment and control of electrical focus.

Fig. 4-27A shows a typical circuit for the 6326 tube when grids 3, 4, and 5 are tied together. The same arrangement is used for tubes of the 7038 variety, except that pin 3 is not used. Fig. 4-27B shows a typical circuit when grid 3 is controlled separately.

For film pickup, an average high-light illumination of 50 to 300 foot-candles is required on the faceplate of the vidicon for minimum lag and best black-level uniformity. For this range of illumination, the target voltage usually ranges between 20 and 40 volts.

For direct pickup, a good picture can be obtained with a constant high-light illumination of not less than 10 foot-candles, or its equivalent when color filters are used, on the faceplate. At 10 foot-candles, the target voltage usually ranges between 40 and 70 volts. With the relatively low light levels encountered in direct pickup, the lag and black-level uniformity will be inferior to the lag and uniformity obtained under the higher light levels and lower target voltages used in film pickup. It is also characteristic that lag or smearing of moving images increases at low values of signal-output current.

In most cases, the illumination level and/or target voltage is limited or adjusted so that the peak signal-output current does not exceed a value of 0.3 to 0.4 microampere. In order that the signal-output current may be known at all times, the camera is provided with a microammeter in the target circuit of each vidicon, or a calibration pulse of the proper magnitude is fed into the input of the video preamplifier to indicate peak signal current.

The maximum amount of illumination on the photoconductive layer is limited primarily by the temperature of the faceplate. This temperature should never exceed 60°C and preferably should be maintained within an operating range of 25°C to 35°C for most satisfactory performance.

In most recent color cameras, the magnetic-focus vidicon has been eliminated in favor of an electrostatic-focus, magnetic-deflection type. An example of this type of tube is the 8134 (Fig. 4-28). This tube is 1 inch in

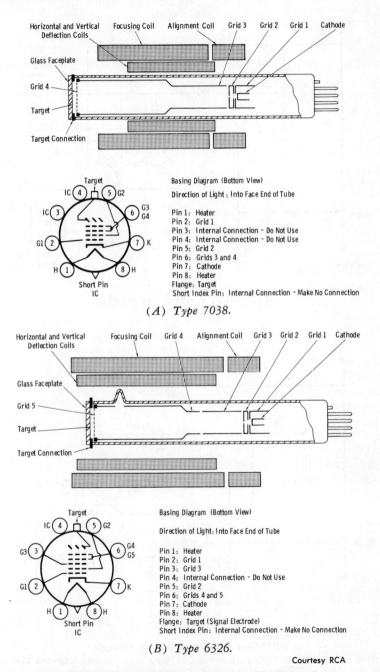

(A) Type 7038.

(B) Type 6326.

Fig. 4-26. Basic grid arrangements of vidicons.

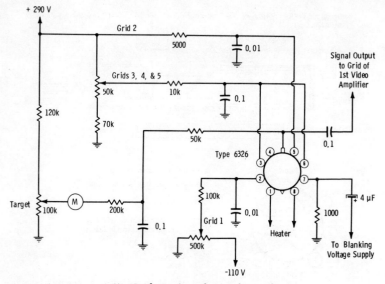

(A) Grids 3, 4, and 5 tied together.

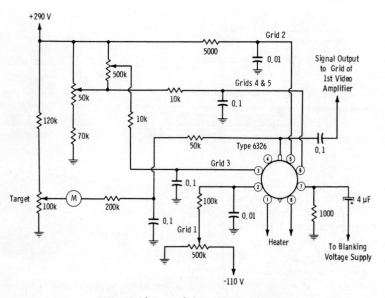

(B) Grid 3 used for vernier focusing.

Courtesy RCA

Fig. 4-27. Circuit connections for type 6326 vidicon.

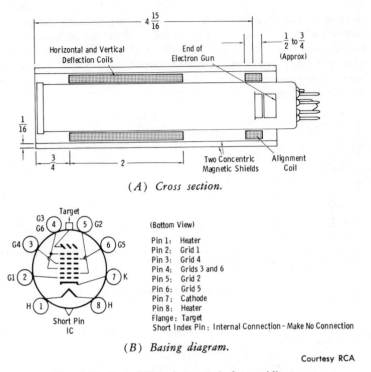

(A) Cross section.

(B) Basing diagram.

(Bottom View)

Pin 1: Heater
Pin 2: Grid 1
Pin 3: Grid 4
Pin 4: Grids 3 and 6
Pin 5: Grid 2
Pin 6: Grid 5
Pin 7: Cathode
Pin 8: Heater
Flange: Target
Short Index Pin: Internal Connection - Make No Connection

Courtesy RCA

Fig. 4-28. Type 8134 electrostatic-focus vidicon.

diameter, the same as the previously discussed 7038 and 6326. An additional design feature of the 8134 is a separate connection for grid 5, which facilitates beam-landing adjustment for the most uniform signal output from the entire scanned area of the target. Grid 6 is connected internally to grid 3 for overall optimum performance. Control of alignment fields for optimum focus uniformity is required with the 8134. Also, magnetic shielding must be used to prevent external fields from impairing the resolving capability of this tube.

Many other vidicons of the electrostatic-focus, magnetic-deflection type use the same base connections as those in Fig. 4-28. For example, Types 4493, 4494, and 4495 are 1-inch-diameter, electrostatic-focus, magnetic-deflection vidicons designed especially for use in the chroma channels of suitably designed color-TV cameras in live-pickup service. The 4493 is intended for use in the red channel, the 4494 in the green channel, and the 4495 in the blue channel. The low heater-power and deflection-power requirements of electrostatic-focus, magnetic-deflection vidicons are advantageous in systems with multiple camera tubes.

Another type of vidicon is the 8480 (Fig. 4-29), which is 1½ inches in diameter and also employs the electrostatic-focus, magnetic-deflection arrangement. The diagonal of the useful scanning area on the photoconductive surface is 1 inch compared to 0.62 inch for a 1-inch vidicon. Note that the pin arrangement for the 8480 is slightly different than for the tubes previously described. The 8480 normally is operated with a useful signal output current of 0.5 to 0.6 microampere, or about twice that of the 1-inch vidicon.

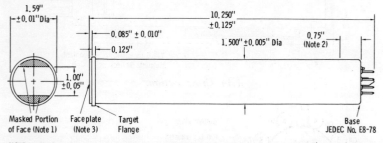

NOTE 1: Straight sides of masked portions are parallel to the plane passing through tube axis and short index pin.
NOTE 2: Within this area the minimum bulb diameter does not apply.
NOTE 3: Faceplate thickness is 0.135" ± 0.005".

(A) Physical dimensions.

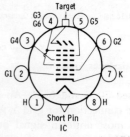

(Bottom View)

Direction of Light: Into Face End of Tube

Pin 1: Heater
Pin 2: Grid 1
Pin 3: Grid 4
Pin 4: Grids 3 & 6
Pin 5: Grid 5
Pin 6: Grid 2
Pin 7: Cathode
Pin 8: Heater
Flange: Target
Short Index Pin: Internal Connection – Make No Connection

(B) Basing diagram.

Courtesy RCA

Fig. 4-29. Type 8480 1½-inch vidicon.

4-6. THE *PLUMBICON* (LEAD-OXIDE) TUBE

The *Plumbicon* pickup tube is similar in characteristics to the vidicon tube, but there are some important differences. The *Plumbicon* uses the same general type of deflection and focusing coils, with similar anode characteristics. It measures 8 inches in length and 1¼ inches in diameter; the conventional vidicon is 6 inches long and 1 inch in diameter.

The *Plumbicon* is a photoconductive device, with each picture element on the photocathode forming a small capacitive charge; one "plate" of the "capacitor" is at the positive potential of the signal electrode and the other "plate" (gun side) is floating. The beam sweeps across the gun side under the influence of the focusing and deflection fields. Since the amount of illumination fixes the amount of charge that leaks through the photo layer, the gun side of the layer contains a positive-potential pattern corresponding to the image focused by the camera lens. Electrons are deposited from the scanning beam onto the layer until the surface potential is reduced to the cathode potential. The resulting capacitive currents of the individual picture elements constitute the video signal.

From this description, we can see that the basic action is identical to that of the vidicon. The important difference is in the properties and construction of the photosensitive element. The vidicon employs antimonious sulfide or selenium, whereas the photosensitive element of the *Plumbicon* consists of three layers. The middle layer is relatively thick and consists of lead oxide (PbO) acting as an intrinsic semiconductor. The outer and inner layers are thin; the material is doped into a p-type semiconductor on the gun side and an n-type semiconductor on the signal-electrode side. This forms a p-i-n diode (i for intrinsic) connected in the reverse direction; the p-type layer is toward the cathode, and the n-type layer is biased by the positive signal-electrode potential.

Table 4-4. Comparison of Pickup Tubes

Resolution (Uncompensated Response*), Typical Values (Percent of "Flat Field")

TV Lines	Plumbicon	Vidicon	I.O.**
100	100	94	100
300	62	48	72
400	45	28	46
500	32	17	28

*The "uncompensated response" of a pickup tube does not fix the possible resolution in the final picture. This is determined by the amount of "aperture correction" possible to use for a given tube, limited by noise level.
**Lens adjusted at knee of curve.

Sensitivity (Typical Operations)

Plumbicon	I.O.
Monochrome	
300 μA/lumen	Latest: 200 μA/lumen Type 5820: 90 μA/lumen
10-12 foot-candles on scene, lens at f/2.8 (comparable to I.O. at f/5.6 for same depth of field).	70-120 foot-candles on scene, lens at f/5.6.
Color	
100-150 foot-candles on scene gives fully saturated colors at f/2.8. Y-channel S/N ratio better than 40 dB.	200-300 foot-candles on scene, lens at f/5.6 with Type 4415-4416 I.O.'s, 4416 in blue channel. S/N ratio 32 dB.

The intrinsic region contributes largely to the high sensitivity of the *Plumbicon*. The conductivity is low in this region, but the electrical field strength is high. Thus, all the discharged (liberated) carriers in the intrinsic region contribute to the photocurrent when the target potential is high enough.

The *Plumbicon* has much lower dark current than the vidicon, and this dark current saturates at about 0.004 microampere regardless of how high the target potential is raised. Basic characteristics of the *Plumbicon* are compared with those of other tubes in Table 4-4.

NOTE: Operational adjustments for all types of pickup tubes are discussed in Chapter 10.

Courtesy RCA

Fig. 4-30. RCA TK-60 monochrome camera.

4-7. THE PICKUP-TUBE YOKE

The pickup tube mounts within the yoke assembly. For monochrome cameras (Fig. 4-30), the focus knob causes the entire tube and yoke assembly to slide on rails behind the camera lens. Color cameras (such as in Fig. 4-7) have fixed-position yokes, and focusing occurs within the zoom-lens assembly.

Fig. 4-31 shows the yoke assembly and the provisions for air cooling the pickup tube. In the particular yoke illustrated, a thermostat is mounted internally for the purpose of automatic temperature control of the image orthicon.

Courtesy RCA

Fig. 4-31. Yoke assembly of TK-60 camera.

The first subject to be considered is the action of the various coils positioned around the pickup tube. These coils are the focusing coil, the alignment coil, the horizontal-deflection coils, and the vertical-deflection coils.

The focusing coil encircles the tube from the emitting end of the electron gun to the face. Its purpose is to concentrate the electron stream into the narrowest diameter possible for good resolution of picture content. The action of the focusing coil is commonly termed magnetic-lens action; it may be visualized better by reference to Fig. 4-32. Recall from basic theory that an electron has an electrostatic field of force about it, converging upon the charge (Fig. 4-32A). When the electron is set in motion, the electrostatic field is distorted, creating an electromagnetic field at right angles to the electrostatic field. In Fig. 4-32B, the magnetic field of the focusing coil is shown. If the electron enters the field so that its magnetic lines of force are perpendicular to the focus field (that is, in a straight line through the coil), no interaction occurs except that required to keep the electron in a straight path. In Fig. 4-32C, the initial direction of the electron is such that its magnetic field is not perpendicular to the focus field, and interaction occurs to "line up" the direction of travel. There is no interaction between perpendicular magnetic fields, but when the angle shifts from 90°, interaction occurs. Thus, the electrons are focused into a beam by the coil.

It is now possible to examine more clearly the mechanics of the scanning function. If strong bar magnets are positioned near a cathode-ray tube as in Fig. 4-33A, with an N pole at the bottom and an S pole at the top of the tube, the beam is deflected to the right. Similarly, when the deflection sawtooth current is passed through the horizontal-deflection coils in one direction, an N pole is at the bottom of the tube and the lines of force are running upward. In Fig. 4-33A, we are looking toward the electron gun from the target, and the beam is therefore traveling toward the

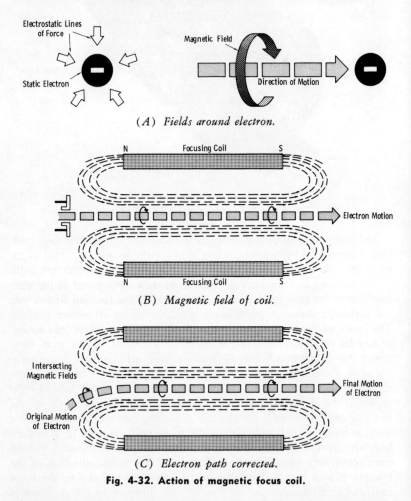

(A) Fields around electron.

(B) Magnetic field of coil.

(C) Electron path corrected.

Fig. 4-32. Action of magnetic focus coil.

observer. The left-hand rule should be used to find the direction of magnetic lines of force about the beam: Grasp the beam of electrons in the left hand, and point the thumb in the direction of flow (toward the observer); the fingers indicate the direction of the magnetic field, in this case clockwise. Therefore, the lines of force to the left of the beam add to the field from the coils, and those to the right of the beam (since they run in opposition) tend to cancel the field from the coils. Consequently, force is exerted on the left side of the beam, deflecting it to the right.

Fig. 4-33B illustrates the action when the sawtooth current through the horizontal-deflection coils reverses in direction, therefore reversing the direction of the magnetic field. The lines of force on the left of the beam now cancel, and those to the right of the beam add. Force is now exerted on the right of the beam, deflecting it to the left.

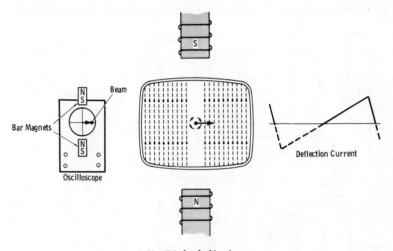

(A) Right half of trace.

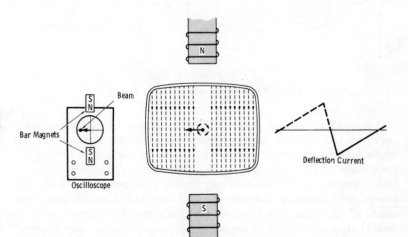

(B) Left half of trace.

Fig. 4-33. Horizontal deflection of beam.

Fig. 4-34 illustrates the interaction of the magnetic fields for vertical deflection of the scanning beam. Again looking from the target toward the electron gun, consider the electron beam coming toward the observer and the deflection current causing a magnetic field in the directions shown. In Fig. 4-34A, the beam is moved to the bottom of the screen and then rapidly back to the center. In Fig. 4-34B, the current reverses, and the beam is carried beyond center to the top of the screen (retrace completed). The current then begins to decrease, sweeping the beam back toward the center of the screen. At the center, the current reverses again, and the scan continues.

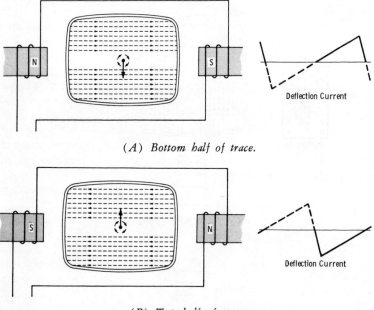

(A) Bottom half of trace.

(B) Top half of trace.

Fig. 4-34. Vertical deflection of beam.

It should be possible now to visualize the details of the entire scanning sequence. The scanning for a complete frame is started at the upper left of the image, and sweeps back and forth on alternate (odd) lines to the lower right corner of the area. At this time, the *field retrace* (vertical retrace) returns the beam to the upper center of the raster. The beam now sweeps the remaining alternate (even) lines until it reaches the lower center of the target. The *frame retrace* (since two fields have been completed) returns the beam to the top left corner to start the same sequence over again. *Field* retrace and *frame* retrace are produced from the same type of waveform; the terms simply designate the position at which the retrace starts and ends.

The foregoing should be made clear by close study of Fig. 4-35. Numbers on the scanning raster are correlated with those on the horizontal-sawtooth waveform shown. Point 1 is the start of the last line of the first field of a complete frame. The *left* side of the raster (looking toward the electron gun) is scanned by current in the horizontal coils contributed by the *damper* tube. At the center of the picture (point 2), the sweep begins to be affected by the current in the horizontal coils contributed by the *driver* tube, and the *right* side of the target is scanned. Point 3, therefore, represents the end of a full line of horizontal scan. In this example, it also indicates the end of the field, and the field retrace occurs. This means that point 3 on the vertical sawtooth has been reached, and vertical retrace begins. It should be observed that the beam does not return immediately to the top center, but sweeps back and forth horizontally a total of 12 to 15 lines. This interval is blanked as discussed previously so that the beam does not actually cause any spurious charges on the target. At point 4, the second field is started. This is the *right* side of the image, and commences on the portion of the horizontal sawtooth contributed by the *driver* tube. Point 5 is then the start of horizontal retrace to return the beam to the left side of the image, and the sequence is repeated. When the bottom half line of the second field is reached, frame retrace occurs, and the beam is returned over an interval of another 12 to 15 horizontal lines to the top left corner to start another frame.

We have assumed the field retrace to start at the lower right and the frame retrace to start at the lower center. It is, however, possible for just the opposite to be true; the field retrace could start at the bottom center and the frame retrace from the bottom right. The sequence depends on whether the vertical-sweep oscillator happens to be triggered by an even-numbered or odd-numbered vertical-sync pulse. Remember, however, that retrace between fields and frames is always one-half line apart, regardless of the position of the beam when retrace starts. Also, retrace may start anywhere along the bottom line.

The field rate is 60 per second (nominal value), compared to the line rate of a nominal 15,750 per second. Since a frame constitutes two fields, the frame rate is 30 per second. There are 262.5 lines in each field for a total of 525 lines in a complete frame. Since the vertical-blanking period occupies about 21 lines per field (42 lines per frame), there are actually only about 483 *active lines* per frame.

Remember in this connection that the above discussion applies to the vertical-blanking signal that is *transmitted* in the composite TV signal to blank the receiver kinescope during the vertical-retrace period. The blanking pulse for the pickup head is *not* transmitted; it is applied only to the camera and is shorter in duration than the transmitted blanking pulse. In this way, the longer vertical-blanking interval at the receiver allows a safety margin for good "trim" of picture edges. The receiver therefore has a vertical-blanking interval equivalent to 40 horizontal lines. The camera

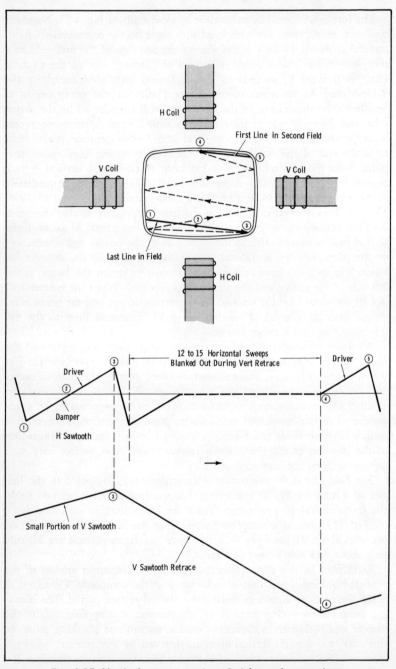

Fig. 4-35. Vertical-retrace portion of pickup-tube scanning.

blanking at the studio must be slightly less than a 25-line interval. The shaping and timing of these various signals is analyzed in Chapter 6.

For the convenience of the reader, the entire scan sequence for an I.O. tube is reviewed here. Because of the importance of understanding this sequence, it is suggested that any step that does not seem clear should be reviewed in the previous text.

1. The heater voltage is applied to bring the electron gun to operating temperature so that electrons are emitted from the cathode surface.

2. The electrons are drawn toward strong positive potentials on a path toward the target. They are limited by the potential of grid 1. Their velocity is determined by the potential of grid 2. The exact alignment of their travel is influenced by the current through the alignment coil at the emitting end of the gun.

3. Current in the long focusing coil (which surrounds the entire tube and yoke assembly) concentrates the electrons into a thin beam for the purpose of scanning small areas of the target. The potentials of grids 4 and 5 assist in this beam formation. Grid 3 has negligible effect here.

4. At the start of a frame scan, sawtooth currents in the horizontal- and vertical-deflection coils create a cross magnetic field that positions the beam at the upper left corner of the image. As the sawtooth current in the horizontal coils increases in one direction (because of the action of the damper and driver tubes in the deflection generator), the scanning beam moves to the right across the image. Under the influence of the changing current in the vertical-deflection coils, the spot is given a slight downward displacement.

5. On the target, positive charges are formed that correspond to the varying degrees of light in the televised scene. As the aperture traverses each line, these charges extract electrons from the beam. The number extracted at any point depends on the amount needed to neutralize the charge, or, in other words, on the degree of brightness at that point. As the beam that is deflected back toward the electron multiplier (return beam) thus varies in strength, it is said to be amplitude modulated in accordance with the brightness variations of the image on the target.

6. The beam continues along the line until it reaches the right edge. At this time, the current in the horizontal coils rapidly reverses direction, changing the direction of the magnetic field and therefore deflecting the beam back to the left side of the target. Because of the slightly downward trace of the scanned line, the beam returns to the alternate line below (interlaced scanning). At the instant this retrace deflection starts, a blanking pulse triggered from the camera blanking generator is received on the target screen (or on any electrode, such as the cathode, to cut off the beam). This pulse drives

the target far enough negative that the entire forward beam is repelled, corresponding to maximum black level of the signal. Thus, the retrace beam never actually reaches the target.

7. The above sequence is repeated throughout the field to the bottom of the image scan area. At this time, the current in the vertical-deflection coils reverses direction, changing the direction of the magnetic field these coils produce. The beam is thus deflected to the top center of the image over a time duration equivalent to about 12 lines. During this time, the image is blanked by the vertical-blanking pulse, which keeps the return (signal) beam at black level.

8. Scanning of the second field now starts, and the beam falls into the remaining alternate lines to complete the frame, which constitutes a whole picture.

It should be obvious from the study thus far that the shape of the sawtooth current in the deflection coils is highly important. The current should rise linearly at a definite time rate for absolutely linear deflection. Since both resistance and reactance are present in the deflection coils, modified voltage waveforms must be applied to them to achieve a real sawtooth current waveform. Fig. 4-36 illustrates the principle under discussion. A sawtooth wave applied to a pure resistance (Fig. 4-36A) will produce sawtooth current, since no reactance is present. If, however, a sawtooth wave-

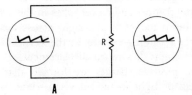

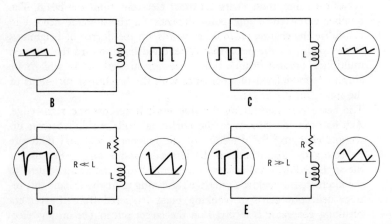

Fig. 4-36. Voltage and current waveforms in RL circuits.

form is applied to a pure inductance (Fig. 4-36B), the self-inductive properties create a back emf that impedes the start of the current, and then when the applied voltage decreases tends to maintain the current; the result is distortion of the applied sawtooth into a rectangular wave as shown. Conversely, a rectangular waveform applied to a pure inductance will become close to a sawtooth in form from the same action (Fig. 4-36C).

In the horizontal-deflection coils, the inductance is large in comparison to the resistive component, but the resistance cannot be discounted in considering the shape of the necessary applied waveform. Fig. 4-36D shows a typical waveform applied to the horizontal-deflection coils. Such a waveform has a rectangular component along with a sawtooth component, in the proper ratio to produce a pure sawtooth current. Fig. 4-36E shows a typical waveform as applied to the vertical coils, since at 60 Hz the yoke may be made largely resistive. In this case, a sawtooth is used along with a small rectangular component.

For the reasons just outlined, all horizontal- and vertical-deflection amplifiers have some means of shaping the deflection waveform. Shaping controls in the pickup head usually are marked "H" or "V" "Saw" or "Linearity," and are used to vary the shape of the applied sawtooth over narrow limits. The main shaping is determined by the design of the sweep circuits themselves.

Fig. 4-37 illustrates the principle of forming a sawtooth wave from a rectangular input pulse such as the camera driving pulse. Since the transistor is not forward biased by dc, it is normally cut off. During this time, capacitor C2 is charging toward the negative potential from the "constant-current" supply at the collector. (A constant-current source is simply a high-impedance source such as the collector of another transistor.) Upon application of a positive-going pulse at the input, the transistor is caused to saturate, shorting saw-forming capacitor C2 to ground and rapidly discharging the capacitor. Capacitor C1 and resistor R1 form a differentiating

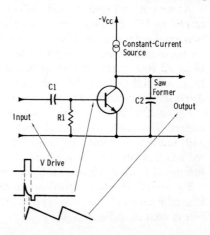

Fig. 4-37. Basic action of a sawtooth generator.

circuit (short time constant) so that the retrace interval (interval in which C2 is discharged) ends before the trailing edge of the applied drive pulse.

A point to remember concerning the control of sweep linearity is that the damper circuit in the horizontal-sweep amplifier contributes to the left side of the picture, and the driver (sometimes called the output stage) contributes to the right side. This is common in all magnetic sweep systems at the studio or in the receiver.

NOTE: Direct currents of the proper amplitude and polarity must be passed through both the horizontal- and vertical-deflection coils to provide centering of the scanning beam. These currents are controlled by horizontal-centering and vertical-centering potentiometers, as shown in Fig. 4-38.

Fig. 4-38 shows the basic approach to achieving identical scanning rasters in a three-tube color camera, when all three tubes are of the same type. The deflection coils, including their associated size and centering controls, are driven in parallel from a common source, as shown here for both the horizontal and vertical circuits. The adjustable resistors labeled "Q" in the horizontal circuits permit adjustment of the R-L ratio to achieve identical current waveshapes in the parallel paths. This is the term used in earlier cameras; later cameras employ the more familiar term "linearity controls."

To obtain satisfactory registration of all three images, a characteristic known as *skew* must be controlled. The drawing at left center in Fig. 4-38 illustrates two rasters with a skew error of one raster relative to the other. Control of this effect is provided by cross-mixing a small amount of vertical sawtooth into the horizontal sawtooth current. As shown here, skew controls are provided only in the red and blue deflection circuits. The green tube and yoke assembly are used as a standard against which the other two are adjusted.

As we can well realize, when different tube types are employed (for example, an I.O. for luminance and vidicons for chroma), different deflection amplitudes must be obtained. Fig. 4-39A shows a typical arrangement for horizontal-deflection circuits. The master size and linearity adjustments affect all four channels; they generally are set for the monochrome tube. Blue, red, and green (BRG) deflection is then obtained through a waveshape and attenuation network such as the one shown.

Fig. 4-39B shows a typical simplified arrangement for the four-tube camera. The green size control becomes the master for all chroma tubes, with individual controls provided for red and blue. Linearity controls for red and blue are variable resistors in series with the deflection coils. (Remember that all of this is for the purpose of obtaining registration readily.)

Fig. 4-39C shows a typical arrangement for vertical deflection in the four-tube camera. The same type of arrangement for the individual channels is used as in Fig. 4-39B.

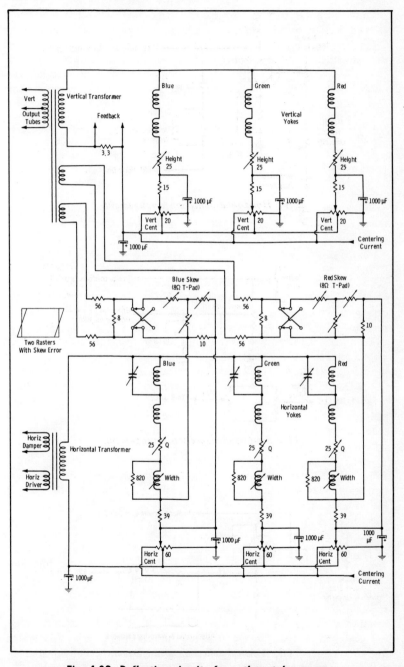

Fig. 4-38. Deflection circuits for a three-tube camera.

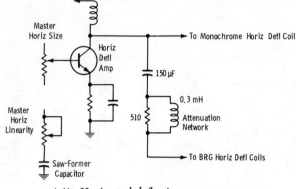

(A) *Horizontal-deflection arrangement.*

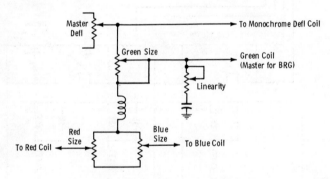

(B) *Simplified deflection arrangement.*

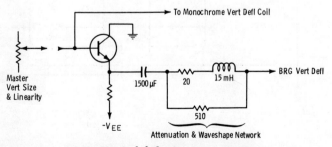

(C) *Vertical-deflection arrangement.*

Fig. 4-39. Deflection principles for dissimilar tubes.

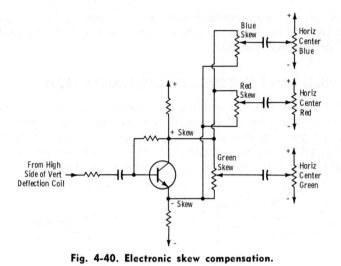

Fig. 4-40. Electronic skew compensation.

Fig. 4-40 shows the latest development in electronic skew control. Contrast this with the older method of Fig. 4-38. The polarity-reversal switches in the T pads of Fig. 4-38 are not required in the circuit of Fig. 4-40 because of the opposite phases obtained from the phase-splitting transistor. In a four-tube camera, skew controls normally are provided only in the three color channels to match the luminance-tube yoke as a standard. In three-tube cameras, the green channel is taken as a standard, and only red and blue skew controls normally are provided.

Some cameras do not contain electronic circuitry for skew correction. A mechanical adjustment is provided in the yoke assembly to permit rotation of the horizontal coil relative to the vertical coil. Other color cameras provide both mechanical and electrical adjustments.

Focus coils for electromagnetically focused tubes are in series across a common regulated focus-current supply so that identical focus fields are obtained (Fig. 4-41). In four-tube color cameras employing three vidicons in the color channels, the vidicons normally are electrostatically focused and do not require focus coils. In this case, the focus-current supply

Fig. 4-41. Focus coils in series.

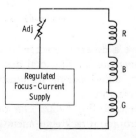

is used only for the monochrome tube. Otherwise, two supplies would have to be provided because of the different current required and the need for individual current adjustment.

4-8. PULSE-DRIVE SYSTEMS FOR THE CAMERA HEAD

Most modern color cameras employ a *timing pulse* to make possible automatic pickup-tube drives and blanking that are properly timed with incoming sync, regardless of camera-cable length (up to about 2000 ft). See Fig. 4-42. The basic function is to sense the delay through the individual camera chain (including the luminance delay in the encoder) and

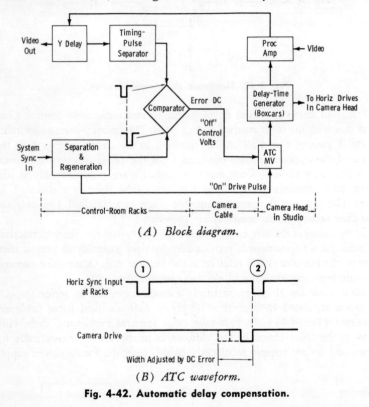

(A) *Block diagram.*

(B) *ATC waveform.*

Fig. 4-42. Automatic delay compensation.

derive a dc error voltage to start camera drives and blanking to exactly compensate for this total delay. (NOTE: The physical location of the circuitry differs among manufacturers and even among models of the same manufacturer.)

Refer to Fig. 4-42A. One output of the horizontal-sync separator in the control-room rack equipment goes down the camera cable to an automatic-time-correction (ATC) multivibrator. This multivibrator turns off after about two-thirds of a line (Fig. 4-42B), depending on a dc control voltage from the compensator. The trailing edge of the ATC-multivibrator pulse initiates a delay boxcar. The trailing edge of the delay pulse starts a timing-pulse boxcar, the output of which is inserted in the camera video (during blanking time) and fed back to the encoding equipment in the control room. This pulse is separated from video and fed to the other end of the comparator, where its trailing edge is compared to the leading edge of sync. The resultant dc error signal provides the off control for the ATC multivibrator.

By using the leading edge of the timing pulse to derive horizontal drive and blanking, the camera signals are advanced relative to incoming sync timing. The amount of advance depends on the corresponding cable length and Y delay. This is to say that the pulse in the camera head corresponding to a given video line is initiated by the sync pulse corresponding in time to the preceding line.

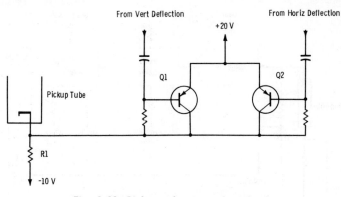

Fig. 4-43. Pickup-tube protection circuit.

If a failure occurs in the timing-pulse loop, the picture will shift horizontally, and the front porch will not be correct. Since a line-to-line function is involved, an intermittent condition can cause erratic shifts of portions of the lines in the raster. Faulty camera cables or connectors can cause this condition. If the camera horizontal-drive multivibrators are of the driven type requiring pulses derived from the timing section, lack of scanning will activate the pickup-tube protection circuits and disable the camera.

Failure of either horizontal or vertical deflection of the scanning beam would confine the beam motion to a thin line with resultant burning of the target. Therefore, the camera employs a protection circuit that cuts off the beam when either the horizontal or vertical deflection fails.

There are many forms of protection circuits; Fig. 4-43 illustrates one form. The pickup-tube cathode returns through R1 to −10 volts. During active line scans, samples of the deflection voltages hold the bases of Q1 and Q2 sufficiently positive to prevent conduction. Thus, the pickup-tube cathode is at about −10 volts, and there is beam current. Loss of either deflection voltage will cause the associated transistor to saturate. Since the collectors are common, the +20 volts at the emitters appears at the pickup-tube cathode, and beam current is cut off. Also, during retrace (blanking) time, the deflection voltages go sufficiently negative to cause the transistors to saturate. Thus, both blanking and pickup-tube protection are supplied by the same circuit in this specific arrangement.

4-9. VIDEO PROCESSING

A typical video path for one channel of a three- or four-channel color camera is shown in Fig. 4-44. The sequence of functions is different for every manufacturer, but the basic function of each block is identical. These functions can be outlined briefly as follows:

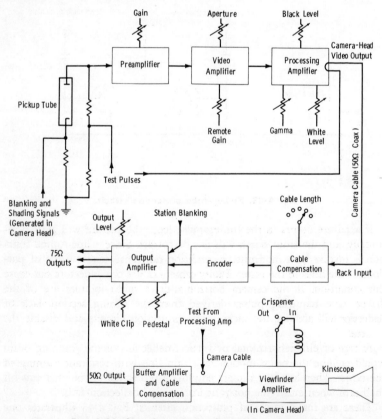

Fig. 4-44. One channel of color-camera system.

Preamplifier. This stage normally is used only for vidicons or *Plumbicons,* to convert the video signal current from the pickup tube to a sufficiently large amplitude to equal that from an I.O., or to a value for satisfactorily processing the signal.

Video Amplifier. This stage normally is the first amplifier for an I.O. tube. In this case, the stage includes remote control of gain, and aperture correction. Aperture correction compensates for the finite spot size of the scanning beam in the pickup tube. In a four-channel camera employing an I.O. and three vidicons, it often is included only in the monochrome (I.O.) channel, since the three chroma channels are narrow-band. (NOTE: Some cameras employ aperture correction in the color channels as well as in the luminance channel.)

Processing Amplifier. This stage (often called a "proc amp") involves voltage amplification to boost the signal voltage to a level suitable for gamma correction, gamma correction to predistort the signal to correct for kinescope characteristics, and manual level sets for black and peak white.

Cable Compensation. This function corrects for camera-cable losses at all frequencies across the video passband, for cable lengths of 200 to 1000 or 2000 ft.

Output Amplifier. The output amplifier provides multiple 75-ohm outputs for distribution of the video signal to the system, and a 50-ohm output to feed back to the viewfinder through the camera cable. It also contains circuitry to limit peak black and peak white signal excursions to some preselected value under overload conditions.

Buffer Amplifier. This stage provides isolation for signal feed to the viewfinder, and compensation for the viewfinder coaxial-cable portion of the camera cable.

Viewfinder Amplifier. In this section, the video signal is amplified to a level suitable to drive the display kinescope. Also, it normally includes a switchable *crispener* circuit as a focusing aid for the cameraman.

From Fig. 4-44, it may be noted that shading signals may be injected at the pickup-tube cathode in addition to blanking pulses. These signals are added because, in practice, portions of a scene may appear darker than other portions even though all parts are evenly lighted and of the same relative reflectance values. This effect is caused by the accumulation of tolerances in tube output sensitivity (from side to side or top to bottom), deflection circuitry, optics, etc.

Controls on a shading device (Fig. 4-45) provide a means of setting both the amplitude and phase of the correction waveforms. Only horizontal sawtooth (Fig. 4-45A) and parabola (Fig. 4-45B) are shown, but vertical sawtooth and parabola also are provided. All shading signals for each channel (including monochrome in a four-tube camera) are mixed together to form a composite shading signal. (For clarity, the dc emitter returns have been omitted in Fig. 4-45A.)

The parabolic waveform is derived by integration of the sawtooth waveform. An integrator circuit is simply a combination of series resistance and shunt capacitance; it is therefore sensitive only to the low-frequency components of the signal.

A horizontal-sawtooth shading signal provides darker shading toward the left of the picture or, by reversing the phase, toward the right. Vertical-sawtooth shading may provide progressively darker shading toward the

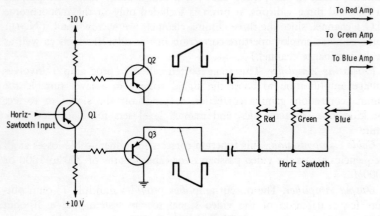

(A) Separate amplifiers.

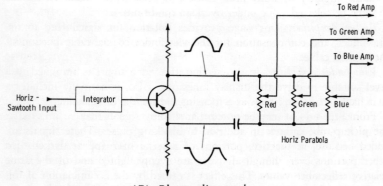

(B) Phase splitter only.

Fig. 4-45. Basic shading-generator circuits.

top of the picture, or, by reversing the phase, toward the bottom. Use of the other waveforms provides a means of controlling the shading at any point in the frame.

NOTE: Shading signals are not always introduced at the pickup-tube cathode. In some instances, they are inserted in a following video-amplifier stage in the chain.

The input amplifier and preamplifier of Fig. 4-46 provide current amplification for low-level signals from a vidicon or *Plumbicon* tube and couple the tube to the low-impedance input of a video amplifier. The input amplifier, which uses a field-effect transistor (FET), is a separate unit that is designed for mounting on a yoke assembly, thereby minimizing input-circuit capacitance. The output from the FET amplifier provides the input signal for the preamplifier module.

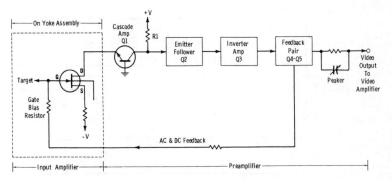

Fig. 4-46. Typical low-level video circuit.

When used together, the input amplifier and the preamplifier function as a signal amplifier with low-noise current gain. The video output is passed through a peaking network in the preamplifier module to compensate for input source capacitance that would otherwise cause high-frequency attenuation (and phase shift) of the video input signal from the pickup tube. Ac feedback from the preamplifier to the input amplifier is employed to cancel the capacitance loading effect on the transistor elements; dc feedback is employed for bias stabilization.

A field-effect transistor with characteristic high input impedance and low noise figure operates as a current amplifier in the input amplifier. A resistor in the input gate circuit provides a path for the degenerative video-signal and bias-stabilization feedback. The signal current from a vidicon target, or other constant-current device, is passed through the gate biasing resistor. The signal voltage developed across the resistor is applied to the gate (G) terminal of the FET, setting up space charges within the transistor that vary with the signal. These varying space charges modulate the amplitude of the current between the drain (D) and source (S) terminals. The result is amplification of the input signal current.

The output from the field-effect transistor is coupled to the cascode amplifier (Q1) in the preamplifier. Transistor Q1, a grounded-base amplifier with low input impedance and high output impedance, provides neutralization of the degenerative capacitance effects between the drain and gate materials of the FET. The video signal output from Q1, developed across resistor R1, is applied to the base of Q2, an emitter-follower amplifier with a high input impedance and a low output impedance. The video signal at the emitter of Q2 is amplified and inverted by Q3, and applied to a feed-back-pair amplifier, containing transistors Q4 and Q5.

Solid-state aperture correction, like its vacuum-tube counterpart, uses a delay line with one end terminated and the other end open. The problem with transistors is that an effective open circuit is hard to obtain in an arrangement similar to the conventional tube circuit. Fig. 4-47A illustrates

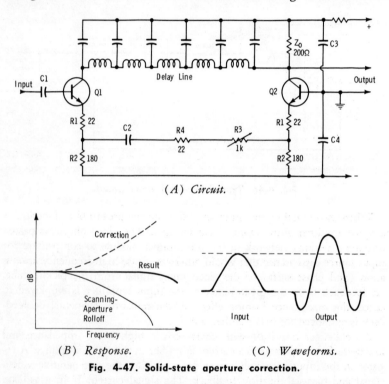

(A) *Circuit.*

(B) *Response.*   (C) *Waveforms.*

Fig. 4-47. Solid-state aperture correction.

how this problem is solved. The collector of Q1 provides a sufficiently high impedance to act as the open circuit for the delay line, in this case at the sending end. The line is terminated in its characteristic impedance at the collector of Q2. The input signal is divided into two paths: the delayed signal from the Q1 collector, at the Q2 collector, and the undelayed signal from the Q1 emitter. Since the collector signal is delayed, the undelayed signal of reverse polarity at the output is, in effect, anticipatory. Now note that the undelayed component appearing at the collector of Q2 will travel back through the delay line, reflect from the relatively high (unterminated) impedance at the Q1 collector, and return to the Q2 collector as a *second* component of reverse polarity. Thus, this single delay line supplies both an *anticipatory* component and a *following* overshoot on transitions, resulting in symmetrical aperture correction. Fig. 4-47B shows how the aperture correction is made to compensate for the scanning-aperture rolloff. Fig. 4-47C shows the sine^2 pulse response indicating "phaseless" aperture correction. Conventional peaking circuits introduce phase distortion that must be compensated by a phase-correction stage. Adjustment potentiometer R3 (Fig. 4-47A) is the correction-amplitude adjustment, since it sets the ratio of main signal current to aperture-correction current.

Although we will find considerable variation in gamma-correction circuits, nearly all work on the principle of diode conduction-threshold adjustments, as shown basically in Fig. 4-48. Some gamma circuitry works only in the gray-to-black region; multiple diodes are used so that the transfer curve can be made to "break" at different conduction levels. The circuit of Fig. 4-48A employs both black stretch and white stretch (Fig. 4-48B) so that we can visualize clearly the principle involved.

When the white-stretch control is rotated clockwise (direction of arrow), a point is reached at which X1 becomes forward biased. Since white is a positive-going signal at the emitter, this voltage fixes the point of "bend" in the white region. When X1 conducts, the emitter resistance is partially bypassed, reducing degeneration and increasing gain (see the white-stretch curve in Fig. 4-48B).

Similarly, as the black-stretch control is adjusted clockwise, the back bias is reduced on X2 and it conducts. Note that the black signal is a negative-

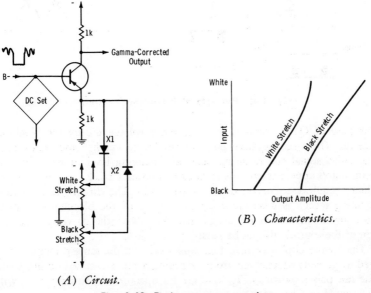

(A) *Circuit.*

(B) *Characteristics.*

Fig. 4-48. Basic gamma correction.

going signal at the emitter. The gain increases in the gray-to-black region, stretching blacks (see the black-stretch curve in Fig. 4-48B).

Any stage in which a black-signal dc reference must exist has a clamper used for this dc set. This function is represented at the base of the transistor in Fig. 4-48A. At the beginning of each video line, the base line of the signal waveform at the transistor base starts from the negative reference dc established by the clamper reference supply. (Review Section 3-7 for a discussion of clamping circuits.)

The principle of black-level control is shown in Fig. 4-49. Negative-black video is combined with negative blanking pulses at the Q1 base.

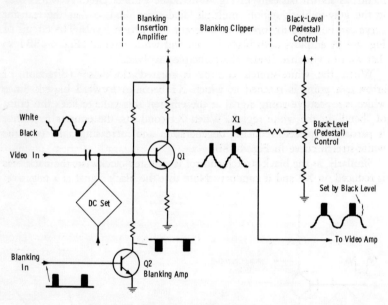

Fig. 4-49. Principle of black-level control.

The base of Q1 is clamped to a reference voltage during the blanking interval. The large negative pulse drives Q1 to cutoff (npn transistor). The video signal and resulting pedestal (blanking level) formed by collector cutoff appear as positive excursions (positive-black) at the collector. These are impressed on the diode clipper (X1). The clipping level (hence blanking level) is determined by the setting of the black-level control, since any amplitude of the signal that rises above the dc voltage on the arm of the control will not be passed by the diode.

The "white clip" shown in Fig. 4-44 works on the same principle. It is used to prevent white levels from exceeding a predetermined output level. The test pulses shown in Fig. 4-44 are for operational adjustments, as will be covered in Chapter 10.

4-10. THE RCA TK-42 COLOR CAMERA CHAIN

Fig. 4-50A is a pictorial layout of the RCA TK-42 color camera chain, and Fig. 4-50B is a block diagram of the rack and control console positions. Fig. 4-51A illustrates a typical rack installation for four TK-42's and two TK-27 color film chains. Fig. 4-51B is a view of the operating-console positions for the rack equipment of Fig. 4-51A. The TK-42 camera head was illustrated in Fig. 4-7.

All camera-head setup controls are located on a panel under a hinged cover at the rear of the camera. The operating console therefore includes only horizontal and vertical centering, white and black balance, master chroma, white-level, black-level, and "sensitivity" (iris) adjustments.

The separate luminance channel in the TK-42 adds the fine detail to the picture; it is comparable in function to the black plate used in four-color printing. Use of the separate luminance channel provides a monochrome signal that is produced directly by an image orthicon tube rather than by a matrixing process. Therefore, the picture on a monochrome receiver or monitor is independent of the accuracy of registration of the red, green, and blue channels. (Three-tube color systems have no separate tube to produce this monochrome signal. Rather, a luminance signal is developed by matrixing the three color-tube outputs. This signal is then combined with the I and Q signals to form the composite color signal. In practice, the choice between three-tube and four-tube systems is largely one of individual preference, and RCA therefore also supplies a three-color chain designated the TK-44.)

Circuits to generate special camera-test pulses, circuits for I.O. alignment, and a color-bar generator are built into the camera equipment. There is provision on the control panel to permit either manual or automatic operation of several functions.

Control of the four channels proportionately and simultaneously is accomplished by automatic white- and black-level controls, maintaining the proper modulation and picture colorimetry. This is done to maintain control of picture brightness and color quality.

The camera monitor features a single-waveform display (familiar to monochrome operators). This is made possible by special circuits that accomplish nonadditive mixing of Y, B, R, and G signals. (This is described in Chapter 5.)

A cooling and heating system maintains optimum temperatures for the four pickup tubes, to help prevent variations in camera performance. Aging of components and ambient-temperature changes are compensated for by feedback stabilization circuits. Critical voltages and currents are maintained at proper values; some are stabilized to less than 0.1-percent variation.

The input signals required for the camera are sync, blanking, subcarrier, and burst flag. Horizontal and vertical drive pulses derived from sync are

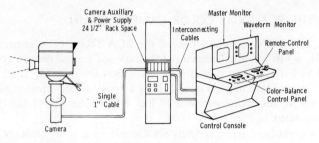

(A) Physical arrangement.

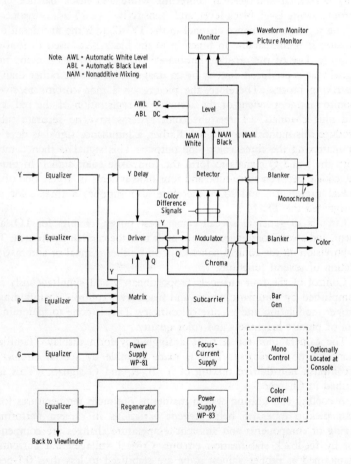

(B) Block diagram.

Courtesy RCA

Fig. 4-50. RCA TK-42 four-tube color chain.

(A) Rack equipment.

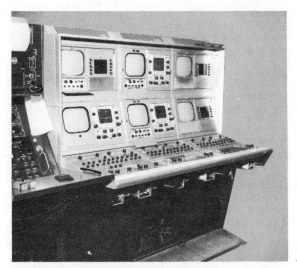

(B) Operating consoles.

Fig. 4-51. Color-camera control-room equipment at WTAE-TV.

generated within the camera. Also, video output is automatically corrected for colorplexer and camera-cable delay, eliminating the need for pulse-distribution amplifiers and delay lines for other cameras in the system.

4-11. THE AMPEX BC 210 TWO-TUBE COLOR CAMERA

The Ampex BC 210 camera, illustrated in Fig. 4-52, makes use of a two-tube system. As demonstrated by Fig. 4-52A, a relatively small-diameter camera cable is used. This cable carries two coaxial cables, two twisted pairs, and ten small-gauge wires. It is less than one-half inch in diameter and weighs about 0.15 lb/foot, compared to about 1 lb/foot for conventional color cable. As the reader would suspect, this camera finds use in portable operation, although it is used in studios also. (The camera is, in fact, an outgrowth of a back-pack color camera for field use.)

Both tubes are *Plumbicons*. The luminance tube has comparatively high sensitivity because the reflectance of the beam splitter in the green region is reduced. The chrominance-channel response is accurately shaped by optical red and blue filters in a filter wheel and by the pickup-tube response.

Fig. 4-53 shows a simplified block diagram of the BC 210. A two-way light splitter behind the zoom lens channels the light into two paths. One path goes to the luminance tube, the other (through a revolving red and blue filter wheel) to the chrominance tube. The filter-wheel motor is locked to vertical sync by means of a digital feedback servo. The wheel revolves at 450 rpm (7.5 revolutions per second), and the number of red and blue segments results in 30 red and 30 blue fields per second. To meet NTSC and FCC specifications, there must be 60 fields each of red and blue. Therefore, each field is used twice, once as the original signal, and again as a delayed signal displaced in vertical position by one line. This is the function of the field switch, which is a solid-state processor resulting in a 262.5-line (one field) delay. Green is derived by subtracting proper amplitudes of red and blue from the luminance (Y) signal.

The sync generator in the camera provides lock to the main studio color sync generator regardless of cable length; proper timing is maintained with over 3000 feet of cable. In addition, the power level is automatically adjusted for changes in cable length.

The viewfinder monitoring switch allows checking of luminance and chrominance, as well as registration, which is observed by red and blue subtraction from the Y signal.

4-12. THE MARCONI MARK VII

The Marconi Mark VII (Fig. 4-54A) is a camera that employs four *Plumbicons*. Also shown are a typical control console for this camera (Fig. 4-54B), and the associated rack equipment (Fig. 4-54C), both as installed at WBBM-TV.

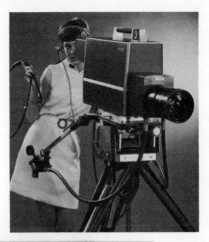

(A) Overall view.

(B) Side view.

(C) Control panel.

Courtesy Ampex Corporation

Fig. 4-52. Ampex BC-210 two-tube color camera.

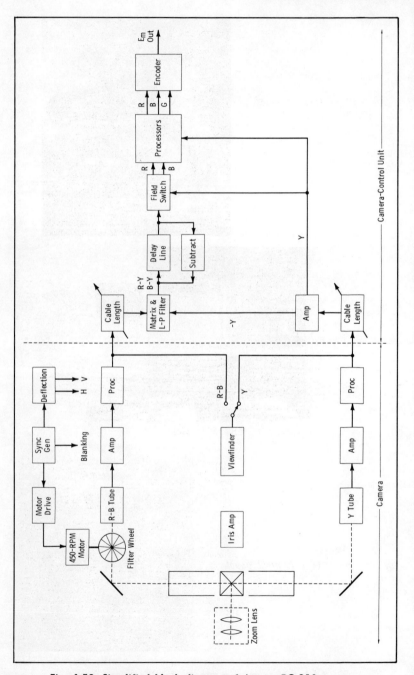

Fig. 4-53. Simplified block diagram of Ampex BC-210 camera.

Courtesy Ampex Corporation

(A) Camera.

(B) Control console.

Courtesy WBBM-TV

(C) Rack equipment.

Courtesy WBBM-TV

Fig. 4-54. System using Marconi Mark VII color cameras.

Heavy current feedback is employed to obtain stable performance in registration and other factors under varying line voltages and temperatures.

Although some image lag can occur in the *Plumbicon,* the problem is not as severe as the "target lag" under low light-level conditions for the vidicon. The lag in the *Plumbicon* is a "beam lag," which requires careful setting of the individual beam currents of the tubes. Since no knee exists for the *Plumbicon,* automatic white-level control is important.

NOTES: (1) Basic operational setup adjustments and operating techniques for the different types of camera chains are covered in Chapter 10.

(2) Some circuitry common to both studio and film chains, such as nonadditive mixing (NAM), solid-state encoders, and color-bar generators, is covered in Chapter 5.

(3) Other cameras, such as the Philips PC-70 and PC-100 are covered in Chapter 11, "Mobile and Remote Telecasts." These cameras lend themselves well to field applications, but are also popular in studio use.

EXERCISES

Q4-1.   In modern color-camera optics, why do you normally find blue reflected first?

Q4-2.   When you say a lens has a 5:1 range, what does this mean in terms of range of angles?

Q4-3.   What is the field-of-view range at 20 ft for an I.O. camera with a lens of 40-mm to 200-mm variable focal length?

Q4-4.   What is the field-of-view range at 20 ft for a *Plumbicon* camera with a lens of 18-mm to 180-mm variable focal length?

Q4-5.   What would the maximum diameter of a circle of confusion be for a *Plumbicon* camera?

Q4-6.   Compare the depth-of-field characteristic of a *Plumbicon* camera with that of an image-orthicon camera.

Q4-7.   What does the problem of Q4-6 assume with respect to operating parameters?

Q4-8.   A color camera is "starved for light" because of light-transmission losses in the variable-focal-length lens and color-splitting optics. Why, then, are neutral-density filters required in certain optical paths?

Q4-9.   Where do you install a range converter, and what does it do?

Q4-10. You know that the longer the focal length of a lens is, the less "speed" it will have because of practical limitations. Why, then, in Table 4-3 can you obtain a maximum of $f/4$ with a 4-20 inch lens, but only a maximum of $f/8$ with the 3.2-15 inch lens?

Q4-11. On what principle (photoemissive or photoconductive) does the image-orthicon tube operate? Does this differ from the vidicon or *Plumbicon?*

Q4-12. What is the purpose of a field mesh in an I.O. pickup tube?

Q4-13. Is operating temperature an important factor for any pickup tube?

Q4-14. What is the main disadvantage of the vidicon for live pickups as contrasted to film-camera use?

Q4-15. What is the purpose of the yoke in a TV camera?

Q4-16. How is electrical centering of deflection currents achieved in a TV camera?

Q4-17. Where does the spectral response of the average camera pickup tube peak in the visible light region?

Q4-18. Does incandescent lighting help or hinder the spectral response of the average pickup tube?

Q4-19. Are the deflection yokes for all channels in a color camera in series or in parallel?

Q4-20. Are the focus coils in a three-channel image-orthicon camera in series or in parallel?

Q4-21. What is skew, and how is it corrected?

Q4-22. What should be the reflectance value of reference white in a scene for color pickup?

Q4-23. If you see controls marked "Q" on a color camera, what function do they perform?

Q4-24. In a four-tube camera, is cable compensation used in all four channels?

Q4-25. In a four-tube camera, is aperture correction used in all four channels?

CHAPTER 5

The Film Camera Chain and Circuitry Common to Studio and Film Chains

In this chapter, we will consider the application of the television camera to film and slide pickup, and we will study the special type of film projector required for television systems. Also, we will discuss additional circuitry applicable to both live and film chains, such as nonadditive mixing for monitoring color chains, color-bar generators (which normally are included in the rack equipment for adjustments), and solid-state encoders.

5-1. THE TV PROJECTOR

The fundamental properties common to any film projector are illustrated by Fig. 5-1. Standard 16-mm sound film (Fig. 5-1A) is 0.630-inch wide by about 0.006-inch thick. The base is acetate and is noninflammable. The picture frame is 0.284-inch high by 0.380-inch wide; each picture is separated from the next by a 0.006-inch opaque line; there are 40 frames to the foot. While the picture frame is pulled down intermittently and allowed to rest momentarily at the gate (Fig. 5-1B), the sound track is caused to roll steadily over the sound drum and photocell assembly. Audio is 26 frames ahead of the corresponding picture frame. The exciter lamp supplies light to the photocell through the sound track, and thus the photocell receives fluctuations of light according to the sound recorded on the track. The photocell in turn feeds an audio preamplifier. The picture projection lens places the image on a field relay lens, or directly on the pickup-tube surface when only one projector feeds one camera.

The basic problem in the use of motion-picture film for television broadcast is the requirement for a 30-frame/second rate for TV. Long ago, the motion-picture industry standardized a 24-frame/second rate, and all sound motion-picture film is on this standard. Projectors for television must convert the 24-frame/second rate into a 60-field, 30-frame/second rate, to avoid the presentation of incomplete information during some of the scanning periods (Fig. 5-2).

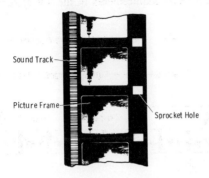

(A) 16-mm sound film.

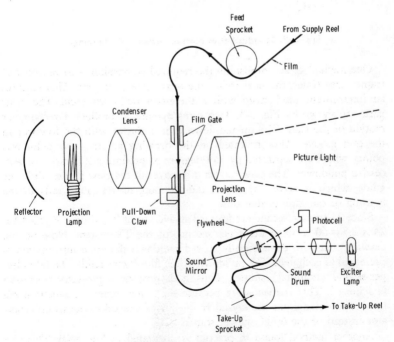

(B) Parts of projector.

Fig. 5-1. Fundamental film-projector components.

The vidicon is an excellent *storage* device. (Storage is the ability to retain momentarily a photoelectric charge image corresponding to the film image projected on the light-sensitive surface.) It is known that if the projector light-application time is at least 30 percent of the total time, and the same information is presented to successive scanning fields, a vidicon tube has no critical phase requirements in the TV system.

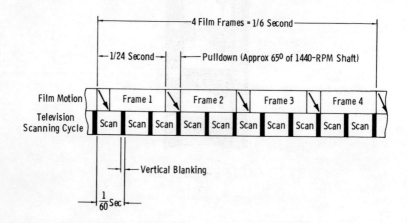

Fig. 5-2. Standard film motion versus TV scanning.

One method of accomplishing the required conversion is to project one frame three times, the next twice, the next three times, etc. This requires an intermittent mechanism with a three-two pulldown cycle. The basic parts are shown by Fig. 5-3. Two cams operate to produce a rectangular motion of the claw. The up-and-down cam is shown with the follower in the cam groove. Also indicated on the drawing are the two pull-down points, which are separated by 144 degrees to produce a 2:3 ratio for successive pulldowns. The claw travels in grooves in the claw guide, while the guide, which is driven by the in-and-out cam, moves along rails in the housing of the intermittent unit.

Since 24 frames/second are scanned an average of 2½ times (Fig. 5-4A), $24 \times 2.5 = 60$ scanned fields per second for the TV system. Note on the drawing that with an unequally spaced pulldown the same information is presented to each field. The pulldown of the film is covered by the 60-pulse-per-second (nominal) shutter. Thus, nonsynchronous projector operation is achieved. This is mandatory because the color camera is synchronized by a countdown from the crystal frequency of the color-subcarrier generator instead of the 60-Hz power source.

Another method found in practice is illustrated in Fig. 5-4B. This is a fast pulldown projector with a 120-pulse-per-second shutter. Significant times in this method are:

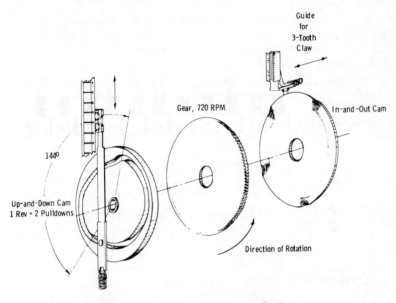

Fig. 5-3. Intermittent-motion mechanism.

Pulldown time = 0.00463 s = 4630 μs
1/60 sec (1 TV field) = 16,660 μs
Light-application pulses = 0.00277 s = 2770 μs
Two light pulses per field = 2 × 2770 = 5540 μs
5540 μs = 33 percent (approx) of one television field. This is 66 percent of a frame.

The 120-pulse-per-second shutter produces five light pulses for each film frame, four per television frame, or two per television field. Note therefore that one film frame is equal to 2½ television fields, to obtain the necessary conversion from projected film rate to the television field rate.

The film projector normally uses a 1000-watt projection lamp. At 66-percent illumination per frame: (1000)(0.66) = 660 watts effective. Of course, much of this light is lost in the following beam-splitting optics for color cameras.

The 2″ × 2″ slide projector normally employs a 300-watt projection lamp. Thus, when film projectors and slide projectors are multiplexed into one camera, the film projectors are equipped with rotating neutral-density discs (remotely controlled) so that the film camera can be fixed in setup adjustments. In practice, differences in lamp-reflector efficiencies must be considered. Even with proper light-source intensity match on open gate, varying film densities require some "gain riding," but they should not require readjustment of the target or beam controls.

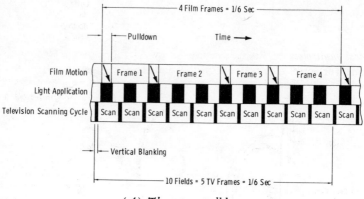

(A) Three-two pulldown.

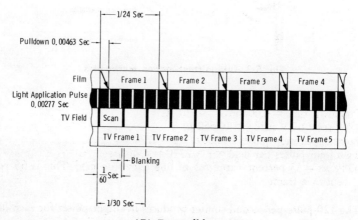

(B) Fast pulldown.

Fig. 5-4. Film scanning for TV pickup.

In monochrome practice, projector-lamp voltages were adjusted for the desired light output to obtain intensity match with another source. In color, extreme caution must be exercised because of color temperature of the lamp; normally, it is not possible to use less than 105 volts on the projector lamp; lower voltages result in a predominant red output. Color balancing the film camera is discussed in Chapter 10.

5-2. FILM ISLANDS

In practice, it is normal to use only one camera for three or four separate film sources. Each installation of this type is termed a *film island*. Fig. 5-5 illustrates a typical installation of three film islands; each camera accommodates two motion-picture projectors and one slide projector. (One island is partly obscured at the left of the photograph.)

The use of one camera for multiple film sources is made possible by an *optical multiplexer*. This is a device for switching several image sources to one (or sometimes two) television film cameras. There are two main types of multiplexers, (1) the "no-moving-parts" type using semimirrors or prisms, and (2) the moving-mirror type, which may be further subdivided according to vertical or horizontal movement and number of inputs and outputs.

A semimirror passes a portion and reflects a portion of the light that strikes it. The *reflecting* surface of such a mirror can be detected by placing the fingernail against the surface. If a space exists between the fingernail and its image, this is the *back surface* of the semimirror. The *front surface* of the semimirror is identified when the image meets the fingernail. This is the surface that must face its respective film projector. If the back surface is mounted toward the projector, "ghost images" appear in the picture because of the difference in optical path lengths that results from the thickness of the mirror.

A typical multiplexer for monochrome cameras is shown in Fig. 5-6A. Light from the slide projector goes through semimirrors C and B and is

Fig. 5-5. Film islands at WTAE-TV.

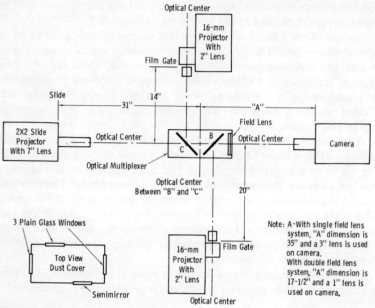

(A) Semimirror type.

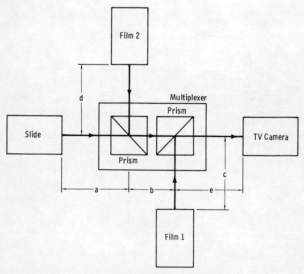

(B) Prism type.

Fig. 5-6. Typical optical multiplexers for film islands.

focused on the field lens. The motion-picture projectors also are focused on the field lens, with the light from each reflected by the semimirrors as shown. Note that the same optical path length is maintained for both film projectors by the spacings indicated.

The optical multiplexer normally may be used in a fully lighted projection room. A cover is used to provide protection from dust. A window in the cover adjacent to mirror B is also a semimirror to achieve equal light levels on the field lens from each projector.

Fig. 5-6B illustrates the prism type of optical multiplexer, which is usually more stable than the individual mountings of Fig. 5-6A over a longer period of time. The path lengths for film projector 1 ($c+e$) and film projector 2 ($d+b+e$) must equal the length of slide-projector path $a+b+e$. Measurements are made from the projector film gate to the field relay lens. The standard height of the optical path above the floor is 48 inches.

Multiplexers of the type shown in Fig. 5-6 allow "supering" of film sources on the same camera by controlling the projector light sources with neutral-density wheels controlled from the remote operating position. Unfortunately, this type of multiplexer has poor light transfer due to optical losses, and the moving-mirror type (with front-surface reflection and no "transparency") usually is employed for color cameras.

The multiplexer in Fig. 5-7 has provision for four inputs and two outputs. Both outputs have color capability, to achieve either preview or

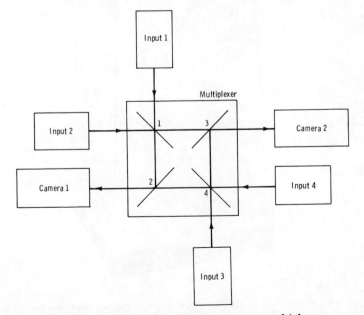

Fig. 5-7. Principle of four-input, two-output multiplexer.

Optical Assembly

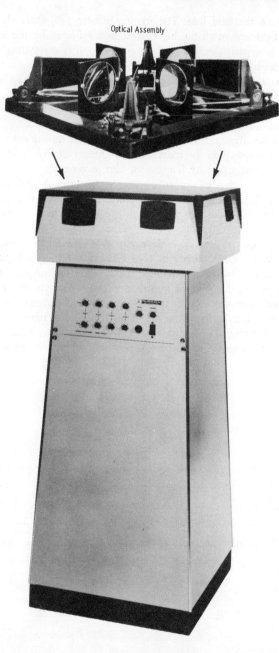

Fig. 5-8. TeleMation Model TMM-212 optical multiplexer.

color back-up capabilities. Ideally, the control system should allow assignment of input-output combinations to the same or separate remote control positions, thus permitting the island to be operated as a single unit or as two separate islands ("dual-island" operation). Note that for input 1 to feed camera 1, mirror 2 is up (in operating position) and mirror 1 is down. For input 1 to feed camera 2, the control system places mirror 1 up and mirror 3 down, etc.

Fig. 5-8 illustrates a multiplexer in which four movable mirrors are used to permit optical switching of any of four film and/or slide projectors into either of two cameras. A precision-machined aluminum optical base plate is used as a rigid mounting for the drive motors and mirrors. The entire optical assembly "floats" on a three-point mounting so that external stress will not impair optical alignment. Each mirror is mounted on two ball-bearing pillow blocks. These blocks, which are an integral part of the base-plate casting, determine mirror alignment in vertical and horizontal planes.

A fiberglass hood protects the optical assembly. Since this hood is supported by the pedestal base (not the assembly itself), it can be used to support objects such as monitors or oscilloscopes without distorting optical alignment.

A built-in 24-volt power supply provides voltage for operation of the multiplexer control circuitry. A switch panel mounted on the multiplexer permits each of four optical sources to be switched into either of two cameras. The control circuits permit camera 1 to pre-empt camera 2, whether controlled locally or from a remote position. This prevents accidental switching of the "on-air" source and allows camera 2 to preview any of the three unused sources.

The control and logic circuitry uses bistable relays to prevent unintentional switching during momentary power failures or when control is transferred. "Local" and "remote" buttons on the local panel permit exchange of control, and the optional remote panel is equipped with a button and indicator light to permit the control-room operator to take over at any time.

5-3. THE FIELD LENS: CENTER OF COLOR OPTICS

The field lens collects the light from the various film sources in the form of a real image. All film projectors are focused on the field lens, and all channels in the color camera are focused on the field lens. Since the image here is a real image, it can be observed by placing a screen (such as a white card) at the field-lens position. The focal plane of the field lens is at the *center plane* of the lens.

In order to have the same effective f number for a relay system, all projector lenses must match in angular throw for the particular film source used. This is to say that 16-mm and 35-mm (or $2'' \times 2''$ slide) projectors must have matched lens complements for a given field-lens system.

Example 1: We have on hand 16-mm motion-picture projectors fitted with 2-inch projection lenses (focal length = 2 inches). It is desired to find the focal length of the lens in a $2'' \times 2''$ slide projector necessary to match the angular throw of the existing 16-mm motion-picture projectors.

First, determine the angle of throw of the 16-mm projector. (Review Fig. 4-1, Chapter 4.) From the multiplexer data in Table 5-1, we can determine that the diagonal of the useful aperture size of 16-mm film is approximately 0.475 inch. Then, for F = 2 inches:

$$\theta = \tan^{-1} \frac{0.475}{2 \times 2} = \tan^{-1} 0.118 = 6.72° \text{ (approx)}$$

The total angular throw (not important except as a check, in these computations) is $2\theta = 13.4°$. Since we have determined that $\tan \theta = 0.118$ for the 16-mm projector lens, it is only necessary to find what focal length is necessary for the lens of the $2'' \times 2''$ slide projector to match this value.

From the multiplexer data table (Table 5-1) we determine that the diagonal (w) of the useful aperture of a $2'' \times 2''$ slide is about 1.4 inches. We know that

$$\theta = \tan^{-1} \frac{w}{2F} = \tan^{-1} 0.118$$

Therefore:

$$0.118 = \frac{w}{2F} = \frac{1.4}{2F}$$

and

$$2F = \frac{1.4}{0.118}$$

Then:

$$F = 1/2 \left(\frac{1.4}{0.118} \right) = 1/2 \ (11.9) = 5.92 \text{ inches}$$

Table 5-1. Multiplexer Data

Material for Proj	Aperture Size (Inches)	Diagonal (Inches)	Required Magnification for Vidicon Only	Approx Required Magnification for $3'' \times 4''$ Pattern On Field Lens
16-mm Film	0.284×0.380	0.475	1.3	10.55
35-mm Film	0.612×0.816	1.0	0.61	4.9
$2'' \times 2''$ Slide	0.844×1.125	1.4	0.44	3.55
$4'' \times 5''$ Opaque	3×4	5	0.125	1.0

Therefore, for practical purposes, we determine that the focal length of the $2'' \times 2''$ slide projector should be 6 inches. This focal length will properly match the angular throw of the 16-mm film projector equipped with a 2-inch lens.

A simpler computation is possible by merely comparing the ratios of diagonals, which is $1.4/0.475 = 3$ (approx). Then $2 \times 3 = 6$ inches, the focal length required for the slide projector when the lens on the 16-mm motion-picture projector has a focal length of 2 inches.

Example 2: Given a $2'' \times 2''$ slide projector with a 7-inch lens, match a 16-mm projector lens. For the slide projector:

$$\theta = \tan^{-1}\left(\frac{1.4}{2 \times 7}\right) = \tan^{-1} 0.1 = 5.7° \text{ (approx)}$$

The total angular field is 2θ, or $11+°$.

For the lens of the 16-mm projector to match, you know that $\theta = \tan^{-1} w/2F$, and $w/2F$ must equal 0.1. Since w for 16-mm film is 0.475:

$$0.1 = \frac{0.475}{2F}$$

$$2F = \frac{0.475}{0.1}$$

$$F = 1/2\left(\frac{0.475}{0.1}\right) = 1/2 \ (4.75) = 2.38 \text{ inches}$$

The nearest standard focal length is 2.5 inches.

For a quick computation (as noted in Example 1), the ratio of diagonals is $0.475/1.4 = 0.34$, and $7 \times 0.34 = 2.38$, or a 2.5-inch focal length in practice.

Example 3: Match a 35-mm film projector to the $2'' \times 2''$ slide projector and 16-mm projector of Example 2. Since w for 35-mm film is approximately 1 inch:

$$F = 1/2\frac{1}{0.1} = 1/2 \ (10) = 5 \text{ inches}$$

Thus, the focal length should be 5 inches for the 35-mm projector, to match the conditions of Example 2.

After the angular throw has been matched for all projectors, the approximate distance from each film gate to the field lens can be determined by making use of the fact that the necessary focal length of a projector lens is approximately:

$$F = \frac{mD}{(m+1)^2}$$

where,

F is the focal length in inches,
D is the distance in inches from the film gate to the field lens,
m is the magnification required (Table 5-1).

The necessary distance (approximate) between film and image for a given focal length is found by solving for D in the above formula:

$$D = \frac{(m+1)^2}{m} F$$

In practice, the actual distance is slightly greater than that indicated by the formula. The direction of adjustment is therefore *closer* to the multiplexer unit by 1 inch or more.

The most usual combination of lenses in practice is a 7-inch slide-projector lens with 2.5-inch lenses on 16-mm projectors or 5-inch lenses on 35-mm projectors.

5-4. FILM-CAMERA OPTICAL ALIGNMENT

Study Fig. 5-9. This is a typical layout for modern four-vidicon color film cameras. With fixed-prism optics of this sort, there is always one chan-

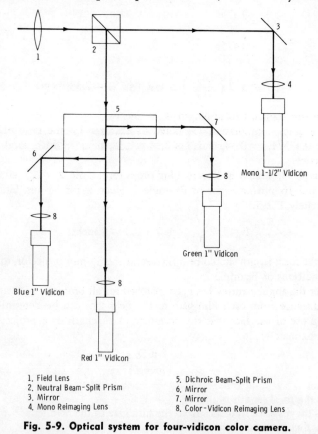

1. Field Lens
2. Neutral Beam-Split Prism
3. Mirror
4. Mono Reimaging Lens

5. Dichroic Beam-Split Prism
6. Mirror
7. Mirror
8. Color-Vidicon Reimaging Lens

Fig. 5-9. Optical system for four-vidicon color camera.

nel that is the "standard"; it is factory adjusted in optical alignment, and only the vidicon positioning behind the lens is variable in order to obtain proper focus at the proper image size.

Note that the "standard channel" in this case is the red vidicon path. The neutral prism and the dichroic prism are fixed. What does this mean in initial setup? Simply that the projectors and the multiplexer positioning must be made and adjusted to obtain a properly oriented image on the face of the red vidicon. The variable elements are then the front-surface mirrors (3, 6, and 7).

The field lens normally is equipped with a mask that outlines the necessary image size. Throw distance determines this size. Usually, the film-gate outline must overlap the mask by about 1/16 inch all around. The proper size must occur with the image focused in the center of the field lens. We determine this by first holding a white card against the front of the field lens (between the mask and face of field lens) and focusing the projector lens for sharp focus at this point. Then we hold the card tightly against the rear of the field lens and note in which direction we must turn the projector-lens focus to again obtain sharp focus. When we split the difference, we have proper optical focus. (NOTE: Always have the projector lens stop "wide open" to obtain a shallow depth of field.)

The projector adjustments for the above procedure are distance (throw), vertical and horizontal positioning, and tilt adjustments for optical axis alignment. The optical axis alignment is not correct until we can stop down the projector lens and the camera lenses to the smallest "dot" opening, and obtain the dot of light dead center on the small opening. We do this first for the standard channel, then adjust the front-surface mirror angles to obtain the same condition for the remaining channels. *Caution:* Some manufacturers have tooled these mirrors such that special equipment is required to change mirror positionings. Where this condition exists, *be certain* that the field engineer responsible for the initial installation checks this alignment. Unless the dot of light can be made to fall exactly at the optical dead center, less than ideal results will be obtained.

When all projectors have been positioned to obtain focus at the field lens with the proper image size, and the dot of light is properly centered, the final optical adjustment is movement of each vidicon toward or away from its individual reimaging lens to obtain sharp focus with the lens focused on infinity. Normally, all camera lenses are operated "wide open" (maximum stop opening). For these initial adjustments, projector lenses are also wide open to obtain a very shallow depth of field for critical optical focus. For operation, the projector lenses should be stopped down one or two stop openings to allow a safety margin for varying film characteristics (buckling, warping, base-thickness variation, etc.).

NOTE: Electrical adjustments and color balancing of the film chain are discussed in Chapter 10.

5-5. THE COLOR-BAR GENERATOR

Any color-camera chain (either live or film) normally includes a color-bar generator that is used to set up the encoding system. A switching system is used to select either the color bars or the camera signals. In Fig. 5-10, the technique for a four-channel camera is shown. When the color-bar test position is selected, luminance is "derived" monochrome, as covered in Chapter 2. In normal operation, the luminance is "true" pickup from the luminance tube. In three-channel cameras, luminance is always derived, whether the switch is in the color-bar test position or the normal operating position.

It is important to understand at the start that any signal inserted at the "red" input of an encoder will activate the red gun in the color picture tube. Similarly, the green and blue inputs of the encoder activate their respective picture-tube guns. The color-bar signal is universally used in encoder setup and adjustment procedures.

A typical color-bar generator (Fig. 5-11) employs three basic multivibrators (one for each primary color), which trigger one another from

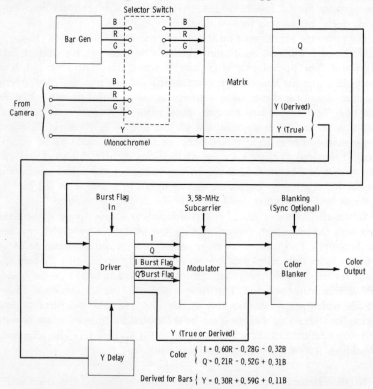

Fig. 5-10. Integration of color-bar generator into camera chain.

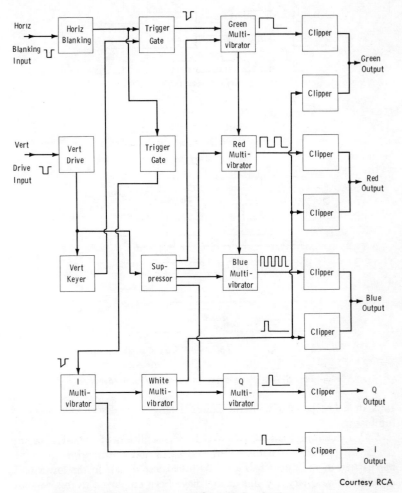

Fig. 5-11. Block diagram of RCA WA-1 color-bar generator.

the trailing edges of their respective outputs. The waveform timing for the color-bar sequence is shown in Fig. 5-12A. The active part of each line period is divided into seven bar intervals, and the circuit is so arranged that each multivibrator is on for four of the seven intervals and "off" for the remaining three. All of the multivibrators are on for the first interval, so white is produced as the first bar. In the usual arrangement of the device, the green multivibrator produces only one pulse (four intervals wide) per line, the red multivibrator produces two pulses (each two intervals wide) per line, and the blue multivibrator produces four pulses (each one interval wide) per line. The combination of these signals re-

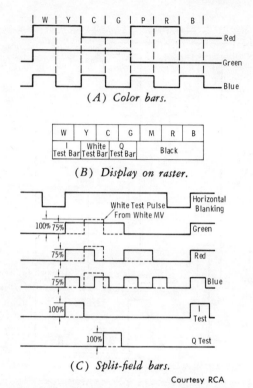

(A) Color bars.

(B) Display on raster.

(C) Split-field bars.

Courtesy RCA

Fig. 5-12. Waveforms in color-bar generator.

sults in a bar pattern with the colors arranged in descending order of luminance.

Added flexibility can be achieved by the addition of a 60-Hz keying circuit to separate the pattern into two horizontal bands, with the color-bar pattern in the upper half and additional test signals in the lower half. Among the auxiliary test signals that have been employed in this way are ordinary monochrome or color pictures (usually of test patterns), step signals for linearity adjustments, and white pulses the amplitude of which may be adjusted independently of any white bar in the basic color-bar pattern. In some color-bar generators (as in this one), special test pulses, which may be injected by separate circuits directly into the I and Q channels of an encoder to facilitate phase adjustments, are generated.

In the generator we are observing, the 60-Hz keying is obtained from the vertical-drive input. (The standard line-rate pulses are started from the trailing edge of horizontal blanking.) Thus, this generator is capable (by selector-switch function) of providing a "split-field" color-bar pattern on the color monitor, displaying standard color bars on the top of the raster and a white bar together with the I and Q test bars on the bottom (Fig. 5-

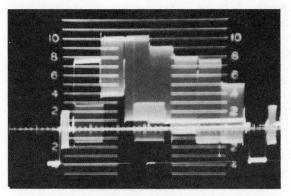

(A) Split field, wideband response.

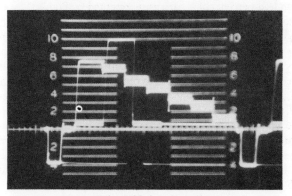

(B) Split field, low-pass filter.

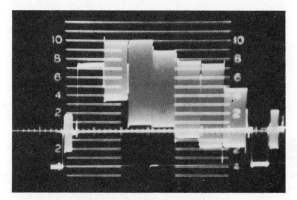

(C) Bars without split field.

Fig. 5-13. CRO displays of 75-percent color bars.

12B). The pulses from the I and Q outputs are termed as they are simply because the signals feed the "I" input and "Q" input of the encoding system for tests and adjustments.

The timing chart for the split-field presentation is shown in Fig. 5-12C. In this instance, all bars (except the white test pulse) are shown reduced to 75 percent of their nominal input value. Thus, the first "white" pulse (upper half of raster) is actually a "gray"; it is at 75 percent of reference white level. However, the special white test pulse (lower half of raster) remains at the 100-percent reference level.

Fig. 5-13A shows the resulting CRO display of the split-field pattern (75 percent), with blanking and sync added. Fig. 5-13B shows the same signal passed through a high-low cross filter in the low position, showing more clearly the luminance levels. The highest bar is the reference white level, or 100-percent amplitude. Note that for 75-percent bars there is no overshoot of chroma beyond this 100-percent (reference white) level.

One-hundred-percent color bars, as represented by Fig. 2-18 in Chapter 2, require considerable overmodulation of the transmitter. Whenever the overall system is checked in practice for performance measurements, the color bars are reduced to 75 percent of full amplitude. This more nearly represents the maximum saturation of colors the transmission system will be required to handle from live or film cameras. But keep this in mind: The colors displayed at 75-percent amplitude cannot be called desaturated. Study the pulse-timing diagrams, and note that for all bars (except gray or white) no more than two encoder inputs exist at any one time. For a color to be desaturated, a mixture with white must exist; this means some amount of all three colors must be present. Therefore, although the bars are reduced in amplitude, they are still correctly termed "fully saturated color bars at 75 percent of full amplitude."

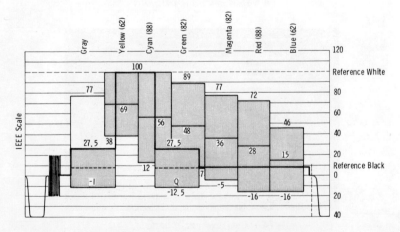

Fig. 5-14. Standard split-field color-bar signal.

Fig. 5-13C is the CRO presentation of 75-percent color bars when the split field is not used. Some bar generators in camera chains do not provide the split field, and it is not necessary for camera setup. The value of the split-field presentation is in system checking so that a reference-white luminance level is always present to distinguish between loss of overall level and loss of chroma level only.

Fig. 5-14 illustrates the standard split-field, 75-percent color-bar signal in terms of IEEE units, when 1 volt = 140 IEEE units. The peak-to-peak chroma values in IEEE units for each color bar are indicated at the top of the chart. Practicing engineers become thoroughly familiar with this signal in overall system checkouts and in checking the incoming network lines.

5-6. NONADDITIVE MIXING (NAM)

See Fig. 5-15, a simplified block diagram of the RCA TK-27 color film chain. It is relevant here to take a brief look at nonadditive mixing (NAM) techniques, since many modern color chains employ this quite important feature. Note that in this discussion, Y is direct or "true" luminance, and Y' is derived luminance.

Remember that the three-channel camera derives luminance from the three primary-color channels. In the four-channel camera, the three color tubes operate on color brightness only; the luminance channel provides the "true" luminance information, which *can* be different from chroma luminance. Therefore, to obtain proper color balance in camera setup, proper luminance-to-chroma ratio, and proper "gain riding" of signal level, the NAM monitoring signals are provided.

First of all, review the "Receiver Matrix" subsection of Section 2-15 (Chapter 2). The NAM circuits are fed from a "receiver matrix" based on the principles discussed in that subsection, with this fundamental difference: B, R, and G are derived as follows:

$$(B - Y') + Y = B$$
$$(R - Y') + Y = R$$
$$(G - Y') + Y = G$$

This is to say that a matrix from the I and Q circuitry prior to modulation obtains the color-difference signals on principles outlined in Chapter 2, and then adds the "true" luminance from the Y (monochrome) channel (Equations 2-6, 2-7, and 2-8, Section 2-15). Of course, we realize that when the color chain is operated in the test position on color bars, the luminance provided is the derived value, since the luminance tube is not contributing to the signal (nor are any of the color tubes). So now the monitoring facilities in NAM position will be looking at a signal the same way the color receiver sees it.

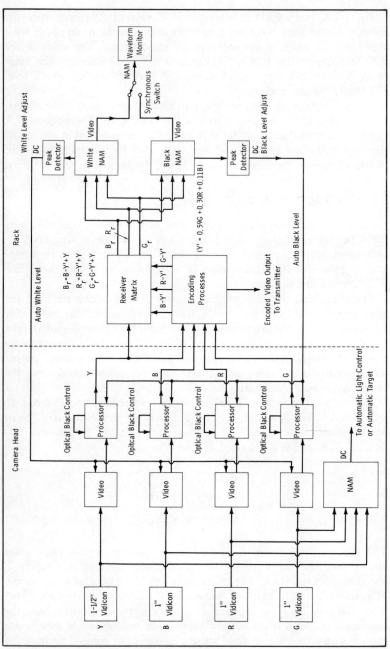

Fig. 5-15. Block diagram of RCA TK-27 color film chain.

Study the NAM gated detectors of Fig. 5-16. The red, green, and blue signals just described are applied to the detectors as shown in the drawing. Note that the negative detectors are pnp, whereas the positive detectors are npn. Assume for the moment that the gates are saturated, and the emitters of all transistors are held at essentially zero (ground) potential. Further note that positive-going pulses will have no effect on the pnp transistors, and that negative-going pulses will not affect the npn transistors, because of the respective base-emitter junction-bias conditions.

Now assume that the red signal momentarily has the highest amplitude. The negative excursion will turn Q2 on, and the resultant negative pulse at the emitter (hence across R1) will hold Q1 and Q3 off, since their bases are not as negative as the base of Q2. We can see that only the red negative-going signal will appear across R1. This is to point out that whichever signal has the largest negative-going amplitude at any instant will appear across R1, holding the other two transistors off.

Similarly, the signal with the largest positive-going swing will appear across R2, holding the other two positive-detector transistors off. This is the basic principle of NAM; only that signal with the largest amplitude is passed, and hence is not added to the other signals.

Now look at the gates, Q7 and Q8. They are held in saturation by the currents through R3 and R4, until negative pulses arrive at the gate bases

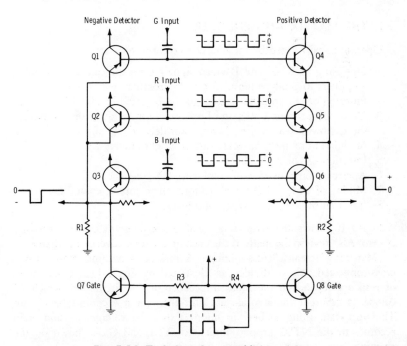

Fig. 5-16. Technique for nonadditive mixing.

to turn them off. Since these pulses are from opposite sides of the NAM multivibrator, one gate is on when the other is off. The pulse durations are such that three to four lines of NAM white are on, then three to four lines of NAM black are on, repeating throughout the entire field. We can see that it depends upon video polarity at the point of detection which gate is monitoring "white" and which gate is monitoring "black." The important point is that NAM is monitoring the signal of highest level in any of the channels at the time of sampling. This is one form of electronic switch.

The output of the NAM "switch" is fed to the CRO waveform monitor, which sweeps at a standard 1/2-horizontal rate. The effect as seen on the waveform monitor is a superposition of two signals, maximum black and maximum white. To the video operator, these signals appear just the same as the signals from a monochrome camera; thus, he is able to adjust the color-film controls in the same way that he would for a monochrome camera.

Note also from Fig. 5-15 that white and black detectors associated with the NAM circuitry enable automatic white and black control of the video signal. The Optical Black Level mode is achieved by gating the black-level detectors on the individual processor modules so that they respond to an unilluminated portion of the vidicon (provided by an optical mask), and peak black is automatically referenced to this.

5-7. THE MATRIX AND ENCODER

Operations performed by the encoder (colorplexer) are:

1. Matrixing of R, G, and B video signals from the camera processing amplifiers to produce luminance and chrominance signals.
2. Filtering of chrominance signals to the required bandwidth.
3. Delay compensation of Y and I to correct for the delay resulting from the narrowest-bandwidth (hence most-delayed) Q signal.
4. Modulation of the 3.58-MHz carrier by chrominance signals.
5. Insertion of color sync burst.
6. Aperture compensation of luminance (Y) signal.
7. Mixing of Y, I, and Q signals to form the complete color signal.
8. Insertion of composite sync (optional).

Fig. 5-10 shows the basic functional block diagram of the encoding system. Also review the math of the encoding system covered in Chapter 2.

"Matrixing" means "cross-mixing." A matrix is nothing more than a cross-connected voltage divider, as illustrated by Fig. 5-17. Actual values of resistors depend on the input impedance of the circuit fed, which, of course, is different for a solid-state unit than for a vacuum-tube circuit. The important thing to bear in mind is that the voltage division is in accord wth the NTSC proportionment of Y, I, and Q, as shown on the drawing.

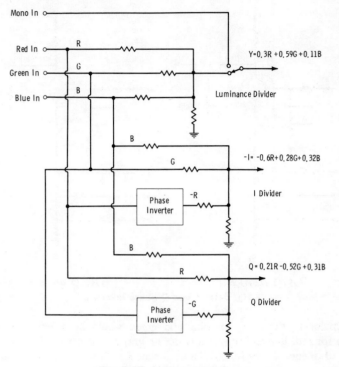

Fig. 5-17. Matrix network.

There are just two other characteristics to explain here. First, note that the luminance (Y) divider is switchable between the matrixed RGB signal and a separate monochrome input. In a four-tube camera, R, G, and B are matrixed to form I and Q only; the luminance information is supplied by the luminance pickup tube. However, many such cameras are switchable so that the operation can be changed to a three-tube function in case of failure of the monochrome channel. In this case, the monochrome (luminance) information is obtained from the matrixed RGB signal (derived luminance).

The other characteristic to explain is that a "minus I" signal is shown at the I-divider output. Note from the basic proportionment of a "plus I" signal (Fig. 5-10) that *two* phase inverters would be required to produce it. Since the −I signal is simpler to produce (one phase inverter), this is the form usually employed.

See Fig. 5-18 for more detailed drawings of the specific I and Q networks. The I phase inverter (Fig. 5-18A) inverts the phase of the red signal injected at the base; since blue and green are injected at the emitter, no phase inversion takes place for these components. Hence, the video signal is −I at the collector.

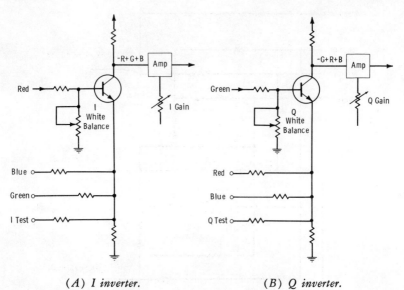

(A) I inverter.   (B) Q inverter.

Fig. 5-18. I and Q phase inverters.

Obviously, the gains for blue and green would be different than the gain for red. Remember that a white or gray signal means that all inputs are of identical amplitude. Thus, when $R = G = B$, the white-balance control (actually a red-signal gain control) is adjusted so that all subcarrier is cancelled in the I output.

Note also that since the I-test pulse is injected in this stage, it becomes a −I-test signal. Since this pulse occurs either on a split field (without RGB present at the same time) or by itself (also without RGB present), and since it is *not simultaneously matrixed* with the luminance channel, it has *zero luminance* and is inserted on the blanking pedestal as shown by Fig. 5-14. It is matrixed such that with normal input level, its amplitude at the encoded output is the same peak-to-peak value as a properly adjusted sync-burst amplitude. It is filled with subcarrier just as are all of the actual color bars; only white (equal input amplitudes) contains no subcarrier. All of this is evident from Fig. 5-14.

The reader should now be able to correlate this signal processing with that of the Q phase inverter (Fig. 5-18B), and note that the output here is a +Q signal. In this circuit, green is the inverted signal and is adjusted for Q white balance.

Following the matrix, we amplify, control the individual gains, and shape the bandwidths of the video signals.

Fig. 5-19 illustrates the Y, I, and Q video signals prior to subcarrier modulation. The "standard color-bar signal" is assumed here at 100-percent levels. We recall that this signal has a sequence that is of descending lumi-

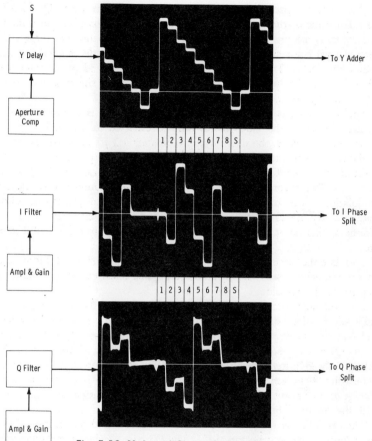

Fig. 5-19. Y, I, and Q signals for color bars.

nance value. The reader will find it a great help to memorize the sequence of color bars as they are numbered in Fig. 5-19. This is:

1. White
2. Yellow
3. Cyan
4. Green
5. Magenta (Purple)
6. Red
7. Blue
8. Black

The next space is sync, which is optional. When sync is inserted in the encoder, it is normally inserted following aperture compensation of the Y channel, and just prior to the Y delay line.

Note that on space 1 (white) the value of both the I and Q video is zero. The same is true at black level. Each of the other color-bar video intervals contains the proper ratio of I and Q to the luminance value for 100-percent-level color signals. Note particularly the −I video signal. For the yellow bar (space 2), the I video goes negative by −0.32 of the reference white level at that point in the circuitry. A + I video signal would go to +0.32 for the yellow bar.

The Y channel is not limited in bandwidth, as this is automatic at the transmitter where the video bandwidth is restricted to 4.2 MHz. The Y signal must, however, be delayed approximately 1.2 μs to equal the delay of the Q video channel. (All of this was covered in Chapter 2.)

Carefully review Chapter 2; be sure to understand Figs. 2-27, 2-33, 2-34, and 2-35. This theory covers the fundamentals of vacuum-tube subcarrier modulation. Since transistors do not employ multigrid principles, solid-state circuitry differs radically from corresponding tube circuitry. We must assume at this point that the reader has a basic background in solid-state theory.

Correlate the circuit of Fig. 5-20A with Fig. 2-33, Chapter 2. From here on, we will need to study slowly and carefully; be certain to understand each step in this progression.

We will now discuss Fig. 5-20B. Note that the "I flag out" ("flag" is burst-keyer pulse) of Fig. 5-20A goes to the base that receives +I video, and that the "Q flag out" goes to the base that receives −Q video. Transistors Q1 and Q2 form a bootstrapped feedback pair, with the base of Q1 held (in the quiescent operating state) by the voltage-breakdown point of −5.6 volts of the zener diode at the emitter. Base resistor R1 couples a dc voltage from an automatic carrier balance to be discussed in connection with the circuit of Fig. 5-21.

The Q1 collector is dc coupled to the base of Q2, and the emitter of Q2 is dc coupled to the collector of Q3. The Q3 base receives the I subcarrier input. Transistor Q3 is the modulator, in this case acting as a switch. The video (and burst flag) at the collector of Q3 is switched to ground at the subcarrier rate. In the RCA modulator, the subcarrier sine wave is converted to a semisquare wave so that Q3 becomes a switch with a 50-percent duty cycle, with on and off times clearly defined on each alternation of the subcarrier cycle.

Thus, the video at the modulator collector has a dc component, which is the collector potential when Q3 is a closed switch to ground. This voltage is between 0.5 and 1 volt, depending on the type of transistor. In the RCA system, it is +0.7 volt. In any system, it is a "designed-in" and known level.

The Q circuitry (Q4, Q5, Q6) is identical to that just described. Now note carefully the polarities. At the base of Q7, −I video and +Q video appear. We recognize Q7 and Q8 as another bootstrapped feedback pair with phase inversion. Thus, +I and −Q video appears at the Q8 emitter to be coupled into the mixer stage (shown in Fig. 5-22). The gain of Q7-

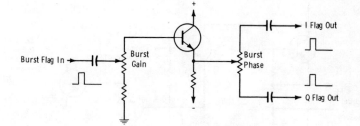

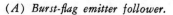

(A) Burst-flag emitter follower.

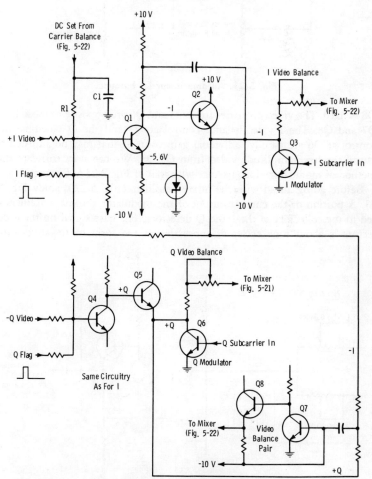

(B) I and Q modulators.

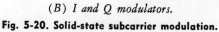

Fig. 5-20. Solid-state subcarrier modulation.

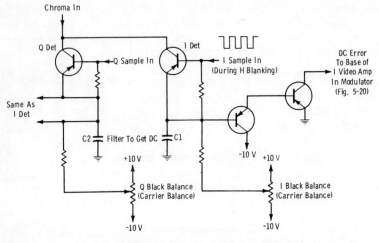

Fig. 5-21. Automatic carrier balance.

Q8 is fixed. The video polarities from Q3 and Q6 are opposite to those from Q7 and Q8. The I video-balance control at Q3 and the Q video-balance control at Q6 are simply individual gain controls that may be adjusted to cancel the fixed-amplitude video from Q7-Q8. We can now correlate this method of modulator video balance with that of Fig. 2-27A, Chapter 2.

Before going ahead with the mixer amplifier (Fig. 5-22), study Fig. 5-21. A portion of the chroma output of the modulator chroma amplifiers is fed to the collectors of the I and Q detectors. The base signal on these detectors is about nine cycles of subcarrier in the form of negative-going

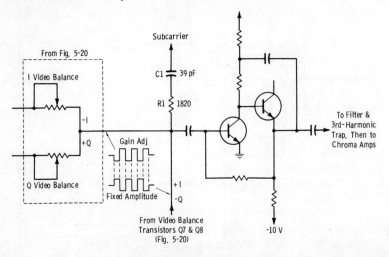

Fig. 5-22. Mixer circuit.

pips gated on during horizontal blanking. This period is therefore during picture black, at which time no carrier should exist. Since these are pnp transistors, the negative pips serve to tie the collectors to the emitters, and at these instants, any amount of chroma at the collector charges filter capacitor C1 or C2 to the amplitude and polarity of the signal. Thus, if the black-balance controls (carrier-balance controls) are adjusted with no signal input (black condition), the application of video will transfer the circuitry to an automatic carrier-balance action. Let us see how this occurs.

Note that a dc amplifier in each detector is tied into its respective base resistor of Fig. 5-20. The current through this resistor varies in accordance with the dc error voltage from the detectors. Therefore, the dc level at the video-amplifier output to the modulator collector varies in such a direction as to cancel any subcarrier existing during the blanking interval. To complete the visualization of this process, go now to Fig. 5-22, the mixer amplifier.

The mixer amplifier (another bootstrapped feedback pair) receives four signals, as follows:

1. The modulated I signal.
2. The modulated Q signal.
3. The output of the video-balance amplifier.
4. A subcarrier signal that is shifted in phase and fixed in amplitude by C1 and R1. The reactance of C1 causes the voltage to lead the input voltage by the amount of circuit delay in the modulator function. This provides a fixed subcarrier reference that directly offsets the carrier unbalance caused by the positive dc component of the video-amplifier output. So we have a fixed carrier balance consisting of the effect of the dc video component cancelled by the phase-compensated carrier component, and the adjustable carrier balance, which becomes automatic with the application of video. All four of the signals, in proper amplitude and phase, must be supplied to the mixer for correct encoder functioning.

It was mentioned previously that the output polarity of the I matrix (a − I signal is more simply produced) is of no concern since this is conveniently taken care of in the modulator circuit. We were introduced to the phasing of the subcarrier drives to tube-type balanced modulators in Chapter 2 (Fig. 2-35). Now refer to Fig. 5-23. The I (in-phase) subcarrier is fed to its associated amplifier; the subcarrier is delayed by 90° for the Q feed. Note the difference in this method from that of Fig. 2-35. In Fig. 2-35, the cosine axis (for the I modulators) was derived from a *leading* voltage caused by the reactance of capacitor C. In the arrangement of Fig. 5-23, the I subcarrier (0°, or in-phase subcarrier) is *delayed* by 90° to give the component required for Q modulation.

Note in Fig. 5-23, from conventional transformer theory, how the following requirements are satisfied:

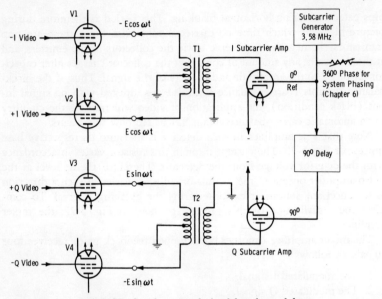

Fig. 5-23. Q subcarrier derived by phase delay.

E cos ωt goes to the tube with +I video.
−E cos ωt goes to the tube with −I video.
E sin ωt goes to the tube with +Q video.
−E sin ωt goes to the tube with −Q video.

Note in the case of Fig. 5-23 that −I video is on the V1 grid, whereas +Q video is on the V3 grid. Do you see an alternate method of accommodating the correct phasing? Transformer T1 could be used in the same arrangement as T2 if the grid feeds from the I phase splitter were reversed: −I video to the grid of V2 and +I video to the grid of V1.

If the reader has followed this discussion with comprehension, he will be able to analyze any subcarrier-modulation circuitry he is likely to encounter in practice. To recapitulate, see Fig. 5-24. In Fig. 5-24A, we have the conventional representation of an in-phase vector and a 90°-delayed quadrature vector. In Fig. 5-24B, the −I and −Q vectors have been added. In Fig. 5-24C has been added the result of the burst flag inserted into the +I and −Q video during horizontal blanking (recall Fig. 2-34, Chapter 2). Fig. 5-24D is the same as Fig. 5-24C; the difference in appearance is that the entire vector diagram has been rotated counterclockwise so that B − Y corresponds to 0°.

See Fig. 5-25. The I and Q modulator outputs are passed through low-pass filters to remove subcarrier harmonics. If the subcarrier is converted to a square wave, an additional third-harmonic trap is used because of the strong odd-harmonic content.

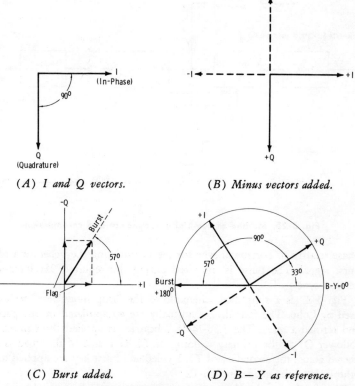

(A) I and Q vectors.   (B) Minus vectors added.

(C) Burst added.   (D) B — Y as reference.

Fig. 5-24. Subcarrier phase diagrams.

The monochrome and chrominance adders have paralleled output circuits for addition of luminance and chrominance. The waveform at the chroma adder is that of both I and Q.

Following Y, I, and Q addition, a driven clamp circuit normally is used to establish line-to-line black-level control. The output amplifier usually drives two or more outputs; negative feedback is employed in this stage to reduce distortion and provide a low sending-end impedance for the transmission lines. Unlike the case of previous circuitry for the encoding process, conventional clamping and amplifier designs normally are found in either tube or solid-state equipment.

A solid-state encoder that does not require an external burst-flag input and that contains a color-bar generator providing either full bars or split-field bars is illustrated in Fig. 5-26. Input signals required are video, sync, blanking, and subcarrier.

In this encoder, burst flag is generated internally from composite sync. The circuit produces a rectangular pulse of suitable width and proper

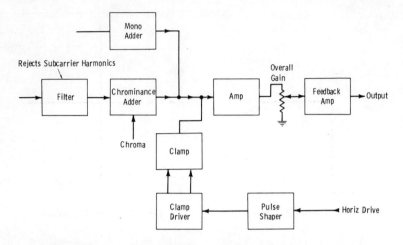

Fig. 5-25. Method of combining luminance and chrominance.

phase relative to horizontal sync for use in initiating the generation of the burst signal. The pulse is inhibited during the required 9H interval of vertical sync.

Fig. 5-27 is a simplified schematic of the "ring modulator" technique used by Cohu. The four diodes normally are encapsulated in one package and termed a *quad*. The 3.58-MHz subcarrier is applied through emitter follower Q1 to the primary winding of T1. One side of the quad is connected across the secondary of T1. I video and burst key are applied to the other side of the quad through T2.

The fundamental subcarrier frequency is suppressed (balanced) by equalizing the offset currents by means of the current in Q2. Control R1 is adjusted so that no subcarrier exists for black (with no video) and is termed "I black balance." This is a *carrier balance* control.

Application of I video unbalances the quad so that subcarrier can flow in T2. Since the carrier fundamental is suppressed, only sidebands exist. The instantaneous amplitude of the sidebands is proportional to the instantaneous amplitude of the I signal.

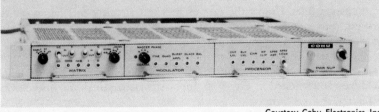

Courtesy Cohu Electronics Inc.

Fig. 5-26. Cohu solid-state color encoder.

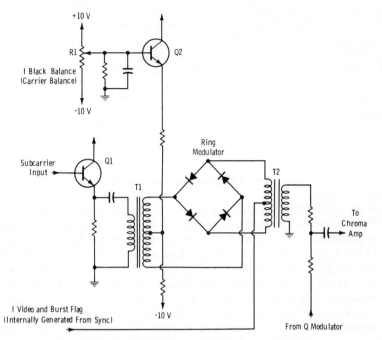

Fig. 5-27. Simplified schematic of I ring modulator.

5-8. SPECIAL CIRCUITS

Special circuits used by some color cameras are (1) "contours out of green," and (2) "vertical aperture correction." First, we will consider the contours-out-of-green principle, with which the reader should become familiar.

Sharpened edges (both vertical and horizontal) produced from the green channel of a three-tube camera are fed into the red, green, and blue channels. This is done to provide the same tolerance to misregistration that a fourth tube provides. Remember, however, that this applies to a monochrome receiver displaying a color program; misregistration will not result in a loss of sharpness on the monochrome receiver, but will cause color fringing (hence, loss of sharpness) on color receivers.

See Fig. 5-28. Enhanced contours in the image from the green tube correct all three channels and become the derived luminance signal. Since the enhanced contours come from a single tube, these contours cannot be degraded by registration errors. The delay of the green signal and the contour signal relative to uncompensated red and blue signals is corrected automatically in the normal registration procedures.

The contour signal is enhanced by employing both vertical and horizontal aperture correction. We know that horizontal aperture correction

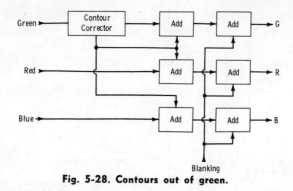

Fig. 5-28. Contours out of green.

is limited by the bandwidth of the system, including the home receiver. Vertical aperture correction (being a low-frequency correction) produces a noticeable increase in sharpness not limited by bandwidth. Briefly, it enhances a vertical transition from black to white by slightly darkening the preceding line and whitening the succeeding line in the transition. Therefore, we would expect to find some means of delaying a line of video information, as is shown by Fig. 5-29. The video is delayed 1 line (1H) in block 1, and another line in block 2. The signal delayed one line is the main uncorrected signal. This signal goes to a subtractor, an adder, and a 100-

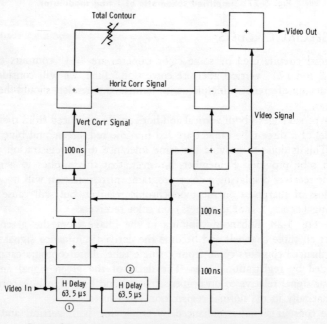

Fig. 5-29. Vertical and horizontal aperture correction.

nanosecond (one picture element) delay line. Thus, this signal serves as the input to the horizontal aperture corrector. The same unit results in both horizontal and vertical aperture correction. The second line delay (block 2) is used to achieve the functions described in basic research supplied by the Philips laboratories:

For large areas in which each line is like every other, subtracting signal from adjacent lines is like subtracting a signal from itself. To get more equalization, more adjacent-line signal is subtracted, and less overall large-area signal remains. This means that changing equalization would also change gain. To avoid this problem, a detail signal, which has no large-area information, is made. This is done by subtracting enough adjacent-signal information to completely cancel the signal when adjacent lines are alike. The contour signal is a signal containing *only* vertical and horizontal transition detail. This contour signal is then added in any desired amount to make the main signal without changing gain.

EXERCISES

Q5-1. In television practice, why are film-projector lamps always of higher wattage rating than slide-projector lamps?

Q5-2. In a field-lens film-camera system, why are all film sources that project onto a common field lens matched closely in angular throw?

Q5-3. What is the difference between a virtual image and a real image in an optical system?

Q5-4. Can projection-lamp voltages be used to "color balance" film cameras?

Q5-5. Should the projected image be focused at the front or the rear of the field lens?

Q5-6. Why should a projector lens be opened to the widest f stop for initial setup and check of optical focus?

Q5-7. What is the most critical check for optical alignment of a multiplexed color film camera?

Q5-8. What is the initial step in "standardization" of a color film camera to assure color balance?

Q5-9. What is the difference between derived luminance and true luminance?

Q5-10. When color-camera encoders are properly set up for correct burst phase, does this assure proper "flesh tones"?

Q5-11. Do "75-percent bars" provide fully saturated color bars?

Q5-12. What is the main advantage of split-field color bars?

CHAPTER 6

The Synchronizing Generator,
Pulse Distribution, and
System Timing

The synchronizing generator, commonly called "sync generator," consists of two basic sections—timing and pulse shaping. The simplest timing relationship that can be used is a 15,750-Hz line frequency obtained from an oscillator and a 60-Hz pulse derived directly from the power line for the field frequency. This form of timing generator is employed in a number of closed-circuit industrial systems, since the signal is not intended for regular broadcast. However, in order to produce the field interlace required for standard broadcast transmission, it is necessary that the field frequency be derived from the master-oscillator frequency. When two independent generators are used (as in the industrial system), a form of *random interlace* results, since the frequency relationships are not locked. Also, the actual number of lines constituting a single frame varies with the amount of drift between the line and field frequencies.

6-1. WHAT THE SYNC GENERATOR MUST DO

Fig. 6-1 provides a capsule review of what the sync generator must do, and it should be referred to during the following discussion. From waveforms N and P, note that field 1 has a full line (H) from the preceding horizontal sync pulse to the first equalizing pulse, whereas field 2 has a half-line interval. This is the requirement for odd-line, interlaced (two-to-one) scanning, and the master oscillator must, therefore, have twice the line frequency so that either H or $\frac{1}{2}$H pulses are available. (NOTE: In this description, field 1 refers to the field that has a full-line interval before the 9H vertical time, and field 2 refers to the field that has a $\frac{1}{2}$-line interval as

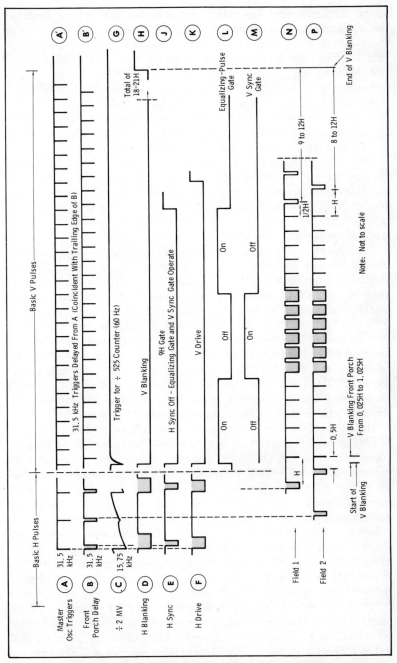

Fig. 6-1. Basic pulse timing of the sync generator.

shown. Also remember that field 1, therefore, has ½H spacing from the last equalizing pulse to the first H sync pulse of field 2, and field 2 has H spacing from the last equalizing pulse to the first H sync pulse of new field 1.) This characteristic is useful in setting the vertical-blanking duration described later.

Waveform A in Fig. 6-1 represents the 31.5-kHz triggers from the master oscillator; these triggers initiate the leading edges of waveforms B, C, D, and F. The width of pulse B determines the front-porch interval, since the trailing edge of this pulse times the H sync information (waveform E). Some generators use a tapped delay line for this purpose.

Although a considerable difference exists in methods used by various manufacturers of sync generators, the *timing* is as shown in Fig. 6-1. The master-oscillator frequency, which is divided by 525, triggers the leading edge of vertical blanking, vertical drive, and the various gates shown. Delayed triggers (waveform B') still provide the coincidence timing of H sync during the long 9H vertical interval.

The importance of timing accuracy and allowable tolerance may be realized from the following examples. Consider the observation of a thin vertical line corresponding to a frequency of 4 MHz. If the timing on alternate lines of the raster is shifted by as much as one-half cycle at 4 MHz, fine detail is lost. Since a complete cycle at this frequency occurs in 0.25 microsecond, a half cycle occurs in 0.125 microsecond. This means that the *overall* (sync generator and monitor or home receiver) allowable variation of timing between successive horizontal pulses is 0.125 microsecond for a 4-MHz detail. This is roughly an accuracy of 1 part in 600 for overall horizontal timing. For the sync generator alone EIA standards allow approximately 0.0008H in an averaging process of 20 to 100 lines, or about 0.05 microsecond (less than one-half of 0.125 microsecond).

The tolerance on vertical timing is even more severe. Since a difference of only one-half line exists for alternate fields, complete loss of interlace occurs with a vertical timing error of ½H (about 32 microseconds). In practice, the overall vertical difference in line spacing must be less than 10 percent to preserve the illusion of perfect interlace. This is approximately 3 microseconds out of the vertical interval of 16,667 microseconds, or less than 1 part in 5000.

Before we leave the basics of interlace, it is pertinent to review briefly the real function of equalizing pulses. They do not, as the name implies, equalize the charge on the vertical-sync integrating capacitor between the H and ½H fields. Fig. 6-2 should make this clear. The time from t1 to t3 represents H (at start of field 1), and the time from t2 to t3 represents ½H (at start of field 2). If the first vertical sync pulse occurred at t3, the integrating capacitor would start charging with a 2:1 voltage difference between fields, resulting in a much earlier firing of the oscillator for field 2 than for field 1. However, if a 3H waiting, or equalizing, interval (t4) is allowed, the following condition prevails:

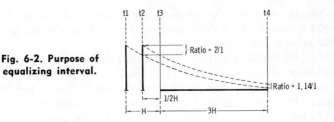

Fig. 6-2. Purpose of equalizing interval.

Field 1: H + 3H = 4H
Field 2: 0.5H + 3H = 3.5H
Ratio: 4H/3.5H = 1.14/1

Thus, by the time the first vertical-sync pulse occurs at t4, the charge on the integrating capacitor is sufficiently equalized between the even and odd fields to fall within the requirements of interlace, assuming proper adjustment of the receiver vertical-hold control.

Obviously, pulses must be present during this interval to maintain horizontal sync. These pulses must be at twice the horizontal frequency so that equalizing pulses 1, 3, and 5 supply triggers for field 1, and equalizing pulses 2, 4, and 6 supply them for field 2. The trailing edges of alternate vertical-sync pulses (serrations) then step the horizontal-sync oscillator during that interval. Equalizing pulses are made one-half the width of horizontal-sync pulses so that no shift in the ac axis occurs at the transition between the line frequency and the double-frequency equalizing pulses. Preventing this shift in axis is important to attenuate the inherent 30-Hz component of horizontal sync introduced as a result of alternate fields being displaced by one-half line.

The sync generator provides the electronic coordination of the entire TV transmission system. This means that the receiver and monitor picture tubes start scanning a particular line at the same time as the camera scan, sweep the line at the same rate, start the retrace from right to left at the same time, blank this retrace at the same time, and return the beam from bottom right to top left at the same time for the start of the next field. The receiver and monitor kinescopes have their own deflection and blanking in synchronism with those at the camera.

It is understandable, then, that the sync generator is necessarily rather complex in function. It must generate not only precisely *timed* pulses, but precisely *shaped* pulses as well. A thorough understanding of the sync generator is not beyond the scope of the TV operator if he will take the pains to study. The engineering and development of circuit designs are not important to the operator. What *is* important is familiarity with the pulses in the composite TV signal, their purpose, and how the sync generator forms and mixes the various pulses for combining with the picture information from the camera.

First, let us review the functions required in coordination of the television system. Some of these functions are performed at the station, and some occur in the receiver.

The photosensitive surface of the camera pickup tube must be scanned from upper left to lower right, as outlined previously, so that the picture is transmitted in elements of instantaneous value. The pickup head contains its own deflection and blanking mechanisms. Since these mechanisms determine where the scanning beam is directed at any given time, they must be timed from one coordinating "headquarters"—the sync generator. Therefore, horizontal and vertical *driving pulses* are supplied to the camera sawtooth deflection generators from the sync generator. (Remember that some recent camera chains do not use drive pulses as such from the sync generator. Composite sync is used, from which camera driving pulses are formed.) *The camera deflection generators cause the scanning beam to trace across the target; the driving pulses trigger the retrace.* Since these driving pulses do trigger the time of retrace, they are also used to trigger the blanking so that the retrace lines are made invisible on the monitor or receiver screen.

Consider now the receiver in the observer's home. An electron beam (which is being modulated in intensity by the video information) scans the kinescope screen under the action of the receiver deflection generators. Therefore, at the same time that a driving pulse is fed to the studio camera, a sync pulse is fed to the receiver, together with a blanking pulse to drive the kinescope tube into the cutoff (black) region. The sync pulse initiates the kinescope retrace, and the blanking pulse makes the retrace invisible to the viewer. Such action occurs in both the horizontal and vertical scanning functions.

As a starting point for our study, we may list seven general types of synchronizing signals delivered to the overall system by the sync generator (not considering color at this time), as follows:

1. *Horizontal-Driving Signal:* This signal consists of short-duration rectangular pulses at the horizontal-scanning frequency of 15,750 Hz fed to the sawtooth generator in the camera to trigger the horizontal-retrace voltage. These pulses also trigger the blanking generators in the camera so that the target of the pickup tube is driven negative and repels the scanning beam so that the signal corresponds to black, thus blanking the retrace action. Although formed at the same *time* as the horizontal-sync pulse, which is part of the signal radiated from the antenna, the driving pulses are *not* transmitted, but are fed only to the cameras.

2. *Vertical-Driving Signal:* This signal consists of rectangular pulses that have longer duration than the horizontal-driving pulses and that occur at the vertical frequency of 60 Hz. These pulses also are fed only to the camera, and they trigger the camera vertical-sawtooth gen-

erator. They are used also to trigger the camera blanking generator, thus blanking the vertical return trace of the pickup tube.

3. *Kinescope Horizontal-Blanking Signal:* This signal consists of rectangular pulses at the horizontal-scanning frequency of 15,750 Hz. They are added to the transmitted video signal for the purpose of driving the receiver kinescope into the cutoff region for retrace blanking. These pulses are transmitted at the "maximum black" level, and form the "pedestal" voltage upon which the horizontal-sync pulses are transmitted in the "blacker-than-black" region.

4. *Kinescope Vertical-Blanking Signal:* Pedestals of longer duration than the horizontal-blanking pulses are transmitted at the same black level at the vertical frequency of 60 Hz. Their purpose is to drive the kinescope tube into the cutoff region, blacking out the tube during the longer vertical-retrace time. (NOTE: Horizontal and vertical blanking is a single composite output of the sync generator.)

5. *Horizontal-Synchronizing Signal:* Short-duration 15,750-Hz pulses, slightly narrower than the horizontal pedestals upon which they are properly positioned in time, are used to trigger the receiver horizontal-sweep generator at the end of each line. These pulses occur at the same time as the horizontal-driving pulses for the camera.

6. *Vertical-Synchronizing Signal: Serrated* pulses of longer duration than the horizontal-synchronizing pulses are generated at the vertical frequency of 60 Hz. They are properly placed in time on the vertical pedestals, or blanking pulses. The vertical-synchronizing pulses are used to trigger the vertical-sweep generator at the receiver, initiating the return of the beam from the bottom to the top of the screen at the end of each field.

7. *Equalizing Signal:* This signal consists of short-duration pulses also placed on the vertical pedestals; these pulses immediately precede and follow the vertical-sync pulses. They are used to achieve uniform firing of the receiver vertical retrace, and they prevent *line pairing* that otherwise would result.

NOTE: Horizontal sync and vertical sync containing the equalizing pulses form a single composite output of the sync generator.

Now it is possible to discuss in detail the composite TV waveform illustrated in Fig. 6-3. The relative levels and timing of both the horizontal and vertical pulses, as well as the allowable tolerances, are indicated. The necessary shaping is shown also in this figure.

Waveforms 1 and 2 in Fig. 6-3 show the horizontal and vertical pulses much as we have discussed them in this section. At the left are the last few lines of a field, showing the horizontal pedestals and sync pulses. When the bottom of the picture (end of a field) is reached, the vertical blanking takes over. The equalizing and vertical-sync pulses are constructed on this vertical pedestal. The equalizing pulses assure that the receiver integrating

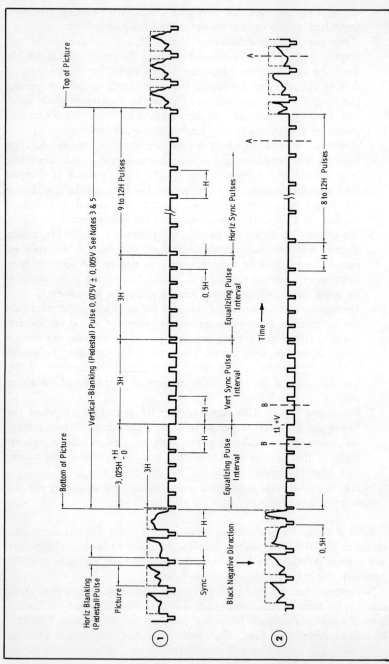

Fig. 6-3. Details of transmitter

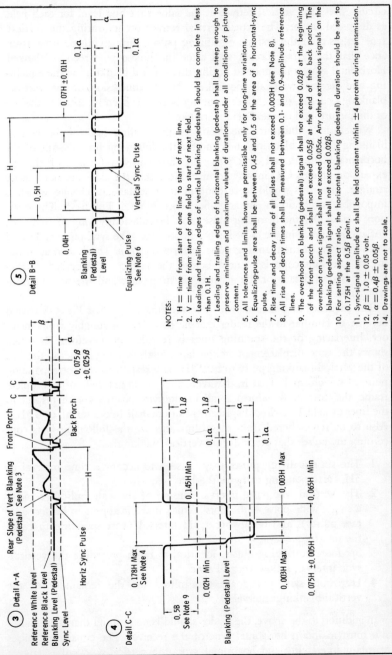

NOTES:

1. H = time from start of one line to start of next line.

2. V = time from start of one field to start of next field.

3. Leading and trailing edges of vertical blanking (pedestal) should be complete in less than 0.1H.

4. Leading and trailing edges of horizontal blanking (pedestal) shall be steep enough to preserve minimum and maximum values of durations under all conditions of picture content.

5. All tolerances and limits shown are permissible only for long-time variations.

6. Equalizing-pulse area shall be between 0.45 and 0.5 of the area of a horizontal-sync pulse.

7. Rise time and decay time of all pulses shall not exceed 0.003H (see Note 8).

8. All rise and decay times shall be measured between 0.1- and 0.9-amplitude reference lines.

9. The overshoot on blanking (pedestal) signal shall not exceed 0.02β at the beginning of the front porch and shall not exceed 0.05β at the end of the back porch. The overshoot on sync signals shall not exceed 0.05α. Any other extraneous signals on the blanking (pedestal) signal shall not exceed 0.02β.

10. For setting aspect ratio, the horizontal blanking duration should be set to 0.175H at the 0.5β point.

11. Sync-signal amplitude α shall be held constant within ±4 percent during transmission.

12. β = 1.0 ± 0.05 volt.

13. α = 0.4β ± 0.05β.

14. Drawings are not to scale.

input-signal waveform.

circuit "fires" the vertical retrace at the same time and level for each field, as discussed earlier. The actual vertical retrace occurs during the vertical-sync pulses after the first six equalizing pulses, thus giving positive assurance that the retrace occurs well after the kinescope screen is blanked. Horizontal sync is maintained during this vertical blanking by the equalizing and serrated vertical-sync pulses. At the conclusion of the vertical-blanking pedestal, the receiver scanning beam has been returned to the top of the picture, and the next field is started.

Waveform 3 illustrates the details of the trailing edge of the vertical blanking and the blanking and sync for the following line. Details of the horizontal-blanking and -sync pulses are shown in waveform 4, which is an expansion of area C-C in waveform 3. Note that the sides are not exactly straight, since this would correspond to an instantaneous change, which is impossible to obtain in any amplifier. (See Note 4, Fig. 6-3.) The duration (width) of the blanking pulse must be sufficiently long to cover the horizontal retrace in the most inefficient receiver. The sync pulse is delayed from the start of the blanking pulse to form the *front porch,* so that blanking of the kinescope is assured before the retrace is triggered. Waveform 5 shows details of the sixth equalizing pulse and the first vertical-sync pulse.

The sequence of pulses shown in waveforms 1 and 2 of Fig. 6-3 should become so familiar to the reader that he may draw it roughly from memory. Interlacing of the scanning lines is revealed in waveform 2, which shows the time displacement of the horizontal-blanking pulse at the left of the vertical-blanking pulse to be 0.5H with respect to the corresponding pulse of waveform 1. That is, if waveform 1 is the first field of a complete frame, the lines of waveform 2 fall in alternate positions with respect to the lines of field 1 (a time displacement of one-half line interval, or 0.5H). Also to be remembered from these drawings are the following facts concerning the pulses that occur during vertical blanking:

1. The six equalizing pulses may be seen to occupy a time interval of 3H, corresponding to twice the line frequency.
2. The vertical-sync-pulse interval consists of six wide pulses (actually a single wide pulse with "slots" which make it appear to be six separate pulses), also occupying a 3H interval (corresponding to twice the line frequency).
3. Six more equalizing pulses, identical to those preceding the vertical-sync interval, follow vertical sync.
4. Horizontal-sync pulses are transmitted for the remainder of the vertical-blanking interval.

In addition to the above, the reader should keep in mind that these pulses also must maintain horizontal sync of the receiver. The equalizing pulses maintain interlacing and horizontal sync, the vertical-sync pulses trigger the receiver vertical retrace and maintain horizontal sync, and regular horizontal-sync pulses follow the last equalizing pulses. It is because of this need for maintaining horizontal sync during the vertical-blanking interval that the vertical-sync pulse must be split up, or serrated. The "slots" are approximately equal in duration to the horizontal-sync pulses, and since they are of twice the line frequency, *every other* serration maintains the receiver horizontal-sweep oscillator in proper sync. Since the equalizing pulses are also of twice the line frequency, *every other* equalizing pulse maintains horizontal sync.

As shown in waveform 5 and Note 6 of Fig. 6-3, the equalizing pulse is one-half the width of the horizontal-sync pulse. If the equalizing pulses were made the same width as the sync pulses, the integrating capacitor would charge over a longer period of time during the influence of the equalizing pulses, and premature firing of the receiver vertical-sweep generator could result.

As already pointed out, it is important to maintain the horizontal-sweep oscillator in perfect sync during vertical blanking; otherwise, the horizontal sweep would be out of sync after each field. The receiver horizontal-sweep oscillator is driven from a differentiating circuit that accepts pulses at the line frequency (15,750 Hz) and above. Therefore, in order that the horizontal-sweep circuits be driven during the vertical-blanking and -sync time (which recurs at a low frequency, 60 Hz), the vertical-sync and equalizing pulses are at twice the line frequency as described. The action of the receiver during this time is illustrated in Fig. 6-4. In Fig. 6-4A, the action of the differentiating circuit on the usual horizontal-sync pulses may be observed; the leading edge of each pulse triggers a discharge through the succeeding stage to drive the sweep generator and cause the retrace of the scanning beam. Fig. 6-4B shows how horizontal sync is maintained during vertical blanking.

At this time, the reader may wonder why the horizontal-sweep oscillator does not "sync" at the 31,500-Hz frequency of the equalizing pulses and vertical-sync serrations. The answer is that the receiver horizontal-sweep oscillator is essentially a "free-running" type, and in order to "lock in" its operation, it is necessary for the "lock-in" voltage to be very close to the natural frequency of the oscillator. Thus, if the free-running oscillator is adjusted close to 15,750 Hz, frequencies of double this rate have no effect on its operation, whereas frequencies of 15,750 Hz will lock the oscillator in. In this way, *every other* pulse constructed on the vertical-blanking pedestal will affect the firing of the oscillator, thus maintaining lock-in at the horizontal frequency.

6-2. AFC CIRCUITRY

All sync generators use some form of automatic frequency control (afc). The master-oscillator frequency normally is controlled by setting the front-panel frequency-control switch to one of the following positions:

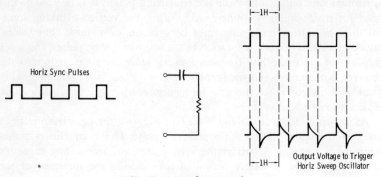

(A) Horizontal-sync pulses.

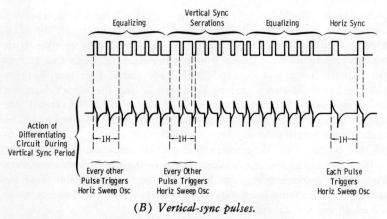

(B) Vertical-sync pulses.

Fig. 6-4. Receiver sync function.

1. Genlock
2. 60-Hz (Line lock)
3. Off (Free-running)
4. Crystal
5. External (Such as the count-down from a color-frequency standard)

The following description applies to the RCA TG-2 generator.

For monochrome telecasting, the generator normally is run with 60-Hz afc. This function may be understood by following the simplified schematic of Fig. 6-5. The afc discriminator (V16) compares the count-down existing at the vertical-sync gate with the 60-Hz wave derived from the power line, and develops a dc voltage that controls the master oscillator by means of the afc control tube. Tube V16 conducts only during a relatively short interval (3 lines, or 190 microseconds, compared to a half-cycle interval of 8333 microseconds at 60 Hz), charging capacitor C1 to the value of

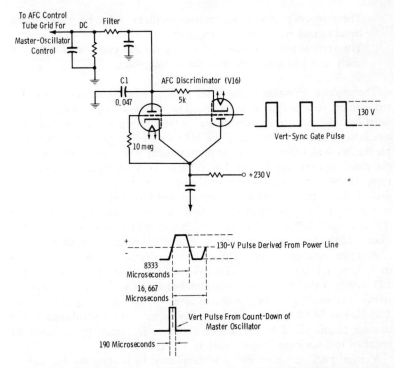

Fig. 6-5. Afc action to maintain constant master-oscillator frequency.

the line-frequency pulse existing at that instant. As you can see from Fig. 6-5, the afc voltage developed depends on the position of the vertical pulse relative to the power-line pulse. The circuit tends to hold this pulse position at the point where the power-line waveform crosses the zero axis. In this way, the master oscillator is stabilized at a frequency exactly 525 times the local power-line frequency. This generator also employs a variable phasing control to shift the phase of the 60-Hz line reference voltage so that proper synchronization of motion-picture projectors with short-application shutters (as used with iconoscope camera chains) may be obtained. The iconoscope type of camera is obsolete, but it is used in a few isolated cases.

When it is necessary to superimpose local material on network or remote video, the *genlock* position is used. In this position, the incoming signal is brought into a stabilizing amplifier where the supersync region is stripped from the composite signal and used as triggers for the local sync-generator control. The video-only portion of the signal is then fed to the local switcher, where it can be handled as any local camera signal (noncomposite input). Two conditions must be met to lock the local sync generator to that at a remote point:

1. The *frequency* of the local master oscillator must be made exactly equal to that at the remote generator.
2. The vertical pulses of the local sync generator must be brought precisely into *phase* with those at the remote point.

The genlock operation of the TG-2 generator functions as follows: The stripped-off remote sync is amplified. The horizontal component is separated by a differentiation network and is then amplified and clipped. Positive horizontal pulses, instead of the 3H vertical-sync gate pulses described for the line-lock function, feed afc discriminator V16. In genlock operation, the charging capacitor (C1 in Fig. 6-5) is removed from the ground return. At the same time, the grid of the afc amplifier receives a horizontal-drive pulse from the local sync generator and produces a sawtooth voltage that is applied to V16 in place of the 60-Hz power-line wave. This time, the dc control voltage fed to the afc control tube serves to lock the frequency of the local sync generator to that of the remote generator.

A typical sync generator uses four counting circuits with division ratios of 7, 5, 5, and 3, respectively. The combined product of $7 \times 5 \times 5 \times 3$ is 525, which is the number of lines (total of active and inactive lines) per frame. The product of the number of lines (525) and the field frequency (60 Hz) is 31,500 Hz, or the frequency of the master oscillator. A fifth counter circuit, dividing the master oscillator frequency by 2, yields the required line-scanning frequency of 15,750 Hz.

Vertical phasing for genlock is accomplished by causing the 7:1 counter and all following counters to miscount until the local field-rate signal is lined up exactly with the remote signal. Remote vertical sync is separated by an integration circuit, then amplified. A negative pulse from the plate of a genlock vertical-phasing tube and a positive local vertical-sync gate pulse appearing at the plate of a 3H amplifier are applied to the genlock phasing amplifier. This tube is biased so that it conducts during the positive local vertical pulse until the negative remote vertical pulse is brought into position to cancel the local pulse. As long as the tube is conducting (vertical pulses not in coincidence), the 7:1 counter (4500 Hz) miscounts, causing the fields to "slip" until exact coincidence is achieved.

6-3. THE COLOR-SYNC TIMING SYSTEM

Fig. 6-6 shows a highly simplified block diagram of a typical sync-generator system for color. In older equipment, blocks 1 and 2 are separate units external to the main sync generator. In later equipment, the subcarrier generator (color standard) and the burst-flag generator are integral parts of the main unit.

The color standard normally contains a precision 3.579545-MHz crystal mounted in a thermostatically controlled oven. The output from this unit is a sine wave of about 2 volts peak-to-peak in 75 ohms. The count-down

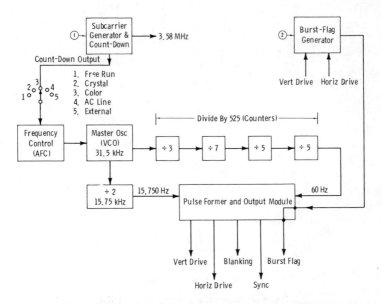

Fig. 6-6. Basic functions of sync generator for color.

from this master oscillator is fed to the "color" position of the timing section, or to the "external" input of a separate sync generator when used (covered in Chapter 2).

The burst-key (burst flag) generator was covered in Chapter 2, but let us look briefly at a modern solid-state version of this unit. Fig. 6-7 presents typical circuitry. The horizontal pulse is amplified and inverted by Q1. The amplified pulse is integrated by the network made up of R1, R2, and C1. Note that the Q2 emitter is returned to −10 volts, and the Q1 collector voltage holds Q2 off *until* the integrated pulse at the base becomes more positive than the emitter voltage. When this occurs, Q2 saturates and its collector goes essentially to −10 volts. So the Q2 collector is at zero volts prior to the pulse (cutoff) and swings to −10 volts for that portion of the base pulse more positive than −10 volts. Adjustable resistor R2 (delay control) sets the rise time of the integration, as shown on the accompanying waveform. Therefore, the Q2 collector output pulse is delayed a proper amount for the *breezeway* interval (interval between the trailing edge of sync and the first cycle of the color burst).

Transistor Q3 is connected in a "boxcar" circuit in which the base pulse width (and therefore the output pulse width) is determined by the C2R3 product. Width control R3 is adjusted to obtain the proper number of cycles in the burst interval.

A typical transistor gate, which provides a pulse 9H in duration for burst inhibit, is shown in Fig. 6-8. Note here that when either transistor

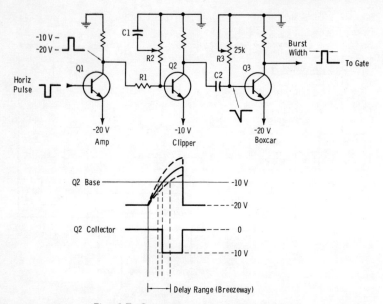

Fig. 6-7. Generation of burst-key pulse.

is saturated, the common collector is at ground (zero) potential. Note also that *both* transistors must be at cutoff for the collectors to reach the supply voltage.

In the absence of input pulses, Q1 is saturated and Q2 is at cutoff. The collectors are at zero (ground) potential. When the positive burst-width (burst-key) pulse arrives at the Q1 base, this transistor is cut off. Since Q2 is already at cutoff, the collectors rise to the negative supply voltage. Therefore, an inverted key pulse is passed to the output across R_L. This action continues until the 9H eliminate pulse arrives at the Q2 base. For the duration of this pulse, Q2 is saturated and the collector voltage remains at ground, eliminating the burst-key pulses from the output.

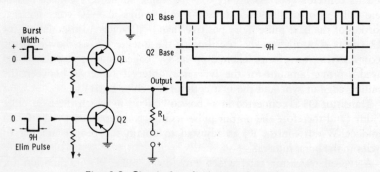

Fig. 6-8. Circuit for eliminating burst key.

NOTE: All synchronizing generators inherently employ a 9H gating pulse in the formation of composite sync. In modern color-sync circuits, we will find that the burst flag is generated in the main portion of the sync generator without need of additional equipment. With older generators, a separate unit is employed.

In describing color genlock, we will assume the reader is familiar with normal monochrome genlock functions as previously discussed. Briefly, these are:

1. The frequency and phase of locally generated horizontal- and vertical-sync pulses are compared with those of corresponding pulses received from a network or remote pickup.
2. From this comparison are derived correcting signals, which reflect the frequency and phase difference between the local and remote signals.
3. The correction signals are applied to the appropriate circuits in the local sync generator so that its output pulses are in exact phase with the incoming pulses from the remote source.

Now assume that the remote signal is a composite color signal, and it is desirable to lap-dissolve or superimpose a local color signal during the program. Since the local sync generator (in genlock) is already locked in frequency and phase to remote sync, the task remaining is to lock the local color-subcarrier oscillator to the exact frequency and phase of the remote subcarrier burst.

The color-genlock circuitry extracts the burst from the remote signal and, through a reference phasing network to obtain proper phasing with local color vectors, forms a sawtooth for use in a phase-detector circuit such as that of Fig. 6-9. In this particular type of system, a sampling pulse is formed from the local subcarrier burst. This pulse is applied to the base of Q1. Diode X1 suppresses any negative-going signal that might appear at the Q1 collector, and prevents ringing from inductive kickback. Thus, only a positive-going sampling pulse appears at the Q1 collector to be coupled by T1 to the error-detection circuitry of Q2 and Q3.

The sawtooth formed from reference burst appears at the center tap of the T1 secondary. Transformer T1 splits the phase of the sampling pulse so that pulses of opposite polarity are fed to the Q2 and Q3 bases. Transistors Q2 and Q3 are normally cut off. When the sampling pulses occur, the transistors are driven into conduction an amount depending on the position (determined by frequency or phase of the local subcarrier oscillator) of the sampling pulse on the slope of the reference sawtooth waveform. Balance control R1 permits balancing for zero potential at the center of the saw slope. The detected error results in amplitude and polarity variations of the charge on C1, and these are coupled through Q4 to the frequency-controlling element of the local subcarrier oscillator. We recognize this as

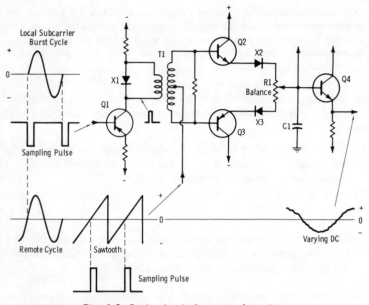

Fig. 6-9. Basic circuit for error detection.

the conventional "pulse sampling" technique in automatic-frequency-control circuitry. In practice, circuits vary greatly in design; the best we can do is acquaint the reader with the principles involved so that he can more readily understand equipment instruction books.

6-4. FCC SYNC STANDARDS IN PRACTICAL FORM

The FCC video waveforms—horizontal-, vertical-, and color-sync signals —are illustrated in Fig. 6-10. It is common practice for the FCC to state all times in terms of H, where H is one television line, or 63.5 microseconds (μs). In practice, times in microseconds are much more useful, so we have included Fig. 6-11 for horizontal pulses (and color-sync burst) and Table 6-1 for vertical pulses. Using this information with your scope enables you to "standardize" your sync generator.

The FCC definition of fields is as follows. Field 1 (even field): The start of field 1 is defined by a whole line between the first equalizing pulse and the preceding horizontal-sync pulse. Field-1 line numbers start with the *first* equalizing pulse in field 1. Field 2 (odd field): The start of field 2 is defined by a half line between the first equalizing pulse and the preceding horizontal-sync pulse. Field-2 line numbers start with the *second* equalizing pulse in field 2.

The FCC states: "The color picture signal shall correspond to a luminance [brightness] component transmitted as amplitude modulation of the picture carrier and a simultaneous pair of chrominance [coloring] components transmitted as the amplitude-modulation sidebands of a pair of suppressed subcarriers in quadrature." Also, "The chrominance subcarrier frequency shall be 3.579545 [MHz] ± 10 [Hz] with a maximum rate of change not to exceed one-tenth [Hz per sec]." The "new" information here is the specified tolerance on the color subcarrier frequency, plus or minus 10 Hz.

The FCC describes the luminance modulation as follows: "A decrease in initial light intensity shall cause an increase in radiated power (negative transmission)."

Table 6-1. Vertical-Pulse Widths

Pulse	Minimum (μs)	Nominal (μs)	Maximum (μs)
Equalizing	2.0	2.4	2.54
Vert Serration	3.81	4.5	5.08
Vert Blanking	1167	1250	1333

NOTES:

1. Vertical-sync pulses are not specified. See detail A-A, Fig. 6-10B. The width of the vertical-sync pulse is set by the tolerance on the width of the leading edge of the last (leading) equalizing pulse to the trailing edge of the first serration.
2. The width of the equalizing pulse must be 0.45 to 0.5 of the horizontal-sync width used.
3. Vertical blanking in terms of H is from 18.375 to 21 lines. Although 21 lines is shown as "maximum" in the above chart, this is the width of vertical blanking maintained by the networks. It allows vertical-interval test signals to be inserted (usually on lines 18 and 19 of vertical blanking) with a suitable "guard band" of blanking lines before the start of active line scan.
4. Horizontal and vertical blanking must be of proper ratio to establish the 4:3 aspect ratio.

"The blanking level shall be transmitted at 75 ± 2.5 percent of the peak carrier level. . . . The reference white level of the luminance signal shall be 12.5 ± 2.5 percent of the peak carrier level. . . . The reference black level shall be separated from the blanking level by the setup interval, which shall be 7.5 ± 2.5 percent of the video range from blanking level to the reference white level."

The above specification is illustrated by Fig. 6-12. The studio scale is shown on the left and the corresponding transmitter scale (percent modulation) on the right.

6-5. CABLE DELAY

The speed of light in feet per microsecond is 983.5 ft/μs. Video and pulse distribution in the television plant is through polyethylene-dielectric coaxial cable, which has a propagation factor of 0.66. Therefore, the rate of travel in such cable is 983.5 × 0.66 = 650 ft/μs. This means there is a delay of approximately 1.5 μs per 1000 feet of cable in the distribution system.

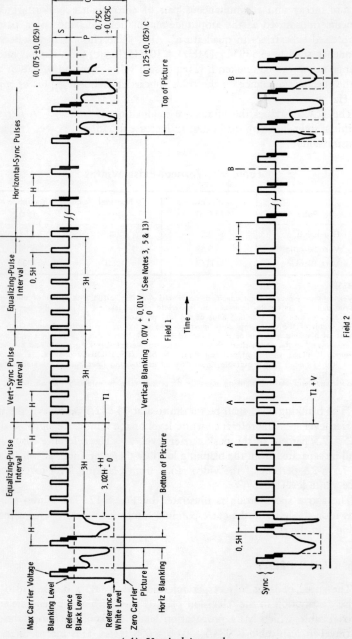

(A) Vertical interval.

Fig. 6-10. Standards for transmitted

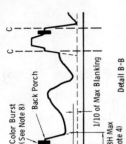

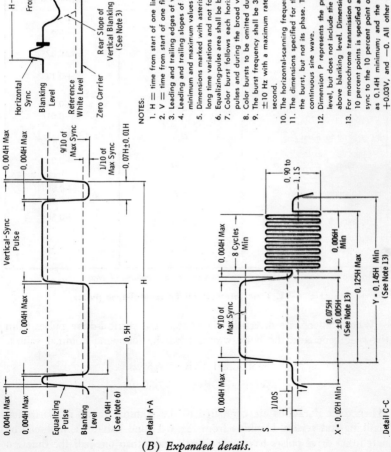

Color Burst (See Note 8)

Back Porch

1/10 of Max Blanking

Detail B-B

Front Porch

H

Z = 0.18H Max (See Note 4)

Rear Slope of Vertical Blanking (See Note 3)

Horizontal Sync

Blanking Level

Reference White Level

Zero Carrier

C — C

C

NOTES:

1. H = time from start of one line to start of next line.

2. V = time from start of one field to start of next field.

3. Leading and trailing edges of vertical blanking should be complete in less than 0.1H.

4. Leading and trailing slopes of horizontal blanking must be steep enough to preserve minimum and maximum values of x + y and z under all conditions of picture content.

5. Dimensions marked with asterisk indicate that tolerances given are permitted only for long time variations and not for successive cycles.

6. Equalizing-pulse area shall be between 0.45 and 0.5 of area of a horizontal-sync pulse.

7. Color burst follows each horizontal pulse, but is omitted following the equalizing pulses and during the broad vertical pulses.

8. Color bursts to be omitted during monochrome transmission.

9. The burst frequency shall be 3.579545 MHz. The tolerance on the frequency shall be ±10 Hz with a maximum rate of change of frequency not to exceed 1/10 Hz per second.

10. The horizontal-scanning frequency shall be 2/455 times the burst frequency.

11. The dimensions specified for the burst determine the times of starting and stopping the burst, but not its phase. The color burst consists of amplitude modulation of a continuous sine wave.

12. Dimension P represents the peak excursion of the luminance signal from blanking level, but does not include the chrominance signal. Dimension S is the sync amplitude above blanking level. Dimension C is the peak carrier amplitude.

13. For monochrome transmission only, the duration of the horizontal-sync pulse between 10 percent points is specified as 0.08H ± 0.01H; the period from the leading edge of sync to the 10 percent point on the trailing edge of horizontal blanking is specified as 0.14H minimum; and the duration of vertical blanking is specified as 0.05V, +0.03V, and −0. All other dimensions remain the same.

Detail A-A

0.004H Max

0.004H Max

Vertical-Sync Pulse

9/10 of Max Sync

1/10 of Max Sync

0.07H ± 0.01H

0.004H Max

0.004H Max

Equalizing Pulse

Blanking Level

0.04H (See Note 6)

H

0.5H

Detail C-C

0.004H Max

9/10 of Max Sync

8 Cycles Min

0.90 to 1.15

0.006H Min

0.075H ± 0.005H (See Note 13)

0.125H Max

Y = 0.145H Min (See Note 13)

X = 0.02H Min

1/10 S

S

(B) Expanded details.

synchronizing waveforms.

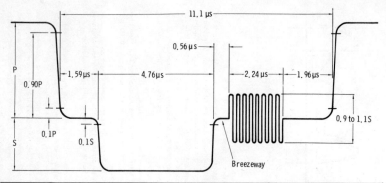

	Nominal Microseconds	Tolerance Microseconds
Blanking	11.1	+0.3 −0.6
Sync	4.76	±0.32
Front Porch	1.59	+0.13 −0.32
Back Porch	4.76	+0.96 −0.61
Sync to Burst	0.56	+0.08 −0.17
Burst	2.24	+0.27 0
Blanking to Burst[1]	6.91	+0.08 −0.17
Sync & Burst	7.56	+0.38 −0.49
Sync & Back Porch	9.54	±0.32

[1]Blanking-to-burst tolerances apply only to signal before addition of sync.

Fig. 6-11. Time intervals for horizontal-sync pulses.

When cameras are driven by horizontal- and vertical-drive pulses from the sync generator, the pulse widths have the following nominal values:

$$\text{Horizontal Drive} = 6.3 \ \mu s$$
$$\text{Vertical Drive} \ \ \ = 680 \ \mu s$$

When these durations are compared to the nominal values of horizontal blanking and vertical blanking from Fig. 6-11 and Table 6-1, it should be noted that drive pulses have only slightly more than one-half the duration of their respective blanking pulses. It is important to understand why this

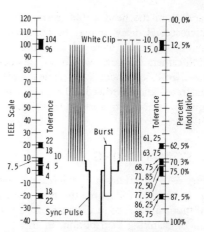

Fig. 6-12. Video-signal amplitudes.

Note: Tolerance value specified at blanking level applies to sync amplitude only. The variation of blanking with respect to the tolerance of setup shown is assumed to be zero.

is so. The 2:1 relationship is particularly important at the horizontal frequency, as is evident in Fig. 6-13. Remember that the transmitted composite blanking and the driving signals for the camera are inserted at the camera-control unit. However, the camera cable may have a length of 1000 feet or more, and allowance must be made for the cable delay, which is roughly 1.5 microseconds per 1000 feet of cable. Since the total path is

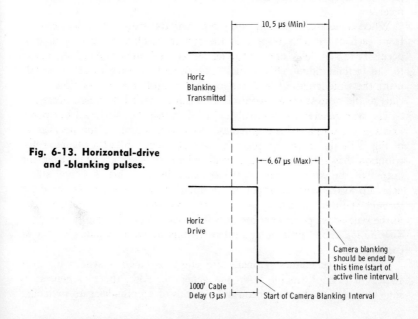

Fig. 6-13. Horizontal-drive and -blanking pulses.

to and from the head (2000 feet total), allowance must be made for a 3-microsecond delay. It may be observed from Fig. 6-13 that if the horizontal-drive pulses were of the same width as the horizontal-blanking pulses, camera blanking would not be ended at the start of the active line interval in cases where long camera cables are employed (unless drive is regenerated and narrowed). This effect is not so important at the vertical frequency, since 3 microseconds is negligible compared to a total of about 1250 microseconds.

6-6. SYSTEM TIMING

Fig. 6-14 illustrates a simple case of noncomposite switching in which sync is inserted following the final switching system. In this illustration, a 900-foot difference in distance exists between two control units and the point of blanking distribution. In this case, the blanking pulse is delayed approximately 1.5 microseconds to control unit 1, and only 0.15 microsecond to control unit 2. If the front-porch-width control of the sync generator is adjusted to obtain a normal front porch in the camera-1 signal after sync insertion, then a switch to the camera-2 signal will result in a lengthened front porch (Fig. 6-14B). Since the receiver retrace is triggered by the leading edge of horizontal sync and the picture is unblanked by the end of horizontal blanking, a lengthened front porch causes the picture area to shift to the left. Similarly, if the front porch is adjusted for normal on the camera-2 signal, a switch of the camera-1 signal will result in a narrowed front porch, and the picture will shift to the right on receivers.

When camera-control units are more than 100 feet (0.15 microsecond) from each other with respect to the system blanking distribution, it is necessary to add delay lines to the nearest control units to equal the delays to the farthest units. This is most conveniently accomplished by simply using the same length of feed line to every control unit. Excess cable in the runs to the nearest camera-control units can be coiled up when necessary.

The sync generator is normally adjusted so that the standard front porch exists at the output of the generator. Now assume that in the installation of Fig. 6-14 all camera-control units have 100-foot cables for blanking distribution. Assume also that all control units have 100-foot cables from the output of the control units to the switcher inputs. This is a total signal delay up to the switcher input of 0.3 microsecond. Now assume the delay through the total switching system is 0.2 microsecond. The total delay up to the sync-adder point now becomes 0.5 microsecond. Therefore, the *sync-distribution* cable must be equivalent to this delay, or about 330 feet of coaxial line.

Two sync generators normally are employed with a sync-changeover switch in case of trouble with one generator. Also, the two generators sometimes are used simultaneously by means of a pulse-delegate switching

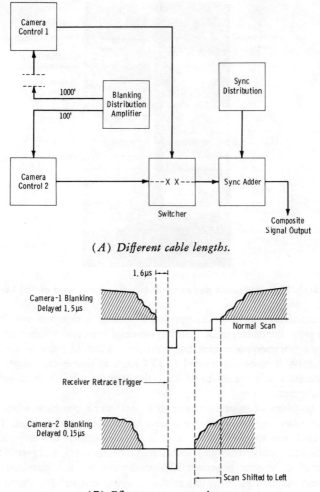

(A) Different cable lengths.

(B) Effect on scan waveforms.

Fig. 6-14. Effect of blanking delay on front-porch width.

system. This is particularly important when genlocking to a remote signal for on-the-air shows while simultaneously tape recording a local show. Video tape recorders require a rock-steady sync input, and the sync-delegate switch removes the possibility of momentary sync instability as occurs in getting into or out of genlock. The tape-recorder sync inputs can be delegated from the generator that is not genlocked to the network or remote signal.

Fig. 6-15 illustrates the installation at WTAE-TV. The photograph shows the two sync generators, changeover switch (which can be oper-

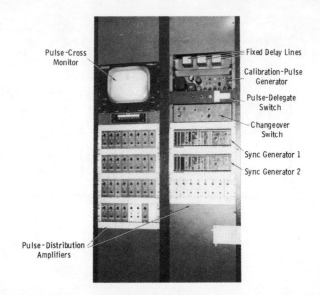

Pulse-Cross
Monitor

Fixed Delay Lines

Calibration-Pulse
Generator

Pulse-Delegate
Switch

Changeover
Switch

Sync Generator 1

Sync Generator 2

Pulse-Distribution
Amplifiers

Fig. 6-15. Sync generators and pulse-distribution equipment at WTAE-TV.

ated remotely from the video operator's position), delegate switch, calibration-pulse generator, pulse-distribution amplifiers, and pulse-cross monitor. The calibration-pulse generator permits a standard pulse of either 0.7 volt or 1.0 volt to be distributed to all CRO's so that every calibration is from the same reference. Use of the pulse-cross monitor is discussed later in this chapter.

The problem of system timing becomes more complex when color camera chains are integrated with existing monochrome systems. This is particularly true for the "older" color systems, which do not employ automatic-timing (pulse advance) techniques (Section 4-8, Chapter 4). Remember that the sync generator is normally adusted for "standard" front porch and standard pulse widths at the generator output. The system must then be properly "timed." The basic problem is illustrated by Fig. 6-16A, in which sync is added following the final switching point. Because of the additional 1.2-microsecond delay (minimum) in the color system, switching from monochrome to color cameras would cause a shift in front porch as indicated by the waveforms in Fig. 6-16A. Lap-dissolving of two such signals would cause a very noticeable "bend" and tearing during the transition. (This discussion assumes the sync delay is proper for the standard front porch on the existing monochrome facilities.)

The solution to the problem is shown by Fig. 6-16B. All monochrome pulse distribution containing horizontal information must be delayed an amount equivalent to the delay of the color system; the color equipment receives undelayed pulses. Output cables to the switcher are maintained

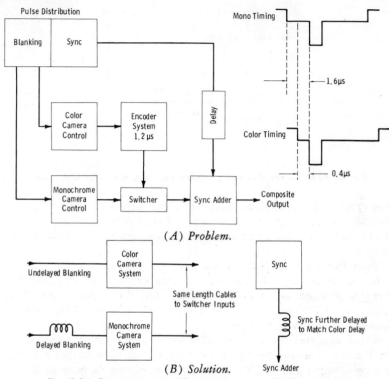

(A) Problem.

(B) Solution.

Fig. 6-16. System timing with monochrome and color cameras.

the same length for each camera. Sync distribution to the switcher is also delayed the additional amount required to maintain standard front-porch durations regardless of signal source. The same problem exists when composite switching is used (sync inserted in each camera chain). The sync distribution must be delayed to the monochrome chains an amount equivalent to the color delay.

System timing is a subject not normally covered in manufacturers' instruction books for specific equipment, even color camera chains. It is therefore important for the engineer to be familiar with the problems involved.

Let us consider the installation of a typical "older" color camera chain, such as the RCA TK-26 color film chain. We will assume that composite switching is used (sync is to be inserted in each camera-chain output), that monochrome cameras are also used, and that new "automatic-timing" color chains are added to the installation.

See Fig. 6-17. We will establish "time zero" as the leading edge of all pulses from the sync generator. In the example shown, the interconnecting-cable and internal signal delays result in a 0.425-microsecond delay up to

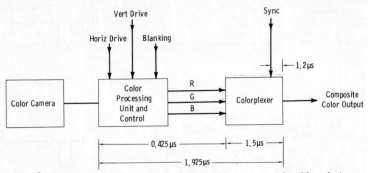

Fig. 6-17. Measured delay of blanking and sync in a color film chain.

the input of the colorplexer (encoder), and there is an additional 1.5-microsecond delay from input to output of the colorplexer. Thus, there is a total delay of 1.925 microseconds from sync generator to color-system output.

Note that from the point of sync insertion (just prior to the Y delay line) to the output, a 1.2-microsecond delay exists. Thus, assuming that the sync generator is adjusted to standard, sync distribution to this chain must have $1.925 - 1.2 = 0.725$ microsecond of additional delay prior to insertion to maintain front porch.

Fig. 6-18 shows the basics involved in integrating the specific camera of Fig. 6-17 into an existing monochrome system. This results in proper system timing at the switcher input for all signal sources. Note again that in the particular case involved, sync timing into the color camera chain is delayed because of the different delay in this chain between blanking and sync. All monochrome chains are delayed in timing of blanking, sync, and horizontal drive (all pulses containing horizontal repetition rates).

Important: Often overlooked is the problem existing when "new" color camera chains (automatic-timing cameras) are integrated with "old" color chains. The automatic-timing camera has an actual measured delay of just 0.1 microsecond between sync input and output. This is a considerable difference from the usual 1.2- to 1.9-microsecond delay in the older color camera. The newer camera must have "standard monochrome timing" of pulses at the input within specified tolerances. Therefore, in adding a new color camera to an installation with older color cameras, *the new system must be driven from the same delayed pulses as monochrome cameras.* In addition, burst-flag distribution to the new cameras must be delayed in the same manner to maintain proper breezeway.

6-7. SYSTEM PHASE FOR COLOR SYNC

Even when the system has been timed for proper horizontal-pulse coincidence at the input of the switcher, we have only scratched the surface

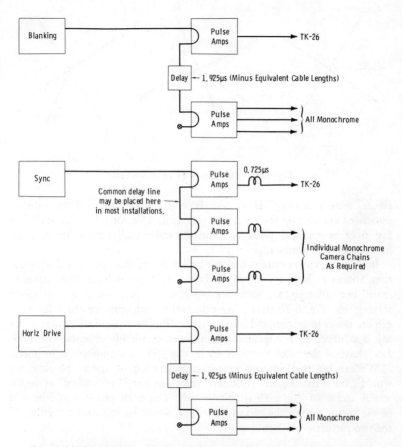

Fig. 6-18. Integration of color film chain with monochrome cameras.

of the subject of color operation. For mixing, lap-dissolving, or use of special-effects amplifiers in which two color sources are displayed simultaneously, the color-sync bursts *must have precisely the same phase* at the switching-system input. Switching systems are delay-compensated internally to provide the same time delay regardless of the switching path (mix, lap-dissolve, special-effects insert, etc.), as covered in Chapter 7. So the problem of color-sync phasing resolves itself into obtaining proper phase match at the switcher input.

See Fig. 6-19; it makes no difference which of the sets of vectors (A, B, or C) the receiver or color monitor "sees." In this case, provided burst phase is correct (which is true for the individual sets of vectors), proper color rendition will be achieved. The synchronous demodulators in the color receiver simply reference to burst phase, so there is no problem at

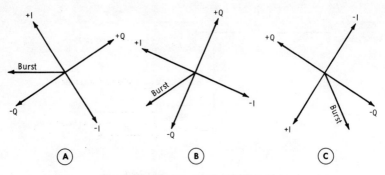

Fig. 6-19. Three sets of color vectors.

all in "system phasing" if we simply switch one camera chain with its associated encoder on the air at any one time. (Vector sets A, B, and C of Fig. 6-19 have proper phase relationships individually, but different absolute phases at a common point.)

But the television system will not be limited to this type of color operation. Sooner or later (if not at the beginning), we will be called upon to "mix" two color signals, such as a lap-dissolve operation or special-effects mixing, etc. Fig. 6-20 shows the principal adjustments involved for each camera chain in system and burst phasing. Proper individual camera phasing is achieved by the quadrature and burst-phase adjustments. The *absolute* phase of the color vector is set by the 360° system-phase adjustment.

We can see that without proper consideration of system phasing, we would have a thoroughly "confused" color receiver if it "looked" at vector sets A and C of Fig. 6-19 simultaneously. The burst phase would be that of the vector sum of the mixed bursts and would be incorrect for either of the two pictures.

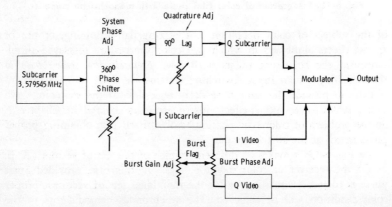

Fig. 6-20. Phase adjustments in color camera chain.

The amount of phase shift of the burst in a coaxial cable can be found easily. One cycle at 3.58 MHz occupies a time interval of $1/(3.58 \times 10^6) =$ 0.28 microsecond. Therefore, at 3.58 MHz, 360° corresponds to 0.28 microsecond. The delay in coaxial cable is about 1.5 microseconds per 1000 feet, or 0.0015 microsecond per foot. With this information, we can set up the proportion:

$$\frac{360°}{0.28} = \frac{x}{0.0015}$$

from which,

$$x = 360° \frac{0.0015}{0.28} = 1.9° \text{ per foot}$$

Therefore, the delay is practically 2° per foot at 3.58 MHz.

The subcarrier into each encoder can be shifted over a 360° range in each unit for the purpose of matching all color chains through a common switching system. Thus, in Fig. 6-21, the cables connecting the encoders and switcher 1 can be any length provided they are the same. You simply adjust the subcarrier phase shifters of encoders 2 and 3 to match the vectors from encoder 1 at the input of switcher 1. This is very conveniently done with a vectorscope by using the "A shared with B" display and matching the absolute burst phases to each other. If there is only one main switching system, the cables can be *any length* (not necessarily the same), since the 360° adjustment will allow absolute phase match.

If there is more than one switching system, it is necessary to use identical-length coax cables from each encoder to the additional switcher so that this switcher output will have identical phases from all units. Note that this does not mean the same *system* phase as from switcher 1. It only means that all color chains will be phased together through switcher 2. Since

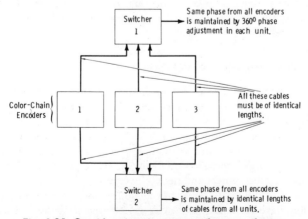

Fig. 6-21. Considerations in system color-sync phasing.

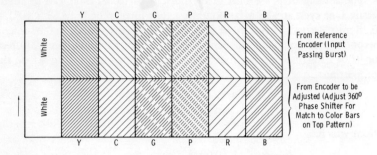

Fig. 6-22. Use of special-effects unit for phase comparison.

there is about 2° of phase shift per foot of cable at 3.58 MHz, we can see that to maintain less than 1° of phase difference the cables must be no different in length than six inches. Of course, if it is ever necessary to mix two or more switcher-system outputs, we must have identical-length cables in *all* encoder-switcher interconnections.

The "color-approved" type of special-effects system (Chapter 7) provides a satisfactory color-match (system phase) facility when a vectorscope is not available. The burst interval is on for one channel only; sync interval is carried by the other channel. Thus, if we feed the "reference encoder" signal (color bars) to one bank and use the "split-field vertical wipe" as in Fig. 6-22, the encoder to be adjusted (also with color bars) can be "wiped up" about half way as shown. Since the color monitor is seeing only one burst (that from the "reference" encoder to which all others are to be matched), we can adjust the 360° phase shifter of any additional encoder for a perfect color-bar match. This means that the "absolute phase" for both encoders being observed is identical at the switcher input, and signals can be mixed without color contamination. Be sure the output levels from the encoders are identical so that you are not confused by brightness differences in colors.

6-8. SYSTEM PHASING DEVELOPMENTS

It should be apparent from our study thus far that total system phasing can become an exacting and time-consuming task that does not readily allow integration of new equipment without additional timing complications. The task becomes much more difficult when we "double up" studios, that is, when we combine one studio output with a production of a second studio ("piggy-back" operation). Obviously, this type of operation occurs in large production centers and network centers.

A number of different approaches to timing and color phasing in large installations have been taken to provide greater operating flexibility. Briefly, these are as follows:

Zero Studio Delay (Andersen Laboratories): In this system, a delay is located at the studio output. This delay, equal to H or 2H minus the studio delay, causes the studio to appear to have no delay.

The Coded-Pulse System: This system requires an encoder that converts the output from a standard sync generator into a single pulse train containing the necessary information for the recreation of all timing pulses. At each destination, a decoder converts this pulse train into the standard pulses required by the particular picture source. For adjustment of studio and systems timing, two facilities may be supplied: (1) a phase-shifting network whereby the studio burst phase is adjusted to match the system burst phase, and (2) a variable delay line to set studio pulse timing to match the station pulse timing. (Only the latter is required for monochrome-only operation.)

The Natlock System: In this system, error signals are coded as audio tones that are generated by comparison equipment at the mixing point. These signals are used for automatic control of the timing and phase of each picture source. The error signals may be carried by telephone-bandwidth circuits in remote or network operations.

6-9. THE PULSE-CROSS MONITOR

Two basic types of pulse-cross monitors are used, the interlaced and the noninterlaced. In both types, the horizontal sweep is delayed so that the horizontal-blanking and -sync pulses appear near the center of the raster. To display the entire frame (interlaced presentation), it also is necessary that the vertical sweep be delayed to place the vertical blanking and sync in the normal active line area of the monitor. Both fields (interlaced) are displayed so that the entire 37- to 42-line vertical-blanking interval is visible. If the monitor vertical-deflection rate is changed to half rate (30 Hz), a single field is displayed with half the number of pulses (noninterlaced presentation). In either case, expansion of the vertical sweep normally is used to allow more critical observation of the pulses.

Fig. 6-23 shows the pulse-cross presentation of a line-output signal on an interlaced monitor. In this case, the video polarity is inverted so that sync is in the white-going direction. Note the convenience as a quick-reference check for the widths of horizontal front porch, sync, and blanking, equalizing pulses, and vertical sync. Vertical blanking is checked conveniently by counting the number of blanking lines. Some stations construct graticules with normal pulse widths marked after an accurate check of the generator with an oscilloscope.

The pulse-cross monitor is extremely useful both as a continuous monitor and as a servicing tool. A 9 × 1 switch panel is used at station WTAE-TV to allow selection of signals from a number of points to feed the monitor, but this switch panel normally is left in the "stand-by generator" position. This enables continuous monitoring of whichever generator is in

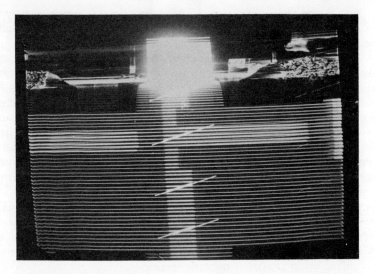

Fig. 6-23. Interlaced pulse-cross display (composite signal).

the stand-by position (composite sync only), as shown in Fig. 6-24A. Fig. 6-24B is an expanded view of this presentation with identification of pulses.

The "cross" is formed at the position in line with horizontal sync. The reader can readily understand the sequential presentation of the monitor (Fig. 6-24) if he will mentally move field 2 (waveform P) of Fig. 6-1 to the right one-half line so that the horizontal-sync pulses of both fields are in vertical alignment and the horizontal pulse immediately preceding field 2 is in line with the first equalizing pulse of field 1. Now, observing Fig. 6-24B, note that the in-line pulses (those occurring at H intervals) are equalizing pulses 1, 3, and 5 of field 1, and 2, 4, and 6 of field 2, spaced on alternate lines of the display because of interlace of fields. The half-line intervals and the remainder of the presentation should be obvious from following a similar analysis.

With an interlaced type of pulse-cross monitor, loss of interlace, such as could be caused by a vertical-countdown error of 0.5H, is readily apparent, as illustrated by Fig. 6-25. Brightness of the display will be greater than normal for a given adjustment because of the double tracing of identical raster lines.

Fig. 6-26 shows how to interpret a single-field (noninterlaced display) pulse-cross display, as is provided on many master monitors. Bear in mind that equalizing pulses 1, 3, and 5 of field 1 occur at horizontal-sync intervals, and equalizing pulses 2, 4, and 6 of field 1 occur at intermediate intervals. Similarly, vertical-sync pulses 1, 3, and 5 start in line with horizontal sync, but pulses 2, 4, and 6 start at the half-line position.

(A) Display on monitor.

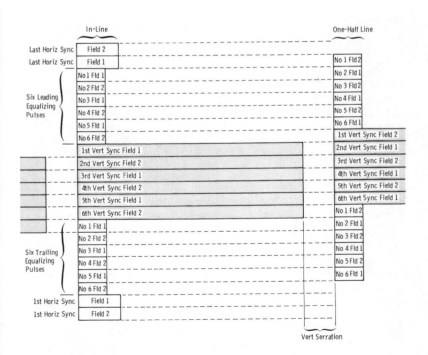

	In-Line		One-Half Line
Last Horiz Sync	Field 2		
Last Horiz Sync	Field 1		No 1 Fld 2
	No 1 Fld 1		No 2 Fld 1
	No 2 Fld 2		No 3 Fld 2
Six Leading Equalizing Pulses	No 3 Fld 1		No 4 Fld 1
	No 4 Fld 2		No 5 Fld 2
	No 5 Fld 1		No 6 Fld 1
	No 6 Fld 2		1st Vert Sync Field 2
	1st Vert Sync Field 1		2nd Vert Sync Field 1
	2nd Vert Sync Field 2		3rd Vert Sync Field 2
	3rd Vert Sync Field 1		4th Vert Sync Field 1
	4th Vert Sync Field 2		5th Vert Sync Field 2
	5th Vert Sync Field 1		6th Vert Sync Field 1
	6th Vert Sync Field 2		No 1 Fld 2
	No 1 Fld 1		No 2 Fld 1
	No 2 Fld 2		No 3 Fld 2
Six Trailing Equalizing Pulses	No 3 Fld 1		No 4 Fld 1
	No 4 Fld 2		No 5 Fld 2
	No 5 Fld 1		No 6 Fld 1
	No 6 Fld 2		
1st Horiz Sync	Field 1		
1st Horiz Sync	Field 2		

Vert Serration

(B) Pulse identification.

Fig. 6-24. Pulse-cross display, sync only.

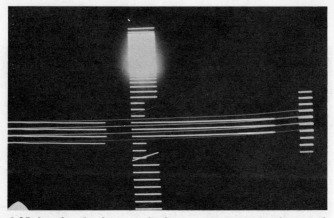

Fig. 6-25. Interlaced pulse-cross display; sync generator has lost interlace.

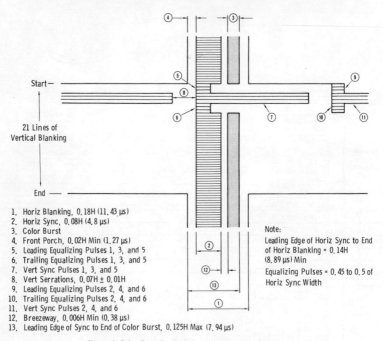

Start—

21 Lines of
Vertical Blanking

End —

1. Horiz Blanking, 0.18H (11.43 μs)
2. Horiz Sync, 0.08H (4.8 μs)
3. Color Burst
4. Front Porch, 0.02H Min (1.27 μs)
5. Leading Equalizing Pulses 1, 3, and 5
6. Trailing Equalizing Pulses 1, 3, and 5
7. Vert Sync Pulses 1, 3, and 5
8. Vert Serrations, 0.07H ± 0.01H
9. Leading Equalizing Pulses 2, 4, and 6
10. Trailing Equalizing Pulses 2, 4, and 6
11. Vert Sync Pulses 2, 4, and 6
12. Breezeway, 0.006H Min (0.38 μs)
13. Leading Edge of Sync to End of Color Burst, 0.125H Max (7.94 μs)

Note:
Leading Edge of Horiz Sync to End
of Horiz Blanking = 0.14H
(8.89 μs) Min

Equalizing Pulses = 0.45 to 0.5 of
Horiz Sync Width

Fig. 6-26. Single-field pulse-cross display.

EXERCISES

Q6-1. Relative to camera-driving pulses, what forms the "trace" and the "retrace" intervals of scanning?

Q6-2. When we speak of horizontal- and vertical-driving pulses, are these separate outputs of the sync generator?

Q6-3. When we speak of horizontal- and vertical-blanking pulses, are these separate outputs of the sync generator?

Q6-4. Are horizontal- and vertical-sync pulses delivered from separate outputs of the sync generator?

Q6-5. Where is the FCC requirement for the width of the horizontal front porch called out in the drawings of Fig. 6-10?

Q6-6. What is the proper amplitude of the color-sync burst?

Q6-7. How many cycles of color-sync burst should be present?

Q6-8. Does the color-sync burst occur following every horizontal-sync time in the complete composite color signal?

Q6-9. In general, what must be done in system timing to integrate color camera chains that do not have automatic timing (pulse advance) with monochrome camera chains?

Q6-10. Can color camera chains that employ automatic timing be integrated with color chains not automatically timed without correcting for timing errors?

CHAPTER 7

Video and Audio Signal Distribution

We will consider in this chapter the video and audio signal routing in typical studio installations. Switching systems, including special-effects equipment, are a necessary part of this study.

7-1. SOURCE AND LINE VIDEO DISTRIBUTION

See Fig. 7-1. All video distribution normally is carried by coaxial cable of 75-ohm nominal impedance. All monitors and distribution amplifiers (DA's) have two paralleled connectors at the input so that the cable can be either "looped through" or terminated. The amplifier input is bridging to 75 ohms so that a number of amplifiers can be fed by the loop-through technique, provided the end of the line is properly terminated in 75 ohms.

Video distribution is carried out at 0.7 volt peak-to-peak (pk-pk) for noncomposite signals, or 1 volt pk-pk for composite signals. Isolated feeds for each path are necessary so that any work required on any specific piece of equipment can be carried out without disturbing the video level for the other destinations. Some video DA's have a single-in, single-out configuration; others have multiple (but isolated) outputs. The DA's of Fig. 7-1 have four outputs for a single input. Modern DA's of solid-state design are usually of the plug-in variety for quick replacement in case of failure.

Fig. 7-1A shows a typical single-source video distribution. In this case, the camera outputs are: one for the master monitor at the control position (switchable between R, G, B, and, for four-channel cameras, Y, or luminance), and two other isolated outputs utilized as shown. Two preview monitors are shown looped through since the signal is not on an air path or recording path. A video DA routes the remaining output to the indicated destinations.

The complexity of source distribution varies greatly with needs. For example, a large production center may provide feeds for each individual signal source to a selectable input of each video tape recorder in the installation.

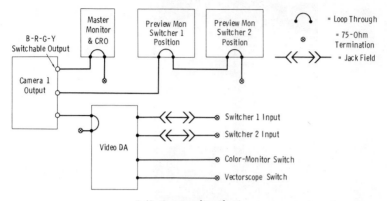

(A) Source distribution.

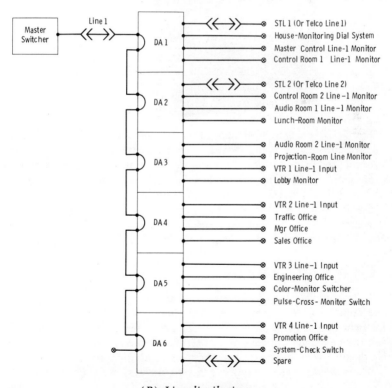

(B) Line distribution.

Fig. 7-1. Typical video distribution.

Video DA's

Patch Fields

Video DA's

House-Monitoring Dial System

Fig. 7-2. Video-distribution racks at WTAE-TV.

Fig. 7-1B is a simplified illustration of typical line video distribution following the master switcher. Again, the complexity varies greatly with the size and production facilities of the particular plant. Many master switchers today are capable of providing three or more separate line feeds as required, with a different program on each line.

Fig. 7-2 is a view of part of the video-distribution racks at WTAE-TV. A number of single-input, four-output solid-state plug-in DA's provide 248 isolated outputs. Each amplifier contains a built-in power supply.

7-2. THE VIDEO SWITCHER

Since most TV stations (even the smallest ones) must use more than one camera source at some time during their daily schedule, some means for switching signal sources must be available. Means of fading out or in and the momentary mixing of two separate video signals for purposes of lap-dissolving or deliberate superimposition of pictures are needed also.

There are three general types of video switchers:

1. *Mechanical push-button switching* with video on the actual switch contacts. The bank of switches is interlocked to prevent more than

one source from being "punched up" at a time. This type of switcher may or may not employ a means of fading or lap-dissolving between channels. It is used primarily (in installations of the past ten years) in portable field units.

2. *The relay switcher,* which employs remotely controlled, rack-mounted banks of relays. The switch banks are not interlocked, since the interlocking function is in the relay arrangement. Modern relay switchers most commonly use a dry-reed relay. This consists of a magnetically activated switch and a solenoid or coil. The switch is completely sealed in a glass tube containing an inert atmosphere, and the coil is wrapped around the glass tube. The reed operates in about 1 millisecond, and switching actions overlap for about 1 millisecond for on-the-air switching. This prevents loss of sync (for composite switching) during the switch from one source to another. This is known as "lap switching," "make-before-break switching," or "form-D switching."

3. *The vertical-interval switcher* that uses solid-state switching plates timed to switch video sources in an interval of a microsecond or two during the blanking time following vertical sync.

In most recent switching systems, the "composite" bank and the "noncomposite" bank are a thing of the past. By using modern mixing and special-effects equipment, it is no longer necessary to require noncomposite signals at the switcher inputs. Some systems accept either noncomposite or composite signals at the inputs; automatic sensing circuits add sync when the signal is noncomposite. Even nonsynchronous sources may be used with automatic circuits that prevent mixing of a signal with a nonsynchronous signal until the original signal is faded completely to black so that the "roll-over" occurs in black (no video signal).

The RCA TS-40 Switcher

The actual switching elements, known as *crosspoints,* are mounted in groups of six. The video input signals to these crosspoints are brought in through small fuses and horizontal copper buses that join all crosspoints to be connected to a given input. Fig. 7-3A illustrates the basic switching configuration from which all TS-40 systems are developed. The 7×1 switcher shown consists of a crosspoint group that can switch any one of six input signals to a common output bus, a latch circuit plate that controls the "switching-off" process in addition to accepting one additional video input signal, and a line-driving output amplifier that recovers gain lost in the resistance isolation of the switching crosspoint. Complex systems are made up of various combinations of these three basic units plus appropriate auxiliary equipment. Fig. 7-3B illustrates a 13×1 switcher, and Fig. 7-3C shows the basic configuration that constitutes a 7×2 switcher. Additional groups may be added to create a switcher with almost any de-

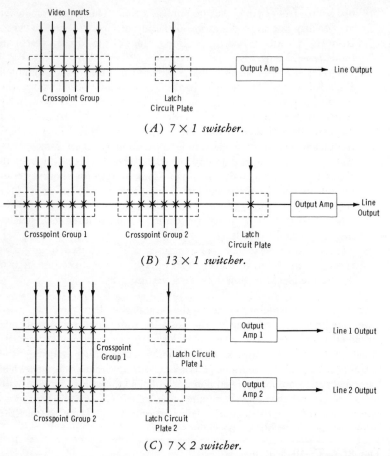

(A) 7 × 1 switcher.

(B) 13 × 1 switcher.

(C) 7 × 2 switcher.

Courtesy RCA

Fig. 7-3. Three ways of combining crosspoint groups.

gree of complexity. The number of groups that may be combined in the manner of Fig. 7-3B is restricted by output-bus capacity limitations. Up to four groups (twenty-five inputs) can be used to feed one output. (NOTE: These groups of six would be added along the *horizontal* lines of the figure.)

The input impedance of each crosspoint is 825 ohms; therefore, a maximum of eleven outputs could be obtained, by paralleling crosspoint inputs, without reducing the input impedance to less than 75 ohms. When fewer than eleven crosspoint inputs are bridged together, resistance is added in parallel with this common input to achieve a 75-ohm input terminating impedance. (For physical reasons involved in construction, it is not usual to bridge more than 10 crosspoint inputs together.)

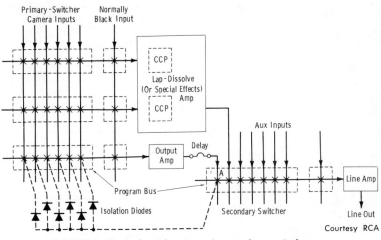

Fig. 7-4. Typical application of secondary switcher.

NOTE: Bridging of crosspoint inputs occurs along the *vertical* lines of the figure. Thus, in Figs. 7-3A and 7-3B, each source encounters only one crosspoint. In Fig. 7-3C, each source encounters two crosspoints. In Fig. 7-4, each source encounters three crosspoints.

Switching systems often involve one main, or primary, switcher section cascaded with a secondary switcher section. Fig. 7-4 shows a typical system in which the program bus incorporates a secondary switcher. This arrangement permits insertion of a delay equivalent to the delay of the lap-dissolve amplifier (or special-effects amplifier) so that the timing of all signals arriving at the output will be the same. This timing is critical when sync is added later, in order that front-porch timing will be uniform. In Fig. 7-4, when it is desired to select a primary input directly on the program bus, it is also necessary to operate crosspoint A on the secondary switcher. Rather than to use two buttons, requiring two separate operations, a single button is used to operate the crosspoint in the primary circuit. A trigger-pulse repeater (TPR) is used to transfer trigger pulses from the push button to crosspoint A. The input of the TPR is connected through isolating resistors to each of the buttons in the primary switcher row that feeds a secondary switcher. The TPR amplifies and clips the trigger pulse to standard shape before feeding it to crosspoint A.

The block diagram shown in Fig. 7-5 represents a color-television studio switching system (noncomposite) providing facilities comparable to those used in relatively large local stations. Note that the system provides both special effects and lap dissolves, and that it has both program and preview output buses. The diagram is greatly simplified in that it does not show the full number of camera inputs that normally would be required, nor all of the isolation amplifiers and other items that may be required in "trans-

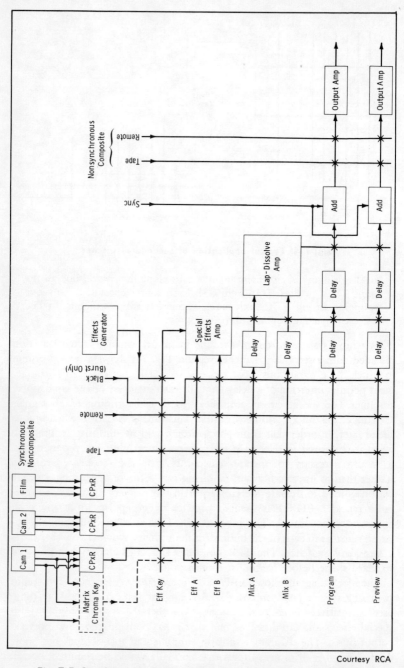

Fig. 7-5. Simplified block diagram of color-studio switching system.

porting" signals from point to point throughout the system. However, it does serve to illustrate some of the basic problems involved in color switching systems.

NOTE: The chroma key matrix and special effects in general are considered later in this chapter. Also note that the "black" input in this case consists of a color-burst signal only. This prevents loss of chroma in the receiver during a "fade to black" of a color signal. This loss would occur about midway through the fading operation, since the amplitude reduction of the color video signal during the fade reduces the attendant color-sync burst at the same time.

Also observe the delay lines incorporated to time the switching system properly. Mix A, mix B, program, and preview buses must be delayed between the crosspoint output and the secondary switching point where the special-effects output path is available. This delay is equivalent to that which occurs in the special effects processing. The program and preview buses must be additionally delayed to the equivalent delay of the lap-dissolve amplifier, as shown.

Fig. 7-6A illustrates the basic elements of a switcher control panel, and Figs. 7-6B, 7-6C, and 7-6D show the three most common types of switching, not including special effects, which are covered later in this chapter. Note that *split-fader* operation is shown in Fig. 7-6A. Normally, the fader handles are "locked" and operated simultaneously.

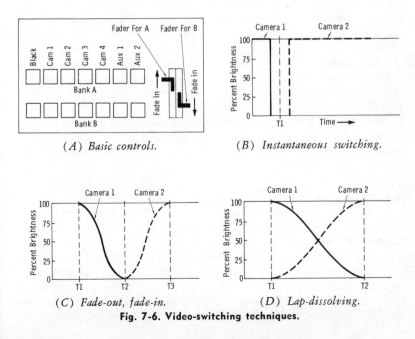

(A) Basic controls.

(B) Instantaneous switching.

(C) Fade-out, fade-in.

(D) Lap-dissolving.

Fig. 7-6. Video-switching techniques.

In Fig. 7-6B, camera 1 and camera 2 are switched instantaneously. In other words, the operator presses the camera-2 push button on the same bank of switches on which the camera-1 push button was depressed. Camera 1 is released immediately, and its signal is replaced with the signal from camera 2. Time interval T1 normally is made to "overlap" so that no signal disturbance occurs during the sync interval.

Fig. 7-6C shows the fade-out, fade-in operation. This can be done in either of two ways. Note from Fig. 7-6A that the fade-in position for push-button bank A is up, whereas the fade-in position for bank B is down. Thus, the usual operation is to have the levers locked together so that, assuming the on-air signal is on bank A, the black button on bank B can be depressed. The fader handles are then simply brought down from the up position so that a fade to black occurs. The alternate camera is then "punched up" on bank A, and the faders are operated back to the up position to fade in the new signal. If, for any reason, it is desirable to have another signal source punched up on bank B (so that the black position cannot be used), the faders can be unlocked and the bank-A fader (alone) operated to the down, or off, position (split-fader operation). In this case, fader A is down and fader B is up, so no signal is transmitted. Then the alternate source can be punched up on bank A and fader A returned to the up position.

In Fig. 7-6D, the camera-1 push button is depressed on one of the banks, and its fader is adjusted at reference brightness. The camera-2 push button on the other bank is also depressed, but its fader is at cutoff and only the camera-1 signal is transmitted up to time T1. At this time, the fader for camera 1 is adjusted toward cutoff, and simultaneously the fader for camera 2 is adjusted toward reference brightness. Thus, between T1 and T2 both signals appear on the screen while signal 1 is decreasing and signal 2 is increasing. At time T2, the camera-2 signal fully occupies the screen, and the camera-1 signal is completely faded out. Again, this is locked-lever operation, which is normal.

In the RCA TS-40 switching system, six crosspoints are mounted on an etched-circuit board to form a crosspoint group (Fig. 7-7). This assembly, fitted with a plug-in connector along one edge, has an output bus that joins all six crosspoints to form a six-input, single-output switching element. For convenience in planning and installing systems, the crosspoint group is supplied as the basic plug-in switching module.

Each crosspoint is functionally equivalent to a relay having two sets of contacts, one for a video signal and one for a tally circuit. It consists, in essence, of a semiconductor-diode switch that is turned on and off by a transistor flip-flop circuit. The circuit is bistable; that is, it will remain indefinitely in either the off or the on position until it is activated externally.

The special voltages required for the transistor circuits are supplied by the WP-40 power supply. In addition, standard 280-volt supplies are re-

Courtesy RCA

Fig. 7-7. Crosspoint frame (crosspoint groups in top row, latch-circuit plates in bottom row of frame).

quired for the amplifier complement, and a 24-volt supply is required to operate tally lamps and auxiliary relays. (The coils of the tally relays operated directly by the crosspoints are powered by the WP-40, but the circuits controlled by their contacts require external power.)

Coaxial fittings are provided at the rear of the WP-40 power supply for the composite-sync input and the output of the trigger-pulse generator incorporated in the supply. This generator consists of a transistor circuit for deriving pulses suitable for triggering TS-40 crosspoints (through the push-button switches on the control panel). The pulses are derived during the vertical-sync interval so that the switching action always occurs shortly after a vertical retrace period, thus minimizing the probability of a vertical roll-over when switching occurs between pictures of widely different duty cycles. The output pulses are at a level of about 30 volts peak-to-peak, and they are conducted by a coaxial cable from the trigger-pulse generator to the trigger circuit plates mounted under the control panel. The pulse rise time is deliberately made quite long to limit the high-frequency energy in the pulses. This permits them to be conducted along ordinary wires from the control panel to the crosspoints without inducing significant cross talk between leads.

In addition to push buttons and fader mechanisms, there are two types of etched-wiring circuit plates mounted beneath the control panels to serve important functions in TS-40 systems. The first of these is the trigger-circuit plate. This circuit is a single-input, six-output transistor amplifier that serves to distribute the trigger pulses generated in the power supply to as many as six rows of push buttons. (Additional trigger-circuit plates may be used for panels with more than six rows of buttons.) Each time a push button is pressed, it connects the corresponding crosspoint to the source of pulses derived from vertical sync. The first pulse that passes through activates the crosspoint, and the complete switching action occurs near the end of vertical blanking.

The second special control-panel component is the trigger-pulse repeater. This device may be used to make any crosspoint a "slave" of one or more other crosspoints so that the "slave" will always be activated when any one of the "masters" is in use. This feature is useful in switching systems that employ delay compensation to keep the total time delay through the system constant no matter which signal path is punched up. A simplified diagram illustrating the function of the trigger-pulse repeater in a system with delay compensation was shown in Fig. 7-4. The push buttons for the secondary switch (shown at the right in Fig. 7-4) may be mounted in the same row as the others in the program bus; this is done so that, functionally, the operator may treat them as part of the same switching bus. No button is required for the crosspoint operated by the trigger-pulse repeater. When any of the crosspoints to the left of the delay-compensation line (actually a length of coaxial cable) is operated, the trigger-pulse repeater produces a pulse to close the second crosspoint automatically. Thus, the circuit is completed through to the output.

The trigger-pulse repeater is actually an amplifier followed by a clipper. Isolating resistors are required at its input to prevent cross talk between the several push-button circuits, and these resistors cause a substantial reduction in the level of the trigger pulses applied to the repeater. However, its gain is sufficient to produce trigger pulses of normal amplitude at its output.

Another basic building block for the TS-40 systems is the latch-circuit plate. This circuit is mounted on an etched-circuit board, which plugs into a frame normally mounted below the crosspoint frame (Fig. 7-7). It performs the same function as the mechanical latching bar in a direct push-button switcher. That is, it trips off the circuit previously turned on each time a push button is operated, thus assuring that each output bus carries only one signal at a time. One latch-circuit plate is required for each independently latched output bus (consisting of up to four crosspoint groups) and is connected to the crosspoints through two buses, designated latch trigger (LT) and latch operate (LO). The latching operation is automatic, requiring no extra connections to the push-button control panel. Each time a crosspoint is actuated by its individual control button, it produces a low-level signal on the latch-trigger bus. This signal is amplified and clipped by the latch-circuit plate and is fed back along the latch-operate bus to all of the crosspoints connected to the same output. The amplified latch-operate signal triggers off whichever crosspoint was previously on. The entire sequence of operation is extremely fast, on the order of 1 microsecond.

The coaxial fittings used for input and output connections are of an unusual design, combining features of standard coaxial jacks and cartridge-fuse holders. The video signals are brought in and out through small fuses that protect the transistors and diodes from damage if excessive voltages are accidentally applied to the interconnecting cables. The fuses are of sufficiently low impedance that they do not degrade the performance of the system.

The interconnections between crosspoint groups and latch-circuit plates are shown in simplified form in Fig. 7-8. The latch-circuit plate includes a crosspoint circuit identical to all the others, but intended to carry the black signal. (In monochrome systems, black may consist of no signal at all, but in color systems it is desirable to provide a black signal containing the color-synchronizing burst, possibly supplemented by a fixed pedestal.) The black crosspoint is connected to the same buses as all the other crosspoints in the same output chain, but its control circuit is interconnected

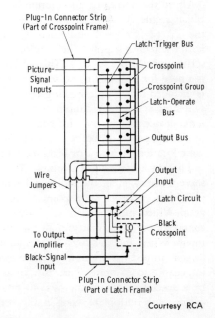

Fig. 7-8. Interconnections between crosspoint group and latch-circuit plate.

Plug-In Connector Strip
(Part of Crosspoint Frame)

Latch-Trigger Bus

Crosspoint

Picture-
Signal
Inputs

Crosspoint Group

Latch-Operate
Bus

Output Bus

Wire
Jumpers

Output
Input

Latch Circuit

LO
LT

Black
Crosspoint

To Output
Amplifier

Black-Signal
Input

Plug-In Connector Strip
(Part of Latch Frame)

with one of the power-supply buses in such a way that the switcher always comes up in a black condition when power is first applied. If it were not for this feature, the bistable crosspoint circuits might be activated in random fashion—some off and some on—when power is applied.

The output signal from a crosspoint is at a relatively low level, because each crosspoint handles a little less than one-tenth of the total signal current applied to the input of the switcher. This current division is necessary to permit up to ten crosspoints to be connected to each input. There also are minor losses involved in the cables required for delay equalization within the system. In order to restore the normal signal voltage level for system distribution, a coupling-circuit plate is used. The coupling circuit consists of a transistor amplifier with an input impedance of 75 ohms and an output impedance of about 1200 ohms.

In a TS-40 system, provision is made to mount the coupling-circuit plate within the output amplifier. This amplifier utilizes etched wiring and is identical to the TA-12 distribution amplifier, except that the coupling-circuit plate has been substituted for a conventional input coupling capacitor. The TA-12 amplifiers are also useful in other applications in which a unity-gain, single-input, single-output isolation amplifier is required. The same basic design is satisfactory for picture signals at a one-volt level, cw subcarrier at two volts peak-to-peak, and pulse signals at four volts peak-to-peak. (NOTE: In most recent TS-40 switchers, the TA-12 has been replaced with all-solid-state plug-in amplifiers, but the function of these amplifiers is the same.)

Up to ten output or distribution amplifiers may be mounted in the amplifier frame. This frame also serves as a housing for such other items in the TA-12 family of equipment as the sync or blanking adder and the heater and bias supply. The sync or blanking adder may be connected in series with the output of a distribution or output amplifier whenever there is a need for adding sync pulses to noncomposite signals. An interlock relay is included in the unit, making the adder suitable for use at the output of switching systems handling both composite and noncomposite signals. This same device may be used for adding a small amount of blanking for "fixed setup" operation. If both sync and blanking addition are required at the same location, two sync or blanking adders may be connected in series.

The block diagram of a simple TS-40 system shown in Fig. 7-9 should serve to tie together the individual functions already described. The actual switching elements (crosspoints) are shown near the center of the diagram. Video input signals are brought in through small fuses and horizontal copper buses that join all the crosspoints to be connected to a given input. When the number of crosspoints connected to a bus is less than eleven, impedance-trimming resistors are employed to adjust the input impedance to 75 ohms. The crosspoints are turned on by pulses generated by a special circuit in the WP-40 power supply and distributed by way of the push-button switches in the control panel. A trigger-circuit plate

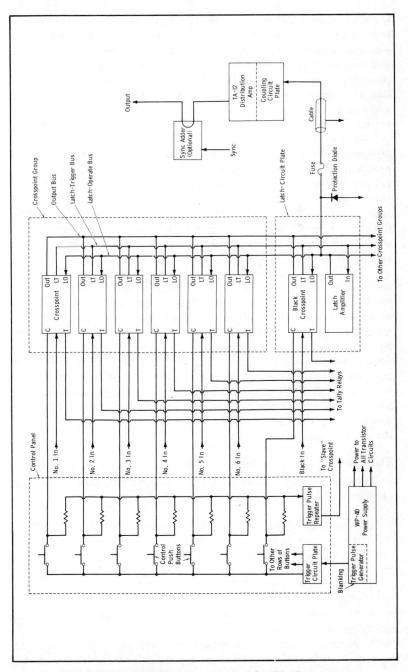

Fig. 7-9. Block diagram of a simple RCA TS-40 system.

mounted beneath the control panel provides a separate output for each independent row of push buttons. A "slave" crosspoint is required to close a secondary switch in systems where delay compensation is employed. In this case, an appropriate trigger pulse is developed by using a trigger-pulse repeater whose input is tied to a group of push-button circuits through isolating resistors. Each crosspoint has an independent tally output, which normally is used to operate a multiple-contact relay to control the several tally lamps and auxiliary circuits associated with each crosspoint.

All crosspoints associated with a given output are joined by three buses, designated output, latch trigger, and latch operate. Up to four crosspoint groups may be joined by these buses. The buses are also connected to a latch-circuit plate, which contains an additional crosspoint for the black-video signal, plus a latch amplifier. The function of the video-output bus is obvious (note that the connection to the output amplifier is brought out through a fuse and a short length of coaxial cable). A protection diode serves to carry the fuse-blowing current in one direction in the event that an excessive voltage is applied to the cable. The crosspoint itself will safely carry the fuse-blowing current in the other direction. The output amplifier is equivalent to a TA-12 distribution amplifier with a transistor coupling-circuit plate installed at its input. A TA-12 sync adder may be placed in series with the output amplifier if sync addition is required. The latch-trigger bus conveys a relatively low-level input signal to the latch amplifier; each time a crosspoint is turned on, it signals this fact by placing a small current on this latch-trigger bus. The latch amplifier then generates from the low-level trigger signal an output pulse of sufficient amplitude to serve as an off trigger for whatever crosspoint was previously on. The latch-operate pulse is conducted to all crosspoints simultaneously, but its action is overridden automatically in the case of the crosspoint to which the control signal is being applied.

The Cohu Switching System

The Cohu solid-state video switcher provides vertical-interval switching of video signals. A list of operational features follows:

1. Accepts either composite or noncomposite signals on any input bus without operator selection.
2. Automatic sync insertion.
3. Input over-voltage protection to ±300 volts.
4. Video input lines automatically decoupled when switcher is de-energized.
5. Control lines are minimized, and separate lines are not required for push-button tally.
6. Nonsynchronous inputs inhibit dissolves and automatically provide fade-to-black/cut/fade-from-black without additional manual operations.

7. Fade-to-black/cut/fade-to-new-source may also be performed by presetting a single mix control. This allows preselection of new source before start of operation.

8. Automatic burst detection and gating prevents mid-fade color drop when dissolving to monochrome.

9. Cut control standard between mix A and B buses, between effects A and B buses, and between preset and program buses.

10. Double re-entry system allows: dissolve to insert or effects, mix to wipe or insert, dissolve to mix, or insert to insert.

11. Auxiliary circuits allow air tally against preset and/or preview buses when desired.

12. Standard air tally circuits provide three sets of independent outputs:
    Set 1. Form-A closure to a common line.
    Set 2. Form-A closure to a common line.
    Set 3. Form-A contacts wired out in individual pairs.

Fig. 7-10 illustrates the control panel for the Cohu system that allows up to 21 inputs. The features of this particular system are as follows:

Inputs:
    21, composite or noncomposite on any input
Outputs:
    2 composite from program bus
    2 composite from preset bus
    2 composite from preview bus

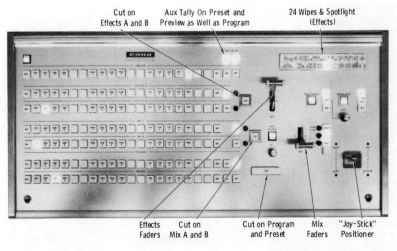

Courtesy Cohu Electronics, Inc.

Fig. 7-10. Cohu 9300 Series switcher control panel.

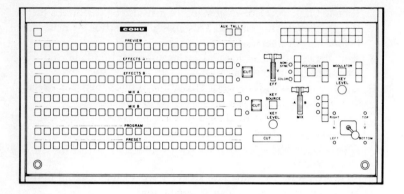

(A) Control panel.

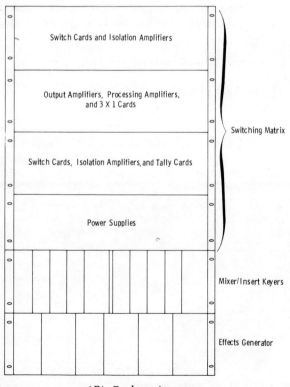

(B) Rack equipment.

Fig. 7-11. Physical arrangement of typical Cohu switching system.

1 each, spare, composite from mix A, mix B, effects A, and effects B
buses
Buses:
7—Mix A, mix B, effects A, effects B, program, preset, and preview

Fig. 7-11A shows details of the control panel illustrated in Fig. 7-10. Fig.
7-11B is a diagram that shows a typical rack-equipment arrangement for
the system.

Fig. 7-12 shows the signal-flow diagram of the Cohu Model 9302 system. This equipment differs from that of Fig. 7-10 only in that 14 inputs
are available rather than 21.

The switch card (Fig. 7-13) has seven identical solid-state switch points
with control and tally logic. Control logic is performed by an integrated-
circuit logic network that accepts signals from the control lines and activates one of the seven switch points. Each switch point is coupled to the
input line in such a way that a constant impedance is presented to the
input isolation amplifier and all other switch points. The integrated circuits of the tally logic provide an output when the associated switch point
is energized and the video is on an output bus.

Circuits of the video processing amplifier (Fig. 7-14) consist of: input
isolation, buffers, sync and clock, amplifiers, white detector, and output
amplifier. Video input signals are sensed, and if they are noncomposite,
sync is automatically inserted. If the input is switched to a nonsynchronous
source, the white detector inhibits the output until after the system has
stabilized; this action prevents noticeable picture roll. Switch timing during
the vertical interval is accomplished by clock pulses, which are derived
from the video-signal sync pulses. The clock-pulse generator and associated circuits provide positive switching and protect against switch contact bounce or random noise spikes.

Fig. 7-15 shows the 3×1 switch/amplifier card. Each card contains
three switch points, control and tally logic, and amplifier circuits. Two of
the switch points, which select delayed or nondelayed signal paths, are
activated automatically by signals from the logic card. The third switch
point provides the double re-entry function and is activated whenever the
mix or effects re-entry push button is pressed. Each input has solid-state
switch points and an isolation amplifier with an FET front end. Integrated
circuits of the control and tally logic provide an output when the associated
switch point is activated.

The integrated circuits of the logic card (Fig. 7-16) sense the control
signals and activate the appropriate automatic switch point. Logic circuits
consist of two functional groups of logic. The first group senses re-entry
control signals and switches the associated crosspoints on the 3×1 switch/
amplifier card. The second group senses the control signals for the program, preset, and preview buses and activates the associated automatic
switch point for the output buses.

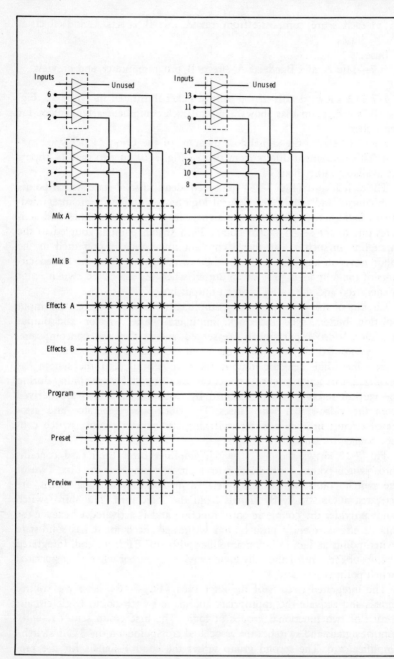

Fig. 7-12. Signal paths in

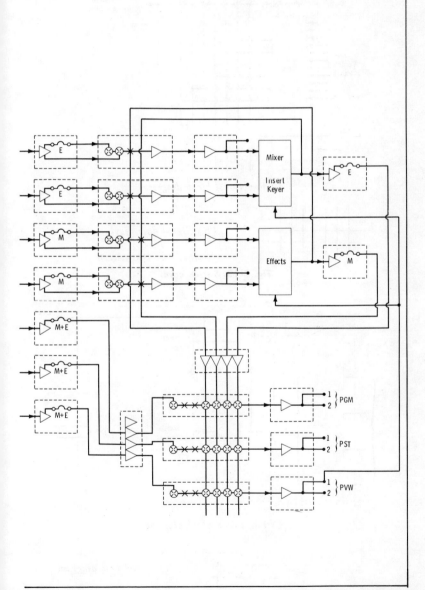

Cohu Model 9302 switcher.

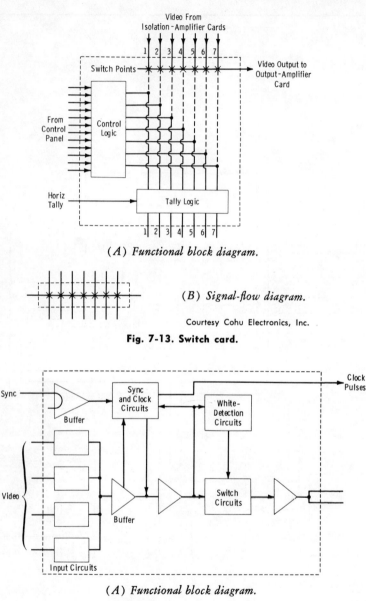

(A) Functional block diagram.

(B) Signal-flow diagram.

Courtesy Cohu Electronics, Inc.

Fig. 7-13. Switch card.

(A) Functional block diagram.

(B) Signal-flow diagram.

Courtesy Cohu Electronics, Inc.

Fig. 7-14. Video-processing-amplifier card.

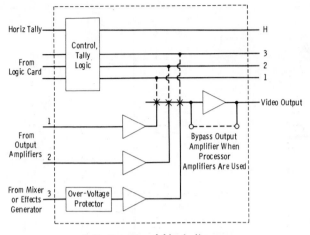

(A) Functional block diagram.

(B) Signal-flow diagram.

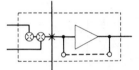

Courtesy Cohu Electronics, Inc.

Fig. 7-15. 3 × 1 switch/amplifier card.

7-3. FUNDAMENTALS OF ELECTRONIC SPECIAL EFFECTS

"Special effects" as applied to television is a broad and varied field. It may be recognized by the reader that some special effects have been mentioned already: lap-dissolving, superimposing of two separate pictures, etc. In the means thus far employed, however, it was necessary to blend whole pictures together. The special effects within the scope of a TV program include not only these methods, but also means of using parts of more than one picture, with a variety of shapes of the boundary between the areas that are combined. Indeed, the scope of special effects is limited only by the imagination and adaptability of the production and technical depart-

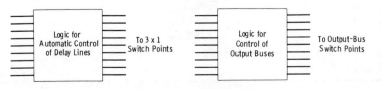

Fig. 7-16. Logic cards.

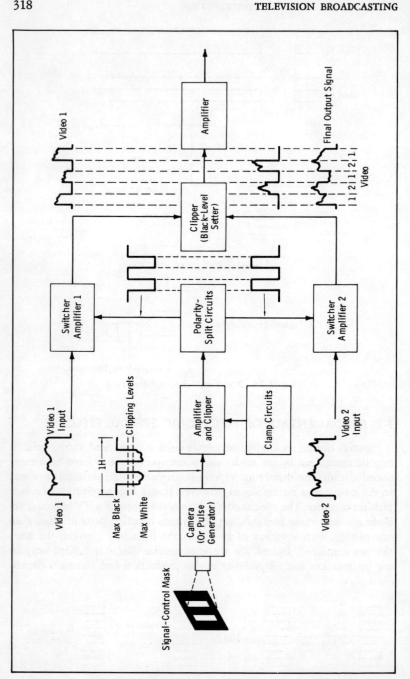

Fig. 7-17. Simplified block diagram of video special-effects amplifier.

ments. Electronic special effects are described in this section. Section 7-4 will cover nonelectronic special-effects techniques.

The action of lap-dissolving, or the gradual fading out of one picture while another is faded in simultaneously, is so common that it is now considered a part of normal operational technique. Superimposing is done less often, as it should be, and is classed as a special effect although it is the most common effect produced by the mixer operator. Since entire pictures are blended together, backgrounds must be watched carefully to avoid undesired effects in the final picture. For example, a dark object that should predominate would be lost to view if the "mixed-in" picture provided a dark background at that section of the raster.

Equipment especially designed to achieve special effects in telecasting usually provides means for blanking out one or more areas of a single picture, inserting another signal into these areas, and changing the separating boundary in any desired shape. Fig. 7-17 will serve as an introduction to special-effects electronics. Note first that the effects "generator" or "control" signal may be either a camera (external key) or a synchronous pulse generator (internal, or self-, key). The latter facilitates changing of shapes by allowing adjustment of the widths and frequencies of the horizontal- and vertical-rate pulses.

The system in Fig. 7-17 functions from the action of a keying signal that is dependent on the shape of an optical mask (or the output of a pulse generator). This signal is used to operate an electrical system that switches from one video signal to another. For convenience in discussing this method, we will assume a mask (or corresponding pulse shapes) that consists of a black background with two vertical white stripes. It will be shown how the video signal selected depends on whether a black or white area is being scanned.

A single-line interval of time may now be traced; note that the video signal from the masking scanner consists of a black line (black positive polarity), a white peak, a black line, another white peak, and another black line. This corresponds to the single-line variation of the scanning signal as it sweeps across the mask. To eliminate spurious noise, the signal is clipped in the following amplifier to produce squared-off tips for proper control. This clipping level is variable.

The resulting clipped signal from the amplifier-clipper is fed to a stage that yields two keying outputs of opposite polarity. One keying signal is fed to video switcher 1, and the other keying signal goes to video switcher 2. These switchers might be connected to separate studio cameras viewing different objects. The switching (keying) signals are clamped during the blanking intervals, as are the video signals. In operation, the keying signals cut off one signal while the other is passed, and vice versa. For example, if video 1 is passed during scanning of the black portion of the mask, video 2 is cut off. Then when white is being scanned, video 1 is cut off and video 2 is passed.

The two composite signals are combined in a clipper stage that sets the black level of the picture. The signal is then amplified to the necessary transmission level and is set at the proper polarity for feeding the studio system.

The simplest special effect is the *horizontal wipe* described by Fig. 7-18. This uses only horizontal-rate keying pulses to key the signal sources alternately. If the faders for the special-effects bus are placed in the center of travel, the pulses are equal in duration. Thus, the pictures are equally split on the monitor as in Fig. 7-18A. As the levers are operated toward the "up" or "down" positions, the relative durations change as in Figs. 7-18B and 7-18C. At the extreme limits of the faders, only one picture occupies

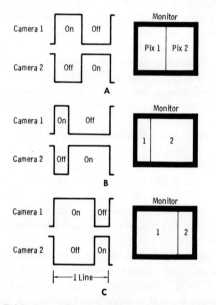

Fig. 7-18. Principle of the horizontal-wipe effect.

the entire screen. In other words, a picture can be "wiped in horizontally" by this technique.

Obviously, a vertical wipe can be made by a similar technique using vertical-rate pulses. If both horizontal-rate and vertical-rate keying pulses are combined, the new picture can be made (for example) to appear first along the top or bottom and one side while the old picture is wiped toward the opposite corner. In common with any wipe effect, this effect can be stopped at any position to provide an inset picture.

The effects control panel of the RCA TE-60B system is shown in Fig. 7-19. (The "joy-stick" positioner at the right is an optional accessory.) There are a number of functions handled by the control panel in addition

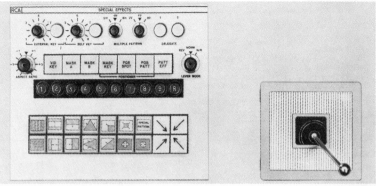

Courtesy RCA

Fig. 7-19. RCA TE-60B effects control panel.

to the obvious ones of pattern and mode selection. The circuitry for pattern clips limit sets, normal/reverse control, self- and external key, video-clip control, aspect-ratio control, and multiple-pattern switching are included in this unit.

A group of 14 buttons (Fig. 7-19) is used for pattern selection, and a group of four corner buttons is used for pattern direction orientation. Each of these button groups is mechanically interlocked. One of the 14-button group, designated SPECIAL PATTERN, activates a group of nine numbered push-button switches. These switches feed the eight pattern-select control wires that run to the waveform-generating and pattern-selection modules and also affect the lever mode operation. This switching arrangement determines the relay closure combinations that perform the actual selection of the pattern waveforms. These switches and their associated control wires are arranged in five groups according to the following functions:

1. Horizontal-waveform selection
2. Vertical-waveform selection
3. Determination of whether the selected waveforms are used individually (horizontal or vertical) or mixed, and the mixing mode
4. Determination of the wipe direction by inverting the pattern selection
5. Mechanical clearing of the special-pattern selection

The pattern-select buttons are arranged to feed the pattern-select control lines through isolation diodes. One of the pattern-orientation switches is always functional on patterns that originate from a corner or side. The polarity control wires of the pattern-select control-wire group are switched to control the polarity of the selected horizontal and/or vertical waveforms. This effectively determines the corner or side of the raster from which the wipe action will take place. Some of the patterns possible are shown in Fig. 7-20.

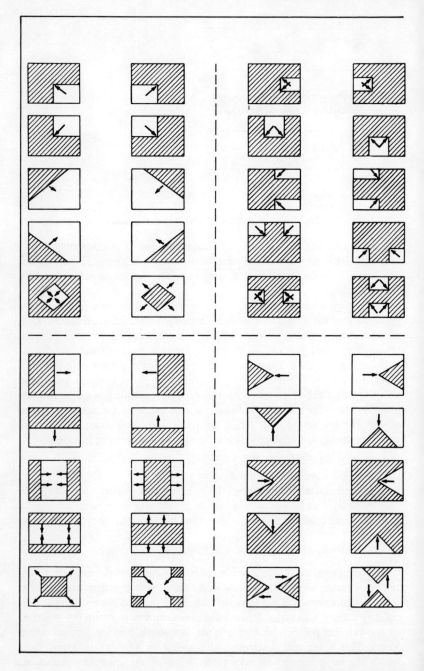

Fig. 7-20. Special effects

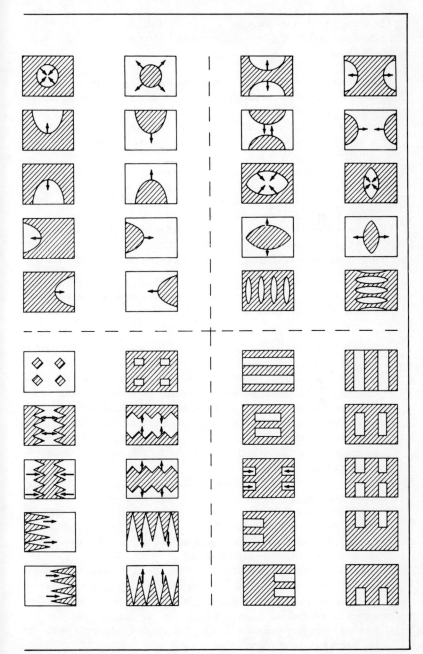

obtainable with TE-60B system.

The mode switches on the effects control panel (Fig. 7-19) are an extension of the mode switches on the control lever assembly. They function as follows:

1. *Pattern-Effects Mode* (PATT EFF): In this mode, the transition from one video channel to the other can be accomplished with various patterns, through use of the control-lever assembly. Some of the more popular patterns are available at single switch positions on the pattern-selector switch. All patterns are available through the special-pattern selection switches.

   The buttons used to select patterns have arrows that indicate direction of pattern travel for the normal setting of the LEVER MODE switch. The corner switch indicating travel from the lower right corner should also be pushed.

2. *Position-Pattern Mode* (POS PATT): This mode allows the pattern used for transition to be positioned anywhere in the raster. Positioning of the pattern is accomplished by moving the positioner control.

   > NOTE: When certain patterns are positioned off center, an illusion of lost identity may occur, and some confusion may result if care is not exercised in pattern selection when this mode is used.

3. *Position-Spotlight Mode* (POS SPOT): This mode is used to give a spotlight illusion in a given scene. For the most advantageous effect, the same video input should be used on both input channels and a circular pattern should be selected. Before the switch is made to the position-spotlight mode, the LEVER MODE switch should be in the normal position, and the fader lever should be on the A limit (away from the operator). This allows the last-on memory to be set for a spotlight effect that will close from the outside in with the appropriate channel spotlighted.

   The relative intensities of the spotlighted area and the surrounding area of the picture are determined by T-pad attenuators, located on the TE-60B connector-plate assembly, which provide 6-dB attenuation in the video B and video D channels. (The "D channel" is the term applied to the B channel in a second mixer when two mixers are used in an effects transfer system.)

4. *Video-Key Mode* (VIDEO KEY): In the video-key mode, it is necessary to select either self-key or external key. The corresponding switches are illuminated at half brightness until one of the video-key modes is selected. This allows selection of the desired video key before selecting the video-key mode. The control associated with the key lamp sets the clip level.

5. *Mask-Key Mode* (MASK KEY): This mode is a video-keying mode that allows the operator to mask the video key with an available

pattern. It is necessary to have the fader levers set on B limit and the LEVER MODE switch at normal when switching to the mask-key mode so that patterns close in the direction shown on the buttons. (The fader lever controls the pattern size in this mode.)

6. *Mask-B Mode* (MASK B): In this mode, only channel-B video is keyed by the special effects; channel-A video is at full level. This mode is used for wiping in lettering or other white outlines.

7. *Mask-A Mode* (MASK A): This mode functions in the same manner as the mask-B mode, except that channel A is keyed and channel B is at full level.

The LEVER MODE switch selects the direction that keying patterns will move:

1. In the normal mode, the pattern always moves in the direction indicated on the pattern-selection buttons, provided the corner selector indicates travel from the lower right. (The corner selector consists of the four switches with arrows showing on them.)

2. In the reverse mode, the patterns always move in the direction opposite to that shown on the pattern-selector buttons.

3. In normal/reverse, the pattern moves in the normal direction from the A limit and in the reverse direction from the B limit.

The ASPECT RATIO control varies the vertical- and horizontal-waveform voltages so that the pattern shape can be varied. Normal setting for this control is at the midpoint of its travel.

There are two switches to control the frequency of the multiple patterns. One selects two, four, or six times the horizontal rate; the other selects two, four, or six times the vertical rate. In the 2V and 2H positions, the two switches do not have any control over the multiple-triangle module that produces triangles at twice vertical and horizontal rate. However, the 4V, 6V, 4H, and 6H positions determine the patterns of relay closure necessary to obtain corresponding outputs.

NOTE: The practical operations of switching and special effects are covered in Chapter 10.

It is important to bear in mind the two basic types of video-signal mixing. (1) *Lap-dissolve* is the normal mixing procedure of gradually reducing the amplitude of one signal while simultaneously increasing the amplitude of the other signal. Thus, a "transparency" occurs in which we momentarily "see through" one picture to the other picture. (2) In nonadditive mixing (Section 5-6, Chapter 5), the signal with the highest instantaneous amplitude at a given point of the resultant picture is the only signal passed. Thus, for example, white lettering can be inserted over lower-luminance signals without "show-through" of the scene in the white lettering. This is the simplest form of video insert.

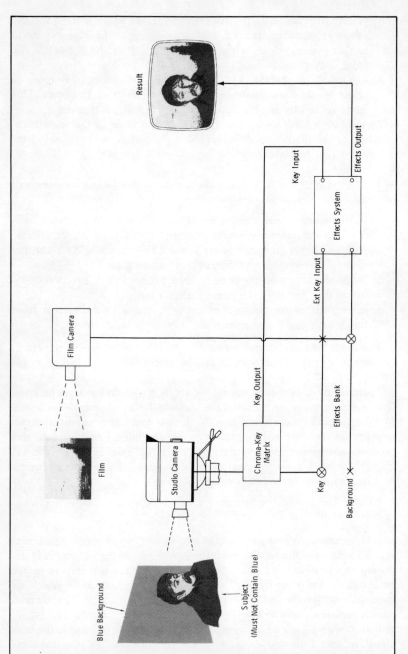

Fig. 7-21. Principle of chroma key.

Thus, a video insert can be either the NAM or the keyed signal (self-key or external key). In addition, the external key can be derived from either brightness (works on luminance differences) or chroma (works on specific hue and saturation). The chroma key is more satisfactory for color operation than a key that works only by luminance differences. In chroma key, the red, green, and blue channels of a specific camera are combined to form a single keying signal. The unit separates a selected saturated color from all other colors.

Fig. 7-21 illustrates a simple example of the use of chroma key. Although not mandatory, a saturated blue flat normally is used as the background for the keyed subject. This provides better control since blue is mainly complementary to flesh tones (opposite in phase to colors making up flesh tones). Thus in this example, the subject must not contain any blue, since the chroma key will create a "hole" in the final picture wherever blue occurs in the studio-camera signal.

In the case of Fig. 7-21, a film camera with a seashore scene is punched up on the background bank of the special-effects buttons. Note that this bank could be either the Effects A or Effects B bank of Fig. 7-5. The resultant picture shows the studio subject inserted into the filmed seashore scene. Since the key is formed strictly from the studio camera, the background scene from the film camera can contain all colors.

NOTE: Operational techniques for chroma key are expanded in Chapter 10, which covers the operation of studio equipment.

An invaluable assist to electronic keying, or "matting," is given by horizontal or vertical *crawl* drums such as the one shown in Fig. 7-22A. Copy is prepared on a special typewriter (Fig. 7-22B). Either horizontal crawl or vertical crawl can be obtained by proper positioning of the drum and proper orientation of the copy. In Fig. 7-22A, horizontal-crawl copy is shown in position on the drum. The drum axis would be rotated 90° for vertical crawl.

The two-foot-diameter drum is made of white translucent plastic. The drum, which rotates on ball bearings, is friction driven by a constant-speed motor that imparts a steady motion to the drum. The housing is constructed of extruded aluminum angle. Aluminum panels, mounted on both sides, support the tubular-steel drum shaft, the drive motor, and a component panel box. Two sets of slide tracks and a mask track are mounted on the front. There are rubber feet for either horizontal or vertical positioning of the unit.

A remote-control box has a neon lamp that indicates that the master switch is on; a motor-speed control; and a three-position switch for forward, off, or reverse motor operation. A push button serves to turn on the lamps and to interrupt the light when desired to obtain emphasis effects. A dimming control adjusts brightness.

(A) Crawl drum.

(B) Typewriter.

Courtesy Telesync Corp.

Fig. 7-22. Equipment for producing crawls.

Black-on-white, white-on-black, black-on-color, or color-on-black presentations are possible depending on copy preparation, which can be outlined briefly as follows:

1. *Positive:* A quick-release carbon is deposited onto a transparent film. The transparent film, now carrying black-on-white copy, is taped to the drum.
2. *Negative:* The carbon is deposited onto a receptive paper. The film ribbon, now carrying white-on-black copy, is taped to the drum.
3. *Color:* To color either positive background or negative copy, a transparent color sheet is selected, taped to clear plastic, and placed in the slide track.

Artwork may be produced by placement of transparent film over an outline map or other artwork to be copied. Tracing is done using India ink with a fine-point pen. Areas may be colored as desired with transparent-ink marking pens.

7-4. NONELECTRONIC SPECIAL EFFECTS

Special effects covered in the previous section involved electronic techniques. We will now consider basic nonelectronic methods.

Special effects include back-projection of slides to provide a background for small objects or entire stage setups, as illustrated in Fig. 7-23. Outdoor scenes may be simulated conveniently by this method. A translucent screen is used for the purpose, and sufficient projection light must be used to prevent the normal key lighting in the studio from "washing out" the projected background. By proper control of front lighting and the use of the ultrasensitive Type 5820 image-orthicon pickup tube, satisfactory results may be obtained.

Also popular in larger stations, such as the key network studios, is the use of back-screen motion-picture projection. As might be expected, one

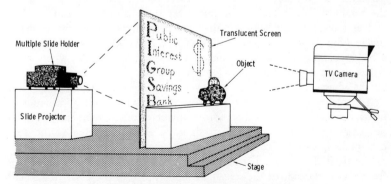

Fig. 7-23. Method of back-screen projection.

of the primary problems is properly synchronizing the projected pictures with the TV camera. At first thought, it might appear that a regular 2-3-2-3 television projector throwing the picture on the screen would be usable. While it is true that this technique is compatible with the studio camera, it has been found in practice that the available illumination on the screen is far below allowable limits. The projector for back-screen projection must be about 18 feet (or more) behind a $9' \times 12'$ screen to fill the picture area, as compared to about 40 inches when the projector is used directly with a TV camera.

This problem can be solved by using a rapid-pull-down 2-3-2-3 type of intermittent mechanism, both a high-intensity carbon arc and an incandescent projection lamp, and a special type of shutter. The major difference, therefore, between a back-screen projector for studio backgrounds and the regular TV projector for the film camera is in the type of projection lens used and the intensity of the projection light source.

Another problem with rear-screen projection is a "hot-spot" near the center of the screen and reduced light intensity at the sides. This problem can be minimized through placement of the projector as far from the screen as possible.

More satisfactory for the average studio is the recently developed "retroreflective front-screen" projection system. Retroreflection means reflection directly back. To produce this effect, millions of hollow spherical glass beads are carefully imbedded in reflective backing material. A ray of light entering a reflective-backed sphere exits *along the same axis*. Since the retroreflective screen returns about 90 percent of the light to the source, the system can be considered 50 times brighter than a 10,000-watt rear-screen setup.

When a 35-mm slide is projected onto a retroreflective screen, the image returns along the same axis (converges to the light source), and it is almost as bright as when it exited the projection lens. The camera does not respond to portions of the image that fall on the performer because the person is not retroreflective and the camera tube "sets up" according to the brightest light, which is the image returned from the screen.

The camera taking lens is either coaxially aligned with the projected image (Fig. 7-24A) or just off-axis (Fig. 7-24B), close enough to obtain a bright picture of the screen image. If you are in a studio where retroreflective front projection is in use, you cannot observe the actual image on the screen unless you are on or very near the camera taking-lens axis.

Advantages of this system over rear projection include elimination of hot spots and corner fall-off, more brilliant colors, sharper pictures, and no wash-out from normal studio lighting. A disadvantage is the fixed camera setup of Fig. 7-24A with the performer at the center of the screen, or the limited camera movement allowable in Fig. 7-24B where the performer is at the side of the screen. A motion-picture projector in a soundproof housing can be used with this system.

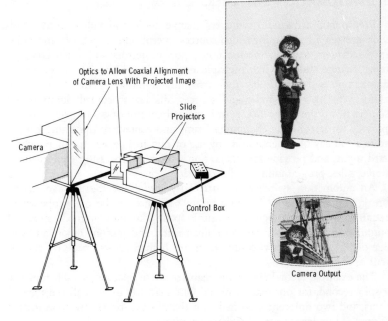

(A) On axis.

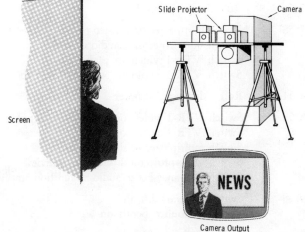

(B) Off axis.

Courtesy Telesync Corp.

Fig. 7-24. Use of front-screen projection.

7-5. AUTOMATION SWITCHERS

Preprogrammed studio switchers change audio and video sources, provide preview of upcoming video sources, preroll film projectors and video tape recorders, start audio tape units, stop projectors and video tape systems, change slides, switch multiplexers, etc. In addition, some expanded automation systems display the remaining duration of the on-air event along with its video and audio sources; display scheduled duration and sources of upcoming events; permit manual control of on-air-event timing; provide for last-minute changing of time and content of upcoming events; store a full day of programming by use of punched-paper-tape or punched-card input; and provide for integration with data-processing equipment in traffic, sales, programming, and accounting departments.

An automation switcher is particularly useful for the average station break, commonly termed "panic period" because of the many operations required over a relatively short time span. The programmed nature of automation switchers eliminates human error, and stations report an average of 95 percent of switching errors are human errors, made by competent operators.

The example below[1] shows the many separate operations, each timed to a split second, for one station break. One man has to give all the instructions, and two different men make all the moves—three-way coordination is required under extremely difficult conditions.

Time	*Event*
03:57:54	Roll film projector 1
03:57:58	*Switch film-projector-1 video on air (60-sec film) *Switch film-projector-1 audio on air Switch VTR-1 video to preview Switch VTR-1 audio to audition amplifier
03:58:51	Roll video tape recorder 1
03:58:58	*Switch VTR-1 video on air (50-sec tape) *Switch VTR-1 audio on air Switch multiplexer to slide projector 1 Switch slide-projector-1 video to preview Switch announce-booth audio to audition amplifier
03:59:48	*Switch slide 85 on air *Switch announce booth on air
03:59:53	*Change to slide 5LS

---

[1]The station-break chart and related text are presented courtesy of Visual Electronics, Inc.

03:59:58   *Change to ID slide
                  Switch live studio-A video to preview
                  Switch live studio-A audio to audition amplifier

04:00:00   *Switch live studio-A video on air
                  *Switch live studio-A audio on air
                  Switch live studio-B video to preview
                  Switch live studio-B audio to audition amplifier

      *On-air operation

Now we will follow the action of an automation switcher in handling the above sequence. This particular break is timed to start at 03:57:58 by the true-time clock in order to finish at exactly 04:00:00. It includes a 60-second film, a 50-second video tape, and three short-duration slides with sound from the announce booth. At 03:57:49, or nine seconds ahead of time, the duration timer in the on-air panel automatically starts in order to alert all personnel to the upcoming initiation of the station break and to preroll projectors or video tape recorders. It counts down at each second—09, 08, 07, etc. At 04 seconds, film projector 1, coming up next, automatically rolls, without a need for personnel to schedule a separate preroll event. The leader appears on the preview monitor. When 00 seconds is reached, the projector, now stabilized, goes on the air. The on-air panel indicates the duration is 60 seconds, the video source is P1, and the audio source is 0 (for audio-follow-video). The count-down of the allotted 60-second duration—59,58, etc.—starts.

Also, as P1 shows on-air, each event in the storage display shifts up one line, and the vacated bottom line is filled by the next event in sequence, which is automatically entered by the prepunched-card reader. This places 50 seconds of video tape from tape recorder 1, audio-follow-video, in the top line; video from this tape recorder is switched to the preview monitor.

When the film duration has counted down to seven seconds remaining, video tape recorder 1 (the next upcoming event) is automatically rolled so that it will have time to stabilize. Its leader shows on the preview monitor.

When the film duration reaches 00 seconds, the video tape recorder is switched on the air, and this source is indicated, with its duration counting down from 50 seconds, by the on-air panel. VTR video is now on the line monitor. As the indication for upcoming slide projector 1 shifts into the top line of the storage display, the multiplexer is switched to the slide projector, and the slide shows on the preview monitor. When the video-tape duration reaches 00 seconds, the slide machine is switched on the air, and the audio is switched to the announce booth. The slides are changed after intervals of 5, 5, and 2 seconds.

New integrated circuits and memory devices have made possible a device known as a television *character generator*. The purpose of this device is to create and display on a television screen alphanumeric characters arranged as titles or news messages. It must include a memory, since it must repeat its display of these characters every sixtieth of a second. The memory lends itself to storing television-program information, and since the basic function of the character generator is displaying information on a monitor, this device lends itself also to displaying upcoming-program information on the operator's monitor (and any other monitor in a television station).

All the information stored in the character generator and used by the switcher is in the coded data form used by teleprinter machines. Thus, the system also lends itself to providing automatic print-out of the station log, including codes to denote discrepancies and variations from the original schedule, as each event goes on the air.

The Visual V7000 digital TV program control system (Fig. 7-25) incorporates the character generator, card reader, video/audio switcher, and teleprinter into a complete system. The storage-display portion of the system provides on a standard TV monitor a display of 12 events plus column headings (Fig. 7-26). This display can be routed to any standard monitors in the station (e.g., projection room, announce booth, switchboard, etc.) to provide constant, up-to-date information concerning on-air timing and upcoming program content. Since last-minute schedule changes are made with a keyboard into this display, these changes are distributed instantly throughout the station without paperwork.

The operator can load the storage display with a keyboard similar to a conventional typewriter keyboard. He may load only one station break of (typically) four events, or he may load the complete display. He may add,

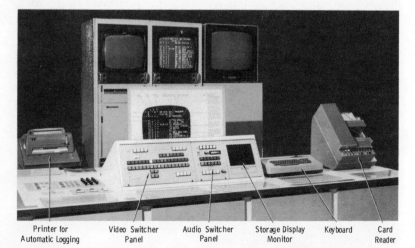

| Printer for Automatic Logging | Video Switcher Panel | Audio Switcher Panel | Storage Display Monitor | Keyboard | Card Reader |

Courtesy Visual Electronics Corporation

Fig. 7-25. Visual V7000 automation system.

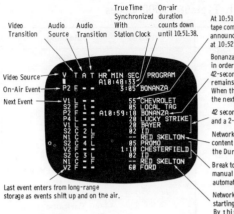

Fig. 7-26. Typical events display for Visual V7000 system.

delete, or change events in the middle of the display at any time while the system is not actually putting events on the air. New events are added at the bottom of the display, and all events move up the screen as the on-air event in the top row ends and goes out of the display.

The format of the display can be changed easily after installation to accommodate different coding, columns, sources, or transitions. The format is stored in a second memory used only for programming the main character memory. This format memory can be addressed by the keyboard, and format changes can be made using the storage display as a readout. Typical codes used in the V7000 system are shown in Table 7-1.

A block diagram of the control system is shown in Fig. 7-27. The storage-display equipment is the heart of the system and consists of three units: the Visual 990 display control unit, Visual 995 keyboard, and Visual SCU switcher control unit.

The display control unit (DCU) can generate any one of 64 characters and locate it in any one of the 512 locations in a TV display of 16 rows of 32 characters or spaces. Information from the keyboard "tells" the DCU which character is to appear in each of the 512 locations. The DCU stores this information in its memory, scans through it 60 times per second, and generates the display as a standard TV signal.

The switcher control unit (SCU) also interrogates the DCU memory and receives data concerning characters and their locations, compares the data with information in the SCU memory, and then operates the appropriate relays in the video/audio switcher. The switcher control unit also inserts data into the DCU memory to perform the countdown function, the upshift function as events go on the air, and the forward count of the true-time indication.

Table 7-1. Typical V7000 Codes

Code	Meaning		Code	Meaning
	Video Sources			**Video Transitions**
P1	16-mm Projector 1		C	Cut
P2	16-mm Projector 2		L	Lap
P3	16-mm Projector 3		F	Fade
P4	16-mm Projector 4		S	Super
S1	Slide Projector 1			
S2	Slide Projector 2			**Audio Sources**
V1	VTR 1		=	Audio Follow Video
V2	VTR 2		1	Announce 1
V3	VTR 3		2	Announce 2
V4	VTR 4 (future)		3	Tape Cartridge 1
			4	Tape Cartridge 2
N	Network		5	Tape Cartridge 3
R1	Remote Line 1			
R2	Remote Line 2			**Audio Transitions**
X1	Spare 1		=	Audio Trans Follow
X2	Spare 2			Video
C1	Camera 1		C	Cut
C2	Camera 2		L	Lap
C3	Camera 3		F	Fade
BC	Black, color		S	Super
BM	Black, monochrome			

The video switcher actually performs much of the automatic operation and is an integral part of this system. The storage-display next-event row operates the video-switcher next-event row after each on-air switch and operates the take button of the switcher when the time to go has reached zero. Also, the video operator can manually operate the single take button of the switcher to take the next event in normal preset-switcher fashion. He is assisted in this regard by the on-air duration countdown, but he is free to take the next event before or after the countdown reaches zero, if the duration inserted in the programmed display was incorrect or he wishes to change the timing for any reason. To make a switch later than the automatic countdown, he pushes the hold button to stop the countdown.

If the next event is a VTR machine or projector, instead of operating the take button the automatic control will operate the single preroll button seven seconds (adjustable) before the end of the on-air event. The countdown display will show seven seconds at this time (although the operator can roll early or late if he wishes). If the next event is a VTR machine, this machine is rolled immediately and is taken on-air automatically seven seconds later. If the next event is a projector, the switcher automatically waits three seconds, rolls the projector, and four seconds later takes the

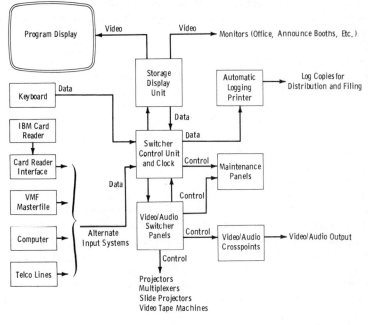

Fig. 7-27. Block diagram of Visual V7000 system.

projector film chain on the air. If the next event on the switcher is a short slide (less than seven seconds), the system will search the next three events in the storage display for either VTR machines or projectors; it will roll these sources and take them on the air at the proper time, after taking the short slides.

If a switch is scheduled to occur at a given true (clock) time in order, for example, to join the network exactly on the hour, this true time is entered in the duration column of the display. When such an event goes on the air, its true time remains fixed indicating when the event will end. Immediately above this time in the display is the clock-operated actual time. The automatic control system compares the actual time with the designated next true-time switch, and seven seconds ahead of time activates the preroll button if required, but if not, continues until coincidence and operates the take button. The slide-change button is operated automatically at each slide event to change the slide projector that is on the air at that particular time. When a slide projector leaves the air, it will be advanced automatically so as to be ready for its next showing.

It is practical to operate this system as a simple preset switcher in which the operator loads up to 12 events at a time into the storage display by means of the keyboard. Stations that carry network programming often have only four or five events per break with long periods of time between breaks. Thus, at a convenient time, an operator can load several breaks covering an hour or more, since the loading process takes only a few minutes.

Some stations, however, prefer to have a full day, a week, or even 30 days of program-storage capacity that automatically loads the storage display as an event goes on the air. Such storage devices also can be computer-loaded, up-dated, and operated with "hard-copy" printers to produce the daily program log. Many stations use IBM cards throughout their operations. In a typical system, for each sales contract a master card is punched with all contract information. Separate cards, which do not include billing information, are also prepared for program titles, unbillable ID announcements, public-service announcements, etc.

To print the program log, appropriate cards for the day are automatically sorted from these files to produce a deck arranged in time sequence. This deck may be sent to the traffic department for editing, rearranging, and having additions made; or a preliminary schedule may be printed in the accounting and printing machine, sent to the traffic department for pencil marking, and returned to be used as a guide for preparing the final log. The final program log is then printed in about ten minutes, the complete process requiring only a fraction of the many hours needed when older methods of preparing a schedule are used.

The same card deck from which the log has been printed is sent to the control room for insertion in the card reader associated with the automatic control system. The card reader reads the deck and feeds information automatically into the storage display as events go on the air. The operator can check the cards before air time by running them through the card reader and observing the storage display during inactive program time.

The use of IBM cards can be tied to other station operations such as billing, availability print-out, invoicing, sales reports, etc. There are many possible arrangements, sometimes including a computer, each tailored to the requirements of the particular station.

Any device that can simplify and perhaps eventually eliminate the requirement for manual logging of the time and program content of every event throughout the broadcast day is an important improvement in operating efficiency. The automatic-logging portion of the Visual system prints out program-log information on a Teletype Model 35 heavy-duty printer. A sample automatic log print-out is shown in Fig. 7-28. On the left is the time of each video or audio switch, provided from the same clock that displays the true time in the storage display monitor. To the right of this time is indicated the on-air event, including the description of the program content, as described in the storage display monitor. Film, slide, or tape numbers can be used instead of word descriptions.

The video-switcher portion of the system can take many forms depending on station requirements. It may be large or small and may be

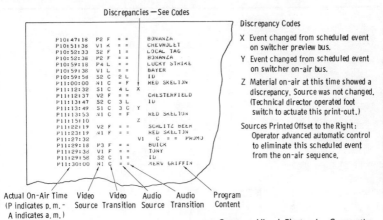

Discrepancies — See Codes

Discrepancy Codes

X Event changed from scheduled event
on switcher preview bus.

Y Event changed from scheduled event
on switcher on-air bus.

Z Material on-air at this time showed a
discrepancy. Source was not changed.
(Technical director operated foot
switch to actuate this print-out.)

Sources Printed Offset to the Right:
Operator advanced automatic control
to eliminate this scheduled event
from the on-air sequence.

Actual On-Air Time Video   Video   Audio   Audio   Program
(P indicates p. m. - Source Transition Source   Transition Content
A indicates a. m.)

Courtesy Visual Electronics Corporation

Fig. 7-28. Sample automatic log print-out.

combined with "production" switching. Special effects may be included and automatically controlled. However, the switcher shown in Fig. 7-29 is typical and includes the features required in many master-control switchers. The switcher illustrated is a Visual Model LS-8C, which provides automatic electronic laps, fades, and supers. It has 17 inputs and three output buses; the outputs are line A, line B, and next-event preview. The uses of these outputs are described briefly as follows.

Lines A and B feed a Visual CMP-1 automatic-transition processing amplifier which, in addition to providing for output clamping and sync addition, provides for the required automatic lap or fade transitions. Although two line-output crosspoint buses are utilized, the control panel contains only a single row of switching buttons for the line output. Two rows of buses are required in order to provide for the automatic transitions. For example, when a lap is called for, the newly selected video crosspoint is activated on the bus not currently being aired. The video signal on the on-air bus is faded down while simultaneously the upcoming video is automatically faded up on the other bus. Similarly, for a fade transition, the bus not on the air is switched to black; a lap transition to black takes place, after which the bus that was on the air originally is switched to the new upcoming video; finally, there is an automatic lap transition to the newly selected video on the original bus. Thus, a fade transition consists of two lap transitions, the first to black and the second out of black to the new video; both of these transitions take place automatically.

The third bus is a next-event preview bus that is fed from the memory in the storage display previously described. The switcher next-event row is fed by the event going into the next-event row in the storage display. The next-event bus feeds a processing amplifier that provides for clamping and sync addition.

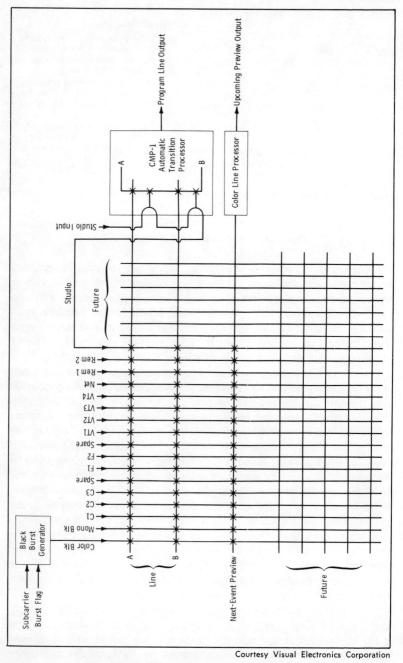

Fig. 7-29. Video-switcher functional diagram.

It is important in any automatic system to be able to bypass the automatic switcher. This permits checkout of the complete automatic system while the station is on the air using one of its studio switchers. A bypass button is installed on or near the station studio switcher, and with it the operator activates the bypass crosspoints that pick up the studio input (Fig. 7-29) and feed it to the program line output. These crosspoints override all other on-air crosspoints, and they are not interlocked so that the automatic system can be operated and the line bus can remain bypassed.

Many stations now have such large numbers of VTR machines and film islands that a video switcher cannot conveniently accommodate them all. Routing switchers often are used to preassign a small number of VTR machines or film chains from a large number of these units. Such routing switchers also can be controlled automatically by codes in the description column of a preceding event.

The audio functional diagram (Fig. 7-30) closely parallels that for video. Two line buses, A and B, are used to provide automatic lap and fade transitions in a fashion similar to that for video. The "super" transition brings the upcoming audio (normally announce booth or a cartridge) over the previous audio, at the same time reducing the previous audio (usually film audio over which the announce tag is made) 10 dB. An audio audition bus is controlled by separate, independent buttons on the control panels.

A monitor amplifier is switchable between the audition output and the line output (Fig. 7-30). The 3×1 crosspoint section is utilized to transfer the studio output or the audition output to the line output when it is desired to feed the studio switcher directly to the line in order to bypass the automatic control for testing.

On the audio switcher, opposite the line and next-event rows of the video switcher are the audio-follow (AF) button and the five audio breakaway buttons. These breakaway buttons are electrically interlocked with the audio-follow button and when activated provide for the energizing of the appropriate audio breakaway crosspoint. If the AF button is activated, the appropriate audio-follow-video relay is energized. The next-event row does not actually control audio relays, however. The breakaway buttons are needed in order to provide for the manual audio preselections.

A custom control panel, Fig. 7-31, normally is required to provide convenient manual override of the automatic show, start, and stop of film; show and slide change of slide projectors; and start and stop of VTR machines. This panel usually is located near the video switcher.

7-6. VIDEO STABILIZING AND PROCESSING AMPLIFIERS

An important link in video distribution along the on-air path is provided by the *stabilizing amplifier* or the *processing amplifier*. Stabilizing amplifiers have been used since early monochrome broadcast days, and

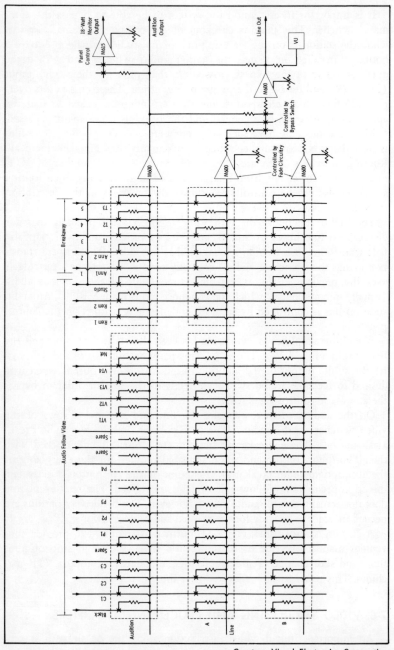

Fig. 7-30. Audio functional diagram.

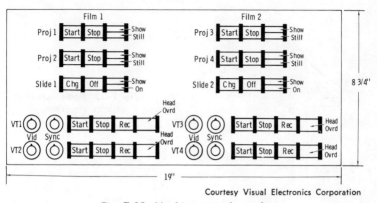

Fig. 7-31. Machine-control panel.

more recently have been modified for color. Processing amplifiers are a later development, and they provide a number of additional correction circuits for modern color transmission.

The Stabilizing Amplifier

The primary function of a stabilizing amplifier is to correct any fault existing in the video signal. Such faults may be hum, switching surges, noise, or sync-tip modulation (Fig. 7-32). The stabilizing amplifier also may be used to insert sync pulses into the composite signal, or it may be used to separate sync from the composite signal for genlock purposes.

Stabilizing amplifiers may be found in the control room as the first amplifier for incoming network or remote programs, and following the switcher unit. This type of amplifier also is found at the transmitter room, where it is used as the first amplifier after the coaxial cable from the studio or after the STL receiver. The unit contains video ampliers, sync separators, sync-insertion circuits, keyers, shapers, and clamping circuits.

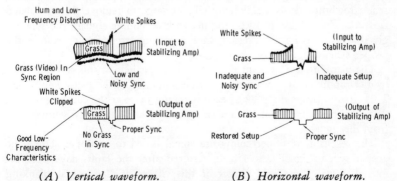

(A) Vertical waveform.   *(B) Horizontal waveform.*

Fig. 7-32. Action of stabilizing amplifier on faulty waveforms.

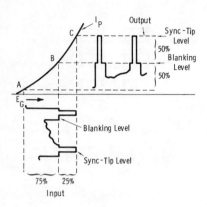

Fig. 7-33. Basic action of sync stretcher.

The first two or three stages usually provide linear video amplification and a means of inserting the composite (horizontal and vertical) sync pulses. Such insertion is accomplished by means of two input stages with a common plate or cathode load. The combined video-and-blanking signal is fed to one tube, and the composite sync is fed to the other tube. The signals are then mixed in the common load. This stage is followed by one or two linear video-and-sync amplifiers.

At this point, *sync stretching* is introduced. Fig. 7-33 illustrates the electrical function of a sync-stretcher circuit. This type of circuit might employ both a Class-A stage and Class-C stage. Over the normal video-signal range up to the blanking level (between points A and B on the combined plate-current curve of Fig. 7-33), the amplification is linear. At the blanking level, the Class-C stage also begins conduction, adding to the total signal from blanking level to sync tips (between points B and C). In this manner, a composite signal of 25 percent sync and 75 percent video at the input may have its sync region expanded as shown at the output. There are several reasons why this function is desirable. For example, if the sync tips should become modulated by pickup of stray noise, sync stretching allows a following clipper stage to remove the modulated tips. Also, in the case of incoming network or remote signals, the sync region is likely to be compressed, thus requiring stretching to restore the proper sync-to-signal ratio.

The blanking level (point B on the curve), is held at this constant level by a clamping circuit that functions independently of the picture amplifier. This circuit clamps the peak of each blanking pulse at the correct point on the amplifier curve and eliminates spurious low-frequency components from the signal.

The resulting amplifier composite signal is fed to a clipper stage. The proper sync-to-signal ratio is thus restored after the faults have been removed. Usually, the grid circuit of this clipper is clamped at blanking level so that the predetermined sync amplitude is independent of variations in the average video signal amplitude. Thus, in this stage also, spurious signals are eliminated from the video and blanking portions of the composite signal.

The output of a stabilizing amplifier consists of three stages. One feeds the line or transmitter, one feeds a monitor bus, and one feeds stripped sync to a sync generator for genlocking purposes.

The sync-stretcher stage is independent of the video signal in the amplifier. This means that keying pulses are derived from the sync portion of the incoming signal or from the sync input itself, properly shaped and amplified, and used to operate the clamping circuits. The keying-pulse shaping circuits that develop the clamping pulses provide a delay in time so that clamping takes place during the portion of blanking signal that follows the sync interval (the back-porch interval of the blanking pulse). Clamping during this interval is more effective than attempting to clamp on sync tips, since any compression of the sync region would tend to defeat the purpose of the clamping circuits.

A color signal differs from a monochrome signal in two major respects, both of which pose problems. First, the addition of the color subcarrier components to the luminance signal causes the resultant color-video signals to extend into the blacker-than-black and whiter-than-white regions. Second, a color-synchronizing burst is placed on the back porch following each horizontal-sync pulse. These characteristics of the color signal give rise to two problems, as follows:

1. *Clipping of subcarrier blacker-than-black excursions:* In monochrome stabilizing amplifiers, the video signal usually is clipped at black level. This removes the sync signal and any noise spikes or signal overshoots that extend into the sync region. The sync signal is regenerated by amplification and clipping in a separate channel and then added back to the video signal. The purpose, of course, is to restore the sync signal to its original wave shape and amplitude; i.e., to remove any distortion introduced during transmission. In stabilizing amplifiers intended for color, some means must be provided for bypassing the burst and subcarrier components around the clipper so that their excursions beyond black level are not clipped off.

2. *Burst distortion:* To insure that video clipping will automatically occur at black level despite changes in signal level or average brightness, the signal must be clamped during the back-porch interval. Since it is during this time that the color-sync burst is transmitted, steps must be taken to prevent the clamp action from causing distortion of the burst.

These two problems, subcarrier clipping and burst distortion, were avoided in the RCA TA-9 stabilizing amplifier by passing the composite color signal through a spectrum-separation network, or crossover filter, in which the subcarrier components are separated from the luminance and

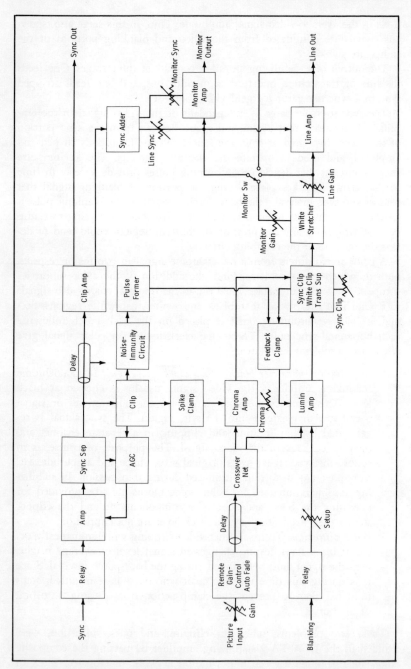

Fig. 7-34. Simplified block diagram of RCA Model TA-9 stabilizing amplifier.

sync signals. Essentially, this leaves a composite monochrome signal that can be processed in the normal manner.

The simplified block diagram in Fig. 7-34 illustrates the major circuit features of the TA-9 stabilizing amplifier. Note that the composite picture signal traverses three paths—for chrominance, luminance, and sync.

First, the input signal is split into two channels, one for picture information and the other for sync. Provision is made for inserting a relay to select either internal or external sync. Use of this relay eliminates the need for the transient suppressor required in many stabilizing amplifiers of older design.

In the sync channel, separation of sync information is accomplished in a high-level clipper. This stage is driven from an automatically gain-regulated amplifier to insure stable and accurate clipping over a wide range of signal-level variations.

A noise-immunity circuit is used between the clipper and pulse former to provide clamp pulses free from the spurious pulses that might otherwise be formed from noise spikes in the incoming signal. The circuit works by virtue of the fact that spurious noise impulses normally are much narrower than the desired sync pulses. The sync signal delivered to the noise-immunity circuit has previously been doubly clipped, so that both the sync pulses and the spurious noise impulses have the same peak-to-peak amplitude. An RC integrating circuit is employed to attenuate the narrow noise pulses greatly so that the sync pulses can trigger the pulse former.

In the picture channel, the signal is again split into two paths, one carrying chrominance information and the other carrying luminance information. The crossover between the channels occurs at the color-subcarrier frequency, 3.58 MHz, with a complete null at that frequency in the luminance channel.

Feedback-clamp and clipper circuits are contained in the luminance channel. The purpose of the feedback clamp is threefold: to maintain clipping at exact black level over long periods of time without readjustment, to set the clipped signal automatically at the proper position on the white-stretcher characteristic, and to provide a high degree of immunity to tube aging and supply-voltage variations. Since the color subcarrier is not present in the luminance channel, sync may be clipped off all the way to blanking level, and back-porch clamping may be performed with full effectiveness without damaging the color burst in the color signal.

Following the clamp stage, where accurate reference level is maintained (for sync clipping, white stretching, etc.), a white-clipper circuit is provided. The purpose of this clipper is to reduce receiver intercarrier buzz caused when the carrier is overmodulated by peak whites. Chroma and high-definition video components may still cause overmodulation, since these components pass through the chroma channel and thus bypass the white clipper. However, the frequency and energy of these components is such that the buzz usually is inaudible.

The chrominance information is passed around the clamp and clipper stages through a two-stage amplifier channel. This allows control over chroma gain and provides proper delay for later recombination of the chrominance signal with the luminance signal. The signals from the chrominance and luminance channels are mixed together and applied to the white-stretch circuit. Here an adjustable degree of amplitude nonlinearity may be introduced to predistort, or compensate, the signal for later passage through equipment that may cause compression, such as a transmitter that does not contain built-in compensation. A switch is provided to bypass this function when it is not needed. The output composite picture signal is formed by addition of the reshaped sync signal to the clamped picture signal.

Differential gain and phase controls are provided to compensate for transmitter characteristics when necessary.

Video-Processing Amplifiers

A video-processing amplifier, or "proc amp," as used in the air-distribution path includes all of the features of a stabilizing amplifier, plus others that are now considered necessary for modern color-station operation. Fig. 7-35 is a block diagram of the RCA TA-19 video-processing amplifier. A brief description of this unit will give the reader an insight into the general requirements of such a device.

Faults in the sync and blanking areas are a cause of worry to many video engineers. Most broadcasters are familiar with the case of the microwave-signal fadeout, in which the signal becomes extremely noisy, and finally the monitors and receivers start to unlock and roll. Normally, it is then necessary to switch back and forth periodically between black and network to determine when the signal becomes good enough to lock receivers again. In this case, the TA-19 will generate normal-amplitude sync and blanking with a composite input signal of only 0.25 volt. It is designed to correct for sync and blanking faults in amplitude, width, and timing and to compensate for missing pulses. The amplifier adds a completely new sync and blanking interval; the "used" input pulses merely serve as a timing reference, and therefore their actual condition is not critical.

Sync, blanking, and drive signals are supplied to cameras and VTR's independently. These pulses may be generated by a common sync generator or by separate sync generators. In either case, the pulse widths and delays might be different. Since the TA-19 generates a completely new sync and blanking signal, the proper widths may be adjusted in the amplifier and will be maintained. The predominant controls are located on the front of the modules for adjustment of blanking width, sync width, front-porch width, and horizontal advance. Additional internal module controls are provided. Individual path delays and variable widths are therefore compensated for in the instrument. Amplitude of sync on the composite output

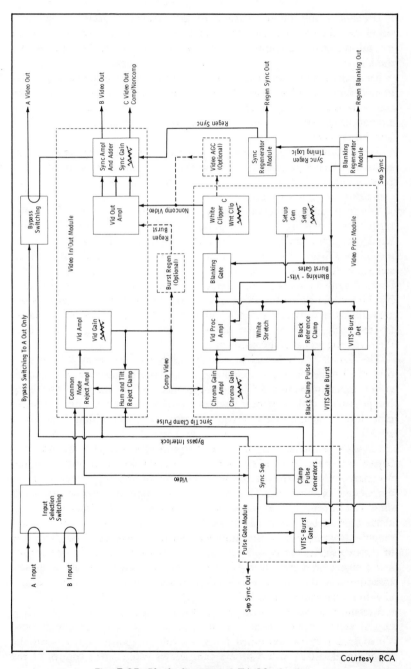

Courtesy RCA

Fig. 7-35. Block diagram of TA-19 circuits.

is dependent only on the settings of the manual controls on the front panel. Should equalizing or horizontal-sync pulses be missing from the incoming signal, the amplifier will "flywheel" and insert these pulses into the signal.

Sense circuits operate to eliminate unwanted signals while passing either monochrome or color. With a color signal, burst, of course, is passed in the horizontal-blanking interval. When burst is not at the input, as with a monochrome signal, the gate closes, and the complete horizontal-blanking interval is processed.

Three 75-ohm video outputs are provided. Two of these, which might be used to feed a program line or for color genlocking, are composite outputs. A third, which might be used to feed a switcher, can be switched composite or noncomposite.

The three pulse outputs consist of separated sync, regenerated sync, and regenerated blanking. Each is available at 4 volts in 75 ohms. The separated sync pulse may be used to reference input sync timing. If it is to be recombined with video, a sync-adding distribution amplifier may be installed, utilizing the noncomposite video output. Regenerated sync may be used to drive monochrome genlock, camera, or VTR. Or, regenerated blanking could be used to drive cameras or VTR's. The regenerated pulses are adjustable in width and delay and are locked to the incoming signal. Consequently, any equipment driven by these signals will be locked to the incoming signal and controlled.

The TA-19 has two bridging inputs selectable by a relay that may be controlled either locally or remotely. A field-effect transistor is employed at the input, providing an input impedance of one megohm (bridging compensated) with extremely low input capacitance and resulting in negligible loading of the source. Frequency response is flat within ±0.2 dB to 10 MHz.

Since the amplifier is normally in the primary signal path, an automatic bypass was built in as a safety feature. Should power be lost, it is not necessary to patch around the unit; transfer of the input signal to the output line is automatic. This automatic switching also takes place under these conditions: (1) if the dc voltage at the TA-19 output stage rises above a prescribed value, (2) if the power supply fails, or (3) if a critical module is removed from the amplifier. During bypass conditions, only one of the two inputs is fed to the output. The second input, which is used for a test signal or remote feed (or any composite signal), is not bypassed. In addition to the automatic operation, bypass also can be switched manually from the control panel.

Remote controls are: sync gain, video gain, chroma gain, setup, white clip, A and B input switching, manual bypass, and unity-variable chroma. Video gain may be controlled for a 1.5-volt composite signal when the input signal is 0.25 volt. Chroma gain allows independent control of the chroma signal to ±3 dB of unity (flat response). Provision is made to switch to unity chroma for a reference (flat frequency response) to indicate the amplitude of input chroma without boost.

At times, it may be desirable to prevent the network vertical-interval test signal (VITS) from passing, especially when a local signal is added for remote keying or any other purpose. For this reason, a switch is provided on the local control panel to permit cancellation of the incoming vertical-interval test signal. The gate is variable in width and delay to provide deletion of a single line or several lines.

One might expect that an instrument capable of achieving all these functions would have inherently high delay. However, this is not the case. The path delay is only 25 nanoseconds, which is the equivalent of approximately 16 feet of video cable. This simplifies the installation in systems that require accurate timing.

The video-agc module is an optional item. A dual detection method and two-step agc action provide precise automatic video-level control. Luminance and chrominance information are not separated, and color saturation and hue are undisturbed over the complete gain range. This system overcomes such problems as fade-to-black, average picture level, and attack-and-release-time errors, at the same time minimizing errors in signals with absence of white references. Adverse overshoots due to lag (common in agc recovery systems) are eliminated. The action is very fast, giving the impression of a correct output level at all times.

Three switch-selectable modes of control are provided in the agc system: Mode 1 provides agc referenced to the vertical-interval test signal. Mode 2 provides agc without external reference signals. Mode 3 automatically references the highest amplitude of both VITS and picture signals.

Controls are provided to adjust the output level manually and to eliminate agc action. A panel lamp indicates loss of agc action.

A separate (optional) module allows independent control of burst amplitude, phase, position, and width. The burst regenerator provides a constant burst output dependent only on the module controls and independent of burst input level. Burst levels may vary slightly with gain changes in the agc. If a precise burst amplitude is required for a particular color system, it is recommended that the burst regenerator be included in the system.

A burst-mode-select control allows switching to incoming burst while "on-air," for phase, amplitude, position, and width comparisons of regenerated burst. Automatic deletion of the regenerated-burst output occurs with a monochrome signal at the amplifier input. A calibrated control is provided as an indicator for burst-phase compensation.

7-7. THE TV AUDIO SYSTEM

The discussion of video in the medium of television is so intense for the average radioman that the all-important "other half" of the telecast—

the audio—is likely to be neglected. This is an unfortunate condition, since the greatest opportunities for improvement of equipment and operational techniques actually exist in the area of TV sound. Long before the television market has reached the saturation level across the country, personnel concerned with TV sound transmission should have achieved a realistic "dimensional" audio.

When pictures are accompanied by sound, the observer must hear. sound that has a satisfactory relationship with what he sees. Although the operator has only a monaural system for sound transmission, he must use every technique available to achieve dimensional effects that are compatible with the monaural channel. It is imperative at this time to become acquainted with problems in acoustics as they exist in the studio, and with equipment used for the sound pickup.

The acoustical requirements of a studio suitable for television shows differ radically from those of strictly aural studios. The primary reasons for this difference are the greater distance from the microphone to the sound source that may be necessary in the TV studio, and the inevitable noise that accompanies movement of performers and crew. Modern radio and recording studios designed for good musical transmission are acoustically treated to achieve sufficient reverberation together with properties of diffusion to prevent boominess. Singers and announcers may then work close to a microphone for clear-cut definition without undue influence on articulation by the acoustical liveness. Such studio design is unsuitable for most television shows, since "action" is considered a must for good video productions. This action necessitates movement not only of personnel, but often of stage props as well.

The basic characteristic of the TV studio, then, is the return to the older-style radio studio overtreated with sound-absorption panels. The average television studio is of much larger dimensions than the average radio studio, and undue reflections from the walls would result in a confusion of unwanted sounds. On shows in which pickup of orchestras or musical groups is included, portable hard *flats* often are placed to the rear and sides of the musical area. Without some means of providing proper reflection of the musical tones, a muffled and "lifeless" quality prevails. Some stations have fed the musical microphone into an artificial-reverberation channel such as a continuous magnetic-tape loop; in this way, any desired degree of liveness may be simulated for normal or unusual musical effects. This practice is increasing in popularity.

Television productions usually call for both fixed and mobile, or *boom-operated,* microphones. On musical shows featuring a vocalist, a microphone mounted on the usual floor stand may be used, since the audience is familiar with this type of pickup on the stage. Often, however, the vocalist "takes off" while singing to roam around the pickup area, adding variety to the video content of the program. This calls for the boom-operated microphone which may be mobile or stationary, depending on the scope of the required movement. This microphone follows the performer, and is not allowed to enter the camera picture area.

The mobile mounting is termed a *perambulator;* it enables the operator to locate the microphone quickly and quietly for proper relationship to the sound source. The rear wheel swivels through 180° and may be clamped to hold a fixed radius. A toggle brake on this wheel may be operated by pushing the tiller back. Operation of a hand wheel adjusts the elevating column so that the column may be raised from about 6½ feet to about 9½ feet. The platform on which the operator stands is raised with the boom. Further elevation, and horizontal movement of the boom, is controlled by a hand rail. Extension or retraction of the boom is accomplished with a hand crank. The telescoping member is counterbalanced by weights adjustable to balance different microphones properly.

The unidirectional microphone is most often used on a microphone boom so that the pickup may be concentrated on the sound source. Movement of the microphone itself (rather than the entire boom) through an arc about its pivot is termed "gunning the mike." The microphone is moved in this respect through 280° by means of the gunning device.

Fig. 7-36 is a simplified block diagram of the audio section of a typical TV control room. The arrangement shown provides the following general facilities:

1. Inputs for seven studio microphones and provision for eight additional microphone lines, which may be patched to consolette input circuits
2. On-air signals for studio, film, video tape systems, turntables, etc.
3. Studio monitoring speaker for cueing or talk-back during rehearsal
4. Intercom circuits as follows: production circuits for program director, stage director, and production personnel; talk-back from audio engineer to microphone-boom operators; talk-back from technical director to camera and dolly operators

The audio console must provide for switching and mixing the audio signals from microphones, film projectors, and remote or network lines. It also is generally used to mix and switch the outputs of one or two turntables, tape systems, etc. The signal from the turntables may be switched into the outgoing line, or fed to the studio speaker for program-background purposes in special cases.

The major difference between audio systems for TV and those for radio is the much larger number of signal sources encountered in normal TV operations. The console of Fig. 7-36 has 15 input channels, but a maximum of 60 input circuits exists. This comes about from the fact that each of the 15 input channels is associated through push-button control with four separate and preselected audio sources. For outputs, "Y" pads (shown by dash lines) permit either a split two-output-line operation (A or B) or an

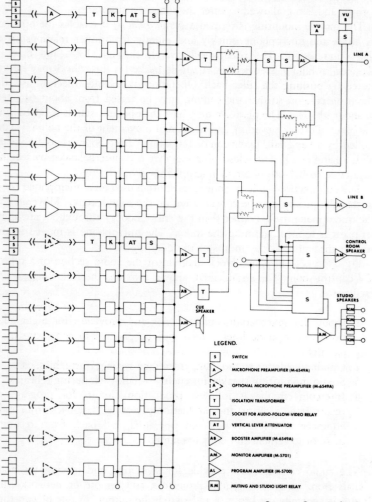

Fig. 7-36. Block diagram of Gates TV-15 television audio-control console.

A + B operation (all input circuits fed to one output line for large productions).

The very important intercom system usually is associated with the TD (technical director, or switcher) position. During both rehearsals and on-air broadcasts, communication must be maintained between the TD, producer, program director, cameramen, camera-control operators, microphone-boom operators, and any assistant production men near the cameras or microphone booms.

To meet the requirements of the intercom and monitoring system, each operator is provided with a double-earphone headset and microphone. The headsets are provided with dual plugs that provide a five-wire connection. This allows one earphone to reproduce the program sound, while the other reproduces the orders and cues given over the intercom network. Cameramen and production men in the studio plug their units into jacks provided at the cameras; microphone-boom operators make use of jacks on the boom. Dolly operators (when used) also wear headphones. Operators in the control room plug their headsets into jacks usually provided on the switcher unit, and at video-control positions.

The intercom control box allows the TD and production director to talk directly with their respective engineering and production personnel, for the purpose of coordinating all aspects of the TV program. One bank of push buttons is for engineering, and one is for production. The TD, for example, can communicate directly with any of his cameramen by pressing the desired CAM (camera) push button on the engineering bank. He also may converse with all engineering personnel simultaneously by pressing the ALL CAM push button. A PL button connects to a private line that may run to the studio or any other designated point.

By means of the production bank, the producer-director can communicate with all assistant production men. He also may talk one-way to all production and engineering personnel simultaneously by pressing the PROG LINE (program line) push button. This connects his microphone into the line that supplies program sound to all personnel.

Film-chain operators in the film room are cued on speakers connected to the same intercom system. Some TD's and producers commonly use the program lines for nearly all instructions, as this gives all personnel an overall view of action and anticipated action.

EXERCISES

Q7-1. What is the nominal impedance of video-distribution cables at the studio and in terminal gear at the transmitter?

Q7-2. What is the meaning of the term "loop through"?

Q7-3. What are the two fundamental methods of switching video signals?

Q7-4. What is the difference between noncomposite and composite switching?

Q7-5. What is the most usual method of timing the various paths in a video-switching system?

Q7-6. Define the term "chroma key."

Q7-7. What is the basic function of the handles on a special-effects control panel?

Q7-8. What is audio-follow-video switching?

Q7-9. What is the difference between a stabilizing amplifier and a processing amplifier?

Q7-10. What is the major difference between an audio control console for TV and one used for normal radio broadcasting?

CHAPTER 8

Television Recording Systems

Television recording techniques are widely varied, and currently there is rapid development in many different types of systems. The methods are basically as follows:

1. Kinescope recording
2. Electron-beam recording (EBR)
3. Magnetic video tape recording
4. Slow-motion and stop-action disc recording

The kinescope recorder has been with us the longest and will be covered first. In spite of inroads of other methods into the video-recording field, the kinescope method has undergone significant improvement in recent years and still meets a definite need, particularly in monochrome recording techniques.

8-1. THE KINESCOPE RECORDING SYSTEM

A great amount of time and many feet of motion picture film a year are used in kinescope recording. This type of recording is made by photographing visual images (displayed on a special kinescope) on motion-picture film in specially adapted cameras. Motion pictures of an image on a picture tube are variously referred to as *kinescope recordings, kinephotos,* or *teletranscriptions.*

Two different methods are commonly used for recording both picture and sound onto a single motion-picture film:

1. *Single-system method:* The picture image and the sound-track image (variable area or variable density) are recorded simultaneously in a single pass through a combination picture and sound recorder. This method usually is employed only when a single recording is to be made.
2. *Double-system method:* Separate picture and sound recorders are operated in synchronism and at constant relative speed. This method

is used for optimum quality of both picture and sound in the final composite release print. It is used most often when a recording is to be distributed for wide use.

In practice, 16-mm cameras are used for kinescope recording. The kinescope upon which the image to be recorded is displayed is a special type of tube employing a flat face, with phosphor and brightness characteristics that allow use of inexpensive film. The circuits associated with such a tube always permit reversal of direction of horizontal sweep (for reasons described below), and provide means so that either positive or negative images can be placed on the screen. The need for this feature will become apparent in the following discussion.

There are several alternate procedures, which may be used in kinescope recording, applicable to either the single- or double-system method. The choice depends on the purpose for which the recording is made, or on the

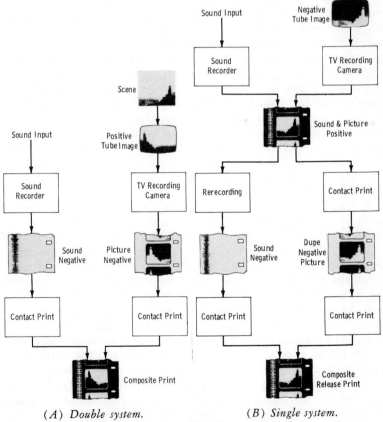

(A) *Double system.*   (B) *Single system.*

Fig. 8-1. Kinescope recording techniques.

time allowable before use of the finished recording. One of the more popular procedures is diagrammed in Fig. 8-1A. In this method, a positive picture appears on the kinescope (just as seen on the monitor or receiver picture tube) and is photographed by the camera. Since the image is positive, the film will have a negative picture. This film is processed either by a local motion-picture laboratory or by the station itself when the developing and printing facilities are available. The composite positive print is then made from the combined sound and picture negatives. The picture print produced in this manner *does not have the standard emulsion position.* (Film is ordinarily placed in the projector so that the emulsion side is toward the lens.) Therefore, *it is necessary to reverse the image, left to right, on the kinescope screen.* Note that the kinescope image is not a negative picture, but simply a picture reversed in horizontal content. This is accomplished by reversing the direction of the horizontal-scanning current for the transcriber kinescope.

When only one copy of a video recording is required, or when time is an important factor in producing a positive print, the method outlined in Fig. 8-1B may be used. A negative kinescope image is obtained by employing an extra amplifier stage or other means. Since the image is a negative picture, the photographic recording may be made on positive film. In this single-system method, both picture and variable-area sound track are recorded on the same strip of film. Development in positive-type developer results in standard emulsion position of the film. If copies are necessary, a "dupe negative" usually is made from the picture and is used in making the final composite release prints. A negative also is made of the sound track, usually by rerecording; this negative is used to print the final positive sound tracks. Rerecording eliminates the necessity for using a duplicate negative of the sound track, which invariably results in serious loss of high-frequency response.

The reader should remember that it is not necessary to use positive prints for film telecasting, since the TV camera employs a polarity-reversal switch so that either positive or negative pictures may be used. Positive prints are advantageous, however, since the film technicians often project the film on a screen in the film control room for purposes of editing, etc.

In video-recording equipment, the engineer is faced with the inverse of the frame-rate problem that occurs in TV projection. In this case, he is faced with adapting the 30-frame-per-second image appearing on the kinescope to the standard 24 exposed frames per second of the recording camera. It is convenient for the discussion here to relate five television fields (5/60 second, or 1/12 second) to two film frames, since 1/12 second (2/24 second) is equal to two motion-picture frames. Therefore, if we omit one of every five scanning traces, we are left with four scanning traces, or two TV frames. This results in compatibility between the two systems.

One way the foregoing conversion is accomplished in practice is by the

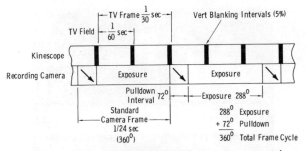

Fig. 8-2. Midfield-splice method of frame-rate conversion.

sequence of operations illustrated in Fig. 8-2, known as the *midfield splice*. If we call the standard film cycle 360°, the "tube cycle" may be seen to equal 288°. During the first 72° of the film cycle, the film is pulled down to advance one frame. The camera shutter is closed during this time, and the actual pulldown time of the film is in the vicinity of 60°. Exposure (shutter open) then takes place for 288° (1/30 second) during another TV picture frame. Pulldown of the film follows to start the next cycle. The exposure time is extremely critical and must be within one-half line of a complete frame (525 lines) to avoid an effect known as *banding*, or *shutter bar*. This is a reproduction on the film of a varying-density region. The exposure time is controlled by either mechanical or electronic shutters.

Fig. 8-3. D. B. Milliken Model DBM-64 16-mm kinescope recording camera.

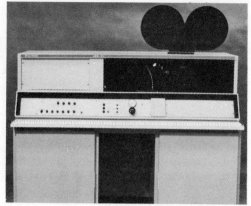

Courtesy D. B. Milliken

Fig. 8-4. D. B. Milliken DBM-R1 video film recording system.

A recently developed kinescope recording camera that does not employ the midfield-splice method is illustrated in Fig. 8-3. An extremely rapid film pulldown is accomplished by compressed air that moves the film through the gate; no cams or claws are used. An air compressor is provided with the control console (Fig. 8-4) and is separate from the console so that it may be installed in any convenient location. The console contains controls for the air supply pressure.

The camera is driven synchronously with the television signal through a camera-drive inverter in the console. A phase monitor provides continuous verification of camera phase synchronization.

Total pulldown and stabilization are accomplished during the standard vertical-blanking interval. On 60-Hz systems, the camera runs at synchronous speed of 24 frames per second. Every fifth TV field is blanked by shutter operation, thus accomplishing the necessary conversion from 30 to 24 frames per second. Pulldown occurs alternately between fields and during the blanked field. This method eliminates the shutter-bar problem and the image shearing that normally occurs during fast panning of the camera. The rotary film transport is designed to eliminate the camera vibration caused by reciprocating mechanisms, thus helping to assure proper interlace.

A special high-resolution picture tube is fitted with an optically flat faceplate for optimum edge resolution and minimum barrel or pincushion distortion. Special blanking circuits are provided to allow either negative-positive recording for multiple copies or direct positive recording for single-copy recordings.

It should be recalled that two basic methods of sound-on-film recording exist, the single system and the double system. In the former system, the sound track is recorded at the same time as the picture, using only a single film. In the latter system, the sound and picture are recorded on separate film, after which they are edited and "married" by joining on one film. It is only natural that the newcomer ask the logical question, "Why bother with two separate films and then go through the process of combining them on one film?"

The reason is just as logical. The reader who is a photography fan knows that a film with a "fast" emulsion (capable of capturing images even of scenes with low illumination) has a comparatively coarse grain. But the sound-track emulsion must be on very fine-grain film (therefore a "slow film") to avoid excessive noise in the sound channel. Thus, in the single system of video recording, a compromise, which never reaches the results possible in the double-system method, must be made. The double system permits a fast film to be used for pictures, and a fine-grain film to be used for the sound. This system also permits more accurate editing, since in the single system it is almost impossible not to lose some picture or sound in the editing process. The sound quality is better in the double system not only because fine-grain film may be used, but also because in the single system the 26-frame lead of the sound track often is insufficient to remove all of the intermittent film motion at the sound drum. The result is a slight flutter on the audio channel, and even a very slight recording of flutter is emphasized upon playback of the film.

The reader should understand what is meant by making a positive from a negative film, and not be confused by the usual snapshot printing process. When motion-picture film is exposed to a "positive" image (just as your eye perceives the scene), the original white areas are dark after exposure and developing, and the original dark areas are light. Naturally, if this film were projected "as-is," a negative image would show on the screen; that is, faces would be dark instead of light. Therefore, in this case, a positive contact print is made, although it is still on transparent material for the purpose of projection. When *reversal* film is used, no extra printing is necessary. This film, when developed and treated a second time, reverses its shades so that portions that were black become white, and vice versa. This is termed a *direct-reversal* positive.

Another film is *release-positive* film. In this case, the image on the transcriber kinescope is made a negative image, and the release-positive film, upon development, will be ready for projection.

Panchromatic film, which is sensitive to all colors through the red portion of the spectrum, is seldom used for kinescope recording. The transcriber kinescope emits a high value of light in the blue region, and only the inexpensive violet-blue-sensitive film is required for recording. Handling and developing of this film may be carried out under red safety lights rather than in total darkness.

It is at this point that a very important operational function should be noted by the reader. To understand the need for this function, it is pertinent to look briefly into the different historical backgrounds of 35-mm and 16-mm film processing. The 35-mm type has been the "professional" film used for theatrical projection, and it has been universally produced by a negative-to-positive method. The standard emulsion position in the projector for 35-mm film is with the emulsion side toward the projection lamp. Just the opposite standard emulsion position has been established for 16-mm film, which in the past has been an "amateur" product usually using a reversal film processed to a positive image. Thus, the correct left-to-right orientation of the image is obtained when the emulsion side is toward the projection lens, instead of toward the lamp. Adding to this problem is the fact that the sprocket holes for 16-mm sound film are along only one edge of the film, and the release print must be such that the correct emulsion position is obtained when the film is properly inserted to engage the sprockets in the projector.

It now may be seen that if the 16-mm film is to have the emulsion side toward the projection lens, the process illustrated in Fig. 8-1A, since it is a negative-to-positive process rather than a direct-reversal process, would result in improper orientation of scenes upon projection. This problem is solved by reversing the image on the transcriber kinescope. Note that this does *not* mean a "negative" picture on the kinescope screen, but that the *horizontal scanning* on the kinescope is *reversed in direction* from normal. The final print will then be such that when the film perforations are on the correct side of the projector, the emulsion will be toward the lens, and proper images will be transmitted. Keep in mind that this is done *for 16-mm film only,* where the emulsion side is to be toward the lens. For 35-mm video recording, the negative-positive process would be correct with normal scanning of the kinescope, since the 35-mm projector is threaded with the emulsion side of the film toward the lamp.

The beginner may ask why the 16-mm film could not simply be threaded just as the 35-mm type is threaded, with the emulsion side toward the lamp. The reason is that all other commercial-type 16-mm film received at the station is already printed for the standard 16-mm emulsion position, and the overall operating procedure would be needlessly complicated. Since perforations are along only one side of the film, the two types could not be spliced together to form a direct sequence of film, as must often be done in telecasting.

Some 16-mm projectors for television use are able to accommodate sound film with the emulsion in either position, toward the lamphouse or toward the lens. In these projectors, the sound lens is refocused automatically for each position of the emulsion. In such equipment, direct-reversal films are threaded with the emulsion toward the lens (as are *Kodachromes* when used), but ordinary negative-to-positive processed films are placed with the emulsion toward the lamp. It is our purpose here to give all these facts to the new or prospective operator, not to confuse him, but to give him the overall picture of the wide variance in equipment and operating techniques. He is then better equipped to grasp the particular techniques that he may encounter on the job.

Whenever time is an important factor and it is known that only a single copy is required, the method of Fig. 8-1B is popular. In this case, a *negative* image is placed on the screen of the transcriber kinescope (usually by throwing a switch that takes the kinescope grid excitation from a stage that has negative video polarity). The picture then may be photographed on a direct-positive film. As a rule, this is a single-system method of recording, in which the picture and variable-area sound track are recorded simultaneously on a single pass (the sound is 26 frames ahead of the corresponding picture frame). Development in a positive-type chemical developer yields a direct positive picture and sound track image, with standard emulsion position for projection and telecasting. If it later becomes necessary to make copies, a dupe picture negative (duplicate negative) may be made to use in printing the final composite release prints. (As was mentioned earlier, the sound track usually is rerecorded onto a negative film for use in printing the composite prints, since use of a dupe-negative sound track loses the highs in the sound spectrum.)

Fig. 8-5 illustrates another alternative in kinescoping. A positive tube image is recorded on reversal film to give a direct positive print, ready for telecasting with standard emulsion position (16 mm only). When copies are made, a dupe negative made from the picture positive is used to print the final composite print. Recording of the sound onto reversal film is not considered good practice. As shown in Fig. 8-5, the sound usually is recorded on a separate negative film that is then used for printing the final composite release print.

The following procedure for determining correct exposure in kinescope recording is presented courtesy of the Eastman Kodak Company: The correct exposure for the film in photographing a kinescope-tube image depends on a number of factors, among which are the type of phosphor used for the tube face, the spectral character of the light it emits, the brightness of the image, the spectral sensitivity of the film, the nature of the subject matter, and the processing conditions for the film. Before any picture tests are made, it is desirable to correlate the brightness level of the tube with the density obtained on the film when the latter has been processed to the recommended gamma value. This is done with a plain raster such as would be obtained by the use of the blanking signal without picture modulation. The brightness of the raster is varied by means of the video-gain control or picture-tube grid-bias control. The beam current may be measured by a microammeter. Since the brightness of the tube is dependent on the power input to the screen, this beam-current measurement serves as a measure of the brightness. A series of exposures may be made by varying the beam current in logarithmic steps. After the film is processed to the recommended gamma value (0.6 to 0.7 for picture negatives, and 2.2 to 2.5 for direct positives), the density may be read on a densitometer and plotted against the logarithm of the beam current. For a negative material that has been developed to a gamma value of 0.65, the

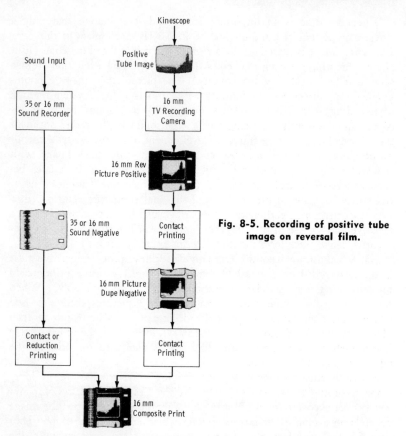

Fig. 8-5. Recording of positive tube image on reversal film.

negative density range normally made use of in the recording of a picture image is from about 0.20 to 1.4 or 1.5. A beam current that gives an intermediate density of around 0.8 to 0.9 might therefore be considered as providing an average brightness level corresponding to that of a picture-tube image that will give an approximately correct exposure. With this beam current as the starting point, a series of exposures over a smaller range may then be made with picture modulation on the tube, in order to arrive at an average exposure value that will be satisfactory for various types of subject matter.

The reader familiar with ordinary photographic techniques will immediately recognize the above procedure as being analogous to amount of light, exposure time, type of film, and processing for still-picture photography. Gamma is the straight-line portion of the film characteristic curve. A film is chemically treated (developed) to a gamma of, for example, 0.7 by using a known solution and development time under a given set of conditions such as temperature, type of film, etc. (A gamma of 0.7 in the negative will result in a final release-print gamma of about 2.3.) For a given value of gamma, the film density depends on the brightness of the scene. (In video recording, the time of image exposure is fixed.) The camera lens, of course, may be stopped down to assure correct focus if brilliance of the tube screen is sufficient. Since the brightness of the kinescope depends almost directly on the beam current, the above procedure uses the value of beam current as the abscissa of a plot on log paper. The density of the film, as determined by the densitometer or calibrated photocell unit, then may be used as the ordinate to form a plot of density versus beam current.

The range of the beam current over the linear portion of the curve should be about 30 to 1 for optimum TV use. This corresponds to a gamma in the final release print of about 2.3. This is the contrast ratio which, in terms of the film, is the *density* ratio. It is the opinion of experienced engineers that the density range of film made specifically for telecasting should be less than in films ordinarily used for the theater. The latter film usually has a density (contrast) ratio of at least 50 to 1.

It should be noted also that for direct-positive film, the development should yield a contrast directly suitable for telecasting. In this case, the film is developed to a gamma of from 2.2 to 2.5.

It probably is obvious even to the general reader that this coverage of kinescope recording has been extremely sketchy. Coverage of this department of a TV station includes a wide range of subjects, including chemistry, electronics, optics, and numerous operational procedures. It would be impossible to include a detailed analysis in this book, which must treat the overall television system. It can only be expected that the element of mystery that may have existed has been lifted somewhat, and that the operator will have gained a basic background of knowledge concerning kinescoping.

8-2. ELECTRON-BEAM RECORDING

The end result of an electron-beam recording (EBR) system is fundamentally the same as that of the kinescope recorder: conversion of video and audio signals to motion-picture film. If we imagine replacing the phosphor-coated face of the picture tube with photographic film, we have the basic idea of EBR. The reaction of silver-halide film to the electron beam is very similar to its reaction to light. In this method, the electron beam scans the film (instead of the phosphor-coated faceplate) to create a latent image in the film; the image is developed by normal film-processing techniques.

Manufacturers of EBR systems such as the one shown in Fig. 8-6 claim several advantages over the kinescope recorder, as follows: In a kinescope recorder, the limited light output available from the phosphor of a cathode-ray tube forces the use of a relatively fast film. To the graininess that re-

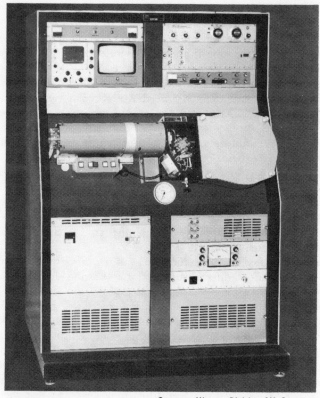

Courtesy Mincom Division 3M Company

Fig. 8-6. 3M Brand Model EBR-100 electron-beam recorder.

sults from the use of a fast film is added the graininess that is caused by
the particulate nature of the phosphor. Uneven shading in the picture is
caused by variations in the phosphor thickness. The resulting picture quality
is less than ideal. These shortcomings are overcome in the electron-beam
recorder by eliminating the phosphor and creating the image directly in
the photographic emulsion.

Fig. 8-7 shows a comparison between raster lines recorded directly with
an electron beam and raster lines recorded on a conventional kinescope

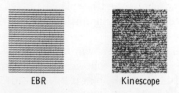

EBR   Kinescope

**Fig. 8-7. Comparison of EBR
and kinescope recording.**

Courtesy Mincom Division 3M Company

recorder. The large grain size and unevenness typical of some kinescope recordings are obvious in the pattern at the right. The corresponding EBR pattern is at the left.

Immediately following is an abstract from a paper presented at an IEEE symposium on broadcasting by John W. Reeds, Jr. It appears here through the courtesy of the Mincom Division, 3M Corporation.

Electron-beam recording introduces some unique new problems. One of these is that the photographic film must be introduced into a high-vacuum system in order to record on it with an electron beam. Another is that the electron beam builds up an electrostatic charge on the film, and this charge tends to deflect the beam from its proper position. In the EBR-100, these problems are solved by evacuating the film magazine and transport to a pressure of about 15 mm. This reduces the requirements on the vacuum seal between the film magazine and the electron gun so that a simple raised land or border surrounding an aperture through which the electron beam passes is sufficient to achieve a vacuum of about 10^{-3} mm at the recording aperture. At the same time, by maintaining the pressure in the film magazine at about 15 mm when a standard commercial positive release film is used, drying out of the emulsion is avoided. Since some moisture is retained in the film, the conductivity of the emulsion remains high enough to dissipate the electrical charge built up by the electron beam. Since the film magazine is pumped to only a modest vacuum, it may be evacuated from atmospheric pressure to the operating pressure in only about 30 seconds.

In recording motion pictures from American standard television signals, it is necessary to convert from the television picture rate of 30 frames per second to the 24 frames per second required by motion-picture standards. This conversion may be done in several ways: Two frames (four fields) may be recorded and then one field discarded; four frames may be recorded and then one frame discarded; or one frame may be recorded, one-half field discarded, and then one more frame recorded (made up of one-half field plus a full field plus one-half field). In order to reduce the speed requirements in the film intermittent drive, and at the same time avoid the flicker problems of recording multiple frames at one film position, the last method was chosen for the EBR-100. The time (approximately 8 milliseconds) during which one-half field is discarded is utilized for the intermittent drive to move the film.

In Fig. 8-8, the frame conversion is illustrated. The first picture on the film is recorded from the first two fields of video (one frame). Then the top half of the next video field is discarded, and the film is advanced to the next frame position during this time. The recording of the next picture on the film is made from the bottom half of a video field, the following interlaced field, and then the top half of the next field. The film is again advanced, and one-half field of video is discarded. The entire sequence is then repeated.

16 mm Film   Television Signal
(24 Frames/Sec)   (30 Frames/Sec=60 Fields/Sec)

Courtesy Mincom Division 3M Company

We can break the electron-beam television film recorder into three general areas: the film transport, the electron optical system, and the electronics. Considering the film transport first, let us look at Fig. 8-9. A coaxial magazine contains both the supply and take-up reels. Film is pulled from the supply reel by a motor-driven sprocket that is servo controlled by the incoming vertical sync. The film goes to the gate area, where there is a conventional intermittent drive, and then to the sound drum and to the take-up reel. If sound is to be recorded, a variable-density optical galvanometer is utilized.

At the interface between the film transport and the electron gun, a vacuum seal must be made on the film. This seal must be adequate to reduce the 15-mm pressure in the film chamber to the approximately 10^{-3} mm required at the area where the image is being recorded. A relatively flat plate having an aperture surrounded by raised lands, chrome-plated and polished, is used as part of a film gate to accomplish this seal. When the film chamber is being threaded, or in the event of film breakage, a valve closes off the electron gun. This valve may be manually opened or closed, and is closed automatically by an interlock in the event of a vacuum failure.

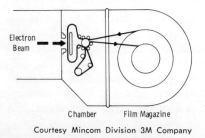

Electron
Beam

Chamber   Film Magazine

Courtesy Mincom Division 3M Company

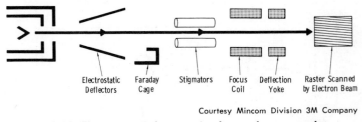

Courtesy Mincom Division 3M Company

Fig. 8-10. Electron optical system in electron-beam recorder.

The electron optical system (Fig. 8-10) uses a single electromagnetic focus coil to image the crossover spot of the electron gun onto the film. A hairpin tungsten filament is used. In order to achieve good filament life, the gun is pumped by two separate diffusion pumps. The first pump maintains a vacuum of about 10^{-3} mm at the image plane, and the second pump maintains a vacuum of about 10^{-6} mm at the filament. An octopole stigmator is used to correct astigmatism of the record spot. An electrostatic deflector allows deflection of the electron beam into a Faraday cage for sampling the beam current. Horizontal and vertical sweeps are electromagnetic deflections.

Since the electron beam is focused to a fine spot, adjacent lines recorded by conventional sweeping of the beam would not come together. On a direct-positive recording, this would leave a clear space between recorded lines, thus making it impossible to obtain high densities desired for projection. If this image were rescanned for televison broadcasting, *moire* patterns between the lines in the image and the scan lines would appear. To prevent these undesirable effects, while at the same time retaining maximum possible resolution, the electron beam is wobbled vertically at about 30 MHz in order to blend adjacent raster lines together.

When the electron beam is not writing raster lines, it is deflected into the Faraday cage, and the signal from the Faraday cage is displayed on a waveform monitor. During recording, the one-half field that is discarded during the film pulldown time provides a signal to allow monitoring of the video being recorded. On the back porch of every horizontal-sync pulse, a reference pulse is inserted. This reference pulse is gated out of the Faraday cage and is used to control the bias of the electron gun for automatic stabilization.

When the electron beam is scanning a raster at the record aperture, a loop antenna surrounding the aperture collects secondary electrons. The signal collected by the antenna is displayed on a television monitor, and thus creates a picture of the object being scanned. For focusing the electron beam, a metal target or screen is placed in front of the beam, or a leader on the film is coated with graphite paint. By reducing the size of the raster being scanned, the display of the target on the television monitor may be greatly magnified. This permits careful observation of the

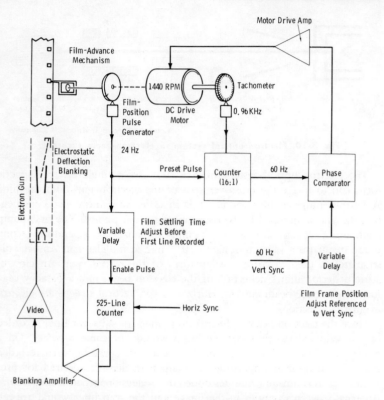

Fig. 8-11. Block diagram of Model EBR-100 electron-beam recorder.

beam focus and astigmatism. During this focusing, the 30-MHz vertical spot wobble is turned off.

The electronics in the electron-beam recorder provide the following basic functions:

1. Synchronizing the film motion to the video input (Fig. 8-11)
2. Counting raster lines to record precisely the right number (Fig. 8-11)
3. Amplifying the video, including switchable high-frequency boost and gamma correction for recording both negative and direct-positive films

Switchable high-frequency boost may be used when desired to improve the subjective quality of the particular material being recorded. High-frequency boost gives sharper resolution at the expense of increased high-frequency noise.

The gamma-correction video amplifiers provide adjustable break points to permit adjustment of proper gray scale in the film. A typical gamma

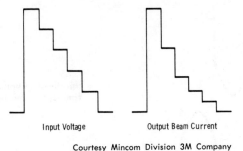

Input Voltage   Output Beam Current

Courtesy Mincom Division 3M Company

Fig. 8-12. Gamma correction for direct-positive film.

correction for recording a positive film is shown in Fig. 8-12. (At the time of this writing, only monochrome recordings can be made by the EBR process.)

NOTE: Several manufacturers are developing laser systems for film recording. The advantage of these systems is elimination of the high-vacuum system required in electron-beam techniques.

8-3. VIDEO TAPE RECORDING SYSTEMS

The quadruplex (four-head) video magnetic tape system is by far the most popular method of television recording at the time of this writing. Television tape recording requires a magnetic-field manipulation that will adequately handle an 18-octave signal range at a satisfactory signal-to-noise ratio. We will investigate problems and the approach to their solutions. The equipment and techniques involved are quite extensive, and a book-length treatment would be required for full coverage of the subject. We can only hope that this section will serve as a worthwhile introduction to the subject for the beginner.

The familiar frequency-to-wavelength conversion, which is a fundamental time-space relationship, is represented in Fig. 8-13. For any given medium, the velocity at which waves travel is determined by the medium or system in question, and the velocity determines how far a point in a wave travels in a given time. On waveform A of Fig. 8-13, point 1 of a passing wave is at reference position T1. One second later (waveform B), the passing wave has advanced to T2, and point 2 is at the reference position. The distance between T1 and T2 is the wavelength, and it also defines one cycle of the wave.

The velocity of sound waves, although influenced by temperature, humidity, and height above sea level, may be taken to be approximately 1088 feet per second. Since the wavelength is equal to the velocity divided by the frequency in hertz, a 1000 Hz tone will have a wavelength (in air) of 1088/1000, or 1.08 feet. Doubling the frequency to 2000 Hz results in a

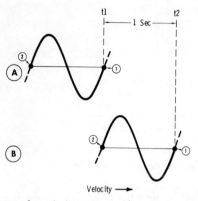

**Fig. 8-13. Fundamental
time-space relationship.**

Velocity ⟶

wavelength just one-half as great. However, an increased velocity would result in a greater wavelength for the same frequency.

The tape velocity of broadcast-type audio tape recorders is either 7.5 or 15 inches per second (in./s). At 7.5 in./s, the recorded wavelength of a 1000-Hz tone is 7.5/1000, or 0.0075 inch (7.5 mils). (One mil is 0.001 inch.) At a tape speed of 15 in./s, the recorded wavelength of a 1000-Hz tone is 15/1000, or 0.015 inch (15 mils).

The gap in the magnetic head must have a certain minimum physical size to place adequate field strength on the tape. The relationship of head-gap size to frequency is illustrated in Fig. 8-14. If the frequency is increased and the tape speed is held constant, the recorded wavelength approaches the physical size of the head gap. This results in zero output because of north-south field cancellation of the signal. Therefore, the high-frequency limit of the system is fixed by head-gap size and tape speed.

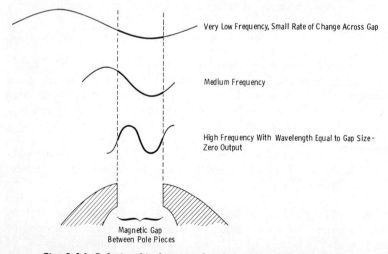

Fig. 8-14. Relationship between head-gap size and frequency.

The strength of the signal induced in the playback head is determined by the strength and the rate of change of the magnetic field recorded on the tape. For a given recorded amplitude, as the frequency is increased (increased rate of change) the induced voltage is increased. As the frequency is doubled (increased by one octave) the voltage is doubled. This voltage increase, measured in decibels, is 6 dB. Therefore, in magnetic recording a 6-dB-per-octave rise in response occurs with increasing frequency.

Conversely, as the frequency is decreased from a given high-frequency limit, a 6-dB-per-octave fall in response occurs. At a certain low-frequency limit, the change rate of the magnetic recording is so low, compared to the gap size, that very little output exists. When the signal output falls to a point at which there is an objectionable signal-to-noise ratio, the lower-frequency limitations of the system have been exceeded.

The frequency limitation of direct magnetic recording is about nine octaves, or 30 to 15,000 Hz in the audio range. If it is desired to increase the upper frequency limit to 30,000 Hz, the lower frequency limit also must be doubled to maintain an adequate signal-to-noise ratio. The result is still a nine-octave range, 60 to 30,000 Hz.

To recapitulate, we have learned that at very low frequencies, there is a very small rate of change across the gap and a correspondingly low induced signal. Also, the smaller the gap, the less the induced signal can be. At higher frequencies, there is a greater rate of change across a given gap width, hence more induced signal.

At still higher frequencies, several additional factors appear. There is a loss in the tape because of tape-thickness factor. Also, as the wavelength approaches the gap width, cancellation of energy occurs. Fig. 8-15A shows the typical response curve (6 dB/octave increase) up to the high-frequency limitations, where the response falls quite rapidly. Note that at some frequencies the response is near the noise level and below and cannot be called usable. The noise level is the result of tape, modulation, and input-amplifier noise.

See Fig. 8-15B. You have learned that response can be expanded into higher-frequency regions by two methods, (1) increased tape speed to produce a longer wavelength for a given high frequency, and (2) decreased gap size. Note that a higher tape speed extends the high-frequency response, but since the wavelength at a given low frequency is made longer, the low-frequency response suffers. Also, reducing the gap size means that less energy is available at the lowest frequency, so again the low-frequency response suffers.

Fig. 8-15C represents the typical equalization curve to correct the magnetic-recording characteristic of Figs. 8-15A and 8-15B. The rising response at the high-frequency end is to correct for the rolloff caused by gap size and tape magnetic-coating factors.

The normal bandwidth of modern studio video equipment is at least 8 MHz. The picture transmitter, however, is limited to approximately

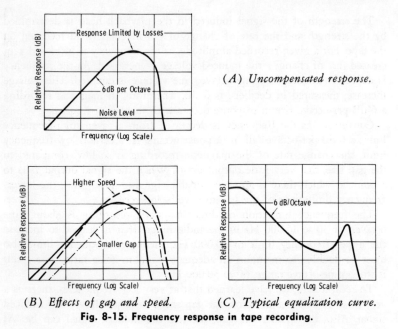

(A) Uncompensated response.

(B) Effects of gap and speed.   (C) Typical equalization curve.

Fig. 8-15. Frequency response in tape recording.

4 MHz by FCC channel assignments and engineering standards. Therefore, the horizontal resolution is restricted to about 320 lines in the home receiver. The complexity of the problem encountered in recording pictures on magnetic tape warranted a compromise in this particular studio gear. Early television tape recorders had response at least to 4 MHz (320 lines horizontal resolution) with a signal-to-noise ratio of at least 35 dB. This meets the minimum requirements of network-signal distribution as specified by AT&T.

The video-signal frequency range for television tape extends from very low frequencies (actually dc) to a practical upper limit of 4 MHz. If the range is considered to be 10 Hz to 4 MHz, a gamut of over 18 octaves is required; this range is not possible with direct magnetic-recording techniques.

The dc component in a standard video signal is *inserted* by means of line-to-line clampers. Essentially, the clamper charges or discharges a coupling capacitor to a dc reference that usually represents the signal-blanking level. This reference level assures that each active line scan starts from the same reference immediately following horizontal blanking.

The extreme low-frequency requirements in video recording are met by employing an rf carrier that is frequency modulated by the video signal. (Fm was chosen over a-m to allow amplitude limiting to be employed for attenuation of extraneous noise.) The carrier frequency represents the dc component, either sync tip or blanking depending on the system clamp

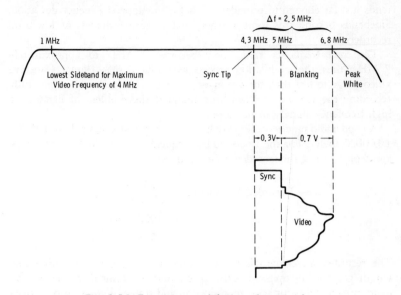

**Fig. 8-16. Frequency modulation of carrier for
low-band monochrome recording.**

reference. Fig. 8-16 shows the modulation characteristic with a carrier
frequency of 5 MHz clamped to the signal-blanking level. The conventional
assumption that the carrier frequency must be at least ten times the highest
modulating frequency is discarded in television-tape applications.

NOTE: There are two standards for magnetic video tape recording, *low
band* and *high band.* The characteristic shown in Fig. 8-16 is for low-band
monochrome operation.

With the carrier frequency clamped at the video-blanking level, it is
standard monochrome practice (for low-band operation) to adjust the
picture-signal gain so that peak white causes an upward frequency deviation of 1.8 MHz; the peak-white signal then occurs at 6.8 MHz. With a
standard video input, 0.3 volt sync to 0.7 volt video, sync tips cause a
downward deviation of 0.7 MHz, to 4.3 MHz. The total deviation of
2.5 MHz is used currently as the 100-percent modulation reference for
monochrome signals, on low-band standards.

When the modulation index is less than 0.5 (frequency deviation less
than one-half the modulation frequency), sidebands beyond a single pair
are practically nonexistent, and the signal approaches a-m characteristics
in this respect. Sidebands occur at the carrier frequency plus and minus
the instantaneous video frequency. The maximum video frequency in the
system passband determines the sideband limits. For a 4-MHz bandwidth,
the lowest sideband occurs at $5 - 4$, or 1 MHz. The upper sideband ex-

tends just far enough to provide a *shelf* for the upper frequency deviation. Sidebands beyond this limit are not used in earlier models of low-band recorders.

The total frequency range now becomes 1 MHz to approximately 7 MHz; thus, the modulation process has reduced the original 18-octave video range to less than three octaves. A practical solution to magnetically recording the video signal has been found if the problem of handling a high-frequency signal can be solved.

A good audio tape recorder may have a head-gap size as small as 0.25 mil (0.00025 inch). For the upper audio frequency of 15,000 Hz and a tape speed of 7.5 in./s, the recorded wavelength is:

$$\text{Recorded Wavelength} = \frac{7.5}{15,000}$$
$$= 0.0005 \text{ inch}$$
$$= 0.5 \text{ mil}$$

The recorded wavelength is twice as long as the gap width. This wavelength/gap-size relationship is the lowest practical limit that permits adequate pre-emphasis for good reproduction and signal-to-noise ratio. (Losses in the magnetic core structure cause the signal to begin to decrease before this point, and pre-emphasis is required.)

The smallest practical magnetic gap for low-band video-recorder heads is approximately 100 microinches, or 0.1 mil. Therefore, the maximum wavelength that will represent 7 MHz (upper shelf limit for low-band recording) must be twice as long as 100 microinches, or 200 microinches. The required tape velocity can be calculated if the minimum required recorded wavelength at the upper frequency limit is known. Since:

$$\text{Wavelength} = \frac{\text{Velocity}}{\text{Frequency}}$$

then:

$$\text{Velocity} = \text{Wavelength} \times \text{Frequency}$$
$$= 0.0002 \times 7,000,000$$
$$= 1400 \text{ in./s}$$

If the tape were actually pulled across the head at a speed of 1400 in./s, 420,000 feet of tape would be required to record one hour of material. Try to visualize the tape transport mechanism required to handle a reel containing this much tape—obviously, this is an impractical solution.

The problem is solved by pulling the tape past a rotating video head at a practical speed of 15 in./s. Insofar as the video signal is concerned, the resultant velocity is more accurately termed *head-to-tape* velocity. The rotating-head principle, illustrated in Fig. 8-17, gives an effective head-to-

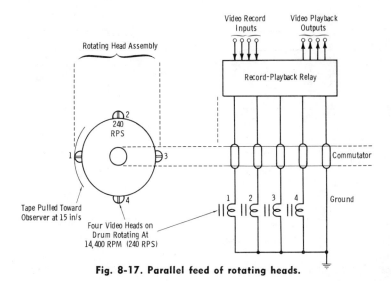

Fig. 8-17. Parallel feed of rotating heads.

tape velocity of slightly over 1500 in./s, with a head-gap size just under 0.1 mil. The fm video signal is fed to the individual heads on the rotating drum-and-brush assembly. Video tracks approximately 10-mils wide are laid down crossing the 2-inch-wide magnetic tape from side to side. Audio is recorded longitudinally along the top of the tape at the conventional 15-in./s rate. A 240-Hz control signal, which indicates the precise position of the rotating head drum relative to the recorded video information, is recorded longitudinally along the bottom of the tape.

To recapitulate again, the direct video signal is not recorded on the tape. The actual signal is in the form of a frequency-modulated rf carrier. Zero deviation of the carrier represents the dc component of the video signal; in Fig. 8-16, the video blanking level is "clamped" at 5 MHz. When a negative-going video signal (sync region) is modulated on the carrier, the carrier is caused to decrease in frequency. When a positive-going video signal (picture information) is applied, the carrier is caused to increase in frequency.

The standard low-band monochrome deviation shown by Fig. 8-16 is repeated at the top of Fig. 8-18A. The sync-tip level is actually at 4.28 MHz (generally rounded off to 4.3 MHz). For low-band *color* recordings, the deviation is reduced, as shown at the bottom of Fig. 8-18A. Because of the strong color-subcarrier component in the video signal, *moire* patterns result if standard monochrome deviation is used. Also, an excessive amount of differential phase and gain of the color subcarrier is present. Therefore, standard deviation for low-band color recording is reduced to 1 MHz for 100-percent modulation, as shown. Blanking is clamped at 5.79 MHz, sync tip occurs at 5.5 MHz, and picture white peaks

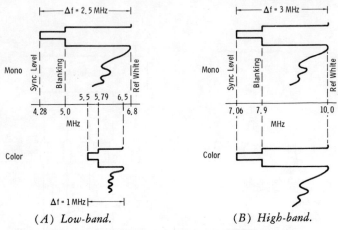

(A) Low-band.   (B) High-band.

Fig. 8-18. Frequency modulation in VTR systems.

at 6.5 MHz. The main characteristic that suffers under these conditions is the signal-to-noise ratio.

Fig. 8-18B shows the standard deviation for high-band recording systems. Note that the deviation for monochrome and color signals is the same in this case. Blanking level is clamped at 7.9 MHz, sync tip occurs at 7.06 MHz, and peak picture white is at 10 MHz. The deviation for 100-percent modulation is then approximately $10 - 7$, or 3 MHz. A number of improvements are possible in high-band systems as compared to low-band systems.

Before recent advances, the head-gap size in video headwheels was 90 microinches. The gap size now has been reduced to 50 microinches. This allows a higher frequency response, as required for high-band recorders. Improved design has resulted in as much sensitivity as existed in the former (wider) gap. Also, both lower and upper sidebands normally are used in modern high-band systems.

The circular mounting that contains the four video heads is termed the *headwheel* by RCA and the *drum* by Ampex. Both of these manufacturers employ horizontal tape transports in some models and vertical transports in other models (Fig. 8-19).

A rotating head contacts the tape through an arc of approximately 120°. Since the head rotates at 240 RPS, 1/240 second is required for 360° of rotation, and one-third of this time (1/720 second) is required to traverse the 2-inch tape (120°). Inasmuch as the tape travels 15 inches in 1 second, it will go 15/720 inch, or 0.02 inch (20 mils), while the head describes its arc across the tape. Therefore, the bottom of each video track is displaced longitudinally 20 mils from the top (start) of the same track.

The orientation of recorded information on the vertical tape transport is shown in Fig. 8-19A. The video heads contact the entire 2-inch surface

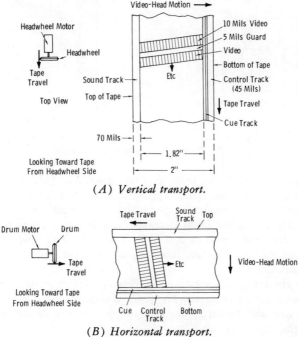

(A) Vertical transport.

(B) Horizontal transport.

Fig. 8-19. Tracks on video tape.

of the tape; however, a 70-mil track along the top is erased for audio information, which is recorded longitudinally to tape travel by a conventional audio head. The video tracks along the bottom of the tape in the area of the control and cue track are not actually erased; however, only that portion used as video information is shown in Fig. 8-19A.

Tapes recorded on either system (RCA or Ampex, vertical or horizontal tape transport) are entirely compatible and can be played back on either system. The illustrated track orientation on the tape would be seen by an observer looking toward the coated side from the rotating heads. The track dimensions for the horizontal version shown in Fig. 8-19B are exactly the same as for the vertical transports in Fig. 8-19A.

The method of holding the tape concentric to the rotating heads is shown in Fig. 8-20A. The vacuum guide holds the tape in place, and the center slot provides clearance for tape *stretch* under head penetration. Fig. 8-20B is the side view of the vacuum guide and the rotating head.

What the television-tape system must do can be outlined briefly as follows:

1. The composite video signal is presented to the four rotating heads simultaneously in the form of an fm signal.

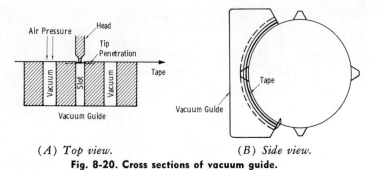

(A) *Top view.*   (B) *Side view.*
Fig. 8-20. Cross sections of vacuum guide.

2. One revolution of the headwheel lays down four video tracks; adjacent video tracks are spaced approximately 5 mils apart.

3. To recover the recorded information, video-head outputs are first amplified, then fed to an electronic switcher that selects the signal from the head that is in contact with the tape. This selection occurs during horizontal-retrace intervals so that switching transients are invisible.

4. The electronic-switcher output is demodulated to recover the video signal from the rf carrier. The video is then processed to "clean up" the sync and blanking intervals for distribution to the television system.

5. For proper synchronization, the tape velocity (15 in./s) must be locked to the headwheel rotation (240 RPS). During recording, stripped-off sync from the incoming video is utilized to obtain a 240-Hz signal (4 times the field pulse frequency of 60 Hz), which is amplified to obtain sufficient power to drive the headwheel. This 240-Hz signal is also converted to a sine wave that is recorded on the control track of the tape.

6. Also during the record mode, a signal proportional to the speed of the headwheel is obtained (by an exciter lamp and photoelectric cell in the Ampex unit, by a magnetic tone wheel in the RCA unit) and processed to yield 60-Hz power for the capstan motor. The speed of the capstan motor has absolute control over the velocity of the tape.

7. In the playback mode, the headwheel again rotates at 240 RPS, with the initial reference being either the power-line frequency or local sync. The 240-Hz control-track signal is phase-compared to the playback speed of the headwheel, and any phase error slightly modifies the 60-Hz frequency applied to the capstan motor. Thus, tape velocity is continuously maintained so that the rotating heads sweep directly across the video tracks originally laid down.

8. The audio signal is recorded and reproduced by conventional magnetic-recording techniques.

Fig. 8-21. VTR room at WTAE-TV.

Fig. 8-22. VTR room at WBBM-TV.

Fig. 8-21 is an illustration of the RCA video-tape installation at WTAE-TV. This installation consists of two TR-70's, one TR-22HB, and one TR-3 (playback only) system. Fig. 8-22 shows the Ampex installation at WBBM-TV. The complete complement of machines includes four VR-2000's (two of which are in the photo) and two VR-1000's (one of which is partially visible in the background).

8-4. THE HELICAL-SCAN RECORDER

Unlike the quadruplex tape-recording system just described (which has tightly controlled standards), the helical-scan system is very diversified. Refer to Fig. 8-23 during the following basic outline.

The capstan, which determines the speed of the tape from supply to take-up reel, may be located after the video head (which is on the scanning drum) to pull the tape through (Fig. 8-23A), or before the video head to meter the tape into the head-drum assembly (Fig. 8-23B). Tape

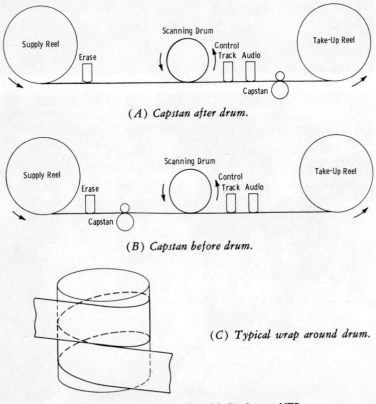

(A) Capstan after drum.

(B) Capstan before drum.

(C) Typical wrap around drum.

Fig. 8-23. Fundamentals of helical-scan VTR.

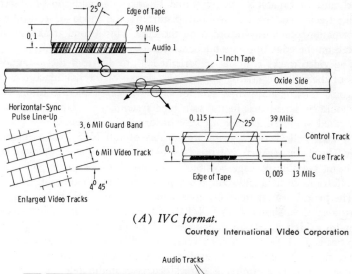

(A) IVC format.

Courtesy International Video Corporation

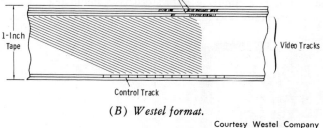

(B) Westel format.

Courtesy Westel Company

Fig. 8-24. Track formats in helical-scan recording.

speed varies, depending on manufacturer, number of heads (one or two), bandwidth requirements, and a number of other factors. The speed is tightly fixed for any one system, usually somewhere between 6 and 10 inches per second.

The fixed erase head erases the entire width of the tape during the record mode only. The tape then passes around the scanning drum (Fig. 8-23C), in which is a rotating shaft carrying one or two video record-playback heads. The signal is fm (IVC uses pim, or pulse-interval modulation), and picture with sync information is laid down in diagonal tracks similar to those illustrated in Fig. 8-24. Next the tape encounters another fixed head where a control track is applied close to one edge. This head stack may also use an extra erase head under the same or opposite edge of the tape to clear a track for audio. Finally, the fixed audio head places either one or two separate audio tracks on the tape.

Some manufacturers place the control track at the top edge; others place it on the lower edge of the tape. In many formats, a narrow strip of the tape must be capable of being erased for the later addition of a second audio track (cue track), without disturbing the video or control-track information previously recorded on the tape. This second track usually is placed on the edge opposite the control track.

The video tracks typically are oriented 3° to 4° from the longitudinal axis of the tape, depending on the basic design of the manufacturer. Each diagonal track contains one field, or 262½ lines, of picture information. It should be noted that helical-scan tape systems use a tape 1 inch wide, compared to the 2-inch tape used for quadruplex recorders.

Modern helical-scan systems are capable of full NTSC color reproduction. The main advantage of this type of recorder is the much more compact and portable package possible as compared to the quadruplex system. Applications in field and mobile use are covered in Chapter 11.

The helical-scan system also finds some use in the TV studio. Fig. 8-25 illustrates the Westel Model WRR-350 studio recorder. This recorder contains a high-speed, small-diameter helical-scanning system that utilizes two heads. Each field occupies several short, diagonal scans across the tape, rather than one long scan as was done in earlier helical systems.

The system is a complete record/playback studio television recorder, designed specifically for high-band color operation. It is supplied for

Fig. 8-25. Studio-type helical-scan VTR.

Courtesy Westel Company

operation on the 525-line, 60-field standard and can be adapted to other standards in common use throughout the world. The machine will record up to 96 minutes on a 14-inch reel of standard 1-inch tape. In 525-line, 60-field systems, longitudinal tape speed is 15 in./s, and the head-to-tape writing speed is 1260 in./s. In addition to the recording of video and control-track information, the studio unit also provides two high-quality audio channels. These can be selectively recorded or erased and rerecorded without disturbing the video tracks.

Electronic circuitry is all solid state and includes a high proportion of integrated circuits. The circuitry is designed for high-band color operation and for the new high-output ferrite heads. Automatic framing, horizontal lock, and certain functions of a processing amplifier are built-in features of the equipment.

The WRR-350 uses five printed-circuit motors that have a high ratio of torque to inertia. Stable tape motion is achieved by servo control of reeling and drive elements together with a tape-tension servo system. As a result, lock-up time is kept to a minimum, and uncorrected time-base stability is well within the "window" of conventional electronic time-base-correction accessories.

Normal setup and alignment procedures are limited to video and audio gain in the record mode, and to initial adjustment of video tracking for playback. Full electronic editing capability is an optional feature.

8-5. THE SLOW-MOTION, STOP-ACTION DISC RECORDER

The need for video disc recording first arose from the desirability of "instant replay" in sports events. To emphasize highlights of a given play, slow-motion or stop-action replay adds interest to the production. The disc recorder also is finding many applications in studio productions for special effects. Individually selected "takes" from the disc can be transferred into a longer production on a standard studio quadruplex recorder equipped with an electronic editing system.

Fig. 8-26 illustrates the basic function of a two-head disc recording system (such as the Visual VM-90). The highlights of this technique can be outlined as follows:

1. The disc consists of aluminum substrates lapped to optical flatness, then electroplated with a thin layer of magnetic nickel-cobalt. This surface is then plated, or *flashed,* with a few microinches of rhodium to provide a mirror-like running surface for the heads and to prevent surface corrosion.

2. The rotating disc is contacted by two record-playback heads, one on the top and one on the underside. NOTE: Some disc recorders, such as the Ampex HS-100, employ two discs and four heads. We will return to this method shortly.

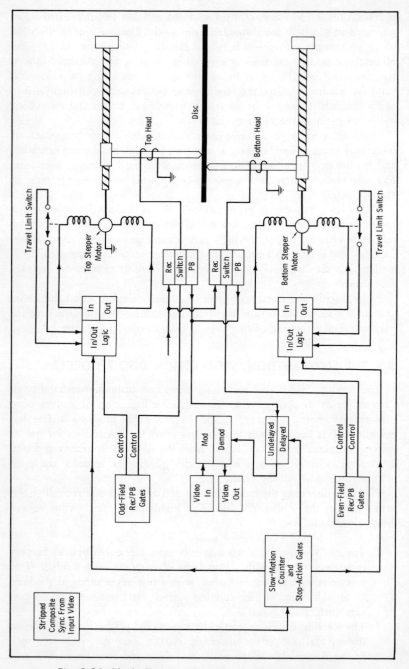

Fig. 8-26. Block diagram of two-head video disc recorder.

3. Individual fields are recorded as separate, circular tracks on the disc surface. In Fig. 8-26, the odd-numbered fields are recorded on the top, and even-numbered fields are recorded on the bottom of the disc.

4. The recording sequence is as follows: Starting at the inside edge of the disc, field 1 is recorded by the top head. Field 2 is recorded by the bottom head. During the time of field 2, the top stepper motor advances the top head *two tracks*. The top head then records field 3. During the time of field 3, the bottom motor advances the bottom head two tracks. The entire process is then repeated.

5. At the end of travel in the first direction (inside-outside), the last two fields are recorded and each head moves back one step, again recording the odd field on top and the even field on the bottom. The steppers then revert to movement to every other track, and information is recorded on the tracks left vacant on the first pass. With this two-head system, all tracks are filled in 30 seconds.

6. For slow-motion or stop-action playback, a field must be used over and over. To produce *interlaced frames,* the video information in a single *field* is first delayed by a half line, then (by means of an electronic switch) undelayed for the next field duration. The information in the two fields is identical, but displaced vertically one-half line on the raster to provide a full interlaced frame. The question usually asked here is why a full *frame* is not repeated rather than one field. The answer is that in scanning the two fields making up a frame, a time differential of 1/60 second exists between adjacent lines at any one point on the raster. A rapidly moving object (such as a football, for example) would produce a very noticeable double-exposure effect in the interlaced frame.

In the Ampex HS-100 system, two 16-inch discs are rotated at a nominal rate of 60 RPS by a common shaft locked to an external vertical-sync reference. One revolution of the discs corresponds to one field, beginning and ending during the vertical-blanking interval.

Four heads are used in the Ampex system, one for each surface of the two discs. Table 8-1 shows the Ampex head sequence. When head A completes field 1, head B starts field 2, etc. When head D completes field 4, head A has been stepped into position to start field 5. Thus, each head

Table 8-1. Ampex HS-100 Record-Mode Sequence

Head	Field 1, 5, 9, etc.	Field 2, 6, 10, etc.	Field 3, 7, 11, etc.	Field 4, 8, 12, etc.
A	Record	Move	Move	Erase
B	Erase	Record	Move	Move
C	Move	Erase	Record	Move
D	Move	Move	Erase	Record

records every fourth field. Note that during the record mode, one head is recording, one head is erasing, and two heads are moving to new positions.

EXERCISES

Q8-1.   Name four basic types of TV recording systems.

Q8-2.   Name three basic methods of magnetic TV recording.

Q8-3.   Name the two basic methods of kinescope recording.

Q8-4.   If a kinescope recording is made on positive film, what is the video polarity on the kinescope?

Q8-5.   Name the three general functional areas in an electron-beam recording (EBR) system.

Q8-6.   For low-band monochrome VTR operation, give the frequency that corresponds to (A) blanking, (B) sync tip, and (C) peak white.

Q8-7.   For low-band color VTR operation, give the frequency that corresponds to (A) blanking, (B) sync tip, and (C) peak white.

Q8-8.   For high-band VTR operation, give the frequency that corresponds to (A) blanking, (B) sync tip, and (C) peak white.

Q8-9.   Does the rotating head assembly of a video tape recorder actually receive video signals?

Q8-10. Describe the basic principle of slow-motion and stop-action video systems.

CHAPTER 9

Studio Lighting

The average radio engineer becomes familiar with Ohm's law early in his training. Even though he can measure voltage, current, and resistance with electrical meters, without Ohm's law his ability to understand radio circuits would be sharply curtailed.

In television, the engineer is concerned not only with Ohm's law, but also with the nature of the medium for which television was created, namely, light. There are certain basic relationships in this medium that should be as well understood as Ohm's Law, in spite of the fact that light meters, video analyzers, and various other instruments may be used to measure the quantities involved.

9-1. THE NATURE OF LIGHT

There is an abundance of radiant energy in free space: radio waves, heat waves, light waves, etc. The difference between such energies lies primarily in their respective frequencies. Light waves occupy a band of frequencies to which the human eye is a natural receiver. Properties of light that concern the TV operator are reflection and scattering of light rays, factors affecting contrast and detail of a picture, the influence of color in composition of a picture relative to equipment limitations, and the difference between *brightness* and *intensity of illumination*.

In a monochrome picture, the *more* light that reaches the eye, the "whiter" the image appears. When light rays strike a white surface, almost three-fourths of the rays are reflected, although not in an orderly fashion. As if they were tiny lamps, infinitesimal points in the reflecting surface borrow some of the energy and scatter the reflected light. A dark reflecting surface reflects less light into the eye, and the image appears "darker."

The phenomenon of *reflection* is best exemplified by the use of a mirror. In this case, light falling on the mirrored surface is reflected at a calculable angle; the angle of reflection equals the angle of incidence. This means that if the original light rays strike the mirrored surface at 40°, the reflected light will leave the surface at 40°. Such surfaces are used in varying shapes as high-efficiency reflectors to control the lighting of the TV pickup area.

The main point to remember, however, is that the eye "perceives" an image because objects scatter the light that strikes their surfaces. Thus, when a beam of light shines through a slit in a venetian blind, unless there are dust particles in the air to scatter the rays of light, the eye does not "perceive" the beam of light except as a bright spot on the floor or wall.

Certain qualities of light are called "color." Again, the difference lies in relative wavelengths or frequencies of *visible* light. Daylight, or "white" light such as unfiltered sunlight, contains all the various frequencies of visible light. "Black" is simply a lack of light. A black substance absorbs many light rays and reflects few. A white substance absorbs few light rays and reflects many. Colored objects absorb light rays of same wavelengths while reflecting rays of other wavelengths. A red glass, for example, makes everything appear red when you look through it. This is not because it turns other rays into the red wavelengths, but because it destroys or absorbs waves of all other lengths except those of the red portion of the spectrum.

The TV operator and producer are confronted with problems in color whether the system is black and white or color. In color systems, various hues in close proximity affect the overall composition. Exactly the same thing occurs in black-and-white transmission. The conversion of the varying colors into the corresponding shades of gray determines the average brightness and contrast of a scene.

The *candle* is the unit of light intensity. One *foot-candle* may be defined as the intensity of light on a one-square-foot white surface illuminated by a standard candle placed a distance of one foot from the surface. (One foot-candle is a comparatively small amount of intensity.) The *lumen* describes the amount of light that falls upon a surface per second. The lumen may be related to the foot-candle as follows: One foot-candle is the intensity of illumination on a surface area of one square foot when perpendicular rays of light flux totaling one lumen strike the surface. This occurs when the standard candle illuminates a one-foot-square white surface located one foot away.

A point source of light radiates light flux in all directions and is said to have spherical radiation. Thus, it helps to consider the amount of light flux (lumens) contained in a sphere of given size. One lumen is the amount of light flux emitted from a source of one-candle intensity and contained within a "unit solid angle." The unit solid angle (termed a *steradian*) is the solid angle that encloses an area on the surface of the sphere equal to the square of the radius of the sphere. Since the total surface area of a sphere is $4\pi r^2$, there must be 4π steradians in the sphere. Since there is one lumen per steradian, the total light emitted in all directions from the one-candle source is 4π, or 12.57, lumens.

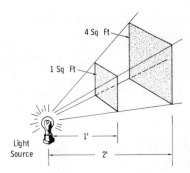

Fig. 9-1. Inverse-square law of light radiation.

Since light does radiate spherically, it spreads out as it progresses from the source. If it is necessary to ascertain the illumination on a given surface at a certain distance from a light source of known intensity, this illumination may be calculated by taking into account the spreading of the light waves. Fig. 9-1 illustrates this effect, called the *inverse-square law*. Rays from the light source shining through a one-square-foot aperture at one foot from the source, will be spread over four square feet at a distance of two feet. Doubling the distance increases the area covered by the square of 2, or 4 times. Since the area covered is much greater, the *intensity* of illumination is *reduced* as the square of the distance:

$$I = \frac{c}{d^2}$$

where,

I is the illumination in foot-candles,
c is the intensity of the source in candles,
d is the distance (in feet) of the surface from the source.

In practice, this basic formula must be modified because it assumes the light path to be normal to the plane of the surface. In an actual situation, however, the illumination must be to one side of or above the TV camera. Since the angle the light path makes with the surface must be accounted for, a cosine factor is added:

$$SI = \frac{c \cos \theta}{d^2}$$

where,

SI is the surface illumination in foot-candles,
c is the intensity of the light source in candles,
θ is the angle the light path makes with a line perpendicular to the surface,
d is the distance (in feet) of the surface from the source.

Example. Assume a test pattern is illuminated by a light source of 1000 candles (power in watts is entirely dependent on type of lamp and luminaire) that is displaced $45°$ from the plane of the test-pattern card, and at a distance of 12 feet.

$$SI = \frac{(1000)(\cos 45°)}{12^2}$$

$$= \frac{(1000)(0.707)}{144}$$

$$= 5 \text{ foot-candles (approx)}$$

NOTE: When diffusers are used in studio fixtures, this principle does not apply.

This is the incident light reaching the test pattern, *not* the reflected light.

There is a decided difference between the *illumination* of a scene and the *brightness* of a scene. It should be realized that, for example, the brightness of a white card illuminated by a 100-candle source is far greater than the brightness of a black card illuminated by the same light source. Therefore, it is necessary to consider the factors affecting brightness, which are the total light flux on the surface, and the reflecting and diffusing properties of the surface (affected by color).

The unit of brightness is the *lambert*. The number of lamberts is equal to π times the number of candles per square centimeter. In the more familiar English system, the number of *foot-lamberts* is equal to π times the number of candles per square foot. The foot-lambert also may be defined as the brightness of a perfectly diffusing surface that emits one lumen of light flux per square foot of surface.

The cosine law must also be taken into account for the brightness of a surface (or portion of the surface, since brightness may vary from point to point of a given surface, as in a test pattern). The cosine law as expressed for brightness is:

$$B = RI \cos \theta$$

where,

B is the brightness at a given distance from the surface,
R is the reflection coefficient of the point under consideration,
θ is the angle between the light path and the perpendicular to the surface,
I is the illumination of the incident rays.

Incident rays are the light rays from the source of light. Reflected rays are the rays of light reflected from the surface that is illuminated by the incident light.

The TV engineer will encounter another factor in reading technical literature pertaining to lighting: the Kelvin (K) rating given to light sources. The precise derivation of the Kelvin scale is quite complicated, and only the practical content of its interpretation is considered here, as applied to problems concerning TV studio lighting.

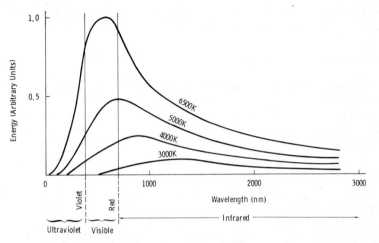

Fig. 9-2. Energy distributions for sources of different color temperature.

The Kelvin scale concerns a factor known as *color temperature*. It is based on the concept of a *black-body,* which is a body that absorbs all incident light rays and reflects none (therefore, a theoretical concept only). If the black-body is heated, it begins to emit visible light rays, first dull red, then red, then through orange to "white heat." Many readers are probably familiar with this phenomenon in the heating of metals. If the metal could be heated enough without melting, a bluish-white radiation would occur. Thus, the higher the Kelvin temperature, the nearer is the approach to the sunlight spectrum. A low Kelvin rating indicates a light source with most of the *visible* radiation in the red region. Most incandescent lamps such as those ordinarily used in house lighting are comparatively low in K rating. Fig. 9-2 illustrates the concept of this scale. "Daylight" is given a rating of approximately 6500 K.

It should be noted that the Kelvin rating has nothing to do directly with actual radiated heat from a light source. For example, a lamp with a low Kelvin rating actually emits most of its energy in the invisible infrared region, which constitutes heat waves. (When light from such a source is raised to a high enough intensity to brightly illuminate a TV studio, the heat is very noticeable to those under the radiation.) The Kelvin rating simply compares the actual distribution of colors, as observed visually. The standard of comparison is the visual emission from a black-body as it is heated to higher and higher temperatures. Daylight is considered cold light, yet it has a comparatively high K rating. Filament-type (incandescent) lamps range from 3000 to 4500 K; certain arc lamps have a still higher K rating, approaching daylight characteristics. The lower-K incandescent lamps are used sparingly in TV studios to eliminate the high degree of heat radiation on the personnel. The operator may find quite a mixture of lamps in use in actual practice. For color operation, the K rating of lamps becomes a very important factor.

9-2. STUDIO LIGHTING

Before we go ahead, it is necessary to become familiar with both the formal and slang terminology of basic lighting equipment, as follows:

Barn Doors: Black hinged panels or shutters (which open and close like real barn doors) attached to the front of a lamp fixture to control the spread of the light beam. Also see *Snoots* (or *Snouts*).

Broad: A type of floodlight used to provide broadside flat lighting. When it has twin lamps, it is termed a *double broad* or *broadside.*

Century Stand: Upright, pole-like stand with clamps. It may be used to hold a gobo, extra small lights, etc.

Deuce: Nickname for a 2-kW lamp. Also see *Junior.*

Dimmer Board: Houses the appropriate switches and controls for fading down and brightening lights or banks of lights.

Elbows: Special hooks used to keep cables off floors.

Flag: Small rectangular gobo. An even smaller gobo is termed a *clip.* Terminology is quite fluid here; a thin flag might be called a *blade;* a round one might be called a *target* or (if very small) a *dot.* A flag with a small hole is called an *ear.*

Flood: *Floodlighting* and *spotlighting* are the two basic types of lighting. The flood is designed to illuminate a broad expanse; it is also called a scoop.

Gobo: Black screen, cloth, or sheet used to block off unwanted light reflections into the camera lens. Also applied to a built-in pattern projector for *Lekolights.*

Inky: Baby spotlight usually mounted on the front of a camera to lend extra sparkle to a person's eyes and teeth. Originally this term was a contraction of "incandescent" (thus, a small incandescent spotlight), but it also is applied in current usage to a small spotlight using the newer quartz-iodine lamp.

Junior: A light not quite as bright as a senior, usually rated at one or two kilowatts.

Keg Light: Small spotlight in a housing that resembles a keg.

Klieg Light: A flood-spotlight (either way). (Trade name of Kliegl Bros. Lighting.)

Leko (Lekolite): Special spotlight shuttered to provide a desired pattern of light. (Trade name of Century Lighting, Inc.)

Luminaire: Light fixture with lamp. (General term.)

Reflectors: Brightly polished concave surfaces used behind a lamp to concentrate the light rays in one direction. The term is also applied to exterior sheets (up to about $6' \times 6'$) used to reflect light into desired locations (such as shadow fill).

Rifle: Spotlight with a very small beam of light used to pinpoint a small detail.

Scoop: See *Flood.*

Snoots (or *Snouts*): Hoods placed in front of lamp housings to limit the area illuminated.

Spiders: The points where cables converge for connection to the current main. They are so termed because they resemble a spider's web.

Spill: Term for excess light that falls in areas not intended.

Spotlight: Highly concentrated beam of light.

Striplights: Lights arranged in strips for lighting backings, windows, and cycloramas ("cycs"). A cyclorama is a semicircular or U-shaped hanging for a background.

Teaser: Large, flat gobo used to block overhead back light from the camera lens.

Additional common lighting-equipment terminology will become apparent in the following descriptions of specific fixtures.

Fig. 9-3 illustrates typical scoops used in TV studios. The scoop provides relatively broad-area lighting, known as floodlighting. Fig. 9-3A shows a 300- to 500-watt, 15-inch scoop intended to provide proper intensities for small-studio, low-ceiling use (10 to 12 feet). It can be used with a color frame and barn doors. This scoop uses incandescent lamps. Fig. 9-3B shows a 16-inch scoop using a 1000-watt quartz-iodine lamp. (The difference between incandescent and quartz lighting is discussed in Section 9-3.) This scoop is used for diffused base and fill lighting. (This terminology also is explained in Section 9-3.) The fixture of Fig. 9-3B delivers a wide range of precisely controlled intensities. These units deliver two to three times as much light as conventional scoops. A "punch scoop" using a 1000-watt quartz lamp is illustrated in Fig. 9-3C. This is a long-range unit, especially effective in theater-type studios when nested on a balcony rail. It is widely used with 4½-inch image orthicons and in color TV for filling in shadow areas with a soft light from a longer-than-normal distance. The *Kliegsun* (Fig. 9-3D) meets studio requirements for the production of a brilliant shaft of light. It uses 1000- to 1500-watt incandescent lamps. An additional front spherical reflector redirects rays from the lamp bulb into a parabolic aluminum reflector, greatly increasing beam intensity and eliminating spill light.

Fig. 9-4 illustrates typical spotlights. Kliegl Fresnel-lens spotlights normally are capable of a beam spread from spot to flood, or approximately 10° to 55°. Fig. 9-4A shows a relatively small 500-750 watt (incandescent) unit with a 6-inch oval-beam lens. The action of this lens is illustrated in Fig. 9-4B. Fig. 9-4C shows the quartz-lamp version, using a 6⅜-inch Fresnel lens in an 8-inch-diameter housing. This unit is focusable from maximum spread to minimum spot. Fig. 9-4D illustrates barn doors. A swivel feature permits 360° rotation of shutters to cut off light at any angle. The light beam may be cut down to a very small size to control spill light.

(A) 15-inch, incandescent.

(B) 16-inch, quartz-iodine.

(C) "Punch scoop."

(D) High-intensity light.

Courtesy Kliegl Bros. Lighting

Fig. 9-3. Examples of scoops.

Fig. 9-5 includes more specialized types of lighting than general flood-lights or spotlights. The Kliegl *Q-Lite* (Fig. 9-5A) uses 500/1000-watt quartz lamps and is designed for use in low studios and other applications where space is a limiting factor. It is used for both base and key lighting. A knob provides adjustable light intensities. Sharp barn-door cutoff permits, for example, framing rear-projection screens with no spill. The *Nook Light* of Fig. 9-5B uses 300/500-watt quartz lamps. It is designed for unobtrusive placement to permit concentrated floodlighting from cramped quarters. Fig. 9-5C illustrates cyc floodlights for color. (For a definition of the term "cyc" see the previous definition of "striplights.") This fixture uses 300/500-watt quartz lamps and is 7 feet long with 8 lights. Because quartz lamps tend to maintain the same Kelvin temperature through the entire lamp life, new lamps may be placed next to old ones without changes in light output or dimmer settings.

Fig. 9-6 illustrates typical slide and pattern projectors. The Kliegl 1676G scenic-slide and effects projector (Fig. 9-6A) uses a 5000-watt incandescent lamp. A forced-air blower system for lamp and slides is included. The projector produces an average of 48 foot-candles (without slide) on a $10' \times 12'$ screen at a throw of 12 feet; it has an approximate one-foot beam spread for every foot of throw. The unit has a telescoping roller caster stand. Fig. 9-6B shows a quartz pattern projector that can be used for effects patterns on a screen or studio wall. With a 1000-watt quartz lamp, it will provide 275 foot-candles with a spread of $7.5' \times 25'$. It is equipped with pattern holders and four-way framing shutters and has an unrestricted burning position. Typical effects patterns available for use with all Kliegl slide and pattern projectors are illustrated in Fig. 9-6C.

Some means for controlling light fixtures from the floor, without the need for scaffolds and ladders, are shown in Fig. 9-7. The Kliegl *Pole-Op* method (Fig. 9-7A) makes use of a three-section pole to permit turns, tilts, and focusing. It is available on all 6-, 8-, 12-, and 16-inch light units. Fig. 9-7B shows the Kliegl 120 TV raising and lowering device, which has a safety collar and hanging hook for a pulley system. It provides 360° horizontal rotation, with a yoke to provide vertical tilt. The pantograph in Fig. 9-7C is a single-spring counterbalanced pantograph permitting raising and lowering, 360° horizontal rotation, and vertical tilt.

The studio lights that must be placed overhead are usually suspended from a network of 1½-inch steel pipe known as the *grid*. The pipe cross-member battens are 4 to 6 feet apart, with adequate clearance to the ceiling to allow electrical raceways, conduits, and sheaves for the wiring, and ventilation ducts. When the studio ceiling is high enough, catwalks may be included around and above the grid to facilitate light adjustment or the hanging of scenery by studio personnel.

The subjects of planning, installation, wiring facilities, and light sources themselves could well constitute a complete book of their own. However, this introductory outline should orient the reader for further study.

(A) 6-inch, incandescent.

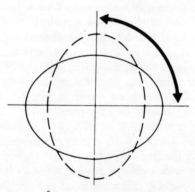

Rotating the lens through a 90⁰ arc positions the projected oval beam as shown above.
The amount of lens rotation exactly governs the degree of vertical or horizontal placement.

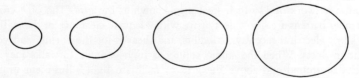

This controlled oval beam always maintains a spread ratio of approximately 3 to 2
regardless of whether flood or spot positions are required.

(B) Action of oval-beam lens.

Fig. 9-4. Examples

(C) 6⅜-inch Fresnel lens.

(D) Location of barn doors.

Courtesy Kliegl Bros. Lighting

of spotlights.

(A) Q-Lite.

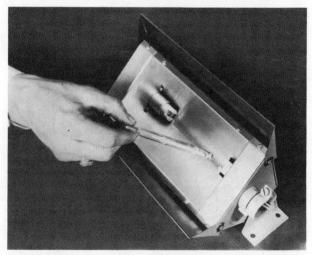

(B) Nook Light.

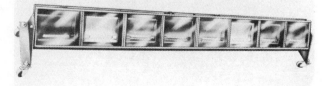

(C) *Cyclorama lights.*

Courtesy Kliegl Bros. Lighting

Fig. 9-5. Specialized lighting equipment.

(A) Slide and effects projector.

(B) Effects-pattern projector.

(C) Typical effects patterns.

Courtesy Kliegl Bros. Lighting

Fig. 9-6. Slide and pattern projectors.

(A) Use of pole.

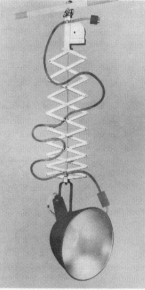

(B) Raising/lowering device.

(C) Pantograph.

Courtesy Kliegl Bros. Lighting

Fig. 9-7. Methods of adjusting light fixtures.

9-3. OPERATIONAL FEATURES OF STUDIO LIGHTING

The basic problem in television studio lighting is this: The light must help furnish the mood called for in the program script, *as interpreted by the "eye" of the TV camera.* The interpretation center of the camera is the pickup tube. Thus it is important to know how lighting effects will be interpreted by the pickup tube. There is a specific kind of situation that each type of camera tube can handle best. There are, however, invaluable generalizations that should be understood before the finer and more sophisticated aspects of pickup tubes are examined. The logical starting point is the basic television process, the monochrome pickup.

Monochrome Pickup Tubes and Lighting

Combinations of different types of lighting sources are used in monochrome primarily to achieve a color spectrum that approaches the daylight range. Recall that some image orthicons have a response well into the infrared range. Infrared waves have a property of penetration into the skin so that men—and some women—appear to have a beard, even though none can be seen with the naked eye. When such tubes as the Type 5655 were used, "minus red" filters, such as the Corning 9788, often were placed over the camera lens. This particular difficulty has been overcome by the development of the Types 5820 and 5826; the 5820 (and its field-mesh counterpart, the 7293) is the most popular pickup tube in use at monochrome studios at the time of this writing.

The amount of incident light normally required for modern studio operation in monochrome seldom exceeds 300 foot-candles. Table 9-1 shows the RCA recommendations of incident light required for the Type 5820 image orthicon. As shown, the most commonly used lens-stop openings are $f/8$ to $f/16$, which give adequate depth of field for most studio pickups. The table indicates a maximum required light intensity of 240 foot-candles. The tabulated incident light indicates the amount necessary at the scene of action, as measured with a light meter (color-corrected) facing the camera lens and perpendicular to the lens axis.

Table 9-1. Incident Light Required for Type 5820 Image Orthicon

Lens Stop	Incident Light (Foot-Candles)
f/1.9	2-4
f/2.8	4-8
f/3.5	6-12
f/5.6	16-32
f/8 ⎫	32-64
f/11 ⎬ Most Often Used	60-120
f/16 ⎭	120-240

The "interpretation" of various kinds of lighting sources by the Type 5820 image orthicon may be seen by observing Fig. 9-8. In Fig. 9-8A may be seen the relative energy of solar radiation and the popular incandescent lamp, as compared to the response of the average human eye. (This may be called the "interpretation" by the human eye and forms the basis of our comparisons.) Fig. 9-8B shows the image-orthicon interpretation of the incandescent lamp, as well as two types of fluorescent lamps, when used without filters over the lens. Fig. 9-8C illustrates the image-orthicon interpretation when a No. 6 Wratten filter is used over the lens system. (The curves in Figs. 9-8B and 9-8C are obtained by finding the product of the image-orthicon and light-source characteristics.) Most tubes have an extended response into the ultra-violet region, which sometimes results in unnatural rendition of skin tones as does infrared radiation. At the time of this writing, the most satisfactory method of handling this problem is the use of a "minus blue-violet" filter such as the No. 6 Wratten. Note from Fig. 9-8C that using this filter with incandescent lighting results in a response very similar to that of the human eye, providing a more predictable scale with which to work in studio lighting from visual observation. This also allows "mixing" of fluorescent with incandescent sources, since the Wratten No. 6 filter attenuates the response in the violet region.

NOTE: The use of fluorescent lights is gradually disappearing in practice. Such lights still are used in limited applications for monochrome, but almost never in color because of the pulse-type response over the blue region.

Incandescent Versus Quartz-Iodine Lighting

The reader should have gathered by now that there are two basic types of lamps used in fixtures for studio lighting: the incandescent lamp and the quartz-iodine lamp. It is important to understand the similarities and differences between these types of lamps.

A color camera must be "balanced" to a given color temperature of light. The normal Kelvin temperature for studio lighting varies between 2800 K and 3200 K, the latter being preferred for color telecasting. If a color camera is balanced on a set using 2800-K lamps, and then moved (without readjustment) to a set using 3200-K lamps, the high lights will turn blue with an unnatural change in skin tones. If the camera is balanced to the higher K temperature and moved to a set with lighting at the lower K temperature, high lights will turn orange, again with an undesirable change in the very important skin tones.

There is no difference between the new-lamp color temperatures of incandescent and quartz-iodine lamps. Either type is obtainable in 3000-K or 3200-K versions. The major difference is in the stability of color temperature and light output with use (aging) of the lamp.

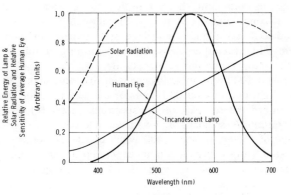

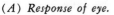

(A) Response of eye.

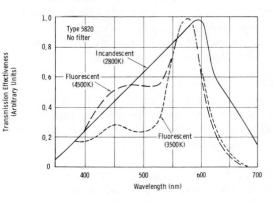

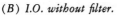

(B) I.O. without filter.

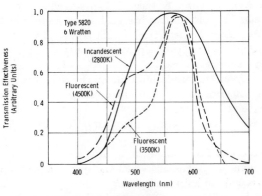

(C) I.O. with filter.

Fig. 9-8. Spectral response of Type 5820 image orthicon.

The quartz-iodine lamp has a tungsten filament and is actually an incandescent lamp, with the same type of continuous spectrum. The difference is that it has a quartz envelope and contains an iodine-vapor atmosphere. This technique keeps the lamp burning at full output and at rated color temperature throughout its life.

A standard 1000-watt incandescent lamp has only about 75 percent of its initial light output after 200 hours of use, and its color temperature is shifted toward the red part of the spectrum (2800 K) from its initial 3200-K rating. A standard 500-watt incandescent lamp undergoes the same type of change in about 400 hours of use. In contrast, a 500-watt quartz-iodine lamp will maintain about 95 percent of its initial light output after 2000 hours of use, with no practical change in the color temperature of the lamp.

Thus, it appears that the more recently developed quartz lamp may eventually replace the commonly used incandescent lamp for studio lighting. In the meantime, the practicing technician may find the two types intermixed, or one type used in one location and another in other locations. It is therefore important to understand the production problems that may be associated with this type of installation.

Operational Terminology

It is always highly desirable for all personnel to "talk the same language." We have previously defined the terminology applied to physical lighting fixtures and accessories. We will now give the terminology currently applied to operational features of lighting technique.

1. High-Key Lighting: A type of lighting which, when applied to a scene, results in a picture that has graduations falling primarily between gray and white. Dark grays and blacks are present only in very limited areas.

2. Low-Key Lighting: A type of lighting which, when applied to a scene, results in a picture that has graduations from middle gray to black with comparatively limited areas of light grays and whites.

3. Key Light: The apparent principal source of directional illumination falling on a subject or area.

4. Base Light: Uniform, diffuse illumination, approaching a shadowless condition, sufficient for a technically acceptable television picture. It may be supplemented by other lighting.

5. Fill Light: Supplementary illumination to reduce shadow or contrast range.

6. Cross Light: Front illumination of a subject in equal amounts from two directions at substantially equal and opposite angles with the camera optical axis and the horizontal plane.

7. Back Light: Illumination from behind the subject in a direction essentially parallel to a vertical plane through the camera optical axis.

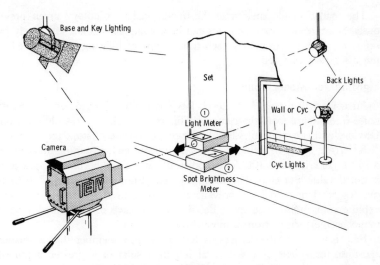

Fig. 9-9. Light measurements in studio.

8. *Side Back Light:* Illumination from behind the subject in a direction not parallel to a vertical plane through the camera optical axis.

9. *Eye Light:* Illumination on a person to produce a reflection from the eyes (and teeth) without adding significantly to the light on the subject.

10. *Set Light:* Separate illumination of background or set other than that provided for principal subjects or areas lighted according to items 3 through 8 above.

9-4. LIGHTING TECHNIQUES

One fundamental of studio lighting must be learned from the beginning; this can be emphasized by the simple diagram of Fig. 9-9. When we speak of "incident light," we mean the light measured with a foot-candle meter on a level with and pointed toward the camera lens from the set area. "Brightness" of any small area of the set is measured with a foot-lambert meter of special design pointed toward the area in question, away from the base and key lighting for the set.

The next basic rule to understand is that light meters are used only as guides for unfamiliar situations. The camera, and the system it feeds, must be the final "judge" of adequate lighting.

Now comes our final basic rule: It is impossible to dictate rules for effective lighting! It is the author's intent here only to offer suggestions and solutions to special problems for the purpose of "getting started." It is hoped that this report of general practice will be of help to practicing personnel concerned with lighting, and will benefit beginners in getting acquainted with the field.

The reader should memorize the operational terminology given previously. Some readers may find this can be accomplished most effectively by referring to this terminology each time a type of lighting is mentioned in the discussion that follows.

Lighting for Monochrome

First, we will discuss light level. Base light must provide an acceptable noise-free picture from the cameras. This is a flat, shadowless illumination. Obviously, this condition is highly dependent on the $f/$ stops required.

Referring again to Table 9-1 (for monochrome), assuming an average stop of $f/11$ is used, 100 foot-candles of base light becomes a practical nominal value. For color operation, a practical nominal value of 250 foot-candles (at lens stop $f/8$) prevails at the time of this writing. We will explore color lighting in more detail after we have first obtained a good monochrome picture from a monochrome camera.

Before we leave the subject of lens $f/$ stop, remember that a smaller aperture, for a lens of given focal length, results in a greater depth of field. Also, with a smaller aperture (larger $f/$ number), more light is required. In some cases, it is desirable from a program-mood standpoint to have background persons or objects deliberately out of focus so that they are subordinate to the central, focused theme. With this requirement, a larger lens opening must be used, or the camera must be moved very close to the focused subject. (If necessary, review Chapter 4.)

Assume for the moment that we have established a base light of 100 foot-candles to start (camera gives a technically acceptable picture). This lighting is far from being "interesting" or "artistic," and therefore it usually is accentuated by various *modeling-light* techniques. These techniques were described briefly under the subheading "Operational Terminology."

It is pertinent at this point to examine the CBS and NBC lighting practices (pertaining to light levels in foot-candles) that were used with monochrome cameras. CBS used $f/$ stops ranging from $f/3.8$ to $f/8$, with $f/5.6$ a nominal value. Base levels were as low as 30 foot-candles with nothing on the set receiving over 100 foot-candles. NBC $f/$ stops ranged from $f/8$ to $f/16$ with a typical base of 65 foot-candles and all other lighting (key lighting, principally) ranging from 100 to 150 foot-candles. Contrast this with the lighting practice of many smaller stations (actually with less overall effective lighting) where base lighting ranged to 200 foot-candles and all other lighting to 350 foot-candles. We will explore the reasons for this discrepancy as we progress. (Remember that, at this time, we are examining monochrome techniques only.)

Key lighting is the "key" to the visual appearance of the scene, since it is the principal factor in determining the camera lens opening. Key lighting should, ideally, provide good scenic illumination from any possible camera angle during the program. Such lighting therefore usually (but not always) is placed at the front of the pickup area at average camera level.

This requires pantographs or floor stands (generally of the dolly type) to provide easy mobility. The technique of key lighting varies considerably among stations. Some lighting engineers for monochrome will not mix "hot" and "cold" light sources to provide key illumination. Others mix these types of light sources, apparently with little difficulty. The fact is that no hard-and-fast rule can be applied here, since the color combinations of the scenery, stage props, and costumes affect the overall interpretation considerably.

In general, key light used to give the effect of a predominant source of light is from one to two times the base-light level, with the base light from the same direction as the key light reduced in direct ratio to the added key light. For this reason, many lighting directors start with key light in establishing camera operating parameters. Base light is then used primarily as fill light.

In practice, mounting positions for key lights are influenced by required camera angles, microphone-boom positions, and the nature of the scene design. Sometimes, these lights are mounted to the side of or slightly toward the back of the set with the beam directed downstage, thus throwing shadows out of camera range toward the front of the set.

The basic problem in key lighting is the number of shadows cast on the rear wall of the set (except, of course, where this effect is deliberately introduced). If key lights on the open side of the set (side toward the camera positions) are kept lower than about 8 feet, microphone-boom shadows are thrown out of camera range into the upper portions of the set. Pattern projectors (Fig. 9-6) sometimes are used to throw break-up designs on the rear set wall to minimize undesired shadows.

Fill lighting is illumination added to portions of the scene or subject to register more detail in the shadow areas. This type of illumination calls for a diffused source, and with key lighting at or near camera level, fill lighting is usually placed overhead. Floodlights equipped with diffusers and beam adjusters to result in a narrow beam instead of a flood spill are often used, or Fresnel-lens spotlights with diffusers may be employed. This lighting fills-in areas that the key lighting cannot illuminate adequately.

Basically, modeling, or accent, lighting may be used to create artistic effects, or to enhance the appearance of the scene or subject. Back lighting is a fundamental type of accent lighting used to provide pictorial separation between an object or subject and the background. Back lighting sets the subject out from the background. The most popular equipment used for this purpose is the Fresnel-lens spotlight, equipped with a barn door that controls the shape of the beam pattern. Back-lighting sources sometimes are placed at an angle of 90° to 135° with the camera orientation, and as low as possible. Accent lighting may include any combination of the basic types of lighting to achieve a desired effect.

Back lighting normally is held to within 1 to 1½ times the base lighting. When this light is measured with the foot-candle meter, the meter should be pointed toward the source of the back light from the central set area, and all other light sources should be "doused" (extinguished). When back lighting comes from relatively high angles, care must be exercised to avoid vertical top light. This causes excessive shadowing under the eyes and nose, and usually results in the video operator's calling for more light on the scene.

It is now becoming apparent to experienced personnel that the call for more light on faces must not always be taken literally. Exactly the same effect can be obtained by correcting any faulty top lighting, or by reducing the intensity of light falling on the background. The camera tube will "set up" on the brightest portions of the scene, and a silhouette of persons in the foreground can well result when a condition of excessive background brightness is extreme. This is the primary reason why, as previously mentioned, unskilled lighting personnel may finish their adjustments with a total of 350 foot-candles or more of incident light on a monochrome camera setup. The monochrome rendition of flesh tones, and the entire gray-scale interpretation of the production, is far better at a maximum of 150 foot-candles for a monochrome camera. The "trick" is to get the proper balance between faces and background, at the proper contrast range for the TV system. We will discuss methods for attaining this objective in practice shortly.

Set light is used primarily on the set wall, or on selected props in the set. It is used to "key" the set itself either as part of scene motivation or simply to control backgrounds properly. With proper adjustments of barn doors on all accent lighting (including key), the set light, also properly barn-doored, can add the final touch to an otherwise unimpressive lighting job.

Effects lighting embraces a wide range of light sources, special accessories, and application techniques. Fire light, window light, cloud effects, "lightning," and a simulated moon are examples of effects possible with lighting sources. Such lighting also may include patterns projected on the studio wall. Projection lamps ranging from 75 watts behind a 3-inch Fresnel lens to 5000 watts behind a 16-inch Fresnel lens may be used in spotlights. An external adjustment handle or knob usually is included to permit a spill-adjustment variation of 5° to 50°. A frame on the front of the spotlight allows the use of any variety or shape of mask, for transmission of any type of light beam desired. Special motor-driven effects discs sometimes are used in front of a spotlight lens for special effects. Thus, for example, shadows of a roaring fire, seemingly originating outside a window, may be thrown across a room.

Now we will consider high-key and low-key lighting. Most studio productions involve high-key lighting, since this is the normal type of lighting for news programs, quiz shows, panel discussions, variety shows, etc. Low-key lighting comes into use largely in dramatic productions or, in special cases, in variety shows.

Low-key scenes are the hardest for the system to handle, and they require considerable skill on the part of lighting directors, scene designers, and video operators. In dramatic productions, a low-key scene normally depicts a somber or threatening mood. Typical low-key scenes have dark backgrounds against which the actors (or dancers in certain variety shows) are lighted in a normal manner. The direction of lighting usually appears to be strong from a specific direction, but this usually requires carefully spaced cross lighting to achieve the desired result from more than one camera angle.

The night scene is a typical low-key application. Here the background is very dark, with attendant problems in excessive contrast unless something such as a moon or other bright object is used to break up the large dark area. In low-key studio scenes, projection of some high lights on the dark walls alleviates this problem and helps to provide a reference for the video operator.

Now we turn to the all-important contrast range for studio lighting and studio sets. We have learned previously in this text that the camera tube has a gamma less than unity and the picture tube has a gamma greater than unity. The average picture tube has a power-law response with an exponent between 2 and 3. A picture tube without excessive stray light or internal scattering of rays can reproduce a brightness range of 40 to 1. This brightness range can be produced by less than a 15-to-1 range in brightness in the actual scene. Employment of gamma correction in the camera reduces this contrast enhancement, but correction can be used only up to the point at which noise in the picture becomes apparent.

We must meet one basic challenge: Props, backdrops, important set areas, and people and their clothing must have reflectance values and contrast ranges the TV system can reproduce accurately. Exceeding the proper contrast range simply means the camera can be adjusted to give good dark-area detail at the expense of washing out detail in the lighter areas, or to give good detail in the lighter areas at the expense of compressing all detail in the darker areas.

See Fig. 9-10. This chart is now in general use and serves as an excellent guide in staging reflectance ranges. White flesh reflects 30 to 40 percent of the incident light. Basic scenery elements under high-key conditions should be allowed to reflect only slightly more; i.e., 50 percent (approximately). For low-key scenes, basic scenery reflectance should preferably be at least 25 to 33 percent as much as for flesh; this means about 10-percent reflectance as a minimum.

Now note a very important point: The specified limits allow only about a 5-to-1 range for basic scenery reflectance. This includes EIA gray-scale steps 2 through 6. Under modeling-light levels, this range may extend perhaps to about 10/1. To insure good skin tones and to allow for camera dependence on *average* scene brightness, it is very important to limit the range for larger areas as shown on the chart.

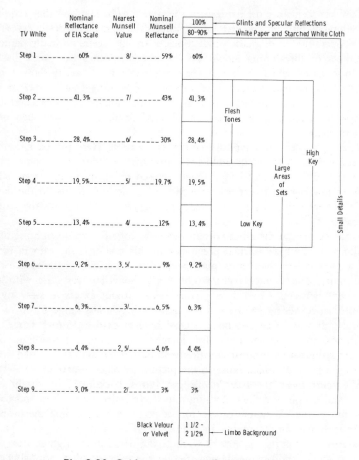

Fig. 9-10. Guide to staging reflectance ranges.

Lighting for Color

Lighting is in the anchor position of any color production chain. The basic function is to supply camera operating requirements while providing as great a latitude as possible in making the scene interesting and varied. The principal points of interest in color lighting are spectral distribution, lighting levels and ratios, and special-effects lighting.

The incandescent lamp of 3000 K or 3200 K is the backbone of color lighting. The major drawback with the ordinary incandescent lamp is that it does not hold its color temperature for very long after it is installed. As the bulbs age, their Kelvin temperature goes down, which means toward red. This characteristic has led to development of the more recent quartz-iodine lamp, which was discussed previously.

Remember that we can balance a color camera to studio light of any color temperature. When the gray scale under this lighting results in zero subcarrier output, the camera is color balanced (Chapter 10). The problem is in keeping all areas, large or small, at the same color temperature. To emphasize this, think of 20 square feet illuminated with 3000-K bulbs, while the remaining 20 square feet of the pickup area is illuminated with 3200-K bulbs.

This brings up the question, "How much variation in color temperature can exist in practice before the camera is no longer color balanced?" The answer is that the *critical* viewer can detect a change of ±20 K; a change of ±100 K has been established as the maximum allowable before the *average* viewer notices the change. The change that results from this variation in color temperature is not noticeable in most of the scenic colors; it is *much more noticeable* in skin tone.

The color temperature of lighting changes about 10 K per volt on a 120-volt circuit. Thus, theoretically, the line voltage can change 10 volts before a 100 K change occurs. The trouble with this bit of philosophy is that as incandescent lamps age, they appear to change in temperature considerably more than this with changes in applied voltage. Again, the quartz-iodine lamp minimizes this problem.

In lighting levels, we begin with the minimum base light required for the particular camera chain (about 250 foot-candles). To this we add back light, key light, and any additional modeling or cross light called for by the production. A total of 300 to 350 foot-candles normally is used on average color productions, with proper contrast range obtained by lighting and control of scenic reflectance values. NOTE: We will explore studio lighting for color cameras more fully in Chapter 10.

Here are some general rules for lighting and staging for color. These rules are almost, but not absolutely, ironclad:

1. Keep incident light uniform within ± 10 foot-candles except where special effects *must* be used.
2. Hold back light to between ½ and 1 times base light.
3. Hold modeling light to a maximum of 1 to 1½ times base light.
4. Try to use small catch lights for eyes and teeth on close-ups. This enhances contrast of monochrome reproduction and is pleasing in color.
5. Background material or lighting should be low in saturation, medium in luminance, and preferably of matte surface.
6. Include in scattered areas throughout each scene some reference white, reference black, and props that provide some reflected energy for each of the primary-color channels to achieve uniformity of camera exposure and shading.
7. Hold the "large area" of each scene to a maximum contrast range of about 5 to 1, corresponding to steps 2 through 6 of the EIA gray scale.

8. Keep specular reflections extremely small. A "hot spot" can ruin an entire picture if this rule is ignored.

9-5. LIGHTING CONTROL

Lighting-control equipment has come a long way since the early days of television. The SCR (silicon controlled rectifier) dimmer forms the basis of most modern design concepts for control of light intensity.

Remote operation usually is employed in lighting-control systems. The use of remote control is mandatory under certain circumstances, such as where presetting of scenes is desired, where two or more control stations are needed, or where remote location of the control console is required. With a remote system, fingertip controls are assembled in a control console. This control center may be located away from the dimming bank in any desired location. At the same time, the power units can be located conveniently near the lighting fixtures so that feeder runs and other wiring may be kept to a minimum.

The diagram of Fig. 9-11 illustrates a typical remote-control system. The control console and preset panel (if required) may be located at a good vantage point away from the dimmer bank. Located between the dimmer bank and the lighting fixtures is a cross-connect circuit-selection panel for the interconnection of lighting loads and dimmers.

A typical rack-mounted SCR dimmer bank is shown in Fig. 9-12. This particular unit contains 18 plug-in SCR dimmer units, available in 3-, 7-, or 12-kW capacities. Each unit employs automatic current-limiting circuitry so that, in case of overload, operation at a reduced voltage results until the overload is removed.

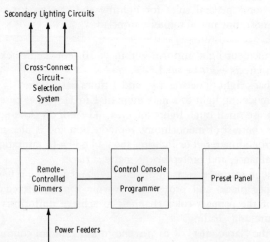

Fig. 9-11. Remote-control system for lighting.

Fig. 9-12. SCR dimmer bank.

Courtesy Kliegl Bros. Lighting

One type of cross-connect circuit-selection system is illustrated in Fig. 9-13. The *Rotolector* is a rotary circuit selector that is used to cross-connect lighting load circuits to dimming or nondimming feeder lines. Fifty of these selectors are shown in the system of Fig. 9-13. Each branch lighting circuit is wired directly to its own selector, which, in turn, is connected to the various dimmer and nondimmer feeders. For example, a 12-point *Rotolector* can be fed by 10 dim and 2 nondim sources, or any other combination. To patch the load to any one of the feeders, the operation is as follows:

1. The handle of the *Rotolector* is withdrawn. This automatically trips the associated circuit breaker before the load contact is broken.
2. The handle is then rotated to the selected position, as indicated by the numbers on the dial, and then pushed in until contact is made.
3. Only after contact is made can the circuit-breaker switch be turned on to energize the circuit. Thus, "hot" connections and dangerous arcing are automatically eliminated.

The magnetic-type circuit breaker built into the *Rotolector* protects the outgoing lighting circuit and, at the same time, can be used as an on-off switch. Each 6½″ × 6½″ unit can be furnished with either 12 or 24 positions and in two sizes to handle either 20- or 50-ampere lighting loads.

Fig. 9-13. Lighting circuit-selection system.

A typical two-scene, four-subscene SCR control console with split faders is shown in Fig. 9-14. The console employs plug-in independent potentiometer controls; this particular console has a capacity of 24 dimmer controls. The scene-1 and scene-2 boards each have independent master faders for the assigned faders. There are two submaster controls for each side of the fader. Each submaster controls, within the selected scene, the potentiometers on the preset panel that have been assigned to it.

With the split-fader control, the following functions can be performed:

1. Fade directly from one scene to another with a single action of the control handles (one-hand operation).
2. Operate the fader-control levers either independently or jointly.
3. Fade in or out of desired scenes completely or partially, as desired.
4. Change individual scene fade-in or fade-out time as desired by fading one scene more or less rapidly than the next one.

Fig. 9-14. Lighting control console.

5. Use either fader control as the master dimmer for the submaster controls on the same side of the fader, thereby making proportional dimming of submasters possible.

The operational technique of dimmers can be highly effective for monochrome or color productions. For monochrome, dimmers also are very useful in quickly adjusting for any desired lighting contrasts. In the case of "hot" (too intense) backlights, or excessive intensity of base light resulting in a "flat" picture, dimmers allow a "fine trim" for proper ratios. In monochrome transmission, light sources can be dimmed to about 25 percent of normal intensity without impairing gray-scale response of scenic colors because of change in color temperature of the lights.

In color, we must be more limited in the use of dimmers in one important aspect, the all-important flesh-tone rendition of faces. This aspect is well covered by the following excerpts from a lecture by Herbert R. More, vice-president and manager of Kliegl Bros. Lighting:

"In color, one tries never to disturb the Kelvin temperature of the lights on the faces but permits unlimited changes in the rest of the lighting. This is actually what is done to create moods in color, and is what dimmers in color are largely used for. For example, to create an evening mood in a garden or on a waterfront, you leave the key and modeling lights aimed at the subject at full Kelvin, but dim all the others until the desired effect is obtained. Since all lights except face lights are being dimmed, more dimmers rather than fewer are required for color than for monochrome. Then, too, the lights that are used with filters to illuminate the cyclorama curtain in various colors are almost always dimmer controlled. This explains why network as well as independent colorcasters demand more dimming capacity than they used for monochrome.

"In summary then, it would seem that the 'don'ts' actually mean 'don't change the skin tones,' whereas the 'do's' acknowledge this fact but want dimmers for the mood and background effects.

"In addition to the above two schools of thought, there are some who insist that they are getting away with dimming faces. Within limits this may be done. Station colorcasting practices have established that a drop of 200 K can be tolerated even on skin tones before the public (not studio engineers) can notice a change. This actually gives the lighting man a great deal of leeway, since the 200-K drop amounts to a 20-volt drop at the lamp socket and a foot-candle drop of 33 percent. This means that if a luminaire delivers 300 foot-candles on the subject's face, it may be dimmed down to 200 foot-candles before the skin tones change objectionably. One must be careful, however, to stay within this 200-K range. Although this breed of lighting man can prove that he is right, many station managers and chief engineers prefer not to take the risk of over-dimming or meddling with the skin tones."

EXERCISES

Q9-1. Is there a direct relationship between light intensity (candles) and wattage of a lamp?

Q9-2. What is the average K rating of luminaires used in modern telecasting studios?

Q9-3. Where are striplights normally used?

Q9-4. What is the basic practical difference between an incandescent and a quartz-iodine lamp?

Q9-5. How is incident light measured?

Q9-6. How is brightness measured?

Q9-7. What is the maximum brightness-contrast range for optimum control of the camera for (A) large areas of a set and (B) small details?

Q9-8. What is the approximate maximum incident light used for (A) monochrome and (B) color studio sets?

Q9-9. Are light dimmers ever used in color studios?

Q9-10. For color lighting, give the optimum ratio relative to base light for (A) back light and (B) key, or modeling, light.

CHAPTER 10

Operation of Studio Equipment

The scope of equipment required in TV control rooms varies widely, depending on the program range of the particular station. An independent or network-affiliated station in a small community might have very limited equipment, such as film and slide chains, with no locally originated programs. At the other extreme, a station originating network programs could have multiple studios with one to six cameras in each, complex associated control-room equipment, a master control area, many film and slide machines, and elaborate distribution facilities for local and network routing of the signals. In such cases, the audio facilities alone may represent triple the cost of all the equipment in a small TV station. Because of this wide range of requirements, the author has chosen for discussion in this chapter a typical setup midway between the extreme cases cited above: a network-affiliated station with one studio, a control room for the studio, and master-control and film-projection facilities. This approach is most appropriate, since every type of control-room equipment may be covered, the only difference in any particular application being in the number of control units involved.

Fig. 10-1 illustrates a typical control-room layout. Fig. 10-1A is a plan view, and Fig. 10-1B shows the relative heights of the various portions of the system. This is an example of the "platform" type of control room that is quite popular in practice. Such a design enables the audio operator and program director to view both the studio action and the control-room monitors. The units immediately in front of the video operator (VO in Fig. 10-1A) are the individual control units for the studio cameras. Those in front of the technical director (TD) are the switcher and mixer units, which usually are operated by the technical director upon instructions from the program director. The audio operator (AO) operates the usual audio switcher and mixer, and the two turntables shown to his left. The program director and technical director have intercom facilities enabling them to converse with the production and technical personnel on the studio floor.

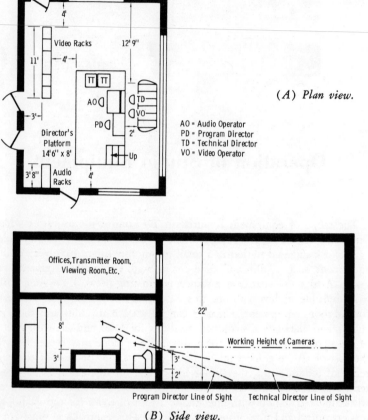

(A) Plan view.

AO = Audio Operator
PD = Program Director
TD = Technical Director
VO = Video Operator

(B) Side view.

Fig. 10-1. Typical studio control-room layout.

The audio operator is provided with similar facilities so that he may instruct the microphone-boom operators. Fig. 10-1B indicates the visibility necessary for the control-room personnel.

Fig. 10-2 is a close-up view of the TD/director section of a console as installed at WBBM-TV. Fig. 10-3 shows an overall view of the operating positions for Studio 1 at the same station.

The video switching position (TD) and director's position are usually (but not always) situated so that visual contact exists with the studio action. However, the video operator's position (at the individual camera-control units) sometimes is situated in a "blind" area somewhere outside the audio-video switching area.

Each studio camera is connected to its respective control unit by means of flexible coaxial cable (within the multiple-conductor camera cable) of about 50-ohm impedance. Each control unit contains a picture tube show-

Courtesy WBBM-TV

Fig. 10-2. Television control room.

Audio Operator   TD   Directors

Courtesy WBBM-TV

Fig. 10-3. Operators in studio control room.

ing the picture output of the camera, and a small oscilloscope to monitor the associated waveforms. It is at the control unit that any necessary adjustments during a show for picture contrast, brightness, or shading are carried out. The video level into the switcher also is adjusted at this point, and is determined by observing the signal levels on the oscilloscope. The control unit for the film camera also is in this group of units, and provides the same facilities for film telecasting as those for the studio cameras. The camera switcher (TD) serves to select a studio-camera or film-camera output, or such effects as lap dissolves and superimposition of any two signal sources. Its associated line monitor shows the picture thus selected and going to the master switcher.

The master switching control (Fig. 10-4) contains a number of push buttons so that the operator at this point may select any of six or more incoming signals (such as from a studio, remote-point coaxial cable, microwave-relay receiver, etc.), and feed them to either of two (or more) outgoing lines. This unit has monitors for preview of all incoming and outgoing signals. The push buttons on this unit operate rack-mounted master switching relays or crosspoints used in the video and audio circuits.

Television shows are born in the minds of clients, writers, producers, and directors to meet a variety of needs limited only by the wide variety of human interests. Equipment concerned with transferring these paper ideas to the minds of the viewers and listeners may be compared to a musical instrument. The finest pipe organ in the world can fill the ears with all the drama of sound of which it is capable only under the touch of commanding and understanding hands. Likewise, television equipment of the finest engineering design can reproduce the original colors, lights, darks, and grays of a scene only under the guidance of commanding and

Courtesy WBBM-TV

Fig. 10-4. Television master control console.

understanding hands. In addition to the necessity of obtaining good image resolution, definition, brilliance, and contrast, the engineer must be conscious of artistic values in both picture and sound. Without that consciousness, he is like an actor who merely reads his lines instead of dramatizing them.

This chapter is dedicated to the purpose of bridging the gap between television theory and the practical use of the equipment. It is designed to answer the all too prevalent cry, "I know the theory, now what do I do on the job?" It is with a humble feeling indeed that the author faces the challenge of analyzing operational problems and their solutions. The fact remains that there is no subject more in need of expansion and so inexhaustible in scope. It is only hoped that the basic foundation given here will prove of service to the trade and inspire future contributions to the field.

10-1. CHECKING THE MONITORS

John R. Meagher, writing for the receiver servicing trade in the *RCA Radio & Television Service News,* has this to say:

"Many TV owners are extremely fussy about having the circles (of a received test pattern) exactly round. Some of them check the circles by holding a small plate in front of the screen, and others measure the wedges to see if they have equal lengths. In some TV areas, this makes life extremely difficult for the television technicians, because it is an unfortunate fact that some stations do not transmit good linearity. Also, the linearity may be different from one camera to another. In one particular city, if the receiver is adjusted so that the test-pattern circle is round on the first station, the pattern from the second station will be egg-shaped vertically, and the pattern from the third station will be egg-shaped horizontally. . . ."

Unfortunately, Mr. Meagher's statement concerning "one certain city" is all too true for a number of "certain cities." How does this come about? If the viewers were to visit the different stations involved, they would find the appearance of the test pattern on the station monitors in good order. It is safe to say that no station *intentionally* transmits nonlinearity in the signals.

The error occurs in setting up and aligning camera chains without first checking and adjusting all monitors in the system against a separate and accurate signal source. It is easy to compensate for nonlinearity of camera sweep by an opposite nonlinearity in monitor sweep. Thus, if one camera chain is adjusted to obtain optimum results on the monitors, and all other cameras are adjusted in the same manner, there is no assurance that the transmitted signal is strictly linear in nature. This practice imposes a severe handicap on viewers of more than one station and on the receiver service technician.

The first logical step is to perform size, linearity, and aspect-ratio checks on the master monitor (usually the line-output monitor) in the control

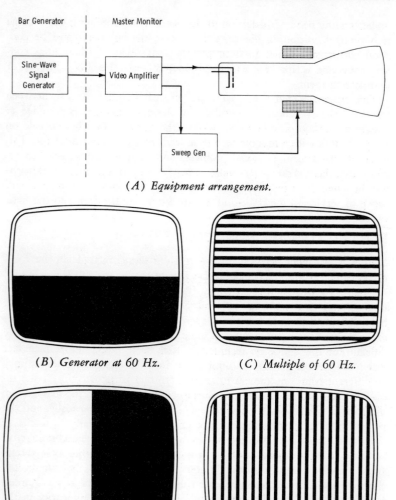

(A) Equipment arrangement.

(B) Generator at 60 Hz.

(C) Multiple of 60 Hz.

(D) Frequency of 15,750 Hz.

(E) Multiple of 15,750 Hz.

Fig. 10-5. Principle of a linearity standard.

room. Individual camera monitors then may be checked against the master monitor. There are two methods of checking picture monitors: using a sine-wave generator, and using a grating or crosshatch generator. The idea in either is to place horizontal or vertical bars (or both) on the screen of the picture tube. In Fig. 10-5A, a stable sine-wave oscillator is connected to the video amplifier of the master monitor. The monitor is operated so that the sync voltage for the sweep circuits is taken from the video amplifier as it is when the monitor is used as a relay-receiver monitor. The sweep is driven from the 60-Hz and 15,750-Hz multivibrators fed from the video amplifier, which in this case is receiving no signal other than the single-frequency sine wave. This sine wave therefore is used simultaneously to modulate the grid of the picture tube and as sync voltage for the picture-tube sweep circuits.

Consider the action if the frequency of the sine wave is 60 Hz. On the positive half-cycle of this signal, the grid of the kinescope is swung in the positive direction, increasing the raster brightness. During this half-cycle time of 1/120 second, the scanning beam, the vertical movement of which is synchronized by the sine wave, has moved about halfway down the screen. The negative half-cycle now reduces the screen brightness. Fig. 10-5B shows the resulting raster, consisting of two horizontal bars, light at the top and dark at the bottom. (If the sweep had happened to sync first on the negative half-cycle of the applied signal, the dark bar would be at the top and the light bar would be at the bottom.)

If the signal-generator frequency is now increased to some multiple of 60 Hz, but still a frequency too low for the horizontal circuit to accept, a number of additional pairs of horizontal bars will appear across the raster. It is observed from the first example that each cycle contributes to the formation of one pair of bars, one light and one dark. Therefore, the number of pairs of bars placed on the raster is equal to the applied frequency divided by the scanning frequency (in this case, the vertical-scanning frequency), minus any pairs that are produced during the vertical-retrace time. If a 900-Hz signal is applied, the number of bars produced is $900/60 = 15$ pairs of bars. However, if the vertical-retrace time is 7 percent of the total scanning cycle, 7 percent of the above number of bars is redistributed back across the pattern too fast to be distinguishable. Since 15 *pairs* of bars are produced for a total of 30, and 7 percent of 30 is 2.1, $30 - 2$, or 28, bars are visible. Half of these bars are white and half are black, so we observe 14 black and 14 white horizontal bars. Fig. 10-5C illustrates a number of bars produced in this fashion; this pattern may be used conveniently to check the vertical-sweep linearity of the monitor circuits. If the bars are evenly spaced all the way up and down the raster, the vertical-scanning motion is of constant velocity and therefore linear in sweep action. If the vertical sweep is not of constant velocity, some of the horizontal bars will be spread out or crowded together, and the monitor vertical-sweep linearity control must be adjusted until the bars become evenly spaced. The linearity control is variously marked v SAW, v LIN, etc., depending on the manufacturer. Some operators have practiced the actual measurement from the leading edge of one bar to the leading edge of the next bar, repeating this process all the way up and down the screen through the use of calipers or other means. Accurately lined transparent masks to fit over the face of the picture tube sometimes are used to spot nonlinearity more conveniently.

If the frequency of the applied sine wave is made equal to or greater than 15,750 Hz, the frequency is sufficient to release and blank the beam during the horizontal sweep. If the applied signal has a frequency equal to the 15,750 Hz of the horizontal sweep, the screen becomes brighter on the positive half-cycle and darker on the negative half-cycle (Fig. 10-5D). Since a few thick bars are not sufficient to check linearity accurately, the frequency is increased in practice to 20 times this rate, or 315 kHz. Then, the number of vertical bars generated is $315,000/15,750 = 20$ *pairs* of bars. This means that 20 white and 20 black bars would appear if it were not for the horizontal-retrace time. If the retrace time is 17 percent of the total scanning time, the number of visible vertical bars is $20 - 3$, or 17 pairs of bars. The effect (for a different number of bars) is illustrated in Fig. 10-5E, and may be seen to provide a convenient means of checking horizontal-sweep linearity. If the bars are crowded at the left, for example, the horizontal-linearity control associated with the damper circuit should be adjusted, since this tube or transistor contributes to the sweep on the left side of the raster.

There may be linearity controls in the grid circuit and the cathode circuit of the damper tube. A control in the discharge-tube circuit affects the linearity of the sweep toward the right side of the raster. As a general rule, these controls are not found on the front panel of the monitor, but are located on the chassis or rear plate. Only the kinescope focus and brightness controls normally appear on the front panel.

The sweep height and width controls on the monitor also affect the linearity of the sweep to some extent. Thus, when the width control is adjusted, the linearity controls usually must be adjusted again. The same is true for a height adjustment. One linearity control normally adjusts the vertical sweep; usually it is a variable resistor in the cathode circuit of the vertical-sweep amplifier stage. This control alters the grid-voltage–plate-current characteristic in such a way that compensation is provided for the shape of the sweep curve.

The interrelation of this procedure for obtaining good sweep linearity with that for obtaining proper aspect ratio may now be seen. The raster on the screen is adjusted to the 3-to-4 ratio, as for example, 6×8 inches on a 10-inch kinescope. As the height and width controls are varied, the linearity controls must be varied also. As an example, adjustment of the vertical-linearity control, being essentially a cathode-resistance change, may also affect the gain of the vertical-sweep amplifier, and therefore the height. This points up the importance of carrying out the master-monitor adjustments with the aid of a bar signal superimposed on the raster so that the effects of each adjustment can be observed.

While this method is used in smaller stations without more elaborate equipment, and is satisfactory with extreme care and understanding, more satisfactory results may be obtained from commercial equipment especially designed for more rapid and accurate checks. The method previously described produces rather thick bars even with comparatively high sine-wave frequencies, and it applies only one group of bars, either vertical or horizontal, at a time.

Most modern sync generators incorporate a grating/dot generator as an integral part of the sync system. The grating-signal output is normally a composite signal with 17 vertical bars and 14 horizontal bars. This fits the normal aspect ratio so that the bars form squares on the monitor. Thus, the vertical-bar oscillator operates at a nominal 20 times the horizontal frequency, or 315 kHz. The horizontal-bar oscillator operates at a nominal 900-Hz frequency.

Fig. 10-6A illustrates the grating pattern from the RCA TG-3 sync generator as observed on the monitor. The extreme left and right bars are almost covered by the kinescope mask. Underswept monitors such as those used on camera controls and as master monitors will, of course, reveal the entire active area against the blanking backgrounds. Fig. 10-6B illustrates the same signal on a monitor with bad vertical linearity. Note the compression at the top and the stretch at the bottom of the raster.

Note that the actual video pulse is at white polarity for good visual display. The width of the vertical pulse is 150 nanoseconds, and the width of the horizontal pulse is two raster lines. The resultant lines are extremely useful in conjunction with a linearity ball chart for camera sweep adjustments as described later in this chapter.

While monitor alignment demands a certain feel gained only by experience, the new operator may obtain just as accurate results by taking a little longer and observing the following general routine. It is imperative that he first become familiar with the purposes of the controls on his particular equipment.

It is assumed that the monitor has been initially set up, with the picture tube properly positioned so that the raster is level, and with the focusing and deflection coils properly aligned. It is further assumed that the sync generator is delivering properly shaped and timed pulses to the system.

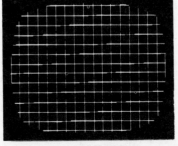   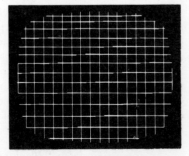

(A) Acceptable linearity.   (B) Poor vertical linearity.

Fig. 10-6. Use of grating signal for monitor-sweep adjustments.

After the equipment has been turned on for about 10 minutes, the raster may be checked to see that it is near the correct aspect ratio of four units wide to three units high. Controls (on the chassis or rear panel) associated with this adjustment are the height control, which affects vertical size, and the width control, which affects horizontal size. In some monitors, these controls are marked V SIZE and H SIZE. In any case, the first adjustment is not critical, since any adjustment of the linearity controls will alter these initial adjustments to some extent. The centering control should be set to place the raster in the middle of the screen.

The focus control on the monitor should now be double-checked for proper electrical focus. The lines of the raster and the bar signal should be as thin as it is possible to obtain by adjustment of this control. The pattern should now be well defined, although the linearity may be incorrect. Assume, for example, that the horizontal bars are crowded together toward the top (Fig. 10-6B). This indicates bad vertical linearity. The horizontal bars also could be "stretched" in the center with crowding at both top and bottom, or crowded only at the bottom. In any case, the vertical-linearity control of the master monitor must be adjusted until even spacing of all horizontal bars occurs. Any bending of the bars indicates the presence of stray magnetic fields.

As previously pointed out, adjustment of the vertical-linearity control may affect the height (vertical size), especially if the necessary correction is large. Therefore, readjust the height control for normal raster size. Any linearity correction now necessary should be a small one, and the process is repeated until correct linearity and correct size of the vertical sweep component are obtained.

If, at this time, the vertical bars are noted to be unevenly spaced, the horizontal-linearity controls must be adjusted. When the operator becomes familiar with his specific monitor, the portion of the raster in which horizontal nonlinearity occurs will provide a clue to which control should be adjusted first: There may be one control that affects overall linearity and other controls that affect the right and left portions. Keep in mind that the control associated with the damper stage in the horizontal-deflection circuit affects the sweep on the left portion of the picture, and controls in the driver stage affect the sweep on the right side. As the linearity is adjusted, it may be necessary to readjust the width control to maintain the proper aspect ratio.

Both aspect ratio and linearity are correct when the grating signal forms perfect squares, provided that 14 horizontal bars and 17 vertical bars are displayed on the monitor and proper system blanking is being used. Note that for 17.5-percent horizontal blanking and 7.5-percent vertical blanking, the following durations apply:

Horizontal Blanking = 11.1 microseconds

Vertical Blanking = 1250 microseconds

10-2. PRELIMINARY OPERATING NOTES
FOR THE IMAGE ORTHICON

The image-orthicon tube is the most complex of the pickup tubes, and it requires somewhat more operator skill and knowledge than the other types. The contents of this section should be studied by both the new and experienced operator so that setup techniques for the image orthicon become meaningful. The more preliminary knowledge the reader can absorb, the more skillful he will become in practice, with added proficiency in meeting unusual conditions.

There are several variations in the construction of I.O.'s. First, there are two basic sizes used for broadcasting, 3 inches and $4\frac{1}{2}$ inches. Second, there are field-mesh and non-field-mesh types. All $4\frac{1}{2}$-inch tubes use a field mesh, which normally is connected to an adjustable voltage. The 3-inch field-mesh tubes usually employ a fixed potential for the field-mesh grid, which generally is tied to the cathode. Other 3-inch tubes do not employ a field mesh. Setup techniques differ between field-mesh and non-field-mesh types of tubes.

A third area of difference is that conduction in the glass targets may be either ionic or electronic. In lime-glass targets, the conduction is ionic. With this type of tube, either an orbitor is used to provide a very slow rotation of the image on the target, or the operator must take care to maintain almost continual slight movement of the camera to prevent *image burn-in*. This is not a serious problem with electronically conducting glass targets, which are used in the "nonburn" I.O.'s. Operational differences exist between the two types of targets.

It should be obvious that the competent video operator must be familiar with the types of pickup tubes in his cameras. For example, the Type 5820 is a non-field mesh, 3-inch I.O. with a lime-glass target. A directly interchangeable 3-inch tube is the Type 7293A, which does employ a field mesh, but which has the same ionic target requiring orbiting. However, the Type 7293A/L has an electronically conducting target which does not require orbiting. This simply emphasizes that the operator should consult the technical bulletin for his particular camera pickup tubes.

During this preliminary investigation of I.O. operating controls, the reader should refer often to Figs. 10-7 and 10-8. This orientation is vitally important to "make sense" of I.O. setup techniques.

Electrical focusing for a sharp picture involves the following: In the image section, focusing is accomplished by means of the graded magnetic field produced by the external focusing coil (Fig. 10-7) and by the electrostatic accelerating field produced between the photocathode (PC) and G6 (Fig. 10-8). In the scanning section, the beam is focused at the target by the magnetic field of the external focusing coil and the electrostatic field of G4. In a field-mesh type of tube, this mesh also aids in the focusing process.

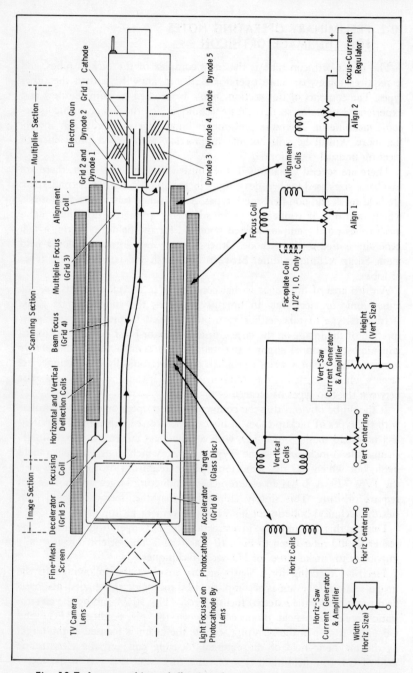

Fig. 10-7. Image-orthicon deflection, focusing, and beam alignment.

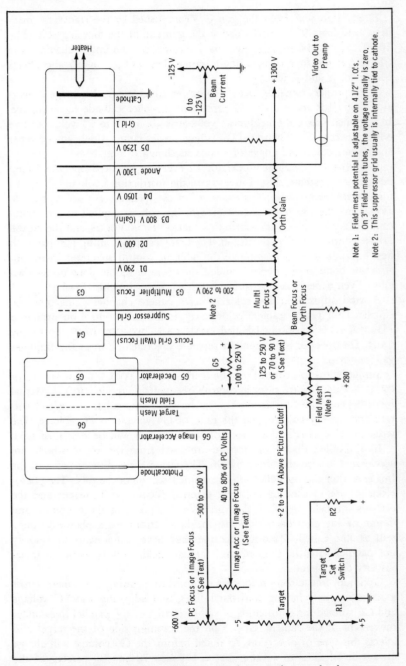

Fig. 10-8. Typical image-orthicon control terminology and voltages.

Beam alignment from the gun is accomplished by the transverse magnetic field from 90° coils located at the gun end of the focusing coil (Fig. 10-7). Proper beam alignment is indicated by minimum shading and "flutter" at the four corners of the raster. The voltage on G5 also affects corner focus and shading.

With proper "mode of focus" (to be described later), the image accelerator (G6) has about 70 to 80 percent of the photocathode focus voltage. In 4½-inch tubes, the field-mesh voltage is about 15 to 25 volts positive with respect to G4. The actual G4 voltage varies rather widely among types and manufacturers of image orthicons, as shown in Fig. 10-8.

IMPORTANT: Every new operator, or every experienced operator facing new camera systems, should investigate the instruction book and diagrams for control terminology. For example, many operators are accustomed to associating the image-focus control with the voltage applied to the photocathode. Some cameras now term this control the PC FOCUS, and the IMAGE FOCUS control actually adjusts G6. The first step in becoming familiar with new equipment is to investigate the table of control functions in the instruction book. If one is not included, the operator should draw up his own table. Even experienced personnel can learn from this initial task.

A vital adjustment for color cameras, where channel gains must be equalized for proper camera colorimetry, is the gain control on dynode 3 (D3). This is variously labelled ORTH GAIN, DYNODE GAIN, or simply GAIN. The operator must mentally separate this function from any amplifier gain controls.

Similarly, the G4 control may be marked WALL FOCUS, BEAM FOCUS, or ORTH FOCUS. In any case, it controls the voltage applied to G4. Actual focus depends on the relationship between the external focusing coil current and the G4 voltage, and the ratio of G4 voltage to PC voltage. The latter ratio is actually the "mode of focus," which will be described next.

It is recalled that photoelectrons released from the photocathode are accelerated toward the target by G6. The target is the divider between the image section and the deflection (scanning) section of the tube. The photoelectrons are focused by both the external focusing-coil current and the electrostatic field of G6. Unless great care is taken in the operation and design of any electron optical system, the electrons (and photoelectrons) will assume a helical motion because they have components of velocity not parallel with the lines of the magnetic field. This is termed a radial velocity component.

Optimum image-section focus is obtained by panning the camera across a scene or chart having horizontal lines, and adjusting the PC voltage and G6 voltage for elimination of "S" distortion of the parallel lines under movement of the camera. Grid 5 on the scanning side of the target also affects this type of distortion. As noted before, the G6 voltage will almost always be adjusted to a value that is between 70 and 80 percent of the PC voltage.

On the scanning side of the target, there exists a phenomenon that is never apparent on diagrams such as Fig. 10-7. The beam path is conventionally shown as making a curve immediately following emission from the gun, and another curve brought about by the influence of G5. In reality, the beam must make five or six complete loops on the path to the target to avoid production of a radial component. The number of loops must be an *integral* number, that is, some number of complete loops. Electrical focus can be obtained on more than one mode (G4 volts), and the proper mode must be used.

The field-mesh tube is more critical in beam alignment than the non-field-mesh type. It usually operates at top performance only at one particular mode of focus. For the 3-inch field-mesh tube, the voltage on grid 4 normally is in the range of 140 to 170 volts. For the 4½-inch tube, the G4 voltage is between 70 and 90 volts. Now look at Fig. 10-9, which shows the circuit of the ORTH FOCUS control in the RCA TK-11 camera. This particular circuit has maximum voltage occurring at maximum counterclockwise rotation of the control. As the control is turned clockwise, the voltage on grid 4 is reduced. The lowest voltage obtained is about 130 volts (a 3-inch tube is used). Thus, in this particular camera, the first mode encountered when starting from the maximum-cw position of the ORTH FOCUS control usually is the optimum mode of focus. On any other mode of focus, a coarse-mesh background may appear in picture low lights; this effect also is noticeable with the lens capped. Also, bad beam-landing effects may be noticed; these include shading in the corners (portholing) and a noisy picture.

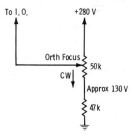

Fig. 10-9. Circuit of beam-focus control.

10-3. IMAGE-ORTHICON ADJUSTMENTS

The handling and operating position of the image orthicon should preferably be such that any loose particles in the neck of the tube will not fall down and strike or become lodged on the target. Therefore, it is recommended that the tube never be operated or carried in a vertical position with the diheptal-base end up or in any other position in which the axis of the tube with the base up makes an angle of less than 20° with the vertical.

The 3-inch-diameter image orthicon is installed in the camera by inserting the base end of the tube through the coil assembly in such a way that the large pin on the annular base (image section) is aligned with the large socket terminal in the coil assembly. In black-and-white cameras, this large terminal usually is at the bottom of the socket, i.e., at six o'clock. In three-tube color cameras, the large terminal usually is located at the top of the socket, i.e., at twelve o'clock. For convenience, a white radial line, or *fiducial mark,* is applied to the faceplate of the tube in line with the large pin. This line serves as a visual check for proper orientation of the installed tube. *Warning:* Forced seating of the tube in an improper orientation may damage it beyond repair.

Installation of the 4½-inch image orthicon is similar to that for 3-inch types. However, instead of annular pins, there are five unevenly spaced contact springs that engage with appropriate contact tracks inside the coil assembly. Visual inspection will show the proper orientation to be used.

The first adjustment is the basic one upon which the remainder of the alignment procedure depends. This is the adjustment for 65 to 75 milliamperes of focusing-coil current; 75 milliamperes is the setting for most 3-inch tubes. The current is measured by inserting a 0-150 mA meter in jacks provided on either the camera head, control unit, or power supply. In the subsequent alignment, no attempt should be made to alter performance by varying this current. This is so because the field developed by the focusing-coil current determines both the orthicon focus voltage (the beam focus at the long cylindrical wall coating, G4) and the image focus (at the photocathode of the image orthicon). Therefore, any change in this current value requires readjustment of all other controls that are normally used in aligning the camera chain. Focusing-coil current usually is variable by a screwdriver adjustment on the power-supply chassis.

Greater focusing-coil current makes the beam "stiffer," decreasing the scan area for a given deflection-current amplitude. Reducing the focusing-coil current increases the scan area for a given deflection current. Therefore, remember that focusing-coil current affects not only electrostatic-focus voltage adjustments, but also picture size for a given setting of the height and width controls.

Most 3-inch I.O.'s require a 75-gauss magnetic field at the center of the scanned area for optimum performance; this field is provided by the 75 milliamperes of focusing-coil current. Most 4½-inch I.O.'s require from 60 to 70 gauss at the center of the scanning area, and around 120 gauss in the plane of the photocathode. This is provided by the design and configuration of the focusing coil and faceplate coil. Most modern cameras employing the 4½-inch I.O. have fixed, highly regulated focusing-current supplies that do not require adjustment.

All I.O.'s should be provided a warmup period of 15 to 30 minutes before alignment for optimum performance is started. Always follow the specific setup procedure shown in the particular tube data sheet or camera instruction book. Different tubes and cameras have different setups, but some general instructions for satisfactory day-to-day operations are outlined below. First, however, there is one basic difference between field-mesh and non-field-mesh tubes that should be understood: In non-field-mesh types, the dc voltages may be applied with the lens capped. If the lens is capped, however, it should be momentarily uncapped during adjustment of the grid-1 voltage to provide a slight amount of beam current. Insofar as field-mesh image orthicons are concerned, *under no circumstances should the lens be capped during application of dc operating voltages.* The lens always must be uncapped and the lens iris must be opened to allow light to fall on the photocathode before application of the dc voltages. This prevents formation of static charges between the field mesh and target, which can cause "sagging" of elements and consequent damage or shortened tube life.

Typical day-to-day I.O. setup procedures for monochrome cameras (color cameras are considered in following sections) may be outlined as follows:

STEP 1. Uncap the lens and open the lens iris.

STEP 2. Apply proper dc operating voltages by turning the camera system on.

STEP 3. Be certain the G1 voltage is adjusted to allow beam current.

STEP 4. Check that deflection circuits (height and width controls) are adjusted to provide a slight amount of *overscanning* of the target area. (Edges of target should show in the four corners.) Overscanning results in a *smaller than normal* picture on the monitor. Since this statement is likely to be questioned by a new student, let us clarify the relationship between scanning size and picture size before proceeding.

The monitor or receiver picture tube has its own normal adjustments of sweep height and width. Receivers are adjusted so that the edges of the raster are beyond the kinescope mask, and therefore invisible. Camera-control monitors and master monitors normally are adjusted so that the raster edges are visible.

If, for example, the camera pickup tube is *overscanned,* the deflection currents cause the beam to sweep out beyond the useful pickup area and include the target ring itself. Thus, the target ring will be visible on the camera monitor. Obviously, the image within the useful pickup area will now be *smaller* than normal. Similarly, if the target is *underscanned,* the image is spread out on the monitor screen, resulting in a *larger than normal* picture. The proper I.O. sweep adjustment is such that the four corners of the picture on the camera or master monitor just barely show. Since most I.O. cameras employ a mask with a diameter of 1.6 inches in front of the pickup tube, these corners are actually from the mask rather than the target itself. Note that to get equal amounts of "corners," both size and centering adjustments are required.

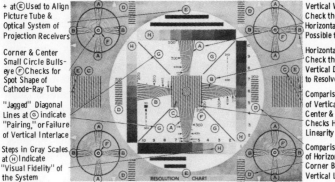

+ at Ⓔ Used to Align Picture Tube & Optical System of Projection Receivers

Corner & Center Small Circle Bullseye Ⓕ Checks for Spot Shape of Cathode-Ray Tube

"Jagged" Diagonal Lines at Ⓖ indicate "Pairing," or Failure of Vertical Interlace

Steps in Gray Scales at Ⓗ Indicate "Visual Fidelity" of the System

Vertical Wedges Ⓐ Check the Number of Horizontal Picture Dots Possible to Resolve

Horizontal Wedges Ⓑ Check the Number of Vertical Dots Possible to Resolve

Comparison of Spacing of Vertical Lines of the Center & Edge Boxes Ⓒ Checks Horizontal Linearity

Comparison of Spacing of Horizontal Lines in Corner Boxes Ⓓ Checks Vertical Linearity

Fig. 10-10. Resolution test chart.

STEP 5. Focus the camera on a resolution chart (Fig. 10-10) or use the *Diascope* (Fig. 10-11) provided with some camera systems. The *Diascope* mounts on the turret in place of one lens and permits use of a $2'' \times 2''$ resolution slide instead of a wall chart. Illumination is from a lamp that receives its power from a remote iris control. With the camera focused on the chart, get initial electrical focus by adjustment of the PC voltage and G4 voltage (PC focus and beam focus) for the sharpest image on the viewfinder. If the image is not aligned with the blanking edges, the entire deflection yoke must be loosened and rotated for a "straight" picture.

STEP 6. Place the target-set switch (Fig. 10-8) in the on position (in reality, this closes the switch around R1). Adjust the target-voltage control until the test-pattern high lights are just visible on the viewfinder set for a

Courtesy Ampex International

Fig. 10-11. *Diascope* used with Marconi Mark IV camera.

high-contrast image. This is the "target-cutoff" point. When the target-set switch is returned to normal, the dc across R1 automatically applies the proper voltage above target cutoff for the particular tube used. For 3-inch I.O.'s, this is generally 2.2 to 2.5 volts above the cutoff value. For 4½-inch tubes, this voltage may be anywhere from 2.3 to 4 volts (or even more) above cutoff, depending on camera and tube manufacturers' specifications.

The beam-current control should then be adjusted to give just sufficient beam current to discharge the high lights. The interrelationship among tube sensitivity, signal-to-noise ratio, and resolution may be used to obtain optimum camera performance for different lighting conditions. The determining parameter is target voltage. At high target voltages, signal-to-noise ratio is enhanced at the expense of resolution. As the target voltage is reduced, this relationship reverses. For a given telecasting session, it is practical and advisable to maintain the target voltage at a single value, because this voltage limits video-gain, gamma-correction, and other adjustments in the camera chain. Furthermore, it is generally advisable to employ the same target voltage for all cameras telecasting a given scene.

The target-voltage control should not be used primarily as an operating control to match pictures from two different cameras. Matching should be accomplished first by individual adjustment of the lens-iris openings, and gamma controls (where used). Small changes in target voltage then may be used to produce picture matching. The target-control voltage calibration (voltage appearing across R1) should be checked periodically to assure that the target-voltage adjustment is correct.

STEP 7. Adjust the distance from the camera to the test chart so that the arrows on the side of the chart just meet the blanking edges. Then adjust the vertical and horizontal size and centering controls for proper aspect ratio. Adjust the horizontal and vertical linearity controls for proper linearity with the proper aspect ratio. Adjustment of these controls requires readjustment of the size controls; a back-and-forth adjustment must be used until proper size with proper linearity occurs. This adjustment will be rechecked with a ball chart in Step 9.

STEP 8. This step has to do with proper beam alignment and involves a number of controls. First, set all shading controls to zero. If a shading-off switch is provided, place it in the off position. Adjust the multiplier-focus ("multi-focus") control (Fig. 10-8) for maximum output with minimum shading. If the two characteristics do not occur at the same setting of the control, adjust slightly toward the point of minimum shading as a compromise. This is best judged by looking at the picture black level on the CRO waveform monitor.

Adjustment of the beam-alignment controls (Fig. 10-7) depends on the type of I.O. used. In the non-field-mesh tube, the dynode aperture and surrounding area blemishes are visible when the electrical focus controls are set for maximum resolution as in Step 5. Adjust the beam-alignment controls until the dynode spot (with lens capped) does not move with rotation of the beam-focus (G4) control, but simply goes in and out of focus. With improper alignment, the white dynode spot moves in an arc and has a "comet tail" when the G4 control is rotated at a fast rate.

With the field-mesh tube, the dynode spot is not visible because of the defocusing of the return beam by the mesh. Most modern cameras employ a *wobbulator* circuit, which applies a 30-Hz synchronous signal (square wave) to G4 when the align-set switch is turned on. This results in a split image of the entire pattern if the beam is not properly aligned. It is then necessary only to adjust the alignment controls to converge the pattern into one well-defined image. This is done first to get the central area stationary; then fine adjustments are made to get best convergence all around the edges. Vernier adjustment of the voltages on G5 and G3 helps to get optimum beam alignment. The final adjustment of G5 is for minimum corner shading and best geometry, and the final adjustment of G3 is for maximum output with minimum overall black-level shading. Note that all of these adjustments have to do with optimum beam alignment. In 4½-inch tubes, the screen-mesh voltage must be adjusted so that the mesh beat is not visible as a pattern in the dark areas of the picture. When a wobbulator is not provided, the alignment controls are adjusted for maximum signal output and so that the center of the image does not move as the G4 voltage is "rocked."

STEP 9. A method for exact measurement of *geometric distortion* (nonlinearity of camera sweeps) is provided by the use of the *ball chart* (RETMA [now EIA] linearity chart) shown in Fig. 10-12A. This aid can be obtained in slide form for film-chain cameras, or in chart form for studio-camera use. The pattern is designed so that the black outlines of the circles are within 2-percent linearity, and the inner portions of the circles describe linearity within 1 percent.

To use the ball-chart method, focus the camera on the chart, and mix a grating signal with the camera signal. Fig. 10-12B shows a kinescope display with excessive geometric distortion from the camera. Fig. 10-12C shows the output of the same camera chain with sweeps adjusted to 1-percent geometric distortion. With zero geometric distortion, the dots, circles, or intersections of the ball chart (or slide) can be made to coincide with the intersections on the grating pattern. Some of the grating-pattern lines will be on the outer edges of the circles when the linearity is within 2 percent. Distortion is measured in distance units (for picture elements)and time and distance figures (for scanning velocities).

Note that the monitor linearity has no bearing on this measurement. If the monitor is nonlinear, the camera sweeps still can be adjusted to obtain camera linearity; then the monitor can be adjusted independently for its own linearity.

If you use a fixed grating signal as obtained from some sync generators, you will not be able to obtain exact phasing of the grating pattern with the ball-chart intersections. (Some grating-signal generators employ a

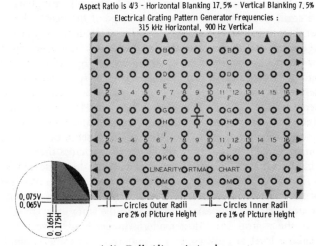

Aspect Ratio is 4/3 – Horizontal Blanking 17.5% – Vertical Blanking 7.5%
Electrical Grating Pattern Generator Frequencies :
315 kHz Horizontal, 900 Hz Vertical

0.075V—
0.065V—

0.165H
0.175H

—|—|— Circles Outer Radii
are 2% of Picture Height

—|—|— Circles Inner Radii
are 1% of Picture Height

(A) Ball (linearity) chart.

(B) Horizontal nonlinearity.   *(C) 1-percent linearity.*

Fig. 10-12. Ball chart used for camera-linearity adjustments.

phasing control.) For the studio camera, you can simply shift the camera or chart position slightly to one side to obtain superimposition. The slide can be shifted slightly in the holder on film setups.

The grating generator must, of course, be adjusted to operate at the proper frequencies to obtain the standard number of horizontal and vertical lines in the pattern to fit the ball-chart intersections. These frequencies are 900 Hz for the horizontal bars (to check vertical linearity) and 315 kHz for the vertical bars (to check horizontal linearity).

After proper size and linearity are obtained, adjust any "high-peaker" control (usually in the pickup-tube preamplifier) for minimum "streaking" of the horizontal bars in the test pattern (Fig. 10-10).

STEP 10. This is a vernier adjustment for field-mesh "antighost" tubes. Position a small, bright spot of light on the edge of the field to be viewed and adjust the G6 voltage so that any ghost that is present disappears as the PC voltage is brought to the level for sharpest focus. Improper adjustment is evident when a light spot that is observed on the right edge of the viewing monitor produces a ghost that appears above the spot, and when a light spot observed on the left edge produces a ghost that appears below the spot. Then readjust the G5 voltage to produce the best compromise between high signal output in the picture corners (minimum shading) and best geometry. Best geometry is indicated by the absence of "S" distortion of straight horizontal lines as the camera is panned across them.

STEP 11. Recheck the target voltage, and then adjust the D3 voltage (dynode gain) so that the camera-head output is at the reference level for the particular camera used. Some of the older camera heads employed an output of around 0.4 volt (peak-to-peak) with the extra gain (to 0.7 volt) made up in the control unit. Most modern camera heads have 0.7 volt of video, blanking level to peak white, at the output. On cameras employing calibration pulses as a reference, the gain of the I.O. is simply adjusted to match the calibration-pulse amplitude. NOTE: For monochrome operation, this level is adjusted with the lens opened to one or two stops over the knee. The knee is found by opening the lens until the picture whites no longer increase. Then, the lens is opened one or two additional stops above this value.

Proper exposure of the image orthicon is required at all times for consistent production of high-quality pictures. The most common error in lighting and exposure control is to overexpose the image orthicon to bring up information in the low lights of the scene. A much better picture can be obtained by filling in the low-light areas of the scene with fill light, rather than by opening the lens and overexposing the image orthicon.

In general, as the light level incident on the image orthicon is increased and the signal output reaches the knee of the light transfer characteristic, picture quality will improve because of an increase in resolution, signal-to-noise ratio, and contrast range. Signal-to-noise ratio and contrast range are directly proportional to the square root of the illumination on the faceplate of the image orthicon, and they increase until the high lights reach the knee of the light transfer characteristic. Any further increase in light level will not materially improve the signal-to-noise ratio but will increase resolution slightly. Operation of the tube with the high lights substantially above the knee allows it to handle a wider contrast range because the whites are compressed without loss of detail and the blacks are raised out of the noise.

Focus the camera on a neutral (black-and-white) test pattern consisting of progressive tonal steps from black to white. Open the lens iris just to the point where the high lights (highest step) of the test pattern do not rise as fast as the low lights (lower steps) when viewed on a video-waveform oscilloscope. This operating point is the knee of the light transfer characteristic. For black-and-white operation, the camera lens then should be opened approximately 1 to 2 stops above the knee for each individual scene. This operating point assures maximum signal, good gray scale, freedom from "black borders," the sharpest picture, and the most natural appearance of televised subjects or scenes.

The camera lens should be adjusted continually to maintain this operating point as the illumination in each scene changes. Operation at this point is especially important for studio pickup in order to obtain the best gray scale in the picture and to reduce the possibility of image retention.

For outdoor and other scenes in which a wide range of illumination may be encountered, the camera should pan that scene that has the least amount of illumination, and the lens iris should be adjusted so that the high lights in that area are just above the knee. The camera will then be able to handle all scenes having higher illumination without requiring lens-stop adjustments. When the camera is to be shifted rapidly from a scene of low brightness to a scene of high brightness, or vice versa, as may take place during panning, the camera should always be set for the dark scene.

IMPORTANT NOTE: Always use just sufficient beam current to discharge the highest high lights in the scene. Use of excessive beam current results in deterioration of resolution as well as poor signal-to-noise ratio in the picture.

When shading controls are used, they are adjusted to obtain the most uniform black level possible in both the horizontal and vertical display on the CRO waveform monitor.

To avoid retention of a scene (sometimes called a *sticking picture*), always allow the tube to warm up properly. Never allow the tube to remain focused on a stationary bright scene longer than necessary. Never use more illumination than is necessary. Image orthicons using electronically conducting glass targets are resistant to "burning" effects and do not need an orbiter, or the necessity of continually moving the camera when it is focused on a stationary scene. Care, however, must be exercised to avoid excessive illumination on the photocathode of these tubes to prevent a change in photocathode characteristics. A maximum allowable photocathode illumination value is specified in the individual data sheets for the specific types.

A persisting retained image generally can be removed by focusing an image orthicon on a clear, white screen and allowing it to operate with an illumination of about 1 foot-candle (1 lumen per square foot) on the photocathode until the image disappears. This value is equivalent to 50 to 100 foot-candles on the screen with an $f/2.8$ lens.

Occasionally, a white spot that does not change in size when the beam-focus voltage is varied, may be observed in the center of the picture. Such a spot, especially if it is visible on the monitor with the camera lens capped, is probably an ion spot. If the spot begins to grow in size with continued operation of the tube, the tube should be removed from service at once, and returned for reprocessing. Continued operation of an image orthicon with an ion spot will eventually damage the target permanently.

New image orthicons should be placed in service immediately upon receipt. They should be operated for several hours before being set aside as spares. Spare image orthicons should be placed in service for several hours once a month in order to keep them free from traces of gas that may be liberated within them during prolonged storage.

10-4. ENCODER ADJUSTMENT TECHNIQUES

Before we can discuss the actual color-camera adjustments, we must get the rack equipment (encoding and processing circuitry) ready to accept the color-camera signal output. That is, setup and adjustment of the color camera must *follow* the encoding and processing setups.

We will not attempt to replace the equipment instruction book in this coverage. What we hope to accomplish is to correlate your previous training with a logical analysis of what each adjustment is supposed to do.

First of all, remember the basic requirements, as follows:

1. *Carrier Balance.* This means that the carrier (color subcarrier) must be cancelled when a monochrome condition exists. A monochrome condition includes black, all shades of gray, and white. Black is the lack of any video signal; for grays and whites, all camera outputs are equal. When the encoder (colorplexer) is adjusted, a known condition must exist. This is why pulses of a known amplitude (color-bar signal) are used in initial adjustments. At this point, stop and think. If we adjusted the I and Q carrier-balance controls on a color-bar signal (or any signal) with whites or grays, we would need to know *for certain* that the matrix circuits (white balance) and the actual encoder input signals were accurate. The logical deduction then is that we must *first* adjust the carrier-balance controls with *no input signals* (black), or with the Y, I, and Q channels turned off. Some encoders have toggle switches to turn these channels off or on as required. When this is not the case, the inputs must be removed entirely, or the color-bar generator module must be removed. Carrier balance is therefore termed "black balance" in some units.

2. *White Balance.* This is the matrix gain balance to achieve proper amplitude relationships among the R, G, and B signals. Again, stop and think. You *can* adjust the matrix gain to cancel all subcarrier on the white pulse, and still have impure whites in the picture. How? Suppose you adjusted this matrix gain with unequal inputs to the encoder. Now, when the camera actually sees a white area and delivers

equal inputs to the encoder, the subcarrier is not cancelled. Therefore, a means often used to set white-balance controls is to use a *tie* switch that connects all inputs together so that the encoding system "sees" the same level (same pulse) from all inputs. When this provision is not made, you must measure each individual video-signal input very accurately and adjust to identical amplitudes. This is clarified in the following procedures.

Now that we have taken the two extremes, black and white carrier cancellation, and mentioned the possible (and most common) sources of error, let us examine a logical step-by-step procedure for setting up the encoding system. This sequence works best in practice, and we will try to show why this is so for each step. In all of the following steps, the oscilloscope is connected at the encoder output, and horizontal drive (or sync) is used as an external trigger for greater trace stability.

STEP 1. *Carrier Cancellation for No Signal (Black).* When switches are provided, turn the Y, I, and Q channels off. This opens only the respective video channels, prior to the subcarrier modulators. With the wideband oscilloscope placed at the encoder output, we will see only a single horizontal line with sync (when sync is inserted in the encoder). See Figs. 10-13A and 10-13B. Adjust both the I and Q carrier balance (the controls may be termed "black balance" in some equipment) for the thinnest possible line (scope set for highest possible gain). This means the subcarrier is cancelled. When there are not individual channel switches, it is necessary to pull the color-bar generator module, or remove the signal-input lines. After carrier cancellation is obtained, restore the video inputs. NOTE: In older tube-type encoding units, there is a switch for the automatic carrier balance. To adjust the carrier balance, place this switch in the out position. Then restore it to the in position after adjustment.

STEP 2. *How to Assure Proper White Balance.* If there is a switch "tie" position (sometimes termed "test" position) that connects the color-bar inputs together so that all channels receive one pulse amplitude, use this position. (For example, in the RCA TK-27 and TK-42 equipment, this is position 2 of the test switch.) If this function is not available, measure

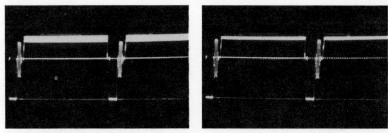

(A) *Carrier not balanced.*   (C) *Carrier balanced.*

Fig. 10-13. Black-balance adjustment.

each of the bar inputs as accurately as possible and adjust the bar-generator gains for exactly equal peak-to-peak amplitudes (usually 0.7 volt—0-100 IEEE units when the scope is calibrated for 140 IEEE units = 1 volt). The procedures now depend on the type of test provision available (bar signals tied together, or not tied).

Procedure with bar inputs tied together. Turn the Y channel off, with I and Q on. (This is automatic in test position 2 in the RCA system mentioned previously.) With Y off, the chroma line should contain no carrier (Fig. 10-14). Adjust the controls for I white balance and Q white balance for the thinnest possible line, as in Fig. 10-14B. Note that this is the same pattern you saw for carrier balance; the difference is that Fig. 10-13 is with no signal input, whereas Fig. 10-14 is with all inputs tied together for the color-bar signal.

The next logical step is to double check the white balance while looking at I and Q chroma (removing the input-signal "tie"), and note whether the subcarrier is cancelled for white. This is the interval immediately preceding the first chroma bar. If subcarrier is present, you know that the signal inputs are not equal in amplitude. In this case, take one of the inputs (for example, blue) as a standard, and adjust the other two inputs (red and green) to cancel the carrier.

Procedure without bar inputs tied together. Now you can understand the necessity for being very accurate in assuring equal-amplitude inputs to the three channels. Without a means of tying all inputs together, this is the only way you can be sure the matrix section actually is "seeing" a white condition. With Y off and I and Q video on, adjust the I white balance and Q white balance to cancel the carrier in the interval immediately preceding the yellow chroma bar. If perfect carrier cancellation can not be obtained, look only at I video and get the best white balance, and then look only at Q video and get the best white balance. This will reveal which matrix is responsible for the inability to cancel the carrier.

STEP 3. *Video balance.* With a standard color-bar signal and with only the I channel on, adjust the I video-balance control(s) to minimize video excursions, as in Fig. 10-15. (Normally, you will not obtain perfect video cancellation, especially when a high vertical-amplifier gain is used on the

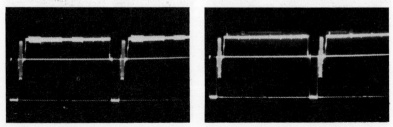

           (A) *White unbalance.*   (B) *White balance.*
                    Fig. 10-14. White-balance adjustment.

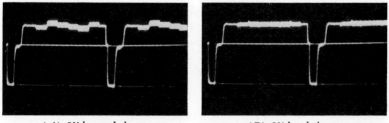

(A) Video unbalance.   (B) Video balance.
Fig. 10-15. Video-balance adjustment.

scope.) Then do the same for Q by turning only the Q video on. *Caution:* Remember from previous study that the video-balance and carrier-balance controls on tube-type modulators, being highly interacting, require a thorough back-and-forth adjustment between carrier balance and video balance. Solid-state circuitry has extremely small interaction between carrier and video balance adjustments.

If desirable, a crossover filter in the low-pass position may be used on the scope for the video-balance adjustment. This will eliminate the chroma and give a clean line that is jagged when the video is unbalanced. Just remember to return the high-low filter to the direct position when checking carrier balance.

STEP 4. *Subcarrier Quadrature Adjustment (Q Delay).* This is normally just a trimmer adjustment that, once set, should never need to be readjusted except in the event of component replacement. (This statement does not apply in the case of older tube-type equipment.) In any event, you must be prepared to check this adjustment at any time.

Vectorscope method. This is by far the simplest, most reliable, and most rapid method. But always be sure the vectorscope is properly self-checking. This means that with the front-panel quadrature and gain-balance controls properly adjusted, the self-checking circles coincide exactly to make one perfect circle. Have both the I and Q channels in the encoder turned on.

With only I and Q test pulses from the color-bar generator, feed the encoder output to the vectorscope. Adjust the vectorscope phase controls to place the I vector directly on the corresponding line of the graticule (Fig. 10-16). To be certain of exact overlay, increase the vectorscope gain to bring the I signal to the calibrated outer circle. The point of the signal vector should be at 303°. Now adjust the quadrature control in the encoder so that the Q-signal vector aligns exactly with the corresponding line on the graticule. Again, you can be very precise by adjusting the vectorscope gain so that the point of the Q vector reaches the calibrated outer circle. The vector should lie precisely at 33°.

Other methods of quadrature phasing. Other methods include special test circuitry that may be incorporated in particular equipment to avoid the necessity for a vectorscope. Naturally, you normally will follow the

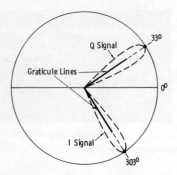

Fig. 10-16. Vectorscope Q phasing.

instruction book for your particular equipment. What we want to do here is to train you to use your knowledge of color fundamentals so that in spite of loss of special test circuitry, etc., you can devise a substitute method of checking.

Refer to Fig. 10-17. Fig. 10-17A shows two vectors of equal amplitude, but phased 180° apart. Now add a vector displaced 90° from the first two vectors (Fig. 10-17B). Fig. 10-17C shows that when a 90° voltage is

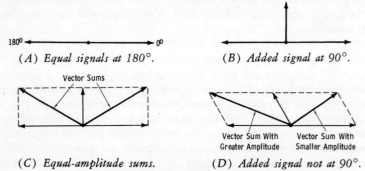

(A) Equal signals at 180°.   (B) Added signal at 90°.

(C) Equal-amplitude sums.   (D) Added signal not at 90°.

Fig. 10-17. Alternate method of Q phasing.

added to two voltages of equal and opposite phase, the new vector sums are equal in amplitude. If the added voltage component is not 90° phased, the vector sums have *unequal* amplitudes (Fig. 10-17D).

All complementary colors are phased 180° from their respective primaries. For example, the vectors for green and magenta are 180° apart. Now observe the representation of the "I Only" chroma in Fig. 10-18. If

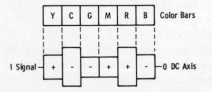

Fig. 10-18. Color-bar I amplitudes.

the Y and Q signals are turned off and only the I signal remains on, we have a convenient signal with which to proceed. That is, in this signal we have intervals of green and magenta adjacent to each other and of identical amplitudes and opposite polarities. This is the condition represented by Fig. 10-17A.

First of all, be certain the I and Q carrier balance is properly adjusted for optimum carrier cancellation. Then obtain a reference amplitude of the green chroma; assume this is 60 IEEE units. Next, deliberately introduce a 90° signal component by unbalancing the Q carrier-balance control in either direction. Adjust this control until subcarrier in the white region is equal in amplitude to the original amplitude of the green chroma, in this example 60 IEEE units. The actual levels that now exist for green and magenta are higher because of the added carrier voltage.

Now, if this quadrature voltage is actually at 90°, the green and magenta chroma signals will still have amplitudes that are equal (although higher than with zero subcarrier unbalance). If they are not equal, adjust the quadrature control in the encoder until they are. Return the Q carrier balance to normal.

STEP 5. *Absolute Levels and Level Ratios.* Use 75-percent bars and have Y, I, and Q all on. Use the IEEE-scale graticule on the scope, and calibrate for 1 volt between −40 and +100 IEEE units. This automatically gives a scale calibration of 0.714 volt between 0 and 100 IEEE units, and 0.286 volt between zero and −40 IEEE units.

Adjust the overall (composite) gain to obtain a white pulse that extends between 0 and 77 units. If the I-Q gain ratio is correct, the tops of the first and second color bars are even, and the bottoms of the fifth and sixth bars are even (see Fig. 5-14 in Chapter 5). Adjust the absolute I and Q gains for 100 units on the first two color bars (yellow and cyan) and −16 units on the last two bars (red and blue). Simply adjust the I gain and Q gain simultaneously (with both hands) until the absolute gains and gain ratio are correct. Adjust the burst gain for 0.286 volt (from 0 to −40 units). (Shift centering of blanking line to measure.) If everything else is adjusted properly, the green bar (third bar) will touch reference black. At this point, such a condition may not exist because a very important adjustment, the burst-phase adjustment, remains to be made. This adjustment can affect slightly the composite amplitude pattern.

STEP 6. *Adjusting Burst Phase.* The burst phase may be adjusted easily, rapidly, and precisely with the aid of a vectorscope. With only I and Q test signals from the bar generator, phase the vectorscope as for Step 4. Then simply adjust the burst-phase control until the burst vector lies on the 180° graticule line (Fig. 10-19). Recheck the burst amplitude if an adjustment of burst phase was necessary. Recheck the composite color-bar signal for correct luminance (overall gain) and absolute and relative levels of I and Q. Fig. 10-20A illustrates a vectorscope, and a typical vector display with proper burst phase is shown in Fig. 10-20B.

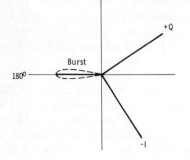

Fig. 10-19. Vectorscope burst phasing.

Again, if you do not have a vectorscope, your particular equipment may be provided with special facilities for setting burst phase. But, as in the case for quadrature adjustment, you can use your knowledge of the color vectors to set burst phase in an emergency. See Fig. 10-21A. When two

(A) Instrument.

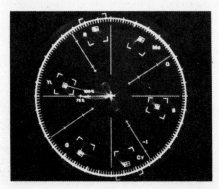

(B) Typical display.

Courtesy Tektronix, Inc.

Fig. 10-20. Tektronix Type 520 NTSC vectorscope.

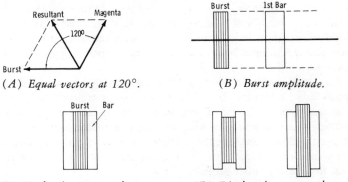

(A) *Equal vectors at 120°.*   (B) *Burst amplitude.*

(C) *Display for proper phase.*   (D) *Displays for wrong phase.*

Fig. 10-21. Alternate method of setting burst phase.

equal-amplitude vectors 120° apart are added together, the resultant has the same amplitude as each of the original vectors. If the burst phase is correct, magenta is actually 119.4° from burst. This angle is close enough to 120° to use in this application, since the resulting amplitude discrepancy is negligible.

Before going ahead with this technique, be *sure* that carrier balance, white balance, video balance, quadrature phase, and I and Q amplitudes have been correctly adjusted. Then proceed as follows:

1. Remove the green-bar input from the encoder. To keep the bar generator functioning properly, terminate the green output in 75 ohms. Turn the Y channel off and the I and Q channels on. The first bar, which originally was white, is now R + B = magenta. This places burst immediately adjacent to a magenta bar in the encoder output.

2. Adjust the burst amplitude to equal the amplitude of the first bar, as in Fig. 10-21B.

3. Adjust the burst delay (flag generator) to move the burst into the center of the magenta bar.

4. Now we have added two equal-amplitude, 120°-phased voltages. If the burst is of correct phase, it will be of the same amplitude as the magenta bar, as in Fig. 10-21C. If it is not properly phased, it will appear as in Fig. 10-21D. Simply adjust the burst phase for equal amplitudes as in Fig. 10-21C.

5. Readjust for proper burst position and burst amplitude. Check for proper breezeway and number of cycles of burst (8 to 10).

10-5. REPRODUCTION OF FLESH TONES

Typical human skin has a spectral reflectance that extends from blue through green (at about the same value but slightly lower in blue), increases somewhat in the yellow region, and then jumps considerably higher in the red region. Flesh tone appears generally yellowish-pink, but varies considerably among individuals. Flesh tones are the most critical to reproduce. The system must be critically balanced to obtain (and hold) natural flesh tones.

This statement can be emphasized by one example with which the reader already may be familiar. If you have ever been fortunate enough to tune in a "test color bar" transmission on your home receiver, you may have observed this phenomenon: Although the hue control was well adjusted for good yellows simultaneously with good reds in the bar pattern, this control needed a readjustment to prevent skin tones on a following program from going either green or purple.

Another example occurs when you adjust the video-tape playback burst-phase control to get proper color-bar presentation on the color monitor during the color-bar test signal at the beginning of the tape. You then (almost always) must change this control at least slightly to get proper flesh tones on the program part of the tape. This does not mean that the actual burst phase as recorded was different for the two sources (although this could happen with careless recording practice). What it usually does mean is that flesh tones are much more critical in adjustments than are the color-bar hues.

We can see the reason for this by returning to Fig. 2-13 in Chapter 2. Visualize flesh tones as covering the region from red-pink through orange-pink and orange to yellowish orange. At the outer perimeter, the band of a particular hue is much broader than it is in the area toward illuminant C. The outer perimeter is analogous to color-bar signals; the inner perimeter is more representative of flesh tones. Although for a given ratio of color-vector amplitude changes the resultant vector is shifted by the *same angle,* a much smaller amplitude change is required to shift the hue in the inner perimeter than in the outer perimeter. This is so because, like the spokes of a wheel, the distance from one color to another is less toward the hub than at the circumference.

Therefore, we must always be concerned with the slightest measurable errors in carrier balance, video balance, quadrature phase, and ratios of Y, I, and Q gains, all of which affect flesh-tone reproduction. So also do differential gain, differential phase, and envelope delay. In other words, good flesh-tone reproduction depends on practically every adjustment we can make in the color system.

Fig. 10-22A shows a reference gray scale against which samples of staging material and paints may be calibrated for live color pickups. The dash-line scale is the *linear* reflectance scale in gamma-corrected camera alignment; visually, it appears to have large steps at the black end because of gamma correction for the 2.2-power transfer characteristic of the average kinescope. The solid line is so chosen that after camera gamma correction (dash line) and encoding, a linear display results on the waveform moni-

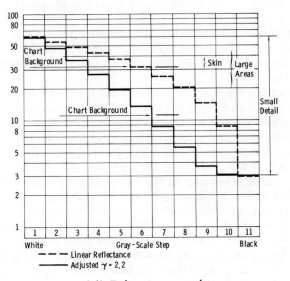

(A) Reference gray scale.

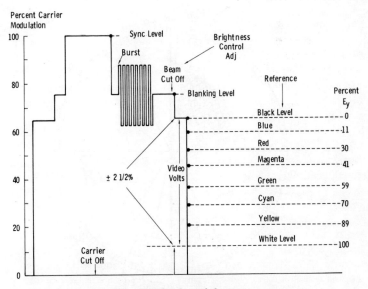

(B) Depth of modulation.

Fig. 10-22. Reflectance and depth-of-modulation charts.

tor (logarithmic gray-scale reflectance chart.) On the picture monitor, then, the solid-line scale has reasonable uniformity of steps, for *linear* reflectance values.

The important point at this time is to note the skin-tone luminance value. The total range of both scales is 20 to 1. The maximum reflectance step is 60-percent reflectance. (Also review the logarithmic reflectance chart of Fig. 9-10, Chapter 9.)

Now correlate the skin-tone luminance range with depth of modulation on the Y (luminance) scale of Fig. 10-22B. Note that, because of gamma correction, the skin-tone luminance level is rather close to (but not equal to) reference white level at 60-percent reflectance on the chart. This should point up the importance of differential gain and phase in this region of studio and transmitter operating characteristics, since the regions toward the black and white extremes are the most susceptible to these errors.

The shaded area of Fig. 10-23 indicates the average range of chroma that makes up proper flesh tones. Note that the +I axis contributes almost exclusively to good flesh tones, and that the Q signals should be properly balanced out for this to occur. Because of its phase relative to burst, the green-magenta complementary line is highly critical. So also is the yellow-blue complementary line. If there is a lack of symmetry between complementary colors, there will be poor flesh tones.

The average flesh tone can be given a specific value; remember this is an "average" value and can vary considerably with individuals. It can be specified as follows:

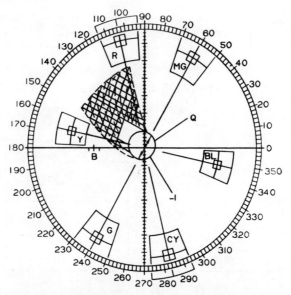

Fig. 10-23. Flesh-tone area on vectorscope.

$x = 0.38$

$y = 0.33$

Luminance Scale = 30-40 percent Reflectance

(60 percent Reflectance = 100 percent Reference White)

This specification is designated in Fig. 10-24. Note that the +I axis contributes almost entirely to this average skin tone. If you focus on an extreme closeup of a face so that it fills the screen, the vectorscope will show a resultant pattern lying somewhere in the shaded region of Fig. 10-23.

We will return to this all-important problem of flesh tones after the preliminary nonsubjective techniques of color-camera setups have been described. We will bypass all arguments concerning the relative merits of subjective and nonsubjective camera adjustments. The following procedure will be followed in this book:

1. Setup and matching of the color cameras in the studio by adherence to strict numerical values as indicated by the picture monitor, waveform CRO, and vectorscope

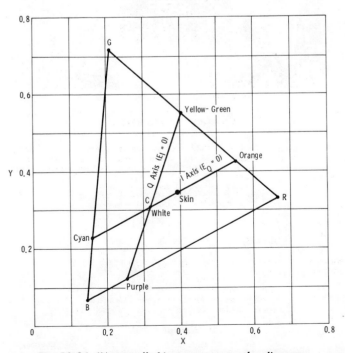

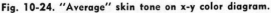

Fig. 10-24. "Average" skin tone on x-y color diagram.

2. Final (and small) adjustments according to operator judgment, using a common color monitor for all sources (subjective technique)

10-6. AMPLIFIER AND IMAGE-ORTHICON CALIBRATION

There are two basic requirements that must be met in a color camera system: (1) The three or four channels must have identical overall amplifier gains from black level to white level, and (2) the three or four pickup

B, R, G Amplifiers
(Luminance Amplifier on Right Side)

Preset Panel (Pickup-Tube Controls)

(A) Controls on camera.

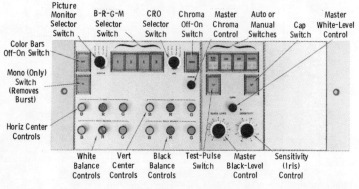

(B) Remote panel.

Courtesy RCA

Fig. 10-25. Controls for color camera.

tubes and associated circuitry must then be adjusted so that the overall gains "track" from black to white levels.

Pickup-tube adjustments may be all at the camera head, all at the control head, all at the control unit in the control room, or some at each location. Fig. 10-25A illustrates the RCA TK-42 camera, in which the preset panel (on the camera head) houses all pickup-tube controls. The vernier centering (registration) and black-white balance (gain) controls are on the remote panel shown in Fig. 10-25B. After setup, only two operational controls are involved, the master-black and sensitivity (iris) controls.

Regardless of the physical arrangement of the setup controls, the principles of adjustment are practically identical for a given type of pickup tube. We will attempt to clarify any differences between four- and three-channel cameras, and among types of pickup tubes, as we proceed.

All channel gains, white levels, and black levels must first be *standardized,* or matched, before color-balancing procedures are undertaken. Modern color chains (both film and studio) provide for this by inserting reference pulses equivalent to pickup-tube beam currents at the preamplifier inputs. By this means, proper amplifier gains, white levels, and black levels can be set prior to pickup-tube balancing.

Fig. 10-26 illustrates the basic principle of amplifier calibration. Assume that we inject a calibration pulse at the output of the camera head. Further assume that this pulse is equivalent to the proper peak output signal of the camera; for example, 0.7 volt peak-to-peak. Then we insert a reference pulse, equivalent to peak high-light current, into the pickup-tube load. This current is normally 0.3 microampere for 1-inch vidicons or 0.6 micro-

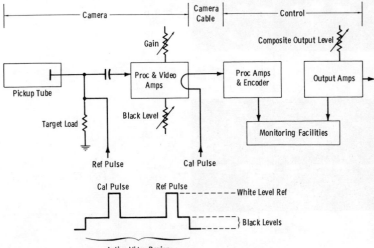

Fig. 10-26. Principles of amplifier calibration.

ampere for the 1½-inch vidicons and I.O. tubes used in the luminance channel. Now if we adjust the camera amplifier gain and black level for the channel involved so that the output for the reference pulse matches that for the calibration pulse, we have standardized the channel gains. If the pickup tube has the electrode voltages applied (but lens capped), the dark current is present for proper setting of the black level.

NOTE: The reader should have a good background in normal I.O. setup procedures before going ahead. If necessary, review the material in Sections 10-2 and 10-3.

An outline of typical color-camera setup follows. Illustrations and waveforms presented are for the RCA TK-42 camera. This system employs an I.O. for the luminance channel, and vidicons for the three color channels. Notes on cameras employing other types of pickup tubes are contained in following sections.

STEP 1. With the test switch in BRGM position (and lens capped), test pulses are applied for initial black and white gain adjustments (Fig. 10-27A). The white level is adjusted to equal the calibration pulse, and the

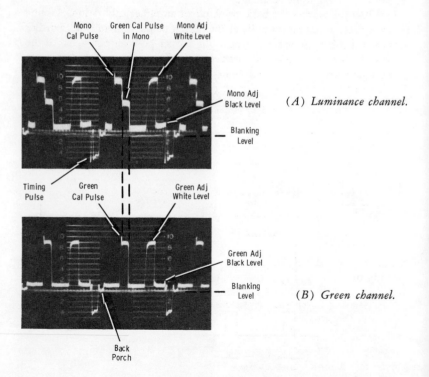

(A) Luminance channel.

(B) Green channel.

Fig. 10-27. Use of test pulses to standardize channel gains.

black level is adjusted just above clipping. The spike preceding the sync pulse is the timing pulse. This pulse is visible because we are looking at the camera-head output prior to removal of the timing pulse in the rack processing equipment.

STEP 2. The monochrome switch is turned off and the green switch is turned on. See Fig. 10-27B. The same process is carried out for the green-channel amplifier as was done in Step 1 for the luminance channel.

STEP 3. The red and blue channels are adjusted for the same reference white and reference black levels as were established for the luminance and green channels.

STEP 4. In any four-channel camera employing an I.O. and vidicons, some means must be used to balance the relative sensitivity of the I.O. to that of the less sensitive vidicons. It will be helpful here to review Fig. 4-8 and Section 4-2 of Chapter 4.

Although only about 20 percent of the light is channeled to the I.O. tube, the remaining 80 percent is severely attenuated by the color-splitting and trim-filter arrangement. Vidicons require that high lights with a brightness of approximately 150 foot-lamberts through an $f/8$ iris be available at the beam-splitter input. The $4\frac{1}{2}$-inch I.O. requires about one-fourth this value (a difference of two full $f/$ stops) to reach the knee of operation.

Fig. 10-28A shows an EIA crossed gray scale (chip chart). Such patterns are available in either $2'' \times 2''$ slides or $24'' \times 36''$ charts. For studio use, the reflectance value is given. For film use, the reflectance density is given. The graduations between steps are logarithmic, as shown by Fig. 10-28B.

NOTE: Some accuracy is sacrificed in printing. For this reason, the corresponding value of Munsell chip is given for each step.

When the chip chart is illuminated with 250 foot-candles of incident light, 60 percent of this light is reflected by the white chips, for a brightness of 150 foot-lamberts. With the iris set at $f/8$, a neutral-density filter is installed in front of the I.O. faceplate (Fig. 10-29) so that the $f/8$ iris opening results in operation one-half stop over the knee. This means that the knee occurs at an opening one-half stop smaller than $f/8$ (between $f/8$ and $f/11$) so that opening the iris to $f/8$ places the high lights a half stop over the knee.

Look at it this way: Assume that the knee of the I.O. is found at an iris setting of $f/16$ for a given scene. With unbalanced sensitivities, the vidicons would require light equivalent to a setting of about $f/8$, or two full stops more, to operate with suitable signal-to-noise ratio, lack of excessive target lag, and best ratio of signal current to dark current. With a suitable neutral-density (ND) filter in front of the I.O., the iris can be opened to $f/8$, and all pickup tubes will be operating at the optimum high-light level. The required value of the ND filter will vary widely with the I.O. installed, even when the same type number is used. The most usual value is

(A) Chip chart.

	Gray Scale Chip Number	Nominal Reflectance of Chip (Percent Relative to MgO)	Nominal Reflection Density	Nominal Munsell Value (Renotation)	Nominal Reflectance Relative to Chip 1
White	1	60.0	0.22	8.05	100.0
	2	41.3	0.38	6.87	68.3
	3	28.4	0.55	5.85	47.3
	4	19.5	0.71	4.97	32.5
	5	13.4	0.87	4.21	22.3
	6	9.2	1.04	3.54	15.33
	7	6.3	1.20	2.95	10.50
	8	4.4	1.36	2.42	7.33
Black	9	3.0	1.52	1.95	5.00
Surround		12.0	0.92	4.00	20.0

(B) Specifications.

Fig. 10-28. EIA crossed gray scale.

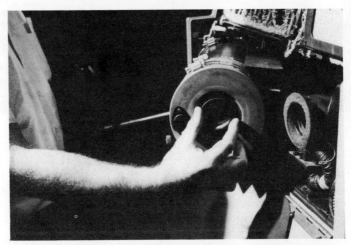

Fig. 10-29. Neutral-density filter in front of image orthicon.

0.1 to 0.3 for a new tube. As the tube ages, the value must be reduced until 0 value is reached. When no ND filter is required, a clear glass is installed to maintain the same optical path length. If the ND glass is omitted entirely, optical focus will be lost at extreme ranges of the zoom control.

The writer has felt it necessary to explain this step in detail to remove any hazy conception of why the operation is required. Here are the actual steps taken:

(A) The camera should be uncapped, optically and electrically focused, and aimed at a chip chart illuminated with 250 foot-candles of incident light (set iris to $f/8$). In the target-set position, adjust the monochrome target control until the first few white steps are visible on the viewfinder. Go back to normal operation. (This step has adjusted for proper voltage above cutoff.)

(B) Open the iris to where the knee occurs. This is the point at which the white level ceases to increase and the adjacent gray levels "pull up" toward white. Then return the iris setting to a point just below the knee (Fig. 10-30A). Note the iris calibration at which this occurs.

(C) Go one-half stop beyond the point at which the knee occurs (Fig. 10-30B). The iris should now be at $f/8$. If it is not, install the proper ND filter (by trial and error) so that this condition exists. The value of the filter should be checked weekly and whenever the I.O. is changed.

STEP 5. Repeat Step 4A. Then cap the lens and apply the monochrome test pulse to get reference white level on the scope. Adjust the scope gain for a convenient calibration, such as 0 to 100 IEEE units for back-porch level to white level.

STEP 6. With the CRO selector in the BRGM position, select the M (monochrome) signal, and uncap the lens with the camera aimed at the

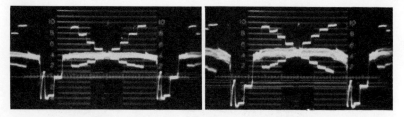

(A) *Just below knee.*   (B) *Over the knee.*
Fig. 10-30. Operating point for image orthicon.

chip chart. If the level from back porch to white chip is more or less than the calibration of Step 5, adjust the dynode gain (D3) for proper level. This adjustment is made with the I.O. at ½ f/ stop over the knee.

Note from Fig. 10-30A that the setup level (picture black above blanking) is near 20 percent instead of between 5 and 10 percent. This is a normal presentation at the output of the camera when the I.O. gamma switch is set on the 0.7-gamma position. The master black-level control is not in the circuit when signals are observed on the BRGM switch positions. Fig. 10-31 shows the corresponding waveform at the output of the rack equipment with proper master black-level adjustment (on the camera-control panel). Note that the CRO display is essentially linear when the camera is looking at a logarithmic chip chart, provided the I.O. gamma is adjusted properly.

Remember that the *amplifier* black-level controls are adjusted on the test pulses. Then on the chip-chart waveform, black level will be somewhat high as in Fig. 10-30A. Proper gray-scale reproduction will be obtained in the final steps when the remote panel is put into service (Fig. 10-31).

STEP 7. Remove any shading signals by whatever means is provided: by pulling the shading module, turning the shading controls to zero, or throwing the switch to the shade-off position. Adjust the I.O. multiplier-focus ("multi-focus") control for maximum video output consistent with best black-level shading. Fig. 10-32 shows the CRO presentation when the multiplier-focus potential is misadjusted. Recheck the black level to be sure it is just above clipping.

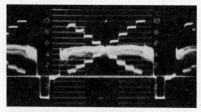

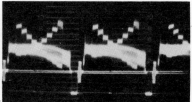

Fig. 10-31. Video waveform at output of rack equipment.   **Fig. 10-32. Waveform with multiplier focus misadjusted.**

STEP 8. Restore the shading signals. Adjust the monochrome horizontal-shading control for equal black levels above blanking. Adjust the monochrome vertical shading for the thinnest base line on the scope horizontal-rate presentation. Recheck the monochrome black level to be sure it is just above clipping.

NOTE: Operators sometimes are concerned by a difference in white level between the left and right white chips of the crossed gray scale. See Fig. 10-33A and note that the left white chip is about 10 IEEE units lower than the right white chip.

First of all, the chip chart must be evenly illuminated. This should be verified by a light meter. It can also be verified by turning the shading off and dollying the camera to get the lower-level white chip into the center of the photocathode area. See Fig. 10-33B, and note that the "left" white chip is now at the same level as the right chip of Fig. 10-33A. This simply means that the I.O. high-light sensitivity varies between left and right of the useful scan area. This variation can be as much as 20 percent and still meet specifications. Improper beam alignment also can result in different high-light levels at the edges.

STEP 9. Check the luminance-tube beam alignment. When a wobbulator is provided (see Step 8, Section 10-3), throw the switch to the "align" position, and superimpose the split pattern on the viewfinder screen. It is also possible to observe the resultant CRO waveform and bring all steps together as closely as possible. Beam-alignment techniques depend on the type and model of camera. If alignment controls are changed, recheck the multi-focus, shading, and black-level controls. Recheck Steps 7, 8, and 9 until the best possible reproduction is obtained. Recheck the dynode gain for reference white level at one-half stop over the knee.

STEP 10. Obtain the best possible focus (still using the chip chart) for the luminance tube by adjustment of all optical and electrical focus controls as outlined previously for the I.O. (The final check will be made with the aid of a resolution chart in a later step.) Then turn the green signal on to super over monochrome. Roughly adjust horizontal and vertical centering for best possible registration of green with monochrome. Do the

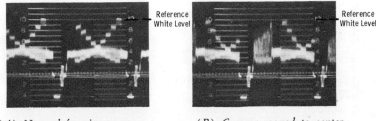

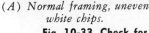 (A) Normal framing, uneven white chips.   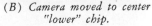 (B) Camera moved to center "lower" chip.

Fig. 10-33. Check for even illumination of chip chart.

same for red, then blue. Final registration checks will be made later. It is necessary at this point only to assure that the vidicons are using the approximate useful scanned area of the tube.

STEP 11. This step starts the procedure for obtaining the same luminance levels from the vidicons as from the I.O. We have already standardized all amplifier-channel gains with the test pulses. It is now necessary to make the pickup tubes track in luminance levels.

Some cameras provide adjustable gamma and *break* controls for the vidicons. The break control provides an artificial knee to prevent the vidicon from increasing in level above the knee of the I.O. Remember that the vidicon does not have a knee on the transfer curve.

Adjust all vidicon break controls to remove any action. With the iris adjusted so that the luminance tube is still one-half stop over the knee (camera focused on the chip chart), look at the green signal only. Adjust the green target voltage for reference white level. (The scope should still be calibrated for this reference white level.) Recheck the black level to be sure it is just above clipping. If the beam appears to be on the verge of discharge on the white chips, adjust the green beam control for more beam current. If the control is at its limit, check the green beam alignment by temporarily raising the target voltage for white bloom-out and adjusting the green alignment controls for maximum beam discharge. Be sure to reset the green target voltage to obtain the reference white level from the green channel. Recheck the black level (it should be just above clipping).

Important: Fig. 10-34 shows a major difference between cameras employing unlike tubes and those using like tubes. When an I.O. and three vidicons are used (Fig. 10-34A), the difference in signal outputs requires a modification in reference-pulse use. Because of the electron-multiplier output of the I.O., the signal output per lumen is greater. The $4\frac{1}{2}$-inch I.O. normally is operated at about 0.6 to 0.7 microampere (rms), which corresponds to about a 20-microampere peak-to-peak current swing on

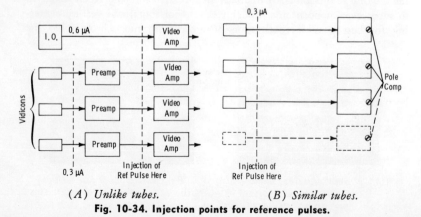

(A) *Unlike tubes.*   (B) *Similar tubes.*

Fig. 10-34. Injection points for reference pulses.

high lights. Therefore, the reference level is at the output of the pre-amplifiers for the vidicons, but at the first amplifier stage for the I.O.

We can see that when the camera is placed in the test-pulse mode of operation, continuity of circuit function is complete except for the vidicon preamplifiers. Therefore, we will find some form of variable or switchable gain controls in the vidicon preamplifiers to obtain the desired reference *signal* voltage from all channels. This is in addition to the vidicon target control.

Plumbicons generally are operated at a peak white current of about 0.3 microampere, the same as for vidicons. Here, of course, all outputs will again be made equal (Fig. 10-34B). The additional control (or fixed circuit) to be found in this case is the *pole compensation* to accommodate the target capacitance of the *Plumbicon,* which requires compensation around 100 kHz in some cameras.

In the type of camera we are now covering (one containing an I.O. and three vidicons), the vidicon preamplifier becomes an integral part of the adjustment for proper level from the tube, since it is not incorporated in the test-signal loop. In most cases, the preamplifier gain should be as low as possible: if the preamplifier gain is too high, the black level usually will be too high for control. It is important to note that both the target and preamplifier-gain controls are involved in setting signal gains of vidicon tubes in this type of color camera.

NOTE: Any vidicon showing black-level shading may have the wrong G5 voltage applied. Check this adjustment as well as black-level gain.

STEP 12. Repeat Step 11 for red.

STEP 13. Repeat Step 11 for blue.

STEP 14. In cameras, such as the TK-42, that employ a timing pulse and registration controls on the camera head, care must be taken to obtain the proper timing relationship in the viewfinder. For the BRGM position at the camera head, the delay through the luminance channel (Y delay line) is eliminated, and the relationship of signal to blanking is different than at the rack-equipment output. In the TK-42, proper timing is obtained by operation of the BARS ON switch on the remote-control panel. The operator must acquaint himself with the registration procedures for each individual type of camera system with which he works.

(A) This step is not necessary after initial installation of the I.O. has been carried out, but is included here for completeness. Swing the I.O. yoke assembly out as in Fig. 10-29 and install the size gauge. Swing the yoke back into position.

(B) Adjust the luminance size and centering controls so that the small circle is tangent to blanking at the top and bottom, and the large circle is tangent to blanking at the sides (Fig. 10-35).

(C) Swing the yoke out and remove the size gauge. Swing the yoke back into operating position and lock it.

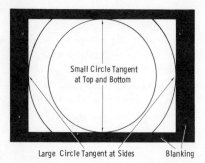

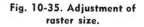

Fig. 10-35. Adjustment of
raster size.

(D) The optical system of the color camera is collimated precisely at the factory. In general, one color pickup tube is the reference; in the TK-42, this is the green vidicon. An optical mask is used in the front of this tube. Looking at the green image, adjust the size and centering so that this image falls just inside the mask dimensions. The camera should be "looking" at a registration chart (Fig. 10-36). Like the chip chart, this aid is available as a 2″ × 2″ slide or a 24″ × 36″ chart.

(E) Now superimpose monochrome and green. Since the green has been centered and sized to the mask, adjust the monochrome to the green. This involves all centering, size, and linearity controls of the two channels. Check all electrical focus controls for maximum resolution.

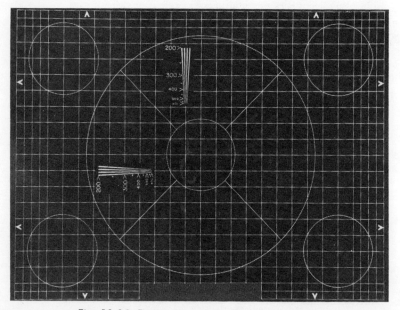

Fig. 10-36. Registration chart for color cameras.

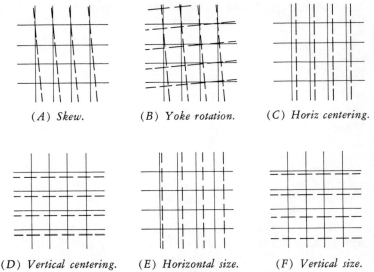

(A) Skew.   (B) Yoke rotation.   (C) Horiz centering.

(D) Vertical centering.   (E) Horizontal size.   (F) Vertical size.

Fig. 10-37. Principles of registration.

Figs. 10-37A and 10-37B emphasize the difference between *skew* and *rotation*. The solid lines represent the "standard" channel, and the dash lines represent the channel being compared with the standard. Skew appears only on vertical lines; rotation appears on both horizontal and vertical lines. Figs. 10-37C and 10-37D show horizontal- and vertical-centering misregistration, respectively. In Fig. 10-37E, centering does not help because horizontal size must be adjusted. In Fig. 10-37F, vertical size must be adjusted.

Registration controls normally must be checked (with the standard registration slide) at least once a day. A little practice should make you quite proficient in this technique.

When adjusting the height, vertical-linearity, and skew controls, observe points on a vertical line through the center of the picture. When adjusting width, horizontal linearity, and yoke rotation, observe points on a horizontal line through the center of the picture.

NOTE: If it is impossible to register monochrome to green by the normal I.O. centering controls, it is necessary to reposition (slightly) the mirror in front of the luminance tube. (This mirror is identified by number 5 in Fig. 4-8, Chapter 4.) An Allen-head locking screw on the mirror-assembly base plate may be loosened, and the mirror may be repositioned to obtain proper centering of the I.O. with the vidicon. Be sure to retighten this screw. After this adjustment, it may be necessary to readjust the focus tracking of the I.O. to the zoom lens. Detailed instructions are a part of all color-camera instruction books, and they differ with make and model number of the zoom assembly.

ADDITIONAL NOTE: A vidicon that cannot be focused sharply when the I.O. is in focus should have its reimaging objective lens adjusted for sharpest focus at this time.

(F) Repeat Step E for red-to-green registration.

(G) Repeat Step E for blue-to-green registration.

STEP 15. Direct the camera at the chip chart again. With the switch in the color position and observing a wideband CRO, adjust gammas for minimum grass in the center of the chip chart. Also remember to keep the flesh-tone concept. Note from Fig. 9-10 (Chapter 9) that on the luminance scale flesh tones lie primarily on steps 2 and 3 of the chip chart. (Thus, where 60-percent reflectance = 100-percent reference white, proper exposure for faces is between 70 and 80 percent.) In the matching of color cameras, it is extremely important to have this area of the gray scale identical for all cameras on a given pickup. Recheck black levels, since gamma adjustment changes black level.

STEP 16. Open the iris another one-half to one stop over the I.O. knee. Adjust the individual vidicon break controls for the same reference white as for monochrome. Open the iris until the least sensitive break control becomes effective. Return the iris to one-half stop over the I.O. knee, and recheck the gammas and black levels for optimum match.

STEP 17. Go to remote-panel operation (Fig. 10-25B). Center all green controls. With the camera lens capped, use test pulses to set the master white level to match the calibration pulse in the test signal. Calibrate the CRO for 1 volt (peak-to-peak) at 140 IEEE units. If the output from the camera rack equipment is not the same (or 0.714 volt noncomposite, which is 100 IEEE units), adjust the VIDEO-GAIN control in the rack.

STEP 18. Uncap the camera and focus on the chip chart. With the switches set for BRGM and M, find one-half stop over the knee by adjusting the sensitivity (iris) control on the remote panel. Set the master black-level control on the panel for proper setup, normally 7.5 percent, or 7.5 IEEE units above blanking level. Make sure that neither black clipping (Fig. 10-38A) nor white clipping (Fig. 10-38B) occurs. White clipping may result from lack of beam, or any white-clipper adjustment in the processing amplifiers. The white clipper should be set to clip at 7.5 to 10 percent over reference white level.

STEP 19. With the switches set for BRGM and G, set the master chroma control for the same level as reference white. This sets the proper luminance-to-chrominance ratio.

STEP 20. Looking at the color output on a wideband CRO, adjust the white and black balance controls for the red and blue channels (green controls are centered) for minimum grass on the steps of the gray scale. Fig. 10-39A illustrates the output from an unbalanced camera; Fig. 10-39B shows the output of a camera that has much better balance. The balance controls are simply individual gain controls for the respective vidicon chan-

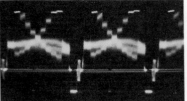

(A) Black clipping.   (B) White clipping.

Fig. 10-38. Waveforms showing clipping.

nels; sometimes they are termed "paint pots" for subjective matching of color cameras as described later.

STEP 21. With the camera looking at the registration chart, use the remote-panel BRGM switch to register the camera.

STEP 22. Be sure that 5 percent minimum black level is introduced. (A minimum-black-level control normally is provided in rack equipment, to assure compliance with FCC rules.) If the master black level is then set too low, clipping of picture blacks (Fig. 10-38A) will occur, but the minimum setup level specified by the FCC is retained.

STEP 23. With the camera aimed at the registration chart, make all final and touch-up adjustments of electrical-focus controls for sharpest detail, high-peaker controls for minimum streaking, and aperture-boost controls for greatest detail contrast possible just under objectionable noise level.

We have now gone as far as possible in setting up the color camera on a strictly numerical basis. The remaining task is color balancing all cameras on a given pickup to one another. This process requires considerable experience in comparison to the nonsubjective techniques just listed.

The subjective method of color matching depends on the judgment of the video operator and on a color monitor that is properly set up. The color monitor should be of the "studio setup" type with a tightly clamped black level. It should have built-in phase reference and check, and a switch that allows monochrome or color display on the kinescope. An input selector should be provided to allow rapid switching between the studio cameras.

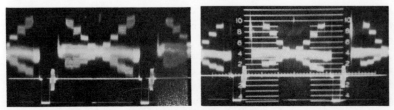

(A) Camera not balanced.   (B) Camera balanced.

Fig. 10-39. Effect of balance controls.

All cameras in the studio should use *one common* chip chart, because of the variance in printing of the gray scale previously mentioned. It is assumed that all cameras have been matched as closely as possible in the preceding steps.

The color-monitor brightness control should be calibrated (with its switch in the monochrome position) by feeding one of the camera chip-chart signals into the monitor, with blacks at 7.5 IEEE units. Adjust the monitor brightness control until the black chips are just above cutoff (are visible). On a *properly set up* monitor, throwing the switch to the color position should result in a pure gray background from the chip-chart signal. If the background is magenta (minus green), adjust the green-channel black and white "paint pots" slightly for neutral background. If the background is yellow (minus blue), adjust the blue paint pots slightly. If the background is cyan (minus red), adjust the red paint pots slightly. There should be no difference in the chip-chart display on the picture tube when the color monitor is switched from monochrome to color and from color to monochrome.

The above procedure should be repeated for each camera on the show. The chip chart should be correctly framed with the arrowheads just touching the raster edges. The chart must be kept in good condition to assure freedom from contamination with any color.

When two or more camera chains have been balanced by this technique, then exposed to subjects in the live set to be televised, it is still possible that minor color variations will be noticed on a common color monitor. Final camera matching is done by switching cameras on a close-up of flesh tones. Pick the camera giving the most pleasing picture and match all other cameras to this one by the "paint-pot" technique. In some cases, it may be necessary to readjust gammas and break controls to get adequate match. Remember that in this purely subjective technique, you are after the most pleasing picture, not the most pleasing waveform! At the same time, it should be evident that this technique requires considerably more experience that matching strictly on CRO level interpretations. The best (and most experienced) operators use both methods.

10-7. LEAD-OXIDE CAMERA SETUPS

We purposely have treated the I.O.-vidicon setup procedure first because the setup of such a camera involves most of the techniques likely to be encountered by telecasting personnel. The reader should understand the preceding sections of this chapter thoroughly, because most of the information is applicable to any type of color camera. Basic examples are electrical focusing for sharpest image, calibration of channel gains and pickup-tube gains, registration, and camera matching.

Obviously, the camera that has lead-oxide tubes in all channels does not have the problem of matching the characteristics of different type tubes, and in this respect it is more quickly and easily adjusted. But other problems result from the unique electrical characteristics of these tubes.

Once a camera is properly set up, it is not necessary to repeat the full procedure outlined in the previous steps. Experience with any particular camera chain soon indicates those parameters that must be checked before each show.

The *Plumbicon**, in common with the vidicon, has no natural knee on the light-to-current output curve. Therefore, gamma correctors and any break controls must be exactly matched between channels and between camera chains. *Plumbicon* cameras normally do not employ break controls; the effective knee is provided by limiters in the video amplifiers, and these limiters must be set correctly to avoid colored high lights.

The basic problem with *Plumbicon* tubes at the time of this writing is that of *beam lag*. This effect shows up as a "comet tail" on high lights when the camera or subject is moved. Since red and blue are the most subject to beam lag (as we will see shortly), the comet tail is usually red or magenta.

Because of the nature of the color-splitting optics, the green tube can have high-light current output three to four times as great as do the red and blue tubes. In some three-channel cameras that derive the luminance strictly from the green channel, the camera is purposely operated in this manner to obtain a noise-free luminance signal. In four-channel cameras, such as the Marconi Mark VII, the luminance signal is provided by a separate tube. In this case, a neutral-density filter is inserted in the optical path for green to limit the high-light current of this channel to not more than twice that for red or blue. This technique more nearly equalizes the charge patterns on the photocathodes of all tubes.

The magnitude of the charge pattern on the photocathode is in ratio to the signal current. Thus, if the green tube has a much higher current than the red or blue tube, the green scanning beam approaches a more highly charged area and is deflected a greater amount than the red or blue beam. This *beam pulling,* or *dynamic misregistration,* results in colored edges.

Beam lag is also a function of the charge pattern. Therefore, it is desirable to minimize differences in charge patterns. We already have mentioned one method, equalizing as much as possible the high-light signal currents from all tubes. There are two additional procedures that can be carried out. The first is well known and is a part of standard operating practice; the second is, at the time of this writing, in an experimental stage. These procedures are:

1. A high light places a given charge pattern on the photocathode of the tube. The beam current can be adjusted to discharge this high

---

Plumbicon is the trade mark of N. V. Philips of Holland for their lead-oxide pickup tubes.

light on a stationary scene. What is not apparent in this static condition is whether the discharge is occurring within just a few fields, or is taking six to ten fields.

If a white object of high-light reflectance value is passed across a dark background (or the camera is panned across a square of 60-percent reflectance against a dark backdrop), there might be a green smear on the leading edge and a red or magenta trail following the high light. The beam currents must be increased sufficiently to minimize this effect. That is, more beam current is required to handle high-light charge patterns under movement than is required on a stationary scene. This procedure is not as simple as it might appear at first thought. Remember that the lag is primarily beam lag; therefore, increasing the beam current beyond some optimum value will increase the lag.

2. The *Plumbicon* usually is operated at a value of target voltage of from 40 to 45 volts. Variation of target voltage around this "normal" operating range has very slight effect on signal output. But the tube *does* have a knee. Signal output rises linearly up to around 20 or 22 volts, and then levels off. The charge pattern produced by the high light is in ratio to the target voltage. Thus, one method of reducing the high-light charge pattern is to reduce the target voltage. Marconi of England has proposed the following procedure:

Starting with the recommended target voltage of 45 volts, reduce this voltage until the signal output has fallen by 5 to 10 percent. Since the high-light charge pattern is decreased, the high-light effect is relatively less severe. Doubts have been raised as to whether the practice of running the tube with low voltage is harmful, but experience indicates that there are no ill effects provided the decrease is limited as above, rather than by setting the target down to a fixed value.

10-8. STUDIO COLOR-CAMERA OPERATIONS

You have become familiar with the basic (static) setup techniques for the color camera chain. You are now ready to use this camera to pick up the studio scene. From here on, you will be confronted by many variables: production whims, lighting techniques and color-temperature variation of lights, lighting and scenic contrast ratios, large-area and small-area luminance and color groupings, etc.

Picture quality, as such, is somewhat subjective. Therefore, it is necessary to tie down "picture quality" as much as possible in terms of basic and essential characteristics that can be defined on the CRO and picture monitor. The final steps (for the *experienced* operator) are purely subjective.

For the moment, forget "color." Our first objective is to concentrate on the luminance information: signal-to-noise ratio (quietness), resolution (detail), gray scale (tonal rendition), and contrast range.

The ultimate capability of the system to reproduce a scene in terms of contrast range and tonal response is determined by the transfer characteristic of the pickup tube(s) and by amplifier adjustments made to cause the pickup to appear as desired on the picture tube. The most troublesome factor here (and this is often the result of artistic efforts of an uninitiated director or producer) is the attempt to exceed the contrast range of the TV sysem. The "system" obviously must include the picture tube. It is our first responsibility to know how to relate contrast range to a *good* monochrome picture; then we can tackle the additional problems of color.

Although the eye is the ultimate receiver, it is a very poor judge of tonal range in any given scene. When lighting is done "artistically" with the eye alone, this tonal range may be anywhere from 100 to 1 to 1000 to 1. As you know, the pickup tube must see a contrast range of about 20 to 1—certainly no more than 30 to 1—and this means the tonal range of small details, not medium to large areas of the set.

Brightness and contrast range are determined by two main factors, scene illumination distribution and scenery reflectance. When you measure incident light, you stand in the pickup area and point the light meter toward the camera. (For color lighting, you should use a color-corrected meter peaking around 555 nm.) After a little experience, you know about what incident illumination you need. But this tells you nothing about contrast range. What you want to know is how bright the various areas—the people, clothing, backgrounds, and staging props—are. So what you need to measure is the reflected light.

Brightness meters, such as the *Norwood Director* or the *Lukeish-Taylor* meter, permit measurement of even small point sources of reflectance. Later versions, such as the GE *Spectra,* are easier to use; you simply aim the instrument at a point, press a button, and read the number on a meter. Earlier designs required balancing the halves in an eyepiece and reading from a calibrated chart. You should investigate all current sources of brightness meters, as the field is expanding rapidly. Some of these instruments have associated color filters so that you can measure the brightness of visible red, visible green, or visible blue.

It is pertinent here to correlate a well-known fact with your judgment of what constitutes proper contrast range from scene to scene, or from high-key to low-key and "limbo" lighting. You probably are familiar with so-called "dramatic" shots, which some producers and directors find very pleasing on the studio monitors (dc restored), only to have a disappointing picture appear on the home receiver (not dc restored). (This statement refers to monochrome receivers without built-in dc restoration, which is common.) As the average picture level (APL) changes drastically, the set without dc restoration washes out the picture.

What does this have to do with color? You might argue that color receivers have some form of dc restoration, which is true. If it were not for this fact, the overall system function would be rather poor, to put it mildly.

The importance of brightness range in color is quite simply that the luminance-to-chrominance ratio is much more tightly controlled when good monochrome practice is followed.

Suppose that you arbitrarily assign zero IEEE units on a CRO graticule to the ac axis, which is average picture level. If you have a clamped-type CRO, you will have the presentation of Fig. 3-31A, Chapter 3. (Also review all of Section 3-7.) The clamper (line-to-line type) holds the signal excursion to a reference despite changes in average picture level. (Incidentally, color-receiver dc restorers are not nearly this good.) If your scope is not clamped, the signal is as shown in Figs. 3-31B and 3-31C.

Most stations have a monitoring CRO that can be operated with the clamps either on or off (examples are the Tektronix Models 525 and 529 monitoring CRO's). This type of monitor can indicate specifically whether the lighting arrangements are exceeding good staging practice for color or monochrome. In this use, place the clamp switch in the off position, and arbitrarily set the blanking level at zero IEEE units, as is common practice. The blanking line should not shift more than 10 to 15 units from scene to scene to maintain tight control over picture quality. Bear in mind that we are not speaking now of contrast range in any given scene; this contrast range will simply determine whether you are sending a linear gray scale, or a gray scale compressed in either the black or white region. Rather, we are talking about the difference between tight shots and wide shots of the same scene, or a complete change of scene from one part of the studio to another studio set, or a change from high-key to low-key lighting (and vice versa).

Low-key lighting can be very tricky, particularly for color. Stop and think about what happens when there is no reference white in the scene. If you are maintaining the same output level now as for high-key lighting, the picture is "heavy in chroma" simply because the peak levels are chroma information—not luminance of a white, but only chroma brightness. It is understandable that picture sharpness is lost, and that you have a strictly "mood" scene without proper luminance-to-chroma ratio.

If the unclamped CRO shows an excessive peak-to-peak shift (black-level shift) on a low-key shot lighted only in the foreground, use sufficient fill light on the backdrop to bring the APL under control and within limits. Even a bright yellow moon in the background is better than a totally dark background. Judicious choice of fill lights or light patterns will maintain good APL control and still allow "mood" lighting. Experience, practice, and experimentation are required.

The first and most fundamental point to bear in mind in tackling the addition of color to the gray scale, is that the system must be compatible. Remember that a large number of monochrome receivers are still in use. The broadcaster must transmit a picture that will be acceptable on *any* receiver, whether a monochrome receiver or a color receiver with the chroma circuits intentionally or unintentionally inoperable.

The monochrome system sees colors only as tones of gray. In order to determine the gray tone for a color, the spectral distribution of the light, the spectral reflectance of the object, and the spectral response of the camera tube all must be combined. Of course, all this is done automatically when you observe the color scene on the monochrome monitor. Yet, in spite of this, beginners are likely to concentrate only on color, with disastrous results on the monochrome receiver.

The basic problem results from a tendency to confuse color contrast with luminance contrast. Two complementary colors side-by-side in a scene can appear visually to be widely different in brightness, and hence to provide good contrast. However, it is possible that on a monochrome monitor the two colors might appear to be of exactly the same brightness.

This will emphasize more than anything else that the color camera is concerned with more subtle subject matter than color bars. For example, yellow and blue are complementary colors. But yellow (fully saturated) has a luminance value of 0.89; blue (fully saturated) has a luminance value of only 0.11, which is comparatively dark. In practice, a very light pastel blue with a great amount of light on it can have just as high a luminance value as yellow. Because yellow and blue are complements, their color contrast is still excellent in spite of the brightness scale. In monochrome, however, you will see no difference in gray scale. It is entirely possible to work out a complete set in highly contrasting colors that will appear in monochrome entirely as one shade of gray.

The point is simply that you must not ignore the monochrome monitor in evaluating a color setup. Even highly experienced color personnel calibrate paint and material to a specific gray-scale number. This is the reason the Munsell value is shown in the chart of Fig. 10-28B. A sample of the paint or material, mounted on a $3'' \times 5''$ card, is held next to the standard gray scale under the studio lighting to be used. If you have the set of Munsell chips from 2 to 8 as shown on the chart, the method is more convenient. Use both the waveform monitor and the picture monitor. The gray-scale number of the chip that gives the same level as the sample then can be marked on the back of the sample for reference in making up the set.

The next important point in color operations is the evaluation of any "large-area" problem. Saturated colors should be confined to *very small* areas; any larger areas should consist of pastels or light colors on matte surfaces. Very important also, both in terms of operating convenience and from an aesthetic viewpoint, is the inclusion of at least a small area each of reference white (60 percent reflectance) and reference black in the scene. Remember also that tight shots are helped by this.

Clothing worn by performers is of sufficiently large area that it becomes an important factor. "White" shirts should be off-white or light blue. Brightness contrast of the clothing must be controlled.

Large-area backgrounds are involved in the basic problem of holding good skin tones. In fact, they have to do with the most apparent problem in modern color programming: It is essential for the performers' background to produce output from *all three* color tubes to prevent skin tones from changing.

Assume there is a large background area of solid blue behind a performer. Adjust the color monitor (if necessary) to obtain pleasing flesh-tone rendition. Now place a backdrop of the complement of blue, which is yellow, behind the performer. Probably, the skin tone will go toward blue; the higher the saturation of the background color, the more pronounced the effect on flesh tones will be.

What happens is quite basic, but often overlooked. A light-blue (which means mixed with white, or desaturated) background is often used because it is the easiest to control in practice, with minimum effect on skin tones. Yellow, if it is reasonably well saturated, is red plus green minus blue. The "minus blue" depends, of course, on the degree of yellow saturation. If you were to look at the three color tubes in sequence, you would find high brightness of red and green (hence high setup of black level) and practically zero blue (hence low black setup). In three-tube cameras, in which the color brightness is matrixed to obtain the luminance signal, the effect on flesh tones is sometimes quite noticeable. You can see what happens. The face becomes a relatively small area against the large-area background. The tubes tend to "set up" so that the small-area skin tone goes toward blue.

RULE: When backgrounds are highly colored (saturated), skin tones tend to go toward the complementary color of the background.

10-9. BASIC CAMERA-OPERATING TECHNIQUES

The basic movements and adjustments of the TV camera are illustrated in Fig. 10-40. Note that the panning handle is an integral part of the friction head upon which the camera mounts (Fig. 10-40A). This handle moves the head left or right for pans (Fig. 10-40B), or up and down for tilt shots (Fig. 10-40C). The friction head contains an adjustment that makes it possible for the operator to choose any desired amount of *drag,* or friction, in turning or tilting.

Fig. 10-40D shows the locations of the turret control and optical-focusing knob. The turret (monochrome cameras only) is rotated by squeezing the turret handle on the rear of the camera and turning it to place the desired lens in front of the image orthicon. The taking lens is on top of the turret after correct positioning. The optical-focusing knob moves the image orthicon and yoke assembly on tracks behind the lens turret. Maximum movement is approximately 2 inches.

Figs. 10-40E, 10-40F, and 10-40G show the basic movements of a mobile dolly or crane-mounted camera. The operator should be thoroughly familiar with the indicated terms. Panning or tilting of the camera on a crane or dolly is accomplished by the same type of panning handle as that shown in Fig. 10-40A.

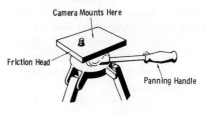

(A) Camera mounting.

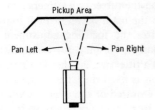

(B) Pan movements.

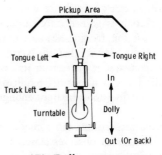

(C) Tilt movements.

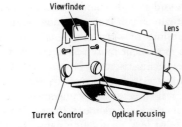

(D) Camera controls.

(E) Dolly movements.

(F) Boom at 90°.

(G) Boom movements.

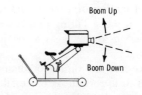

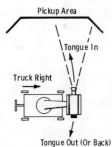

Fig. 10-40. Basic camera operational movements.

In addition to the optical-focusing knob on the camera, the lens itself (on the turret-type camera) may be adjusted by means of a focusing collar on the main barrel of each individual lens. In this way, the operator minimizes the focusing adjustment necessary in changing from one lens to another. He may, for example, adjust his 90-mm lens for focus at 10 feet on a title card, and his 135-mm lens for focus at 25 feet. If the camera position is fixed throughout the show, a minimum of optical-focus adjustment is required of the operator when he changes lenses.

In camera operation as in radio operation, all movements and adjustments must be so solidly a part of the program that the listener-viewer is left completely unaware of the technical aspects involved. The less conspicuous the adjustments, the better is the technical operating technique. It has been said that the camera operator must be adept at mind reading. What is actually meant is that he must be thoroughly "production conscious."

Panning of a camera is probably the most-used operating technique. This movement must be integrated completely with the scenic action, and therefore must actually be *anticipated* by the operator. Although he is aided in this action by instructions received through his headphones from the technical director, the cameraman must ease into the actual operation at exactly the right moment. In order to keep the person properly framed, the pan must *lead* the moving figure. In other words, the moving person must be kept in the same spot on the frame, except in very special cases that will be identified by the director. If the performer's relative position changes, requiring the operator either to catch up or retreat with his camera action, the bobble is noticed at once.

Similarly, the panning must be tapered off in exact accordance with the required movement. The "feel" of such operation is gained only through persistent practice. The operator should adjust the friction-head drag to suit his individual touch. Sufficient drag must be used so that inadvertent movement will not jiggle the head with the consequent distracting jerk of the picture. If, however, the production calls for a "whip shot," requiring a rapid pan (used only on special occasions), the drag must be light enough to allow rapid and smooth operation.

The tilt shot is self-explanatory and is used more often (in a modified form) than the newcomer might suspect. If the reader observes closely the next panel show on the air, he may discover that the view is slightly upward toward the members of the panel. Producers use this technique to add dominance to the central interest of a scene, and to lend a dramatic appeal not evident in straight-on shots. Tilting of a camera during operation must, as a general rule, be done so slowly as to be just barely perceptible to the viewer.

The dolly shot is used for variety or dramatic interest. When the field of interest narrows to a portion of the total scene, the dolly operator or camera operator (depending on the type of dolly) is instructed to *dolly in*. It is always ascertained beforehand that the dolly path is properly lined up so that weaving around obstacles is avoided. Unless sufficient lighting is used to allow a very small iris opening, the cameraman must anticipate his optical focusing so that no defocusing is noticeable when the camera moves in. The same is true when the field of interest again broadens and the *dolly out* order is received. Needless to say, the dolly motion must be absolutely even and smooth both during the movement and upon starting or stopping the dolly action.

A *follow* or *travel* shot is done by dolly movement. In this case, the dolly follows a walking person, or may travel along a showcase in which the sponsor's products are exhibited. Every camera movement must have a definite purpose, since a psychological effect is imparted to the audience for every movement or angle change.

The scanning area of the image-orthicon tube is approximately equal in size to the image in a 35-mm still camera. Camera fans know that the basic lens for a 35-mm camera has a 50-mm focal length. The same is true of the TV camera using image orthicons. A focal length of 50 mm gives an approximation to the angle of view of the human eye. A 35-mm lens gives a wide angle of view, which increases the area possible to be covered but results in much smaller images on the screen. A wide-angle lens is used in such cases as opening shots to give an overall view of the total area to be covered in the immediate future. Also, it can be used to obtain an increased sense of distance.

Observe Fig. 10-41. The horizontal and vertical angles of the most popularly used studio lenses are listed in Fig. 10-41A. Fig. 10-41B shows the approximate width of field for the indicated lenses at 10 feet from the camera (for I.O. cameras). At this distance, the 135-mm lens, with a horizontal angle of 13°, covers about 2.2 feet. The 90-mm lens at 10 feet covers a horizontal field of about 3.3 feet. The 50-mm lens at this distance includes a horizontal field of 6.4 feet, which is approximately equivalent to what the human eye interprets with clarity at a distance of 10 feet. The wide-angle 35-mm lens includes a horizontal field of 9.5 feet at a distance of 10 feet. Fig. 10-41C indicates vertical coverage for lenses with the same focal lengths.

An excellent exercise for the student or new TV cameraman is to plot these angles on a scale drawing of the working area of the studio to be used. (Large-size linear graph paper may be used for this purpose.) The student may assume arbitrary dimensions, such as $40' \times 60' \times 30'$. By using the basic suggestions from Fig. 10-41, one can observe a clear relationship among focal length of lens, position and height, and projected area that may be covered. After considerable practice, these factors instinctively become part of the everyday routine of camera operation.

Zoom lenses on modern cameras generally provide a focal-length range of 18 to 180 mm (0.7 to 7 inches) for *Plumbicon* cameras, which is equivalent to a range of 36 to 360 mm (1.4 to 14 inches) for I.O. cameras. With the zoom lens, the viewfinder in practice becomes the "range finder" for

Lens	Horiz Angle	Vert Angle
35 mm (1½")	50°	38°
50 mm (2")	34°	25.5°
90 mm (3½")	19°	14°
135 mm (5.3")	13°	10°
8½"	8°	6°
13½"	5°	3.7°
17"	4°	3°

(A) Coverage angles.

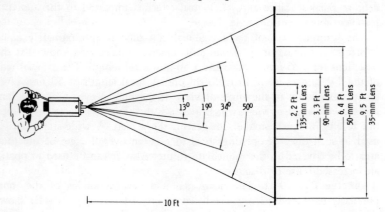

(B) Width of field.

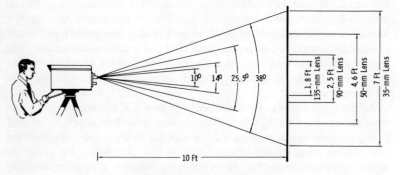

(C) Height of field.
Fig. 10-41. Coverage areas of lenses.

adjustment of the zoom position. Bear in mind the broad gamut of angles provided by the zoom-type lens.

The effect of lens focal length can be thought of in either of two ways: (1) A smaller focal length results in a greater included angle of field, a smaller image, and a greater depth of field. Conversely, (2) a larger focal length results in a smaller included angle of field, a larger image, and less depth of field.

Depth of field, the area which the given lens will cover with fine detail, is governed by the focal length and the iris opening (f/ number). The iris opening, in turn, is governed by the amount and type of illumination, and the scenic content, which determines reflectance values and the range of brightness.

It may be noted, for example, that the depth of field for a 50-mm lens stopped to f/8 and focused at 10 feet is about 12½ feet. All objects from 6'-9" in front of the lens to 19'-3" away will be sharply defined by the lens (when the lens is focused at 10 feet). A 90-mm lens, which has an included angle about one-half that of the above lens, has a depth of field of about 6 feet when stopped to f/8 and focused at 10 feet. In this way, the area of attention is narrowed, and interest is concentrated on a central point. The central objects occupy a larger area on the screen. Objects in the background, which were included by use of the 50-mm lens, are effectively taken away from the picture content. Depth of field also increases with distance from a given lens.

Fig. 10-42 illustrates the practical application of focal length and distance in relation to the area desired to be covered. This illustration represents a panel show involving four members of a panel plus one quiz master. The latter position often includes a place for a guest or participant on the show. This particular example shows how such a production may be handled with only two cameras, although three cameras often are used to increase the latitude of production technique. Camera 1 is shown trained on the entire panel from a distance of 15 feet; the 34° angle of the 50-mm lens results in the proper horizontal coverage area. Camera 2 is shown trained on one member of the panel; the 8° horizontal angle of the 8½-inch lens on this camera is just sufficient to cover one person. This, of course, fills the screen and results in a close-up of the face. Also shown in dash lines is how camera 2 may be panned to the right and used with the 135-mm lens in the lens turret to obtain a close-up of the quiz master's face. Camera 1 may, of course, be panned in any direction, and any of its four lenses may be used to cover any area desired. On a production of this type, judicious use of the eight lenses involved eliminates the need for camera movement, and only fixed tripods with pan heads are required. While one camera is on the air, the other camera operator receives instructions to ready his camera for the following shot, and switching is done in the usual way in the control room. "Ready" and "on-air" lights on the front of each camera inform the participants which camera is in use at any instant.

It should be mentioned that extremely small movements of an actor's head during close-ups, for which little depth of field is required, are greatly exaggerated. For example, in Fig. 10-42, the 8½-inch lens at only

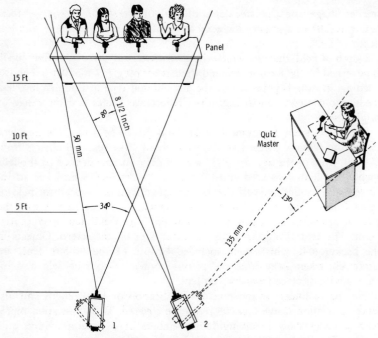

Fig. 10-42. Lens selection for panel show.

15 feet (approximately) has a very sharp angle of field. It is possible for a small movement of a performer's head to cause him to disappear from the picture entirely. Usually, when this effect is noted at this distance, the 135-mm lens is used, and the panel members are spaced widely enough that the 13° angle covers only one person. This points up the need for plenty of rehearsal time on even the simplest of productions.

Fig. 10-43 shows the vertical angles of 50-mm and 135-mm lenses on separate cameras at different heights and orientations. In this specific case,

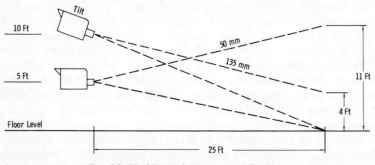

Fig. 10-43. Vertical coverage of lenses.

the camera with the 50-mm lens is covering 11 feet vertically from the studio floor, at a distance of 25 feet. Another camera (on a boom or dolly of the pedestal type) with a 135-mm lens is placed 10 feet high and oriented so as to obtain a tilt shot covering 4 feet up from the floor. This is just one of an infinite variety of combinations for various effects in the picture content.

Production terms, which must be recognized by the TV cameraman, are listed and defined as follows:

Pan: (See Fig. 10-40B.) The cue will be "pan right" or "pan left." Some directors also include the tilt shot under the term pan, as: "pan up" or "pan down." If the direction "head room" is given, the director wants more space above the person or object, and the cameraman tilts the camera up, or "pans up."

Single Shot: The director intends for the shot to include only one person.

Two Shot: The director intends for the shot to include two persons.

Group Shot: The director intends for the shot to include a specified group of persons.

Cover Shot: Also termed "wide shot" or "long shot." The picture should cover the entire scene of action. This is most generally accomplished with the 50-mm lens. If, however, the camera is necessarily close to the whole scene, the wide-angle 35-mm lens must be used.

Close Up: Unless there are directions to the contrary, the close-up should cover the head and shoulders of the specified person. For TV studios of average dimensions, the 135-mm lens is most usually employed.

Medium Shot: Unless there are other directions, the coverage should be from the waist of the specified person up. This most usually requires a 90-mm lens.

Loosen Shot: The director wants more space around the sides of the object, person, or persons. The camera operator must either get farther from the subject or use a lens with shorter focal length.

Tighten Shot: Opposite of loosen shot.

Dolly Back: Move the dolly or pedestal away from the scene of action.

Dolly In: Opposite of *dolly back.*

Tongue Right or Left: Swing the camera boom to right or left as in Fig. 10-40E.

Tongue In or Out: (See Fig. 10-40F.) This direction applies when the dolly or crane is crosswise to the normal direction to the pickup area.

Boom Up or Down: (See Fig. 10-40G.) This action results in vertical movement of a camera that may be on-air, or it may be done to ready the following shot.

When the camera operator receives instructions to "ready number 2 camera for tight shot of singer," he should be able to do just that, with optimum transmission characteristics and with minimum elapsed time.

About the only special effect the camera operator may be called upon to supply is the *blur-out*. This operation is a deliberate defocusing of the image by turning the optical-focusing knob on the camera. Such an effect sometimes is used for transition of time or locale. Either the same cameraman or another cameraman picking up a different portion of the scene may then be called upon to *blur in* the following picture by starting from a defocused image and then bringing the picture into focus by adjustment of the optical-focusing control.

10-10. FILM-CHAIN OPERATIONS

In the film chain, the projector and pickup-tube lenses normally are operated at a fixed iris setting. The video level is then normally controlled by a rotatable neutral-density (ND) disc on the film and slide projectors. Thus, the sensitivity (video gain) control on the film chain is connected to a servo amplifier that operates the ND wheel. In some "automatic" film chains, the sensitivity control sets the automatic target or limiter voltages, and the ND wheel is provided with a separate control.

The color film chain must be accurately installed and optically aligned as described in Chapter 5 before proper operation can be achieved.

Targets are adjusted in the same manner as for monochrome for a certain target current (depending on the type of vidicon) under open-gate conditions. Be sure sufficient beam current is used to discharge the target under this condition, but remember that too much beam current will reduce resolution and possibly cause some image aberrations.

Initial framing (horizontal and vertical sizing) should be done with a ball-chart slide and a grating signal. Beam alignment for each tube should be done simultaneously with sizing, as this affects centering (framing) of the image. Be very exact in this initial framing adjustment, before attempting registration procedures.

You are now ready to undertake laying the images on top of each other to register all channels. Take one channel (normally the luminance channel in 4-vidicon cameras) as the "standard framing," and register one channel at a time to this standard. In 3-vidicon cameras, the standard framing is usually the green channel. Quite often the channel used as the standard incorporates a separate mask in the optical path to assure standard framing. (There are exceptions; the RCA TK-27 optical system is factory standardized on the red channel.) The registration procedure for the 3- or 4-channel film chain is identical to that described for the studio camera. A registration slide normally is used.

Fig. 10-44 shows the usual positioning of pulses in the 4-channel camera. The luminance channel receives the pulses of Fig. 10-26 (the reference pulse is now pulse 3) plus a pulse timed with the green-channel calibration pulse (pulse 2). Pulse 2 is reduced in the luminance channel to 0.59 of unity level (the luminance level of green) for proper matrixing. The

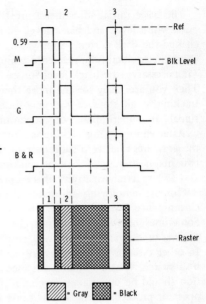

Fig. 10-44. Pulse positions for four-vidicon monitoring setup.

blue and red channels receive only the reference pulse (pulse 3). (NOTE: This last pulse might be designated pulse 5, as in the RCA system, as a result of spacing.) Such an arrangement provides for balancing with test pulses merely by looking at the picture monitor on NAM circuitry. You can see from Fig. 10-44 that when black levels are matched, no "stripes" appear in the number-1 white pulse on the monitor. When white levels are matched, no stripes appear in the number-3 (or number-5) white pulse on the monitor. A CRO presentation will show all white and black levels identical. White-balance controls are simply amplifier gain controls. Black-balance controls usually set a clipper reference voltage for black-level control.

After calibration of channel gains, you have a working standard against which to color balance the pickup-tube operating parameters. The basic requirement, assuming the encoding system has been properly optimized on color bars, is that zero subcarrier must occur for all shades of gray from black to white. This, in turn, demands that the pickup-tube outputs be identical in amplitude (for white and all shades of gray to black) when the optical system is looking at a monochrome gray scale.

In this case, a logarithmic transmission-chart slide (Fig. 10-28A) is used, and the scope monitors the camera output signal. The slide-projector lamp must be operated at nearly the normal line voltage to obtain proper color temperature. Bear in mind also that in this adjustment, you are compensating for any film-base color temperature that might exist, and that this can vary with different film.

The basic initial adjustments are target voltage, to obtain the required peak white level, and black level, since any change in target voltage can change the dark current. Simply adjust the target voltage of one channel at a time to obtain the reference white level for the white chips. Readjust (if necessary) the black level for each channel to obtain reference black. Then you are ready for the more time-consuming procedure of obtaining tracking at all steps of the gray-scale signal, so that all channel output signals are superimposed on all the individual steps from black to white.

After encoding, mistracking is indicated by the presence of subcarrier in the gray areas but zero subcarrier in the white and black areas. There are two major sources of lack of tracking: (1) gamma-correction circuitry, and (2) spectral sensitivity of the optical paths.

Cameras employing adjustable gamma circuits in the luminance and chrominance channels are readily corrected in case of gamma mismatch. Sometimes only the luminance channel is adjustable, with either fixed gamma or no gamma correction employed for the chrominance channels. In other cases, a fixed gamma correction of 0.7, 0.5, or unity is selectable by a switch. Whatever method is used in a particular camera, experimentation should be undertaken to obtain the best possible tracking before the operator proceeds to the second cause listed above. Problems with spectral sensitivity must be corrected by using neutral-density discs in front of one or more of the chroma-tube lenses.

If you have a properly adjusted color monitor (one which is truly black-and-white with a monochrome picture when color burst is present), you can tell immediately where to start in selecting a proper neutral-density filter. For example, if grays are greenish (this is the most common condition), select a neutral filter of about 0.2 and place it in front of the lens for the green tube. You must, of course, readjust the green target voltage to bring the white level back to reference white. Since you have now readjusted the target voltage, you must check for any change in green black level, and adjust it if necessary. If grays have now gone "minus green" (purple), the neutral-density filter is too dense, and a 0.1 filter must be tried. Once the optical paths have been balanced, it normally is not necessary to change filters when vidicons are changed.

When the film camera has been properly registered and tracked, it is imperative in practical film operation to balance whites and blacks of each camera channel for the particular film base and color processing method. For example, a film base can be slightly blue in color temperature. If the color projector can be still-framed, do this on the first frame that has a white and a black area. If the base is blue, the blue channel will have higher setup in the black region than the other channels. Balance black and white gains. This is the *only* way you can keep the flesh-tone concept in telecasting color film. It does require some amount of rehearsal time for previewing the film. But if you are critical in operations, the time is well spent.

A good example of what can happen is illustrated by the following example: You have two studio color cameras, which you have well balanced to one another. The master color monitor shows good flesh tones for the live pickups. Now stop and think a moment. You have balanced and tracked the camera on the scene (essentially) by balancing on a chip chart under the studio lights for the scenes to be used. You roll film for a film clip or commercial. The flesh tones turn greenish or bluish on the monitor. What has happened? Many stations simply balance on the gray-scale slide. This is no guarantee that flesh tones will be reproduced correctly from the film clip, which may have a base with a different color temperature. It is possible to have all encoders showing proper phase setup on color bars, but flesh tones may not be correct from all color sources. This can happen between studio cameras that have not been properly color balanced, or between live and film sources that have not been properly color balanced.

Another basic cause of this trouble is lack of proper color balance between slide projectors and film projectors. You should have the gray-scale slide mentioned previously, and also film loops of the same gray scale for the projectors. Balance the film projectors (if necessary) to each other by slight adjustment of the projection-lamp voltages. Normally, the projection-lamp voltage cannot be made lower than 105 volts; less voltage causes the color temperature to go toward red.

You should obtain two Wratten GL filters, one CC10B and one CC20B. Place one of these behind the lens of the slide projector and start with a lamp voltage of 100 volts. Adjust the lamp voltage and try the two different filters one at a time until the camera stays in balance between the slide-projector and film-projector gray scales. Once good balance is obtained in this way, a minimum of adjustments should be necessary when changes are made from one film to another.

When the operation schedule includes telecasting of film, it is necessary that the director be aware of the 2 or 3 seconds needed for shutter-type projectors to reach synchronous speed. When, for example, short film excerpts are to be used in a show, the leader includes definite frame markings from which the director and operators may judge exactly the required time interval between the order "roll film" and actual "take" of film video. The leader is run through until the designated mark indicates 3 seconds (large numeral 3 on frame) to the first picture frame. The projector is then stopped, and started to roll again 3 seconds before the spot called for in the script. (This time interval usually is about one line of script at normal reading speed.) Film projectors often are operated from the TD position at the switcher panel. When several sections of film on one reel may be used during a program, the TD may observe on a preview monitor the spliced-in sections between the sequences to be used. Thus, when one section is concluded, he switches to the next signal source being used while observing on the preview monitor the run-out of the film projector. He then may stop the projector at the next designated cue mark indicating 3 seconds to the following section.

On short film inserts such as discussed above, it is necessary that the TD and production director be warned of the ending of each section of film. Although in some cases the TD may be sufficiently familiar with the content to know when to cut to the next signal source, most stations use the same type of warning cue as described previously for projectionist use. In this case, the holes are punched in the upper corner of about 4 frames by a special punching device. The first cue is the alert; then the last 4 frames indicate the end of the actual picture content. This avoids the embarrassing possibility of showing the trailer on the film with its jumble of X's on a glaring white screen.

Due to the number of cue marks on many spliced films, some stations do not rely on cue marks at all. When a "stop-down" is to occur in long film shows (such as a feature movie), the projectionist watches for the upcoming white leader. When he sees this, he cues the TD on the intercom to "get ready"; then as the leader approaches the gate, he gives the direct "get out" cue. Other, more sophisticated, systems have a means of inserting a small piece of foil on the film to actuate the next event selected by the automation switcher routing.

Operational cues for film are as follows:

Ready Film: This is the alert to the projector operator and/or TD to expect the direct cue to start the projector running. In installations using more than one projector, the alert usually is more specific, such as "ready film 1," "ready film 2," etc. The numbers designate specific projectors.

Roll Film: (Sometimes given as "hit film") This is the direct cue to start the projector running. Immediately (usually after a 3-second interval), the cue "take film" is given.

Take Film: (Follows order to "roll film") This order usually designates a direct cut to the film control. If "dissolve to film" is given, the lap dissolve is made by the TD. The director usually gives a more specific cue, such as "dissolve to 1," "fade to 1," etc.

Table 10-1 lists running times for 16-mm film.

It is a good idea never to leave film in the projector after use. The portion of the film in the machine is affected by the heat from the projection lamp (even after it is turned off) and by the housing and cams. Film is subject to shrinkage and stretching and should be removed immediately after use and stored in a film can. Many operators keep a moist pad inside the can with the stored film to minimize heat effects, especially if the film room is dry and hot, as is often the case.

Even with proper care of film rolls, normal projector heat causes a certain amount of buckling of old film at the projector gate. This causes a fluttering with consequent changes in focus of the images on the picture tubes. Buckling occurs because the edges of the film in direct contact with

Table 10-1. Footage and Running Time for 16-mm Sound Film*

Footage From 1 Second to 60 Seconds					
Seconds	Footage	Seconds	Footage	Seconds	Footage
1	7.2″	21	12′—7.2″	41	24′—7.2″
2	1′—2.4″	22	13′—2.4″	42	25′—2.4″
3	1′—9.6″	23	13′—9.6″	43	25′—9.6″
4	2′—4.8″	24	14′—4.8″	44	26′—4.8″
5	3′	25	15′	45	27′
6	3′—7.2″	26	15′—7.2″	46	27′—7.2″
7	4′—2.4″	27	16′—2.4″	47	28′—2.4″
8	4′—9.6″	28	16′—9.6″	48	28′—9.6″
9	5′—4.8″	29	17′—4.8″	49	29′—4.8″
10	6′	30	18′	50	30′
11	6′—7.2″	31	18′—7.2″	51	30′—7.2″
12	7′—2.4″	32	19′—2.4″	52	31′—2.4″
13	7′—9.6″	33	19′—9.6″	53	31′—9.6″
14	8′—4.8″	34	20′—4.8″	54	32′—4.8″
15	9′	35	21′	55	33′
16	9′—7.2″	36	21′—7.2″	56	33′—7.2″
17	10′—2.4″	37	22′—2.4″	57	34′—2.4″
18	10′—9.6″	38	22′—9.6″	58	34′—9.6″
19	11′—4.8″	39	23′—4.8″	59	35′—4.8″
20	12′	40	24′	60	36′

Footage From 1 Minute to 60 Minutes					
Minutes	Footage	Minutes	Footage	Minutes	Footage
1	36	21	756	41	1476
2	72	22	792	42	1512
3	108	23	828	43	1548
4	144	24	864	44	1584
5	180	25	900	45	1620
6	216	26	936	46	1656
7	252	27	972	47	1692
8	288	28	1008	48	1728
9	324	29	1044	49	1764
10	360	30	1080	50	1800
11	396	31	1116	51	1836
12	432	32	1152	52	1872
13	468	33	1188	53	1908
14	504	34	1224	54	1944
15	540	35	1260	55	1980
16	576	36	1296	56	2016
17	612	37	1332	57	2052
18	648	38	1368	58	2088
19	684	39	1404	59	2124
20	720	40	1440	60	2160

*24 frames per second—36 feet per minute

the projector mechanism are subjected to more heat than the frame area, which is subject only to comparatively short bursts of light. This results in faster shrinkage at the edges compared to that which takes place through the picture area; as a result, the frame is forced out of shape and flutters in the film gate.

If it is necessary to use old film that has a noticeable amount of flutter, it is sometimes possible to restore a small roll sufficiently to use several additional times. This is done by winding the film with the emulsion side out and placing it in a humidor can with a highly absorbent blotter that has been moistened with water and glycerine. The film is left in the tightly closed can for at least seven days. If 35-mm film is involved, a slight amount of camphor is added to the solution with which the blotter is moistened.

Extreme care must be exercised by the operator in the handling of film, especially in threading the projector. If this operation is done improperly, the loss of the upper or lower loop is likely to occur during transmission. To avoid holes or tears in the sound or picture areas, be sure the sprocket teeth are properly engaged in the sprocket holes before any pressure pads are closed on the film.

Loss of either the upper or lower loop results in an annoying, jerky effect in the projected picture. If the projection becomes very bad, it is preferable to stop the show completely until the condition has been remedied. When the lower loop is lost, the bottom drive sprocket pulls the film off the intermittent pulldown claw, usually damaging the sprocket holes and possibly the rest of the film. There is one emergency operational procedure that has been practiced by highly experienced projector operators, but its use by inexperienced personnel is not to be encouraged. This is the practice of inserting the forefinger just between the sprocket and the lower edge of the gate, and watching intently for the instant when the pulldown claw is retracted, leaving the film momentarily free to be pulled down from the top loop. Two dangers exist in this procedure: First and most important is the danger to the finger from the sharp revolving sprocket teeth, and second is the danger to the film if the claw emerges as the film is being pulled through the gate.

Sometimes foreign matter in the film gate causes the appearance of a fluttering hair at the top, bottom, or side of the picture. If the projectionist is busy, the video operator calls out over the intercom, "Hair in the gate!" The projectionist places an abundant amount of saliva on his forefinger and holds it against the gate side of the film as the film enters the upper section of the threading path. This is the most reliable method of removing "hair" from the gate.

10-11. SWITCHER (TD) OPERATIONS

Many of the basic switching-panel operating techniques were covered in Chapter 7. Direct takes, fades, and lap dissolves, however, have become only a minor part of operating practice involved in modern productions.

The TD, or operator of the switcher-mixer unit, must be thoroughly familiar with production cues as applied to his operational position. The usual terms and their definitions are as follows:

Ready to Fade: A standby cue to the TD to expect the direct cue to fade out the signal now on the air. It is an alert order only, given to "ready" the operator for the action.

Fade Out: The direct cue for action by the TD. He operates the appropriate fader to fade the screen slowly to black. Other terms used by some producers are *fade down, fade to black, go to black,* and *dissolve to black.*

Ready to Fade in One (*or Two, etc.*): A standby cue to the TD to be alert for the direct cue to fade in a specified camera.

Fade in One (*or Two, etc.*): The direct cue for the TD to move the appropriate fader from out to in, bringing the picture up in brightness from a black level.

Ready One (*or Two, etc.*): A standby cue to alert both the cameraman and the TD that the direct cue to take camera one (etc.) is anticipated.

Take One (*or Two, etc.*): Direct cue to the TD to cut to the named camera. This is an instantaneous switch.

Ready to Dissolve to One (*or Two, etc.*): Alert cue to the TD that a direct cue to dissolve is anticipated.

Dissolve to One (*or Two, etc.*): Direct cue to the TD to operate the faders simultaneously, with the push button for the camera to which the dissolve is to be made depressed on the bank being brought on the air.

Ready to Super One (*or Two, etc.*): Alert cue for the TD to anticipate a direct cue to superimpose the designated signal over the one on the air.

Super One (*or Two, etc.*): Direct cue for the TD to add the designated signal to the on-air signal. This may be done gradually, or the push button for the camera being superimposed may be pressed with the associated fader already opened to the required brightness.

Lose One (*or Two, etc.*): Direct cue to the TD to remove the superimposition. This cue also may be *take out one, take out super, go through to two,* etc.

Practically all installations now include special-effects equipment as an integral part of the switching system. The associated control panel usually provides either normal additive mix or nonadditive (peak) mix, insert mode, and variable effects mode as described in Chapter 7.

Fig. 10-45A illustrates the RCA TA-60B control panel. There are two basic switching modes, the additive-mix mode and the peak-mix mode. These and other operational features of this equipment are described in the following paragraphs.

The *additive-mix lap mode* is illustrated by Fig. 10-45B, which indicates locked-lever operation. During a transition from one input to the

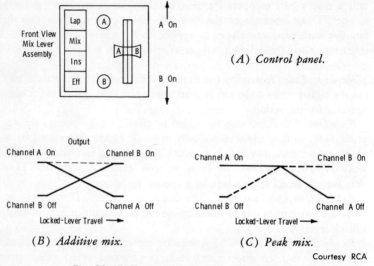

(A) Control panel.

(B) Additive mix.

(C) Peak mix.

Courtesy RCA

Fig. 10-45. Two modes of video switching.

other, the output picture components that are contributed by each input appear to change as indicated. The output signal is the sum of the indicated levels at any point in the lever travel, and is controlled so that it is no greater than the fully on white level of either input signal. The control feature also prevents the possibility of exceeding normal-amplitude white level in split-lever operation even if both levers are placed in the fully on position.

NOTE: The requirement for this lever position—a montage effect using spatially distributed input pictures—is handled properly in the peak-mixing mode.

A fade to black may be made in either of two ways. In one, a black input may be selected on the switcher, and a lap dissolve is made to that input. In the other, with two active picture sources selected on the switcher, the levers may be split, and a fade is made by moving the lever for the on-air channel to its off position.

For color input signals, both must contain color burst to prevent burst fading. Split-lever operation should not be used with color signals.

The *Peak-mix mode* provides a nonadditive mixing operation; it requires that the control levers be locked together. As a transition is made from one input to the other, the picture components in the output appear to change as indicated in Fig. 10-45B. The output signal in this case is dependent not only on lever position, but also on the relative location and brightness of each element in both pictures. Instead of providing a sum, the mixer in this mode instantaneously chooses the brighter of two input picture elements. Thus, as the control lever for input picture B is advanced toward on, the white elements of this picture show in dark areas of picture A and build up until, at the middle position of the lever, the output picture is a composite of all the brighter elements of both inputs. As the lever is advanced further, picture A disappears in a similar manner. The output picture amplitude is never, at any point, greater than the input level. This feature also permits transmission of the montage mentioned in the description of the lap-dissolve mode, and it makes it possible to fade lettering into a picture.

In the preceding example, note that at the middle lever position both pictures are fully on. If one input had been lettering or fine-line art at full white level, it would have appeared to "key" into the background picture. A *white-insert* mode is provided to allow rapid access to this condition without the necessity of moving the levers.

For color-signal operation, the following conditions apply:

1. In any of the mixing modes, both inputs must be in the color mode or contain burst to avoid a burst-fading condition.
2. In the peak-mixing mode, nonadditive mixing occurs during the picture period, but the burst period is gated through undistorted from the appropriate input.
3. In the effects mode, either input may be color, and the mixer automatically takes the burst from the input in which it is present. If both inputs are color, the mixer selects the burst of the signal being wiped in at the middle lever position.

When a nonsynchronous signal is selected on one of the mixer inputs, the switcher must supply a dc sensing voltage simultaneously. If this choice is made in an on input, the mixer amplifier then automatically switches to an internal bypass channel that does not contain clamps, faders, or sync addition. If the choice is made in an off input, the local on signal is not disturbed, and the bypass changeover does not occur until the lever is moved across to the opposite limit. At the same time, an undesirable mix of nonsynchronous signals is locked out by the same automatic circuit, and the lever action results in a direct cut to the remote signal instead of in a dissolve or mix.

A mix transition (either mode) from a synchronous signal to a nonsynchronous signal consists of a fade to black (with normal sync addition) through the full lever travel, and then a direct cut to the nonsynchronous signal.

If the mixer is operating in a special-effects-pattern mode, the wipe transition is the same as that described for the mix modes. In this case, however, cutoff of the nonsynchronous signal during lever motion results in a black output in the appropriate pattern area. Only the synchronous picture can be transmitted, except at the on-limit lever position where a nonsynchronous signal is selected.

There may be occasions when patterns other then those provided by the effects generator are required. Shapes such as a shield, a star, or a keyhole may be produced by supplying a keying signal to the effects amplifier from an external camera source. This may be done by focusing a live camera on a card containing the pattern or by using a slide of the desired pattern with a film camera. To make possible the use of any picture or pattern as the source of a keying signal, an effects-keying bus may be provided in the video switcher, as outlined in Chapter 7.

Fig. 10-46 shows how keying signals are generated externally. The camera is focused on a card that consists of a white background and a black pattern (or vice versa). If the background is white, the black center will be keyed out. If the background is black, the white center will be keyed in. In terms of control-lever operation, if the background is white, the A picture is keyed in. As an aid to remembering this, the following "formula" is suggested: B for black background to key in picture B.

The remote clipping-level control panel is designed for use with externally generated keying signals. Signals from an image-orthicon or vidicon camera are relatively noisy compared to the rectangular keying signals produced by the special-effects generator. When the generator signals are used, the keying gain and clipping levels can be set and forgotten, except for routine maintenance checks. However, any noise on the keying signal causes spurious keying, resulting in ragged edges and background bleed-through on portions of the raster corresponding to those parts of the keying signal for which the noise is strongest. The remote clipping-level control panel provides control over keying-signal gain and clipping level

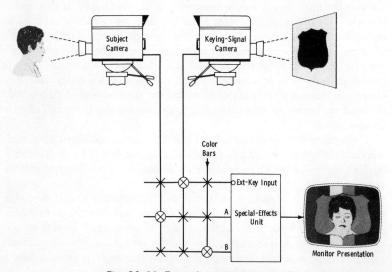

Fig. 10-46. Example of external key.

so that a keying signal having optimum characteristics and waveform can be applied to the amplifier keying circuits.

Remote clipping-level control becomes especially important in traveling matte effects (Fig. 10-47). Here an object or person is positioned before a black background with as much illumination as possible on the subject and as little illumination as possible on the background. (Even the blackest material available reflects some light.) If the signal generated by a camera focused on this scene is applied to both the keying-signal input and to video input A, an area corresponding to the outline of the object or person will be keyed out of whatever video signal is applied to video input B. As a result, the image of the object or person is keyed into the background area.

Subject matter may be moved around at will before the black backdrop, and the keyed area will move about the raster in synchronism with it. The picture applied to the B input serves as the background for the action; the background scene can be provided by a film camera. Thus, a singer can be televised moving about the streets of Paris or on a beach or other appropriate locale. Careful attention must be placed on lighting and the elimination of dark areas and shadows in the scene before the black backdrop; otherwise, spurious keying will result. The remote clipping-level control affords the operator some control over any spurious keying that may arise, but the initial lighting and staging must be carefully done.

For chroma-key operation (review Fig. 7-21 and Section 7-3, Chapter 7), the background is normally a highly saturated blue, well lighted. Actually,

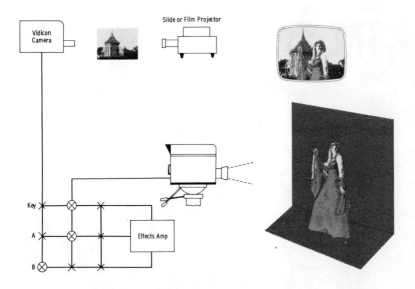

Fig. 10-47. Production of traveling matte effect.

a blue-green (cyan) lies almost directly along the —I axis, which is exactly out of phase with flesh tones. If the backdrop is painted so that a camera signal looking at it shows a "blob" along the —I axis on the vectorscope, flesh tones are quite easily balanced.

10-12. USEFUL SCANNED AREA

It should be borne in mind that camera blanking (from the camera drive pulses) normally has a shorter duration than transmitted blanking. Fig. 10-48 illustrates the usual result. Note that the viewfinder field of view, particularly on the left side of the raster, is more inclusive than that visible on the camera-control monitor or master monitor. To add to this problem, average receiver sweeps are adjusted to extend beyond the picture-tube mask in order to avoid annoying black sides. Therefore, if the cameraman focuses his camera so that an area such as a station-identification card is barely included on his viewfinder, the final result is an incomplete presentation on the receiver.

The solution is to mark off 15 percent of the raster at each side, the top, and the bottom on the master monitor (Fig. 10-49). Then keep all essential information within this area. It is helpful for the beginner to mark off

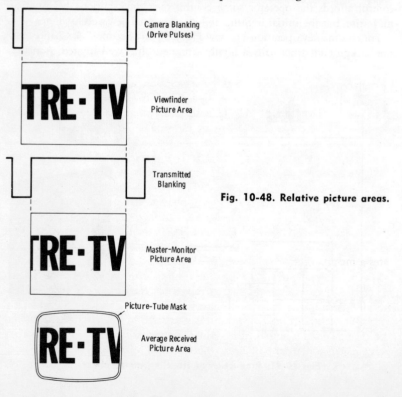

Camera Blanking
(Drive Pulses)

Viewfinder
Picture Area

Transmitted
Blanking

Fig. 10-48. Relative picture areas.

Master-Monitor
Picture Area

Picture-Tube Mask

Average Received
Picture Area

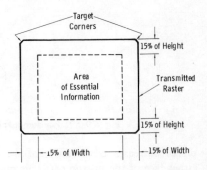

Fig. 10-49. Master-monitor reference for area of essential information.

the camera viewfinder, after careful checks with the master monitor, so that he can hold the essential area within the proper part of the field of view of the lens.

The same problem exists in making up 2" × 2" slides. Always remember that much of the area is "lost," and the area of essential information must be accurately centered horizontally and vertically.

10-13. AUDIO OPERATIONS

In the normal course of a local production, the television sound technician faces just as wide a variety of obstacles to be overcome as does the video technician. Just as in a-m and fm stations, the TV-station audio facilities must accommodate acoustical conditions that are peculiar to the specific installation; no two studios in the world are acoustically alike. In television, this condition is aggravated by the fact that for any particular studio the acoustical conditions change with each program that uses different props and different effective areas of the studio. The TV sound engineer can not "get in a rut"; the daily operations schedule presents a continuous challenge in operational problems. It is imperative that he be familiar with the fundamental properties of microphones as they affect operating techniques.

There are two basic types of microphones, the *pressure* type and the *pressure-gradient* type. Pressure types, in turn, are quite varied, including carbon, condenser (capacitor), crystal, moving-coil (dynamic), etc. There is only one pressure-gradient instrument, the *velocity* type, which incorporates a metallic ribbon suspended in a magnetic field.

Each different type of microphone exhibits a different response pattern (Fig. 10-50). The pattern simply represents the amplitude response for varying positions of the sound sources about the microphone. A *nondirectional* (or *omnidirectional*) pattern is illustrated in Fig. 10-50A. If a sound source of constant intensity is moved around the microphone at a constant distance, the amplitude of the electrical impulses from the microphone will remain constant. In Fig. 10-50B, the response of the micro-

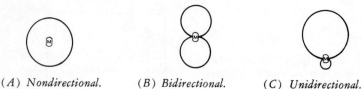

(A) Nondirectional.   (B) Bidirectional.   (C) Unidirectional.

Fig. 10-50. Fundamental microphone patterns.

phone decreases as the angle of the sound source approaches 90° from either face of the instrument. This microphone is sensitive from front and rear and theoretically "dead" at both sides, resulting in a *bidirectional* pattern. Still another fundamental type of pattern is illustrated in Fig. 10-50C; this one is said to have a *unidirectional* response because it is sensitive essentially only to sounds originating in front of the microphone. Briefly, it may be stated that pressure microphones are nondirectional, velocity microphones are bidirectional, and a microphone that is a combination of the two is unidirectional.

Note that these are the patterns that would be measured in an acoustically "dead" room. This condition is equivalent to an open space in which there is absolutely no reflection of sound waves. The effect of room acoustics will be discussed later. It will be shown also that response patterns vary somewhat for different frequencies.

Just as there is a need for a wide variety of response patterns, there is also a need for a difference between the designs for fixed-location or portable use. In addition, there are a number of impedance values for use in the complex fields of radio and sound.

A microphone has either a high-impedance output working directly into the grid circuit of the input amplifier stage, or a low-impedance output using a 30-, 150-, or 250-ohm transformer. This low impedance works into the input transformer of an amplifier. Some microphones built for various applications have a tapped output transformer with a screwdriver-adjustable switch that makes it possible to select either a high-impedance or low-impedance output.

In general, high-impedance microphones are used for general communications, amateur radio, some PA installations, and some home recorders. They are especially applicable if the microphone cable length is less than 25 feet.

The effect of the length of the microphone cable is related directly to the impedance and the application of the sound system. The metallic shielding in a cable forms a capacitance with the inner conducting wires, and this capacitance is added to the capacitance between the wires themselves. This is equivalent to placing a capacitor across the line. The higher the impedance used, the greater is the shunting effect across the line, and the greater is the consequent loss of high-frequency response. Cable length, therefore, definitely must be considered when high-impedance microphones are used. The effect of any length of cable is negligible for low-impedance systems.

Fig. 10-51 shows the signal loss caused by cable length in excess of 20 feet. The curves are based on a good microphone cable, with low capacitance of 0.0007 μF per 20-foot length. For a cable length of 200 feet, the frequency response at 5000 Hz would be 15 dB lower than for a length of 20 feet. Thus, broadcasting stations, recording studios, and special PA installations in which long cable lengths are a necessity always use low-impedance microphones. Low impedance must be used also in any form of communications when the operator works at a point remote from the microphone amplifier, since medium and high frequencies are very important.

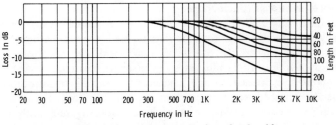

Fig. 10-51. Signal loss versus length of cable.

In the hypothetical instance of the ideal microphone, the instrument should be uniformly directional at all frequencies; however, such is not true in practice. High-frequency waves travel in a beam-like manner compared to low-frequency waves, and a number of design and construction compromises must be made. For example, the use of a nondirectional microphone oriented so as to become semidirectional is possible only for the higher frequencies that are sufficiently deflected by the case. At frequencies of approximately 1000 Hz or lower, the microphone is nondirectional in any position. Such response is useful in reducing high-frequency feedback in public-address installations associated with broadcasts that require a nondirectional microphone.

It is well known that high-frequency response falls off with increasing angles from the zero axis (a line drawn perpendicularly through the center of the diaphragm). Whenever the wavelength of a sound wave is short compared to the width of the diaphragm, several points of unequal pressure exist and cause an irregular movement. Thus, it becomes obvious that for best high-frequency pickup at increasing angles from the axis, the diaphragm must be smaller than the wavelength of the highest frequency considered. In practice, this is impractical because of the extremely low output that would be obtained from such an instrument, and a compromise must be made. It will be noted, therefore, that all microphones are directional

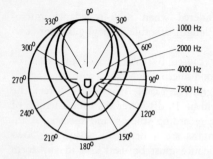

Fig. 10-52. Effect of frequency
on directivity.

at high frequencies, and have wider angles of equal response at the middle
and lower frequencies. A typical frequency-directivity response pattern is
shown in Fig. 10-52. Remember that this is the pattern from the face side
of a "nondirectional" microphone.

This points up a common error with regard to sound pickups, namely,
that any nondirectional microphone may be made directional by pointing
the instrument toward the source of sound, with the back toward unwanted sounds. Fig. 10-52 illustrates the fallacy in this idea. Regardless of
microphone-case orientation, the instrument is equally nondirectional to
all sounds except those at frequencies above about 1500 Hz. Waves at
these higher frequencies are deflected by the housing sufficiently to become attenuated toward the rear of the microphone. To achieve true directivity, the microphone must be "designed that way."

There are several common ways of constructing a microphone with unidirectional response other than the combination ribbon and moving-coil
system. Their basic action, however, is based on the same general principles as the combined unit.

Fig. 10-53 illustrates a method in which a ribbon element, such as is
commonly employed in velocity microphones, acts as a pressure-operated
device. It is suspended in the usual fashion in the magnetic field of a permanent magnet, but it differs from the velocity type in that the face is
exposed to the atmosphere, whereas the rear side is terminated in an acoustic resistance. Thus, the back of the element presents an infinite impedance
to sound, resulting in a microphone that is pressure-operated and has nondirectional characteristics, except at high frequencies.

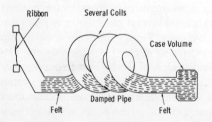

Fig. 10-53. Use of infinite
impedance to produce
unidirectional characteristic.

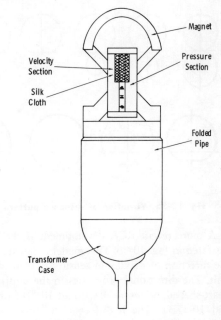

Fig. 10-54.
Unidirectional microphone.

By combining the above principle with that of the pressure-gradient ribbon action, a unidirectional response pattern is obtained. Fig. 10-54 illustrates this method. One continuous ribbon is used, with the upper half acting as a velocity (pressure-gradient) microphone, and the lower half acting as a ribbon pressure device terminated in the rear with a folded pipe (the acoustic resistance). The pipe usually is damped with tufts of felt. Because of the length of pipe behind the pressure section, the velocity of the pressure ribbon leads the pressure in the sound waves at the low audio frequencies. To compensate for this characteristic, a cloth screen is placed in front of the velocity section to introduce a corresponding phase shift in the velocity ribbon. Phase shift at high frequencies is minimized by using the same ribbon element for both units and by using suitable geometrical design of the field structure.

The RCA Type 77-D microphone is similar in design to the type of microphone just described, but it includes a means of varying the acoustic impedance presented to the ribbon. A switch varies the area of an adjustable opening in the labyrinth connector. When the opening is so large that the back of the ribbon is entirely exposed to the air (as in the ordinary velocity microphone), the acoustic impedance is zero, and the response pattern is bidirectional. When the aperture is completely closed, the acoustic impedance is infinite, and the microphone becomes nondirectional. By varying the size of the opening, a great variety of response patterns may be obtained (Fig. 10-55).

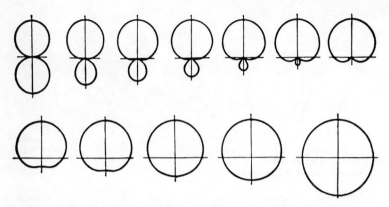

Fig. 10-55. Variation of response pattern by changing aperture size.

A more recent RCA development is the Type BK-5A microphone, illustrated in Fig. 10-56. This model is called a *Uniaxial* microphone because the direction of maximum sensitivity coincides with the major axis of the unit. The directional properties in the midfrequency region are essentially heart-shaped, or *cardioid,* with an 18-dB front-to-back ratio (Figs. 10-57A and 10-57B). The high-frequency directional properties have been improved over those of the Type 77-D, both in pickup angle and in front-to-back ratio. The pattern of the Type BK-5A is not adjustable.

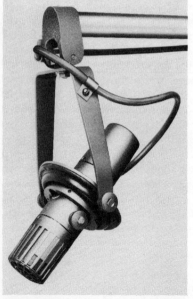

Fig. 10-56. Microphone with cardioid response pattern.

Courtesy RCA

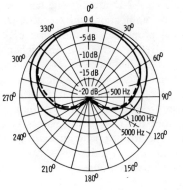

(A) *About vertical axis.*

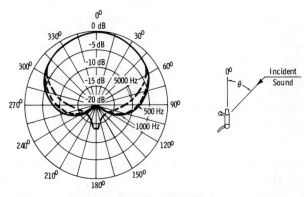

(B) *About horizontal axis.*

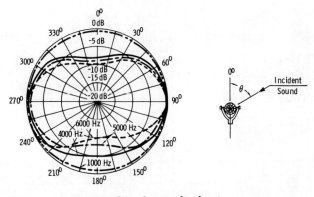

(C) *About longitudinal axis.*

Fig. 10-57. Directional characteristics of BK-5A microphone.

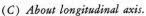

As described previously, the Type 77-D microphone has one adjustable port at the back of the ribbon; in combination with the acoustic labyrinth, this port controls the directional properties. In the Type BK-5A microphone, there are two ports, one placed at each end of the ribbon. The ports are covered with an acoustically controlled cloth to form the proper opening impedance. The movable element is a ribbon with a horn and screen assembly in front and the acoustic connector and labyrinth behind. The port impedance and the acoustic resistance of the labyrinth, along with the physical separation of the front and rear pickup points, give this unit its unidirectional characteristic. The acoustic labyrinth is a pipe about 31 inches long folded into the form of a cylinder. It is damped along its length to eliminate resonances and acts as a pure acoustic resistance over a substantial part of the frequency range.

The electrical circuit of the Type BK-5A microphone consists of the corrugated aluminum-foil ribbon, a line-matching transformer (with connections brought to a terminal board for adjustment to 30-, 150-, or 250-ohm circuits), and a response-compensation reactor. A switch incorporated with this reactor allows selection of three response characteristics, music (M), voice 1 (V1), and voice 2 (V2), which are illustrated in Fig. 10-58. Voice range V1 is often used on microphone booms even for music, especially if a large amount of low-frequency "boom" exists because of studio dimensions and acoustical conditions.

The techniques used for audio pickup depend on the type of show involved. There are three general types of TV shows, as follows:

1. Shows in which all microphones may be visible to the viewer. Such programs as quiz shows, panel discussions, some variety shows, musical productions, formal interviews, news programs, sporting events, night-club pickups, etc., fall into this category.

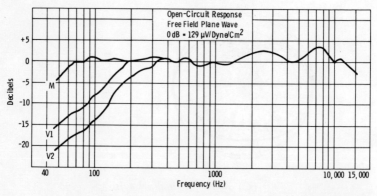

Courtesy RCA

Fig. 10-58. Selectable microphone frequency response.

2. Shows in which some microphones may be visible but others must be kept out of camera range. Some variety programs, such as those that are partly musical and partly dramatic, are a good example of this category.
3. Productions in which all microphones must be kept out of camera range. The most notable example of this category is the dramatic program in which the appearance of a microphone would destroy the atmosphere or mood of the show. This technique also is used in interviews for which an air of informality is to be preserved. Actually, any of the programs mentioned above might call for this method.

Programs in which the microphones may appear in the scene are handled in the same way as are programs in the usual sound broadcasting studio, with one important difference: The microphones must not obstruct the faces of the persons on the show. Microphones such as the one in Fig. 10-59 find wide application as a result of this necessity. When larger microphones are used, they are placed below or above the face. In panel-discussion programs, the microphones are low and sometimes hidden by a projection on top of the desk.

Programs in which some or all of the microphones must be kept out of the picture call for a microphone boom (and boom operator). Typical microphones used for this application are the RCA Type BK-5A (Fig. 10-56) and the Electro-Voice Model 668 (Fig. 10-60). Since the greater distance from the microphone to the sound source emphasizes unwanted sound pickup, a unidirectional response pattern is highly desirable in a boom-mounted microphone. The boom arm must be high enough to be out of the picture, and is usually slanted at about a 45° angle.

Fig. 10-59. Shure Model 576 dynamic microphone.

Courtesy Shure Brothers, Inc.

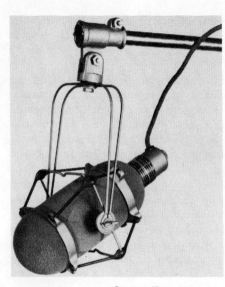

Fig. 10-60. Microphone for boom mounting.

Courtesy Electro-Voice, Inc.

The art of *sound perspective* represents perhaps the greatest challenge the broadcast sound technician faces in making the transition from radio to TV practice. Sound perspective involves matching the apparent distance of sound to the accompanying picture. As an example, assume that the camera is on a close-up of a man and woman seated on a sofa. The man is about to leave, aginst the earnest pleas of the woman. He gets up and walks slowly across the stage to the door, while the woman becomes hysterical. The camera follows the man to the door. Since the viewer is concerned at this time only with the man, it is natural to expect the woman's voice to be left behind as the man walks slowly out of her presence. In this case, even though the man is not saying a word, the microphone-boom operator would do well to follow his movements so that the apparent distance to the woman's voice becomes greater in direct ratio to the movement of the actor. But at this time, there arises a problem that is not apparent on the surface. In monaural reproduction of sound (as contrasted with normal binaural hearing), as the microphone is moved away, the apparent distance becomes increasingly greater than the effect we would experience if we were present "in person." Actually, it is found that the microphone need not follow the man in this example all the way across the room. If this were done, in many instances the apparent distance would be far greater than one room length.

This discussion points up that rehearsals for sound as well as picture must be thorough. No rules may be laid down, since there is no such thing as a "standard studio" acoustically. The desired effect can be achieved only through careful rehearsals. Sound perspective has been discussed here to acquaint the reader with the phenomenon so that, in the event of insufficient rehearsal time, he will be prepared to emphasize the importance of his ear to the success of the show.

Another common problem in television sound occurs when a person rises from a chair, and the area of view is widened from a close-up to include other details. If the person is talking during this transition, the microphone-boom operator must be warned in time to raise the microphone before it comes into view. Then what happens to the sound? Any sudden change in distance of the microphone is quite noticeable to the listener. Consequently, this action must be anticipated far enough ahead of time that the boom may be raised gradually just before the cue to widen the area of view is given. Because of the precision with which this sight-sound relationship must be exercised, many operators tie an inconspicuous length of string onto the microphone so that the string may be noticed on the monitor before the microphone appears in the picture.

Shadows cast by the microphone and boom are an ever-present danger in TV productions, and close cooperation with the lighting engineers is required. Such shadows may be eliminated by using sufficient light on the spot where the shadow occurs to "wash out" the shadow.

When the microphone must be quite close to a camera, it sometimes is necessary to turn off the camera blower to prevent the noise from being picked up. Such occurrences usually last only a short time, and in most modern cameras the blower may be off for a considerable length of time unless the ambient temperature is unusually high.

On unrehearsed shows, in which the continuity of action is uncertain, the boom operator often orients the face of the microphone practically parallel to the floor for wider coverage. The quality of sound under this condition is somewhat inferior, but this procedure often is necessary for adequate pickup.

The "extended reach" often necessary in TV productions is facilitated by such microphones as those illustrated in Fig. 10-61 and Fig. 10-62. The Electro-Voice *Cardiline* Model 642 (Fig. 10-61) is a highly directional dynamic microphone utilizing a combination of the characteristics of cardioid and distributed front-opening designs. This microphone is designed to make possible a working distance which is two or three times that of conventional directional microphones, while offering reduced susceptibility to mechanical shock transfer and wind noise.

Designed for specialized and extended long-range pickup, the Electro-Voice Model 643 *Cardiline* microphone (Fig. 10-62) also combines the characteristics of cardioid and distributed front-opening designs. This microphone provides a cardioid pickup pattern up to 100 Hz and is highly directional over the balance of its frequency range. Integral two-position, low-frequency tilt-down and cutoff filters are included for suppression of room reverberation and retention of "presence." This microphone often is

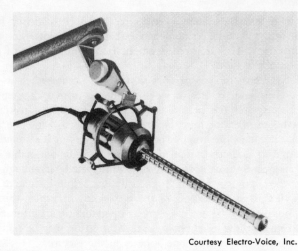

Courtesy Electro-Voice, Inc.

Fig. 10-61. Highly directional dynamic microphone.

used on remote broadcasts. For example, quarterback's signal calls or the "thud" of a punted football are readily picked up in this way to lend "atmosphere" to a football broadcast.

In television, "mugging the mike" (Fig. 10-63) is the rule rather than the exception for many productions. This is necessary to avoid picking up the noise from ordinary "stage business" and unwanted sounds from the many participants and behind-the-scenes workers. In addition, the vocal microphone usually is fed to a PA system for the studio audience. Thus, orchestral pickups for TV require more microphones than normally are used in radio broadcasting. Singers often appear to be "swallowing the

Courtesy Electro-Voice, Inc.

Fig. 10-62. Microphone for long-range pickup.

Fig. 10-63. "Mugging the mike" is common in TV.

Courtesy Shure Brothers, Inc.

mike" when moving about the stage with a hand-held microphone. Therefore, TV-type microphones are designed to handle close-in work without "blasting" and with a smooth overall frequency response.

A common microphone is the lavalier type, which is hung around the performer's neck or clipped to a tie or other clothing. The design must be tailored to compensate for the fact that the microphone is operated off-axis. Fig. 10-64A depicts three locations of the same microphone with respect to the wearer. For purposes of illustration, two things are assumed: (1) the user speaks directly ahead and does not move his head, and (2) the microphone has perfectly flat frequency response.

In position 1 of Fig. 10-64A, the axis of the microphone is aligned directly with the user's mouth, and all frequencies are faithfully reproduced. In position 2, the mouth is at an angle off the axis of the microphone, and the high-frequency response is beginning to fall off. In microphone position 3, against the man's chest, sound is reaching the microphone at an angle even farther off axis, and there is greater drop-off of the higher frequencies. However, two additional changes that affect the reproduction by the microphone have taken place. One, called the *shadow effect,* apparently is the result of the presence of the man's chin in the sound path; this effect causes response to the very high frequencies to start rising. The second effect is acoustical radiation from the man's chest, which reinforces the lower-frequency sounds and causes a rising response at this end of the spectrum. An ideal microphone for this type application, then, is one in which the response characteristics are the inverse of the microphone at position 3. The desired response is indicated by the dash line of graph 3. Notice the similarity of this curve to the frequency-response curve of the RCA Type BK-12A microphone (Fig. 10-64B).

The sound-mixer control console for television differs from the usual one found in radio stations only in that a greater number of microphone in-

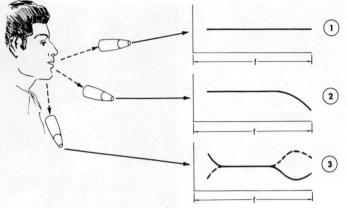

(A) Microphone positions.

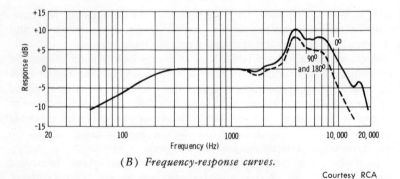

(B) Frequency-response curves.

Fig. 10-64. Lavalier microphone.

puts are available. The console of Fig. 10-65 has eight microphone inputs, or 16 on a preselect basis. Any number of microphones may be assigned to either of two submaster faders for grouping control.

Located on the control panel are the microphone keys and attenuators, turntable keys and attenuators, audio attenuators for the film chain and VTR's, and monitoring and cue-feeding switches and attenuators. A VU meter indicates the level of the audio output from the control room; 100 on the scale corresponds to 100-percent modulation of the fm aural transmitter. For efficient operation, the operator should be familiar with the control and functional circuits of his particular control console, or board.

The turntables often are placed at the side of the control panel and operated by the engineer responsible for the audio mixing. On large productions involving a number of microphones, he usually is assisted by a turntable operator so that his attention may be concentrated on the control of the microphones. A turntable for broadcast use incorporates a filter

Fig. 10-65. Audio console at WTAE-TV.

selector (to accommodate the wide variety of equalization characteristics of recordings and transcriptions), cuing facilities, a fader control in addition to that on the mixer console, and an off-on switch for the motor. The recording is "cued in" by placing the turntable switch in the cue position. When the stylus reaches the sound on the record, the signal is heard either in headphones worn by the operator or on an external loudspeaker cuing system. The operator then backs the record to the start of the sound and turns off the motor. Immediately prior to broadcast, he starts the turntable motor running while he holds the record by hand to prevent its turning. He throws the switch to the broadcast position, and, upon the direct cue, releases the record. This procedure calls for exact cuing of the record on the revolving turntable. The practice is necessary since almost a complete revolution of the turntable is necessary to come up to proper speed. Sound from the record would "wow" if the turntable were simply started revolving at the first sound groove.

Operational cues with which sound engineers must be familiar are as follows:

Open Mike 1 (or 2, etc.): Order to the sound-control operator to open the fader for the designated microphone, or if the fader is open, to close the corresponding switch.

Stand By With Sound: Alert cue to the audio operator or turntable operator to start the turntable revolving while the record is held to prevent its turning. The fader should be open in readiness for sound.

Hit the Sound: Direct cue to the turntable operator to release the record.

Fade in Sound: (Sometimes termed "sneak in sound.") Order to release the record with the fader closed, and then open the fader gradually to the desired level as instructed by the director.

Cut Sound: Cue to cut off sound abruptly. This is accomplished best by using the microphone switch rather than the fader.

Fade Out Sound: (Might be termed "slow fade-out" or "fast fade-out.") Cue to remove sound by use of the proper fader control.

Sound Down: Cue to operate fader to obtain lower sound level. Sound is to be maintained under the principal sound or simply to accompany the picture at a lower audio level.

Sound Up: Cue to operate fader to obtain higher level of audio. NOTE: Some directors do not watch the VU meter and may order the sound up when it already is peaking at 100 on the meter. The operator should realize that he cannot accomplish the desired result by overmodulating the transmitter. Indeed, most installations include a limiter amplifier at the transmitter to avoid such overmodulation. If the period of time is to be very short, the operator may "up" the sound until the meter hits the pin. If the high level is to be extended over a longer period of time, and the director is temperamental, some chief engineers instruct their operators to raise the monitor volume. The technical department must be required to work only within the limitations of the system.

Cross Fade to 2 (or 1, 3, etc.): Cue to the audio operator to fade in microphone 2 (etc.) while the other microphone is faded out.

Segue (pronounced say'-gway or seg'-way): Instruction that another record is to follow immediately at the conclusion of the record that is being broadcast.

10-14. TYPICAL BASIC OPERATING PRACTICES

Organization of basic operating practices is custom tailored to meet the individual needs of each station. The following specific outline is presented here as a reference for practicing personnel, and as an introduction to the field for students and newcomers:

WTAE-TV Basic Operating Practices

System Quality Control

Daily

1. Check overall video system using multiburst test signal.
2. Check overall video system using standard 75 percent, split-field color-bar pattern.
3. Check overall audio system using tone for level setting and noise.

Weekly

1. Check microwave STL and transmitters using multiburst test signal. Record all data per test sheet.
2. Check video tape recorders. Optimize heads and center servo systems.

3. Check cameras, both film and live. Set up heads and check rack equipment.

Monthly

1. Perform complete audio check through entire system (including transmitter). Record frequency, distortion, and noise.
2. Perform video sweep test of transmitters.
3. Check out video switcher systems; measure differential gain and phase.
4. Double-check system color timing to output of each switching system.

Quarterly

1. Run complete proof of performance on transmitters.

Operations

1. Daily operations or discrepancy reports, network quality reports, and news-program reports are kept at studio control by the technicians.
   The official program log is kept by the announcer.
2. FCC maintenance and transmitter operating logs are kept by transmitter personnel.
   Microwave logs are kept at the studio by the technicians.
3. The chief technician (usually the air switcher) is responsible for on-the-spot decisions. The switcher also is director for nonlive programming (operates film projectors and VTR's by remote control).
4. The projectionist loads film, slides, and video tape for playback.
5. The tape operator sets up the video-tape machines for recording sessions, and is the switcher for these sessions. He also is the tape man for news shows.
6. An audio man is used only for live shows, taping sessions, or integrated shows. (Automatic audio-follow-video is in use at other times.)
7. The camera control operator also is responsible for the microwave log and network quality report.
8. The cameramen also do all the lighting.

Film Handling

1. All film for programs and spots is received from the film department on a per-day basis and is aired by the technicians.
2. Slides are stored in the engineering department and are aired in accordance with the program log.

Video-Tape Handling

1. Purchasing, scheduling of reels, and usage charts are the responsibility of the engineering department. (This work is handled by the assistant chief engineer for operations.)
2. Syndicated tapes are shipped, received, and previewed by the film department.
3. Local video-tape shows are recorded, filed, catalogued, and aired by the engineering department.
4. Local recorded (one time or short time) announcements are handled as in 3 above.
5. All other taped announcements are received and stored by the film department and previewed and aired by the engineering department.

Personnel responsible for program operations, news, and engineering operations hold weekly meetings. The assistant chief engineer for operations represents the engineering department at these meetings.

Preventive maintenance at both the transmitter and the studio is done on a planned schedule. The work procedures are specified for each type of equipment by the chief engineer and the assistant chief engineer for maintenance. Ordering of spare and replacement parts is handled by the same personnel.

All other support activities (such as office supplies, etc.) are handled by the engineering office. Remote pickups are surveyed and arranged jointly by the programming department and the engineering department.

EXERCISES

Q10-1. What is the maximum brightness contrast range for optimum control of the camera, and how is this determined?

Q10-2. What is the maximum amount the waveform should drift from scene to scene on an unclamped CRO?

Q10-3. Name the factors that can affect color-picture sharpness.

Q10-4. What is the final check for registration?

Q10-5. When a new color set is used, what is the first thing to check?

Q10-6. Is it possible to have a good color balance on a wide shot, but not on a tight shot in the same scene and with the same camera?

Q10-7. What are the factors pertinent to background effects on skin tone?

Q10-8. How much change in the color temperature of lighting can occur before skin tones show error, and what line-voltage change does this represent?

Q10-9. Are light dimmers ever used in color studios?

Q10-10. For color work, give the optimum ratio, relative to base light, for (A) back light and (B) key, or modeling, light.

Q10-11. What is the proper amplitude of the color sync burst?

Q10-12. In the complete composite color signal, does the color sync burst occur during every time interval corresponding to horizontal sync?

Q10-13. How many cycles of the color-burst signal should be present?

Q10-14. When a receiver demodulates on the $E_R - E_Y$ and the $E_B - E_Y$ axes, are proper flesh tones reproduced?

Q10-15. Which of the chroma axes (I or Q) contributes most to natural flesh tones?

Q10-16. Does the vectorscope show any luminance value?

Q10-17. If you do not have the same system phase for all color sources at the studio, will receivers show a shift in colors when you switch from one source to any other source?

Q10-18. If envelope-delay distortion exists, how does this become apparent on a color picture tube?

CHAPTER 11

Mobile and Remote Telecasts

A wide variety of conditions must be met by the field department of a TV station. Engineers face a continual challenge in providing proper power, interconnection of equipment, orientation for smooth production coordination, and dependable microwave-relay facilities. This very challenge provides the incentive for many engineers to select the field department as the most interesting of all. Certainly it may be said that the field engineer seldom finds himself in a daily routine that never changes.

This chapter covers mobile and portable equipment, and the technical production techniques involved. Many remote pickups and all mobile pickups are relayed by microwave systems. Chapter 12 covers microwave equipment and its use.

11-1. SPECIALIZED MOBILE AND FIELD EQUIPMENT

Fig. 11-1 illustrates how a cameraman can function quite literally as a self-contained, self-powered video cruiser. The Westel WR-201 (record only) helical-scan recorder uses a single-head scanning assembly that is 2.2 inches in diameter. The recorder weighs 23 pounds, including tape and rechargeable nickel-cadmium batteries, and employs an integrated cable-connected vidicon camera head that weighs less than 7 pounds. Fig. 11-1A illustrates the camera-recorder combination with the recorder mounted on a hip pack. Fig. 11-1B shows how the units can be separated (up to 50 feet), and Fig. 11-1C is a view of the recorder assembly.

The use of such an integrated package makes possible television coverage of almost any event, with a minimum of setup time or travel arrangements. Of course the telecast is not "simultaneous," since the tape must be transported back to the studio for playback on the air.

The vidicon camera accepts any standard C-mount lens to fit the field of view desired. The small CRT viewfinder can be switched to operate as a waveform monitor. All operating controls and indicators are located on the back of the camera head for convenient operation.

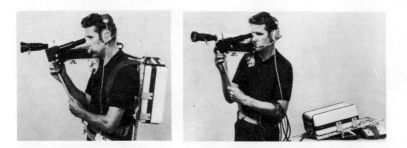

(A) Recorder on hip pack.   *(B) With units separated.*

(C) Recorder cover removed.

Courtesy Westel Company

Fig. 11-1. Portable camera-recorder combination.

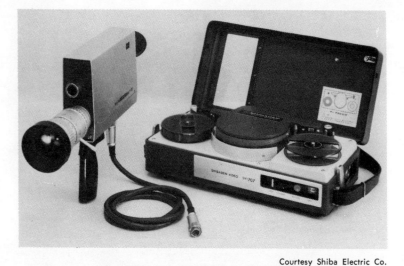

Courtesy Shiba Electric Co.
Fig. 11-2. Audio-video recorder and camera.

The Shibaden Model SV-707U video recorder and Model FP-707 camera are illustrated in Fig. 11-2. The camera employs a 2/3-inch separate-mesh vidicon; all other circuits are built with silicon transistors. Picture framing and focusing are adjustable with the aid of a built-in 1.5-inch electronic viewfinder. ALC (automatic light compensation) and an automatic voltage stabilizer are provided; controls are required for lens focusing and zooming only. A microphone is provided on the front of the camera. The overall frequency response through the VTR is 80 Hz to 10 kHz.

A full-color mobile TV camera is illustrated in Fig. 11-3. The Ampex BC-110 mobile color camera was basically described in Section 4-11 (Chapter 4) as the Ampex BC-210 studio color camera. The camera-control unit, which contains a full EIA sync generator, is carried in the back pack. A multiplexed output is used to enable feed over a single coaxial cable to the base station (which may be in a mobile van up to 2000 feet away) or to an optional microwave video transmitter. When microwave linking is used, a command-link receiver (with a narrow bandwidth of 20 kHz) in the back pack receives synchronizing, iris-control, cue-lamp-control, and intercom signals from the base station.

11-2. CAMERAS FOR EITHER STUDIO OR FIELD

It should be understood that any studio camera, whether monochrome or color, can be used in field telecasting. Obviously, some types of cameras used at the studio are more suitable for remotes than others, because of size, weight, and general flexibility in handling.

Courtesy Ampex Corporation

Fig. 11-3. Portable units of color-camera system.

The Norelco (Philips) PC-70 three-*Plumbicon* camera has been popular in both field and studio applications. The latest Norelco PC-100 camera is small in size and employs digital controls, a multiplexing technique to make possible lightweight cabling, and a one-inch *Plumbicon* in each of the three color channels. It should be mentioned also that the Norelco PCP-90 hand-held camera also uses a digital control system but one that is somewhat different from that of the PC-100. It is available in either cabled or microwave-relay form.

The RCA TK-44 camera employs three channels (lead-oxide vidicons) and is smaller and lighter than the TK-42 described in Chapter 4. Thus, the TK-44 will find more remote applications than the TK-42. (NOTE: A lead-oxide vidicon is the RCA equivalent of the *Plumbicon* tube.) The Visual VP-3 three-*Plumbicon* camera is another color camera popular for either studio or field use. International Video Corporation (IVC) supplies either three-vidicon or three-*Plumbicon* cameras that are suitable for either studio or remote use.

11-3. PRELIMINARY CONSIDERATIONS

Even before the field equipment can be set up and interconnected, a number of preliminary considerations are involved. The type of event to be televised and the physical layout of the locale of origination determine the nature of vantage points available, number of cameras to be used, and general orientation of equipment. For example, in some cases when a mobile-unit truck is available, circumstances will prevent setting up equipment inside the unit. This would occur if the vantage points for the cameras were necessarily more than 2000 feet from the nearest possible location of the mobile unit. This distance is normally the maximum allowable separation between cameras and control units in all present makes of field equipment. Separations greater than this distance usually are not encountered in practical situations, however, and in most cases the control equipment may be operated in the truck.

The first preliminary consideration with which the technical director is concerned is the arrangement with the local wire order chief (Bell Telephone System) for the proper audio lines to be used, where these are necessary. This ordinarily involves the installation of at least two lines. One loop is equalized to at least 5000 Hz to carry the program sound. In some instances, the other loop is unequalized, since it is used primarily for intercommunication between the remote location and the studio, during both preliminary testing (such as microwave-system alignment) and the actual telecast. Intercommunication during the telecast is a necessity when local cut-backs to the studio are involved during the field event. Both lines are run to the prescribed location of the switcher unit; from this point, audio is distributed to the various destinations by means of the intercommunication system.

Several other preliminary arrangements may be necessary. As an example, it may be necessary to construct special operating platforms for cameras and other equipment to provide suitable vantage points. In many instances, this construction must be reported to local building-inspection authorities as required by state laws. Some states require permits before work can be started. If a great amount of electrical wiring is involved, the installation must pass inspection before use. It is important for the field technical supervisor to subscribe to the services of the local building and fire codes so that he is always acquainted with local laws and changes as they occur. Most station managements provide this subscription service for the use of the engineering department.

The adequacy of the power source must be considered. In general, for 117-volt single-phase power lines, about 1000 watts per power supply used in the field setup should be available. Approximately 5 kW of power ordinarily is sufficient for a dual camera chain, and 7.5 kW should be adequate for a triple camera chain. Some pickup points may occur in areas where only dc power is available. In this case, a portable engine-driven ac generator of the required power capability must be used. If the point in question is to be used on a seasonal basis, permanent installation of power lines from such a generator, or from the electric company, may be warranted.

11-4. PROGRAM CONSIDERATIONS

The thought that must be given to the planning and coordinating of field events is emphasized by the intensity of public interest in remote programs. If the reader carefully considers the most appealing programs of his own experience, he will realize that such events as baseball games, football games, golf tournaments, political conventions, news events, and wrestling and boxing matches are near the top of the list. Analysis of viewer interests indicates that remote programs are receiving the widest public acceptance.

For remote broadcasts, the camera range must be broader in scope than it is at the studio, both in angles of coverage and in possible lighting conditions. Extreme care must be taken in planning camera vantage points so that required lens angles can be used. For optimum results in camera-chain alignment, there may be a need for some compromise in the relationship of beam current to target voltage. For monochrome pickups in bright daylight with modern image-orthicon tubes, it is necessary to use neutral-density filters to reduce the photocathode illumination to an amount for which higher-than-normal beam current is not necessary to discharge the high lights. Since the field engineer is concerned with lighting conditions that may change from hour to hour or even minute to minute, it is well for him to keep constantly in mind the relationship between target voltage and beam current. He knows that for best signal-to-noise ratio, both the target voltage and the beam current are related to the average illumination on the photocathode. Camera controls that provide the automatic target set described previously provide a good compromise in the field over a large range of scene brightness. Unusually sharp pictures may be obtained in some instances, however, by a slight readjustment.

In considering such slight adjustments for optimum results (monochrome I.O. cameras only), the following points should be understood:

1. As the average illumination decreases, the target potential may be increased slightly in the positive direction. If the potential is increased excessively, the beam current will have to be increased excessively to discharge the picture high lights, and the signal-to-noise ratio will be decreased.

2. If the average illumination increases (as when the sun emerges from behind a cloud), the target potential may be decreased slightly (made less positive). This adjustment should not be greater than the amount that allows optimum resolution with satisfactory signal-to-noise ratio. If the target voltage is decreased too far, the cutoff point is reached and the picture washes out.

Some monochrome cameras employing I.O. tubes now provide a rotatable neutral-density wheel in the optical path; this wheel facilitates proper picture exposure in field pickups. The target voltage and lens f/ stop may remain fixed.

Weather conditions in the open affect camera performance unless appropriate precautions are taken by the operator. In cold weather, the camera heater should be turned on to shorten the warm-up time. In addition, the heater may be required at intervals throughout the program in extremely cold and blustery weather. If the pickup tube tends to retain an image, the tube may be too cold. If a noticeable loss in picture resolution is apparent, the tube may be too warm, and the heater should be turned off and operation of the camera discontinued until normal operating temperature is restored. Most modern cameras are equipped with thermostatic controls to operate heaters or blowers as necessary to maintain optimum operating temperatures.

NOTE: *Plumbicon* tubes are relatively free from temperature dependence of operating parameters.

In the illustrations to follow, an image orthicon camera is assumed for the angles shown. Remember the multiplying factors for other types of tubes given in Fig. 4-2 of Chapter 4. The I.O. angles are listed in Table 4-1 of Chapter 4. The illustrations that follow serve to show the analysis procedure necessary in typical field pickups involving a lens turret or fixed lenses as used on some hand-held cameras. The examples also are useful in planning for minimum and maximum angles of a given zoom lens assembly.

Tables 11-1 through 11-5 are provided to assist the reader in applying any other type of camera to the illustrations. The "standard" *Plumbicon* covered in Chapter 4 has a useful scanned-area diagonal of 0.8 inch; a later *Plumbicon* is smaller and has the same scanned area as the 1-inch vidicon (Tables 11-1 and 11-2). At the time of this writing, even smaller *Plumbicons* are being developed for portable and field cameras.

The choice of lenses depends on the nature of the event, the distance from the cameras, the type of pickup tube used, and the production techniques. At the studio, a 135-mm lens may be a "medium angle" or "narrow angle" device, but a lens of this focal length may be a "wide angle" lens (for the I.O. tube) in the field, since relatively large camera distances may be encountered. Certain relationships between lens size and distance may be tabulated from experience in the field.

Table 11-1. Lens Angles for 1-Inch Vidicon and *Plumbicon*

Focal Length (Inches)	Horizontal Angle	Vertical Angle
½	54°	40.5°
1	27°	20.25°
2	13.5°	10.1°
3	9°	6.75°
6	4.5°	3.38°

Table 11-2. Fields of View for 1-Inch Vidicon and *Plumbicon*

Distances From 10 to 80 Feet

F*	10' W*	10' H*	20' W	20' H	30' W	30' H	40' W	40' H	50' W	50' H	60' W	60' H	70' W	70' H	80' W	80' H
½	10	7.5	20	15	30	22.5	40	30	50	37.5	60	45	70	52.5	80	60
1	5	3.75	10	7.5	15	11.2	20	15	25	18.7	30	22.5	35	26.2	40	30
2	2.5	1.87	5	3.7	7.5	5.62	10	7.5	12	9.0	15	11.2	17.5	13.1	20	15
3	1.6	1.25	3.3	2.5	5.0	3.75	6.6	5.0	8.3	6.2	10	7.5	11.6	8.75	13	10
6	0.8	0.62	1.6	1.2	2.5	1.87	3.3	2.5	4.1	3.1	5.0	3.75	5.8	4.37	6.6	5

Distances From 100 to 1000 Feet

F*	100' W*	100' H*	300' W	300' H	500' W	500' H	1000' W	1000' H
½	100	75	300	225	500	375	100	750
1	50	37.5	150	112.5	250	187.5	500	375
2	25	18.75	75	56.25	125	93.75	250	187.5
3	16.6	12.5	50	37.5	83.3	62.4	166.6	124.8
6	8.33	6.25	25	18.75	41.7	31.2	83.3	62.4

*F = focal length in inches, W = width of field in feet, and H = height of field in feet.

Table 11-3. Lens Angles for "Standard" *Plumbicon*

Lens	Horiz. Angle	Vert. Angle
35 mm (1½'')	25.75°	19.3°
50 mm (2'')	17°	12.75°
90 mm (3½'')	9.5°	7.125°
135 mm (5.3'')	6.5°	4.875°
8½''	4°	3°

Table 11-4. Fields of View for "Standard" *Plumbicon*

Lens	50' W*	50' H*	100' W	100' H	300' W	300' H	500' W	500' H	1000' W	1000' H
35 mm	21	15	42	31.5	126	94.5	210	157.5	420	315
50 mm	15.75	11.8	31.5	23.6	94.5	70.8	157.5	118	315	236
90 mm	9	6.75	18	13.5	54	40.5	90	67.5	180	135
135 mm	6	4.5	12	9	36	27	60	45	120	90
8''	4	3	8	6	24	18	40	30	80	60

*W = width of field in feet, and H = height of field in feet.

Table 11-5. Nearest Matching Lens

Image Orthicon		"Standard" Plumbicon		1-Inch Plumbicon & Vidicon	
Angle*	Lens	Angle*	Lens	Angle*·	Lens
51.5°	35 mm (1½")	51.5°	¾"	54°	½"
34°	50 mm (2")	34°	1"	27°	1"
13°	135 mm (5.3")	17°	50 mm	13.5°	2"
8°	8½"	9.5°	90 mm	9°	3"
4°	17"	4°	8½"	4.5°	6"

*Actual horizontal angle is shown in each case.

NOTE: The following discussion of lens focal lengths is based on the I.O. camera. The reader should refer to Table 11-5 to correlate this information with any other type of camera.

The 50- or 90-mm lens is used for close-ups of the announcer. These lenses also may be used for opening or overall shots that require very wide angles. A close-up for the 50-mm lens is a distance no greater than 50 feet; for the 90-mm lens it is a distance no greater than 150 feet.

The 135-mm lens or the 8½-inch lens may cover adequately such action as double plays in baseball, depending on the distance from the camera. Camera orientation at most boxing and wrestling matches is such that a 135-mm lens covers the ring sufficiently to necessitate only slight panning. This lens also is popularly used for "cover shots," such as a view of the pitcher and batter with the camera placed behind and above home plate.

The 13½, 17, and 20-inch lenses are used widely for close-ups of plays in football or outfield action in baseball. When it is necessary to use one of these lenses of long focal length on the same turret with a lens of short focal length, the end of the telephoto lens may show in the wide-angle view of the short lens. To avoid this occurrence, the long lens should be mounted diagonally opposite the shortest lens used. When lenses with screw-type bases are used, many stations prefer to use special bayonet-type adapters so that the long lens may be removed quickly for the change from a close-up to a wide-angle view, while another camera is on-air. Whether such adapters are used or not, it is advisable to support a long lens by means of a chain fastened securely to the pickup head or friction head. This is especially important if an accidentally dropped lens would fall into the audience. Of course, the chain must not obstruct proper panning or tilting of the camera.

Sun shades usually are used on field lenses, especially when it is necessary to "shoot" toward the sun or extreme high lights.

For optimum picture quality and good depth of focus, the smallest stop consistent with available light and type of pickup tube is used. A stop too small or too large reduces picture resolution. Too large a stop requires excessive beam current which causes decreased signal-to-noise ratio. The use of long lenses with their fixed stops requires additional alertness on the part of the field camera-control operator. A change in scenic illumination and contrast results in background and signal level changes. This requires constant attention to the video-gain and pedestal controls.

11-5. THE BASEBALL TELECAST

Baseball is one of the more popular subjects of remote telecasts in modern TV programming schedules. Sufficient experience has been gained to formulate a general pattern that has proved most successful in practice, although problems related to the individual field and topography vary considerably.

The fundamental dimensions involved in baseball coverage are presented in Fig. 11-4. The diamond is laid out in a 90-foot square. The pitcher's box is 60 feet from home base, and a distance of 66 feet separates

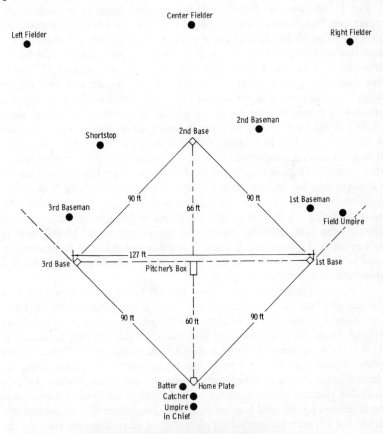

Fig. 11-4. Dimensions of baseball diamond.

the pitcher and second base. The horizontal coverage of the bases may be observed to include 127 feet. The size of the outfield and the distances to the rear and side walls vary among the individual stadia.

The most common area of attention during a baseball game is along a line that includes the umpire, catcher, batter, and pitcher; that is, the major portion of the actual telecast time is centered on this general location. For this reason, it has become almost universal to place a camera behind and above the home-plate area. Whenever possible, this camera (or two cameras) is located on a line directly behind home plate, as illustrated in Fig. 11-5A. This coverage of umpire, catcher, batter, and pitcher usually is referred to as the *cover shot,* and the camera assigned to this purpose (as well as panning other areas) usually is designated Number 1. In Fig. 11-5A, the camera is 50 feet on a direct line behind home plate, and the 135-mm lens coverage is shown. This results in optimum, large-sized images of the center of interest.

Fig. 11-5B illustrates the effect of placing the camera to one side of the direct line. One camera is shown placed 50 feet to the left of the home-plate axis, on a line 50 feet behind home plate. In this case, the reference axis is that line necessary to include the "big three," umpire, catcher, and batter. It is observed that the 135-mm lens with its horizontal angle of 13° would not cover the pitcher. Also, the images of the umpire and two players would be smaller than in the case of Fig. 11-5A because of the greater distance. It is shown that the use of the 50-mm lens with its horizontal angle of 34° would be necessary to cover the same action that could be covered with a 135-mm lens from directly behind the plate. Naturally, the smaller lens gives smaller images, and it is not as desirable as the 135-mm lens at this distance. Thus, it may be realized why the direct-line orientation for the cover shot has become most popular in practice. The only time such a location is not used is when local circumstances make it inaccessible. Observe from Fig. 11-5B that a camera 100 feet to the right of the center axis presents the same problem, except that the images would be still smaller as a result of the increased distance. The off-center positions sometimes are used in conjunction with a camera directly behind the umpire to lend added color and variety during the game.

Just as the horizontal field of view for any given lens is determined by camera position, so the vertical angle and camera height must be taken into consideration. As the height increases for a given lens and distance behind home plate, the vertical field becomes the limiting factor (Fig. 11-6). If the camera considered in Fig. 11-5A, placed 50 feet behind home plate and using a 135-mm lens, is just 12 feet above the playing area, the desired cover shot is obtained with maximum image size. Should the only available position place the lens, say, 25 feet high, the field of view would be as defined by the solid lines of Fig. 11-6. Thus, from this height and distance, the 135-mm lens would supply a shot of an area smaller than the desired cover shot.

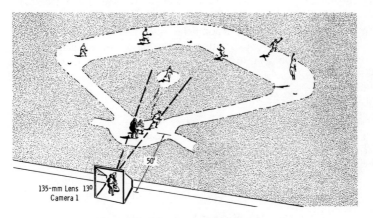

(A) Camera behind home plate.

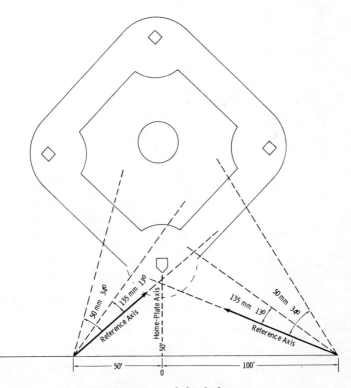

(B) Cameras not behind plate.

Fig. 11-5. Basic horizontal camera angles for baseball telecasting.

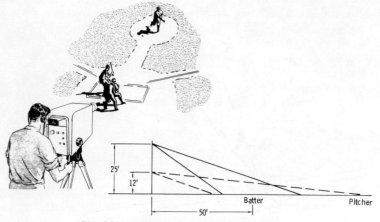

Fig. 11-6. Effect of height on field of view.

The operator will be able to ascertain beforehand the exact lenses needed for specified shots if he plots all available data as in Figs. 11-5 and 11-6. Experienced technicians acquire a sense that permits them to judge these factors by eye after some time in daily routine operations. It is most helpful for the student and broadcast newcomer, however, to practice such graphical analysis so that he gains comprehension even before actual practice on the equipment is involved.

The practice of multiple-camera setup depends on production whims and existing facilities. In most instances of two-camera pickups, camera 2 provides the "color" shots. This means that in addition to following some of the plays in the infield and outfield, camera 2 is able to show the spectators, the dugouts, and views of the field different from those available from camera 1. In three-camera pickups, two cameras may be used in the cover-shot position, and the third camera may be used for color. This type of setup provides the widest production choice of plays and human-interest views. As an example, camera 1 may take the pitcher's delivery, and camera 2 (in the same cover position, but equipped with a 17-inch lens) may follow the ball to the outfield. Should the play result in an infield hit, camera 1 may remain on the batter (now the runner) while he tries for first base. Alternately, camera 3 may be located along the first-base line and show the runner from that position. The latter method is often preferred by sports directors, because it develops the play "into the camera," considered an ideal pickup.

The many special effects possible only with the zoom lens can be used to best advantage by using them in moderation. Also, it is advisable to operate the zoom lens stopped down as much as possible consistent with the prevailing light conditions and pickup-tube sensitivity. When there is a great deal of movement in a sporting event, follow the action in the regular manner by panning the camera; manipulate the zoom action slowly and only when necessary. Too much zooming will distract the viewer's attention from the game being televised. A smooth zoom action can be obtained by grasping the the control lever (or knob) in a normal, relaxed manner and squeezing, not pushing, the control. Apply this pressure slowly so that the beginning of the zoom is hardly perceptible.

Unless the action of the game specifically calls for zooming, try to *under-zoom;* the effect will be all the greater. If a visual impact designed to startle the audience is desired, a fast zoom is permissible. However, it should not be attempted often. It is important to start and end the zoom slowly and gradually. A sudden starting and ending gives an unnatural effect and should be avoided.

Using the zoom lens for baseball telecasting is discussed in the *Back Video Manual* by Capt. W. C. Eddy, at one time manager of WBKB, relative to that station's experiences in Chicago. While written many years ago when the *TV Zoomar* first appeared, his views are quite pertinent to modern techniques. His comments follow:

"Television coverage of baseball in the past has, in many cases, been limited to a visual adaptation of the techniques developed in radio broadcasting. It has been increasingly evident that the earlier accepted use of a camera in fortifying the audio description has introduced confusion into the broadcast, rather than clarifying the description. This is based primarily on the inability of any optical system to adjust itself quickly to both the varying depths of focus and varying angles of coverage.

"The zoom lens solves one of these problems through its ability to change focus without the loss of scene or sharpness. While it is true that a qualified baseball fan might be able to integrate this combined aural description and spotty camera coverage into a continuity of action, it is doubtful that such coverage would make sense to the average layman who does not have a complete knowledge of baseball.

"Officials of the Chicago Cubs have recognized this fact during previous years of experimental coverage of the Cubs' baseball by WBKB. They suggested that steps be taken to analyze and revise television techniques with a view toward creating a production method which would not only correct these deficiencies, but would popularize baseball with the nonfan, as well as satifsy the regular baseball fan and viewer.

"In cooperation with KTLA in Los Angeles, WBKB set up an active test under field conditions at Los Angeles.

"In our analysis of the problem, it was apparent that many of the basic precepts of good showmanship had been overlooked in presentations thus far attempted by television stations covering baseball. Some of these inconsistencies were:

1. All action on the playing field developed away from cameras located behind home plate, so reducing rather than heightening the dramatic impact of the play.

2. The camera placements in common use normally resulted in a wide variation of viewing angles, causing abnormal or complete disorientation of the nonqualified viewer.

3. Because of the limited field of the television lens, normal camera coverage could show only one or two players in detail, thus destroying any visual association with the other members of the team or the playing field itself.

4. It has been a precept in motion-picture switching techniques that the center of interest should not change position on the screen during scene-to-scene transitions. In motion pictures, considerable thought is given to the preplanning of shots in order to retain the focal point of interest in the same relative position during a change of scenes. Television has heretofore disregarded this function.

"In order to develop a satisfactory television technique, it was necessary that as many of the fundamentals of good showmanship as possible be incorporated into the video coverage. Some of these requirements were:

1. Change in the viewing angle from scene to scene should not be allowed to occur.

2. Action of the play should develop toward, rather than away from, the cameras, in order to heighten the dramatic impact.

3. A viewpoint or camera position which would orient the individual player with the team, as well as with the playing field, must be selected.

4. Maximum utilization of good camera techniques and modern equipment must be employed in any successful coverage.

"Coverage of infield plays was the first problem. To satisfy the above-listed requirements, a position alongside the home dugout at third base was selected. It appeared preferable to have this camera quickly adjustable in height so that it could either shoot from the ground level or be elevated to a position from which first base would be visible over the pitcher's mound.

"Various camera heights were tried from the dugout location, and it was established that the lowest camera level possible afforded the most dramatic coverage.

"The second camera position was selected for panoramic coverage of the entire field, with the camera equipped with a zoom lens. In keeping with our plan of establishing a fixed viewing angle, this camera position was chosen in far left field at the foul line, permitting use of the foul line as an orientation reference in establishing the geography of the play.

"With a 340-foot range to home plate, a normal zoom lens could cover the infield satisfactorily and further permit close-up work on any play or player that appeared interesting. It was further evident that the speed of a hit ball would be reduced optically at the start of its trajectory, thus permitting the cameraman to follow the ball into the outfield. Use of the zoom lens then permitted a full detail shot of the play, and following of the delivery of the ball back to the infield.

"The third camera position was selected in the stands in order to provide the psychological advantage of covering the game from the typical fan's viewpoint. Necessarily, this position should be in a spot which provides approximately the same viewing angle as locations one and two. By using this camera for semipanoramic coverage of the foreground, a strong psychological tie-in between audience and player is provided. By utilizing this location as the viewpoint of the announcer, it is possible to provide a highly satisfactory audio coverage of the game.

"Competent use of all three cameras provides sufficient story material at the director's console for satisfactory cutting into a well-balanced and acceptable production which will appeal to the viewer from the dramatic as well as the reportorial standpoint.

"While variations in the locations mentioned can be easily incorporated, the broadcaster must be particularly careful to observe these cardinal points of showmanship:

1. Wherever possible, action should develop into, rather than away from, the cameras.
2. The angle of viewing should not change appreciably from camera to camera in switch shots.
3. Introduction of players and fans into the foreground of the composed shot enhances the psychological effect of the picture, and ties the audience more closely to the fans' reaction.
4. Vertical camera angles should be kept at a minimum on all panoramic effects.
5. Reaction shots of announcers, fans, and players should be used liberally to fill in nonspectacular periods of the game.
6. Close-up switch shots from the dugout camera should only be used in conjunction with the following complete panoramic coverage of the play with the zoom lens, in order to maintain proper orientation of the viewer.
7. The panoramic camera (with the zoom lens) should normally be operated in the unzoomed setting, and in following a ball to the outfield, the cameraman should not attempt to hold the ball in center screen by change of horizon. An unzoomed pan shot should be used until the trajectory of the ball is such that the camera can safely be zoomed without changing horizon.
8. Use of the zoom feature of the panoramic camera should be held to those situations where close-ups are either required or preferable. Continual in-and-out use of this lens destroys its advantages and the impact of the shot.
9. Zoom shots should be accomplished as slowly as possible to preserve satisfactory orientation of the viewer.

"It is evident that the principles of showmanship outlined in this description have equal importance in the coverage of sporting events other than baseball, and that, properly employed, these basic precepts will contribute to the growth of interest in a sport by the television audience, resulting in the creation of new fans at the box office."

In general, the camera sequence, whether accomplished by the zoom lens or by camera switching, follows a specific order of events. The sequence usually starts with two events:

1. *View of Batter Walking to Plate:* This scene is taken with a 13½ or 17-inch lens, or with the zoom lens set for a narrow angle. A close-up image of maximum size is provided.

2. *Pitcher Winds Up and Delivers:* This action calls for a cover shot with the 135-mm lens, or the zoom lens set for a wide angle. The viewer is able to follow the "big four"—umpire, catcher, batter, and pitcher.

When a ball is hit, the skills of the program director "calling the shots," the switcher operator, and the cameramen are tested. Assuming that camera 1 is the cover-shot camera as in 2 above, the sequences of operation for typical plays are as follows (director's orders are enclosed in brackets [ ]):

Infield Hit—Two Cameras: The hit ball is a grounder toward the shortstop. Camera 1 follows the ball by panning, and then stops on the shortstop as he scoops up the ball and throws it toward the first baseman. [Switch to two.] On camera 2, an 8½-inch lens is used to give a medium close-up of the first-base area, to show either an "out" or "safe" play there.

The action described is facilitated by predetermined instructions for the cameramen. For example, the operator of camera 1 may be instructed to keep his lens "on the ball" at all times. The operator of camera 2 may be instructed always to "follow the runner." The program director, watching the monitors, then calls to the switcher for the proper cuts.

Infield Hit—Three Cameras: The ball is grounded toward the shortstop as above. The operator of camera 1 pans the ball. [Take three.] To provide a medium close-up of the shortstop catching the ball, camera 3 has an 8½ or 13½-inch lens (depending on the distance) in the operating position. The shortstop throws toward first base. [Take two.] Camera 2, with a 13½ or 17-inch lens in the operating position, provides a close-up of the play at first base.

Note that this sequence provides close-ups of plays rather than the more general coverage of the two-camera operation. This procedure requires unusual alertness on the part of the cameramen, who may be preinstructed as follows: Number 1, "follow the ball"; Number 2 "always take a close-up of the likely culmination of the play"; Number 3, "always take a close-up (medium angle) of the likely receiver of a hit ball."

For a change of pace to lend variety to the pickup, the director may change these standing instructions at intervals during the game. For example, in the three-camera pickup, during dull moments the operator of camera 3 may be instructed to pick up "color" (such as dugout activity or fans in the stands), and the other two cameras will provide a more general, wider-angle pickup.

Infield Hit—Zoom: Camera 1 "zooms in" to a narrow angle as the shortstop scoops up the ball and throws it toward first base. [Take two.] Camera 2 has (for example) the 13½-inch lens in operating position, to catch a close-up of the play at first base.

Experienced cameramen are always instructed to use their best judgment during active play. An example of this occurs when a player steals base. The director usually designates one cameraman to watch for such activity and catch it when possible.

In the case of a hit to the outfield, the general procedure is the same as for infield hits. Camera 1 follows the ball, and either this camera zooms to the outfield area or a switch is made to another camera (with a 17-inch lens or another zoom lens) to catch a close-up of the outfield activity. Ideally, the second camera should be higher than camera 1, since the relatively low height for a proper cover shot is insufficient for clear coverage of the outfield.

11-6. FOOTBALL PICKUPS

The basic dimensions involved in football coverage are given in Fig. 11-7. The actual playing field is 100 yards (300 feet) long by 50 yards (150 feet) wide. The goal posts in intercollegiate football are 10 yards behind the end lines, making for an overall horizontal area of 120 yards, or 360 feet. Fig. 11-7 includes plots of the horizontal angles and effective area covered for 35-mm, 50-mm, and 135-mm lenses, assuming that the camera is located 150 feet from the nearest edge of the active playing field. Also shown are the nearest and farthest areas of focus, assuming that the camera is 50 feet higher than the playing field. In this specific case, the 35- or 50-mm lens would be used for team orientation or general overall views, and the 135-mm lens would be used for medium close-ups of teams as they line up for scrimmage. A second camera, equipped with a complement of lenses such as 8½-, 13½-, and 17-inch lenses would be used for close-up shots of in-line scrimmage, passing, receiving, kicking, etc.

The person responsible for the technical setup at the field is faced with one primary problem on an initial trial: the direction in which the field is laid out. He is most fortunate in the Northern Hemisphere when the field runs east and west. Then it is possible to set up cameras and equipment on the south side of the field without worrying about shooting into the sun, since the sun will be at least slightly to the south. If the field runs north and south, the choice is a compromise at best. If the west side is chosen (to prevent shooting into the sun toward evening), the southerly location of the sun in the fall and winter months may present problems. If the east side

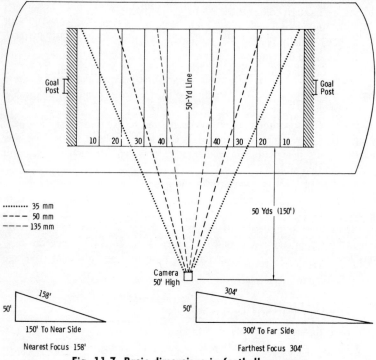

Goal Post

Goal Post

50-Yd Line

10 20 30 40   40 30 20 10

.... 35 mm
- - - - 50 mm
— — — 135 mm

50 Yds (150')

Camera
50' High

158'

50'

150' To Near Side

Nearest Focus 158'

304'

50'

300' To Far Side

Farthest Focus 304'

Fig. 11-7. Basic dimensions in football coverage.

is chosen the source of trouble is doubled, since now the sunest will be
added to the southern brightness. Other factors being equal, therefore, the
proper choice would be the west side of a north-south football field.

Obviously, other factors may influence the choice of camera location.
For example, the stadium may be very high on the west side of a north-
south field, sufficiently high to block any possible direct sun from the late-
evening sky. The source of power and optimum location for the mobile unit
may be on the east side of this field. Thus, the logical choice in this instance
would be the east side of the stadium.

As a general rule, however, the optimum location for cameras is on the
south side of a field that runs east and west, and on the west side of a field
that runs north and south.

Most stations use one or two cameras 100 to 300 feet (or more) from
the edge of the field, and one camera 20 feet or less from the edge of the
field. In one typical three-camera setup, cameras 1 and 2 are 300 feet from
the edge of the field on an extension of the 50-yard line (center of field),
and camera 3 is 20 feet from the field about midway between the center
and one goal post. Camera 1 is equipped with a zoom lens. Camera 2 has
8½-, 13½-, and 17-inch turret-mounted lenses. Camera 3 at the field loca-

tion provides color and "field-level" close-ups when called for. (If all cameras are modern color chains, all have zoom lenses.)

A wide-angle lens (such as a 35-mm, 50-mm, or 90-mm lens, depending on distance) usually is used for the kickoff. Each operator is preinstructed for such possible developments from the line as a pass, kick, or run. Obviously, men familiar with the game make the best cameramen for any sports telecast.

Video tape recorders, of both the conventional type and the slow-motion, stop-action type, are now an integral part of most ball-game broadcasts. The "instant replay" and "stop action" at the climax of the play are nearly always used by the networks (Chapter 8).

The sound portion of sporting events should not be neglected in planning. Nothing is more disconcerting to the sports fan than to see an exciting play developing with almost complete absence of background sounds from the loudspeaker. The cheers (and jeers) of the fans present at the game should be ever-present in the audio, and microphones should be located at strategic points for this pickup. Parabolic-type and gun-type microphones are often used in the field to pick up cheers and game sounds. Without such mixing-in of on-the-spot sound, "atmosphere" is almost completely lacking in the telecast.

11-7. MISCELLANEOUS FIELD EVENTS

To facilitate the drawing of diagrams for analysis of camera setups for the most popular sports telecasts, basic dimensions of the areas of activities are presented as follows:

Boxing and Wrestling: The standard-sized ring is 24 feet square. At least one camera usually is placed at "ringside" for dramatic close-ups of holds or blows. Cameras equipped with the proper complement of longer-focal-length lenses may be placed farther back. Boxing and wrestling matches usually take place under lights that are adequate for modern pickup tubes, except in unusual instances at smaller events. In indoor events of this type, cameras cannot be placed at great distances from the action because of large amounts of haze from smoking spectators. A hydraulic dolly often is used for the main camera at ringside to gain variable angles of shots close to the action.

Ice Hockey: This game also is most often staged under lights, usually indoors. The standard ice-hockey playing area is 200 feet long by 85 feet wide, although some games are played on smaller areas. When games are staged indoors, lighting usually is adequate, but lighting conditions always must be determined in time to plan additional facilities if necessary. Keeping the puck in view of the audience requires excellent panning and well-planned orientation of cameras for most effective shots. Because of the rapidity of action in ice hockey, real close-ups of field action are difficult to attain, and most close-up shots occur at the goals.

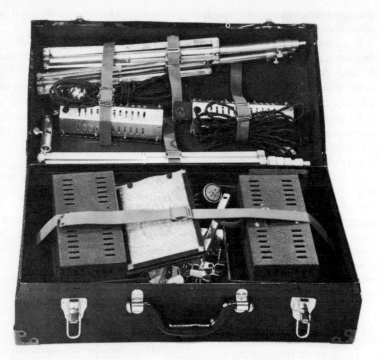

Fig. 11-8. Kliegl Q-6 portable lighting package.

Field Polo: A wide field of play is involved; the standard polo field is 300 yards (900 feet) long by 160 to 200 feet wide. The zoom lens is highly desirable for coverage of such a long, narrow field of play.

Basketball: This game is being televised more and more with growing interest in many states. It usually is played at night with adequate illumination. During daytime games (played only during tournaments in most instances), the camera operators must be careful not to point the lenses toward side windows that are not painted or otherwise darkened. The basketball court is about 50 feet wide by 90 feet long. The length varies slightly from around 84 feet for high-school courts to 90 to 95 feet for college games. The free-throw line is 15 feet from the end of the court, and the rim of the net is 10 feet above the floor.

These court dimensions may be plotted as described previously for baseball diamonds and football fields. The coverage for lenses of various focal lengths then may be determined from the necessary positions of the cameras. One lens should just include the entire court, with sufficient vertical angle to include high balls lobbed toward the basket from a distance. Another lens should include only the free-throw area and basket, with another lens of sufficient length to catch close-ups of such action as toss-ups, etc. Many stations use only two cameras for this sport with very satisfactory results. Naturally, three cameras lend greater variety and a better chance for dramatic close-ups, if the director is experienced enough to warrant their use.

11-8. PORTABLE LIGHTING

As suggested in some of the foregoing descriptions of field events, extra lighting sometimes is required. A portable lighting package is illustrated in Fig. 11-8. This package includes two lights similar to the one shown in Fig. 9-5A (Chapter 9); three lights similar to the one shown in Fig. 9-5B

Fig. 11-9. Kliegl Model R-67/6X portable SCR dimmer.

Courtesy Kliegl Bros. Lighting

(Chapter 9); and accessories such as barn doors, scrims, stands, and cables. All lights are quartz-iodine. The entire lighting package, with required cable, weighs 33 pounds; the $10'' \times 12'' \times 24''$ carrying case fits under an aircraft seat or in a car trunk.

Portable dimmer banks also are available. The unit shown in Fig. 11-9 consists of six 2.4-kilowatt dimmers (14,400 watts total), mounted in an $18'' \times 34'' \times 6\frac{1}{2}''$ bank. Lightweight casters are affixed to the dimmer bank for easy movement from place to place. The control module, mounted as an integral part of the dimmer bank, consists of six controlling potentiometers, an on-off switch, and a master potentiometer. This particular dimmer bank is capable of controlling all the lights in three lighting packages of the type shown in Fig. 11-8.

An even more elaborate system (Model R-67/24X) handles 57,600 watts of lighting. It has two dimmer banks and one 24-channel suitcase control board.

EXERCISES

Q11-1. What is the logical choice of equipment for a delayed broadcast from a helicopter?

Q11-2. Can a "studio camera" be used in the field?

Q11-3. List the basic preliminary considerations for any remote pickup.

Q11-4. Do all *Plumbicon* tubes have the same useful scanned area?

Q11-5. Give the "nearest equivalent lens" (focal length) for (A) I.O., (B) standard *Plumbicon*, and (C) 1-inch vidicon and *Plumbicon*, for a 4° horizontal angle.

Q11-6. If a 50-mm lens is considered a wide-angle lens for the I.O. camera, what is the corresponding wide-angle lens for (A) the standard *Plumbicon* and (B) the 1-inch vidicon and *Plumbicon*?

Q11-7. If you have two I.O. cameras, one using a 3-inch tube and the other a 4½-inch tube, is there a difference in angle of view for a lens of given focal length?

Q11-8. Are zoom lens assemblies ever available on hand-held mobile cameras?

CHAPTER 12

Microwave Relay Systems

Television-relay microwave systems operate in the 1990-2110 MHz, 6875-7125 MHz, and 12,700-13,250 MHz bands. The 7000-MHz region is the most popularly employed band of frequencies for remote pickups and studio-to-transmitter links (STL's).

12-1. BASIC THEORY OF MICROWAVE PROPAGATION

In television-relay practice, the emitted waves are concentrated into a narrow beam to allow exact control in "aiming" the signal toward the receiver, and to obtain a high *power gain*. TV-relay transmitter and receiver antennas take the form of a radiating element surrounded by a parabolic reflector. The microwave transmitting set shown in Fig. 12-1 includes an antenna of this type.

The greater the area of a parabolic reflector is in terms of the wavelength, the greater is the gain. The basic formula for the power gain of a system using a parabolic reflector is:

$$\text{Power Gain} = \frac{4\pi a}{\lambda^2}$$

where,
    a is the effective area of the reflector,
    λ is the wavelength.

Since it is easier to ascertain the *projected area* of a parabolic reflector than the effective area, and since the effective area is approximately 0.65 times the projected area, we may express an approximate relationship:

$$\text{Power Gain} = 0.65 \frac{4\pi A}{\lambda^2}$$

where,
    A is the projected area of the reflector,
    λ is the wavelength.

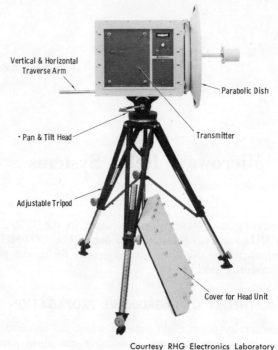

Vertical & Horizontal
Traverse Arm

Parabolic Dish

Pan & Tilt Head

Transmitter

Adjustable Tripod

Cover for Head Unit

Courtesy RHG Electronics Laboratory

Fig. 12-1. Portable microwave transmitter.

This formula results in only a slight error so long as the wavelength is small compared to the diameter of the reflector, as is the case in practice. Thus, for example, an antenna operating in the 6800-7050 MHz range obtains a power gain of 5000 (37 dB) with a 4-foot reflector, and a power gain of 11,500 (slightly over 40 dB) with a 6-foot reflector.

The television operator will encounter the term "half-power beam width" in manufacturer's literature concerning parabolic antennas. For example, a certain 6-foot "dish" may have a half-power beam width of 5°, or a 4-foot dish may be rated for a half-power beam width of 8°. The beam width is defined in terms of the angle through which the dish must be rotated in order to reduce the power available at the receiver terminals to one-half the maximum value. If the dish is aligned with the incoming beam so that maximum power is available, and then the dish is rotated in one direction until the power is one-half this value, twice this angle of rotation (the total angle on both sides of maximum) is the half-power beam width.

See Fig. 12-2. If the operation is in the 7000-MHz (7-GHz) range, a 6-foot parabolic reflector has a half-power beam width of 1.7°. On a 20-mile path, the receiving dish actually is scanning a 1500-foot "aperture" at the half-way point. This aperture would at first appear to be quite broad,

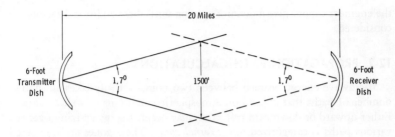

Fig. 12-2. Typical microwave-antenna arrangement.

but if the receiving dish is 1.7° off the central axis, only one-half the maximum power will be obtained.

A 4-foot reflector has a half-power beam width of approximately 3° in the 7000-MHz band. In the 2000-MHz band, a 4-foot dish has a half-power beam width of about 8°. A 6-foot dish results in a half-power beam width of around 5° in the 2000-MHz band. Thus, it may be observed that a higher power gain is accompanied by a narrower beam, calling for careful orientation when the equipment is set up.

Quite often, the operator in the field department of a TV station is called on to give an opinion as to the practicality of relaying a signal from a remote point directly to a pickup receiver at the studio or transmitter location. The distance to the optical horizon is based on the fundamental formula:

$$D = 1.23\sqrt{H}$$

where,

D is the distance to the horizon in miles,
H is the height of the transmitting antenna in feet.

Thus, if the antenna may be placed 100 feet high, the theoretical distance to the horizon (line of sight) is 12.3 miles. However, if it is possible to raise the receiving antenna also, the effective line-of-sight path becomes:

$$D = 1.23 \left(\sqrt{H_t} + \sqrt{H_r}\right)$$

where,

D is the maximum line-of-sight distance between antennas in miles,
H_t is the height of the transmitting antenna in feet,
H_r is the height of the receiving antenna in feet.

If both antennas are at a height of 100 feet, the effective line-of-sight path length becomes 24.6 miles.

Since the preceding basic formulas assume a smooth surface, modifications to allow for the profile elevations along the projected route of signal propagation must be made for practical application. Maps showing elevation contours (height above sea level) at as little as 10-foot elevation intervals are available (from local Geodetic Survey offices.) Unless the proposed pickup point allows distinct line-of-sight service as observed with the eye, the engineer cannot give a definite opinion until these additional factors are considered.

12-2. PROPAGATION-PATH CALCULATIONS

Microwave energy beamed between two points actually takes an infinite number of paths that depend on atmospheric conditions as well as terrain. Either upward or downward refraction may occur. Energy arriving over the various paths is categorized into *Fresnel zones.* These zones are numbered to correspond with circles of different radii centered on the direct line between antennas. The zone with the smallest radius is the first Fresnel zone (Fig. 12-3). Many Fresnel zones exist. Energy in the second and all other

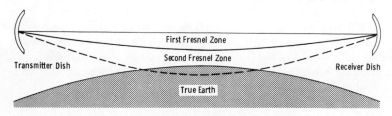

Fig. 12-3. Fresnel zones.

even-numbered zones has a half-wavelength (180°) relationship to energy in the first Fresnel zone. Energy in the third Fresnel zone and all other odd-numbered zones has a full-wavelength, and therefore phase-additive, relationship to energy in the first Fresnel zone.

The primary energy is contained in the first Fresnel zone, and energy contained in the even-numbered zones is phase-cancelling; therefore, it is desirable to obstruct the energy contained in all but the first Fresnel zone. At the same time, the first zone must be provided ample clearance. A value of 0.6 times the radius of the first Fresnel zone normally is taken as the absolute minimum clearance.

The radius of the first Fresnel zone at the point of a major obstruction in the path may be calculated from the equation:

$$R = 72 \sqrt{\frac{AB}{Pf}} \qquad \text{(Eq. 12-1.)}$$

where,

R is the radius in feet,
A is the distance from one end of the path to the point of obstruction in miles,
B is the distance from the other end of the path to the point of obstruction in miles,
P is the total path length in miles,
f is the frequency in GHz.

The following procedure will serve to give the reader a basic understanding of plotting microwave paths for portable setups and STL's.

1. Plot a profile of the transmission path. Graph paper that presents the curvature of the earth on a radius 4/3 times its true value may be used. For limited use, it is more convenient to use ordinary linear graph paper and the data of Fig. 12-4. Paper with ten squares to the inch is ideal for this purpose.

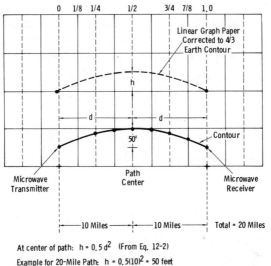

At center of path: h = 0.5 d² (From Eq. 12-2)

Example for 20-Mile Path: h = 0.5(10)² = 50 feet

Fig. 12-4. Method of plotting profile of transmission patn.

2. The path profile and obstructions on the path may be charted from topographic maps. The topographic map gives heights above sea level of the surface of the earth, to which are added the heights of major obstructions. Maps for specific areas may be obtained from the U.S. Geological Survey, Washington, D.C. 20025. For maps of areas west of the Mississippi River, write to the U.S. Geological Survey, Denver, Colorado 80215.

3. The clearance over the tallest obstruction in the path should be at least that shown in Table 12-1.

Take the example illustrated by Fig. 12-5A. The bulge (h) of the earth in feet at distance d1 miles from the near end and d2 miles from the far end of the path is:

$$h = 0.5 \, d1 \, d2 \qquad \text{(Eq. 12-2)}$$

Using Equation 12-2 to find the correction for earth curvature at the point of maximum obstruction:

Table 12-1. Minimum Transmission-Path Clearance (In Feet) Above 4/3 Earth

Path Length (Miles)	1/8 & 7/8 Distance	1/4 & 3/4 Distance	1/2 Distance
5	13	16	19
10	21	27	32
15	29	38	45
20	38	49	57
25	46	59	69
30	54	71	82

$$h = 0.5 \ (12) \ (8) = 0.5 \ (96) = 48 \text{ feet}$$

Using Equation 12-1 to find the radius of the first Fresnel zone at the point in question (point P) for 7000 MHz (7 GHz):

$$R = 72 \ \sqrt{\frac{(12) \ (8)}{(20) \ (7)}} = 72 \ (0.825) = 60 \text{ feet (approx)}$$

Then the minimum clearance is $0.6 \ (60) = 36$ feet.

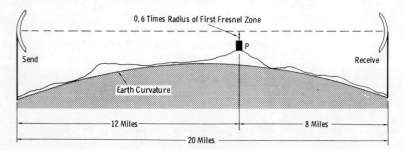

(A) Equal heights.

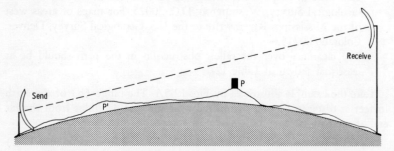

(B) Unequal heights.

Fig. 12-5. Determination of required antenna heights.

Now assume that the height above ground of a structure in the path at point P is 100 feet. We have found that the earth-curvature correction is 48 feet and that the clearance must be at least 36 feet. The sum of these quantities is $100 + 48 + 36 = 184$ feet.

Now further assume that the topographic map shows the sending-end and receiving-end locations to be 1000 feet above sea level and point P to be 1100 feet above sea level. This adds another 100 feet to the "negative clearance" of the dishes. Thus, in this practice problem $184 + 100 = 284$ feet is the required height of the sending and receiving dishes for a straight-line path. In case the installation is an STL, the transmitter location normally includes a tall tower on which the receiving dish can be mounted, and the method of Fig. 12-5B might be used. Note that now point P′ limits how low the sending dish can be mounted for adequate clearance.

Usually, energy in the second and higher-order Fresnel zones is attenuated severely as a result of normal terrain differences. If the path contains variations of magnitude equal to at least 75 percent of the radius of the first Fresnel zone at the center of the path, very little problem need be expected from other than first-zone clearance. If the path is smooth (as over water), the problems can multiply. This is so because reflections are not "dispersed" and can become quite strong.

When accurate topographic maps are not available, accurately calibrated, sensitive altimeters (properly adjusted in accordance with the barometric pressure) can be used; a point of known elevation serves as a reference. Also, aircraft-mounted absolute altimeters may be used. In the absence of these devices, conventional civil-engineering techniques must be employed. This is always recommended when doubt concerning a proposed STL path exists.

Attention to proper first-Fresnel-zone clearance is all that is necessary for the average path, so long as the path length and antenna system are such that the proper *fade margin* is provided. In evaluating microwave-relay performance, the following fundamentals serve as a basic guide:

1. In general, noise will be visible in the picture when the signal-to-noise (S/N) ratio deteriorates to less than 24 dB.

2. Since a microwave beam sometimes is bent and scattered by atmospheric conditions, not only adequate clearance, but also adequate fade margin must be provided. The picture becomes unusable at a signal-to-noise ratio of 8 dB. This 8-dB figure establishes a basis for computing the fade margin.

3. For example, if a 24-dB S/N ratio exists, the fade margin is only 16 dB $(24 - 8 = 16)$. From the graph in Fig. 12-6, a 16-dB fade margin indicates only about 98-percent reliability on a 25-mile path. This would mean total outages of about 117 hours in an average broadcast year, entirely unsuitable for applications of a continuous nature such as an STL. Note that for a reliability of 99.99 percent, the fade margin

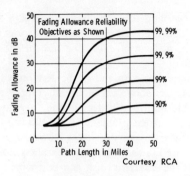

Fig. 12-6. Graph showing fading allowance.

for a 25-mile path should be 37 dB. This requires a signal-to-noise ratio of $37 + 8 = 45$ dB.

4. From Fig. 12-6, compute the allowable fade margin for 99.99-percent reliability, especially if the service is an STL. Assume the path length is 20 miles. This requires a fade margin of 31 dB. Then a minimum S/N ratio of $31 + 8$, or 39 dB, should be indicated by an average of measurements made under average weather conditions. Normally, measurements made daily or nightly over a period of a week or two give a reliable indication of the practical S/N ratio of the system.

You can estimate the signal-to-noise ratio by use of the antenna-system gains and free-space attenuation. Table 12-2 shows the gain of parabolic reflectors as a function of size and operating frequency. Fig. 12-7 shows the system gain at 7000 MHz for each of the designated dish and reflector sizes. (Similar information is obtainable from the manufacturer of a particular microwave unit.) Table 12-3 is a tabulation of the free-space loss in the microwave bands for path lengths up to 30 miles (the maximum distance normally employed for a single hop). Values in the table are based on the formula:

$$A' = 37 + 20 \log f + 20 \log D$$

where,

A' is the free-space loss in dB,
f is the operating frequency in MHz,
D is the distance in miles.

Assume the following data:
Power output of transmitter: 1 watt (0 dBW)

Table 12-2. Approximate Gain (Parabolic Reflector Only)

Dish	dB Gain		
	2000 MHz	7000 MHz	13,000 MHz
4 ft.	25	37	42
6 ft	28	40	45

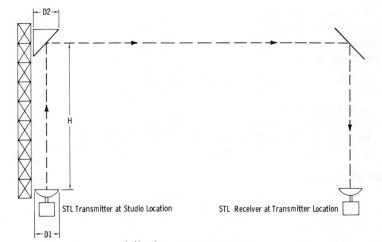

(A) Antenna arrangement.

H Dimension	Gain (dB)		
(Feet)	D2 = 4 ft	D2 = 6 ft	
80	35	39	
100	34	38.5	
120	33	38	
140	32	37.5	D1 = 4-ft Parab.
160	31	37	
180	30	36.5	
200	29	36	
80	36	40.5	
100	35	40	
120	34	39.5	
140	33	39	D1 = 6-ft Parab.
160	32	38.5	
180	31	38	
200	30	37.5	

(B) Approximate antenna-system gain (parabola & reflector)
(frequency: 7000 MHz).

Fig. 12-7. Computation of gain when passive reflectors are used.

Table 12-3. Approximate Free-Space Loss

Path Length	Loss (dB)		
(Miles)	2000 MHz	7000 MHz	13,000 MHz
5	117	127	133
10	123	133	140
15	126	136.5	143.5
20	129	139	146
25	131	141	147
30	132	142.5	148

Frequency: 7000 MHz
Path length: 20 miles
Antenna system gain at each end: 35 dB (total of 70 dB)
The net path loss is:

$$A = A' - G_t - G_r$$

where,

A is the net path loss,
A′ is the free-space loss (from Table 12-3),
G_t is the transmitter-system gain (from Table 12-2 or Fig. 12-7),
G_r is the receiver antenna-system gain (from Table 12-2 or Fig. 12-7).

In this example:

$$A = 139 - (+35) - (+35)$$
$$= 139 - 70$$
$$= 69 \text{ dB}$$

The receiver power input is:

$$P_r = P_t - A$$

where,

P_r is the receiver power input,
P_t is the transmitter power output,
A is the net path attenuation.

The net path loss in dB subtracted from the transmitter power output in dBW (dB above 1 watt) gives the power input to the receiver in dBW. In this example:

$$P_r = 0 - 69 = -69 \text{ dBW}$$

Fig. 12-8 is a graph of the expected signal-to-noise ratio versus power input (dBW) to the receiver for the RCA TVM-1 microwave-relay system. Note that for a power input of −69 dBW, the ratio of peak-to-peak video to peak-to-peak noise should be 38 dB. This is approximately a 56-dB ratio of peak-to-peak video to rms noise. Note also from the graph of Fig. 12-8 that this receiver input power should result in better than a 72-dB signal-to-noise ratio in diplexed sound.

NOTE: Diplexed sound is covered later in this chapter. The relationship of dB to watts is covered further in Chapter 13.

When no direct line-of-sight propagation path from a remote point is possible, it should not be inferred that a microwave relay of a program from that point of origin is impossible. Because of the nature of the microwave beam, it may be bent easily around corners if the proper relative factors are considered. The first consideration is determining a common point that affords line-of-sight paths to both the receiving and transmitting points. In the unusual circumstance when no such common point exists, several

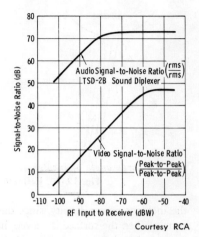

Fig. 12-8. Graph showing signal-to-noise ratio.

points may be chosen to provide a "zig-zag" path between terminals. Fig. 12-9 illustrates the fundamental principle involved in this procedure. The reflecting surface should be flat for maximum reflection efficiency, and it should be oriented so that the angle of beam reflection is proper for the receiver-dish location. Such a reflecting surface may actually be a sufficiently tall steel-constructed building or other structure that may be properly oriented for the purpose. When such a structure is not available, a solidly supported copper screen 6 to 8 feet square may be used, and some stations construct several of these reflectors for just such contingencies. When relays of this type are aligned, it is necessary to install a private line (PL) at the reflector point as well as at the other terminals so that optimum adjustment of all elements in the system may be attained.

The advisability of using some conveniently located *curved* surface as a relay-beam reflector depends on several factors. It may be observed from Fig. 12-9 that the power density reflected from a flat surface is the same as

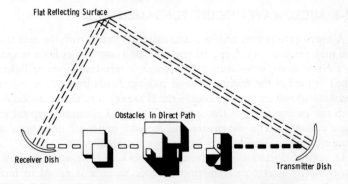

Fig. 12-9. Use of passive reflector to bypass obstacles.

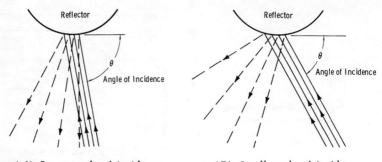

(A) Large angle of incidence.   *(B) Small angle of incidence.*

Fig. 12-10. Reflection from curved surface.

the incident power density, since the angles of reflection and incidence are equal. When the surface is curved, however, the reflected beam diverges an amount that depends on the angle of incidence. In Fig. 12-10A, the effect of a large angle of incidence is shown; the power density is less in the reflected beam than in the original beam. This factor not only decreases the received field strength, but also increases the possibility of troublesome reflections from other objects. These reflections may cause "ghosts" and other phasing effects. As the angle of incidence decreases toward a "grazing" angle, as in Fig. 12-10B, the cross section of the transmitted beam impinges on a greater curved area of the reflecting surface, and the divergence increases, decreasing the power density still further. If such a reflecting surface is very close to the receiving dish, its use may be entirely practical, since sufficient power may be available and reflections are minimized. If such a surface were closer to the transmitter dish than the receiver dish, its use would be entirely impractical. The factors that determine the practicability of this type of reflecting surface may be seen to include amount of curvature (size of circumference of object), angle of incidence, and distance from the receiving dish.

12-3. MICROWAVE-CIRCUIT FUNDAMENTALS

A microwave system employs conventional circuits with the exception of two major items: (1) A special type of tube (normally either a *magnetron* or a *klystron,* or, in solid-state equipment, a terminal diode or lighthouse tube) is used at the transmitter and receiver heads, and (2) a *waveguide* and *buttonhook* are used to couple the rf energy into the parabolic dish at both the transmitter and the receiver. Fig. 12-11 illustrates typical microwave *plumbing.* The wavemeter and monitor cavity are extensions of the same type of waveguide used for the specific application.

In normal operation of conventional tubes with negative grids, a constant-velocity electron stream from cathode to plate is varied in intensity according to the signal voltage applied to the grid. At extremely high fre-

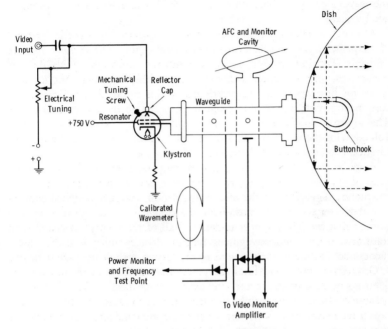

Fig. 12-11. Basic microwave plumbing.

quencies, however, such action becomes practically impossible, because of the transit time of the electrons from cathode to plate. This difficulty becomes increasingly evident in the larger tubes required for handling large power outputs. The electrons that pass the grid during the negative swing of the signal voltage are slowed down, and those that pass through the grid region during the positive signal swing are speeded up. Thus, the electrons reach the plate at different speeds, decreasing the efficiency of the tube in accomplishing any amplification of the applied signal. The klystron tube, however, is designed to use the applied signal to control the *velocity* of a *constant-current* beam, instead of attempting to vary the intensity of a constant-velocity beam.

The reflex-klystron tube has a single resonant cavity, which is designed for the operating band. Electrons bouncing back and forth in this cavity pass through several gridlike structures into a drift space, which is terminated in the reflector electrode. The resonator is operated at approximately 750 volts positive, and the reflector electrode is operated at a negative potential. Thus, the electrons are reflected back into the resonator, resulting in a bunching effect of the electrons, sometimes referred to as *velocity modulation*. The resultant rf oscillations are extracted by a probe in the waveguide and are fed to the coaxial output line through a wideband, coaxial transducer coupling unit. This unit efficiently couples the coaxial output line to the waveguide. The socket through which this line passes is an ordinary octal socket.

Although a reflex klystron is designed for operation in a particular band of frequencies, both mechanical and electrical tuning within this range are available. Turning a screw clockwise (into the cavity) decreases the frequency; turning it counterclockwise increases the frequency. The klystron is set for best operating characteristics by this mechanical adjustment in conjunction with the variable negative potential on the reflector electrode. The klystron oscillator is *frequency modulated* by varying the reflector voltage with the video signal. The modulation sensitivity of a Type 220C klystron averages about 400 kHz per volt of signal.

At radio frequencies above approximately 800 MHz, special techniques must be applied in getting the rf energy from unit to unit or from amplifier to antenna. Therefore, in the microwave region, specially designed connection devices are used for transferring the rf signals. Although it is not important that the TV engineer understand all of the theory associated with transmission lines and waveguides, he should be familiar with the fundamentals so that the major elements of mystery in such devices are removed.

Ordinary transmission lines are practically unusable at microwave frequencies because of the severe attenuation along the line. This loss occurs because of the high series inductive reactance of the line and the low shunt capacitive reactance between the inner and outer conductors. Although this problem has been solved by the development of a special transmission line for transmitters in the uhf range, waveguides generally are used at the microwave frequencies of TV relay operation.

A waveguide is a hollow metallic "tube," usually rectangular (although it may be round or oval) in cross section. Its purpose is to pass energy at high radio frequencies with a minimum of attenuation within the boundaries of the tube. When radio waves are radiated into free space, electromagnetic and electrostatic fields exist at the point of radiation—the antenna. If the frequency is low enough, energy directed into an ordinary transmission line is distributed evenly throughout the conductors as in free space. At uhf and microwave frequencies, however, most of the current is concentrated on the outer surface of the conductor (skin effect), and a high resistance is presented to the passage of the rf current through the line. In waveguides, the dimensions of the tube are such that the concentration of energy takes place in the center, with very little electric field existing at the walls.

Fig. 12-12 illustrates the electrostatic field along two wave paths. Note how the positive and negative portions meet at the walls, effectively cancelling at these points. The positive maxima and negative maxima meet at the center of the tube, adding to the field of force through the center.

The engineer encounters terms such as TE and TM, with various subscripts (for example, TE_{01} and TM_{01}), in literature concerning waveguides. TE stands for the *transverse electrostatic* mode of operation, and

**Fig. 12-12. Electrostatic field
along waveguide.**

Walls of Waveguide

Direction of Propagation →

TM stands for the *transverse magnetic* mode of operation. It should be recalled that all radio energy contains both electrostatic and magnetic fields, which are at right angles to each other. When the electrostatic field is as shown in Fig. 12-13A (that is, across the guide, or transverse to the direction of propagation), the TE mode is designated. When the magnetic field is transverse to the direction of propagation as in Fig. 12-13B, the TM mode is designated. The mode of operation is determined by the manner in which the rf energy is fed into the waveguide. Note that when the rf energy is fed in by a probe or a quarter-wave dipole in the manner shown in Fig. 12-13B, the TM mode results. For each mode, power is extracted in the same way that it is inserted.

The subscripts 0 and 1 designate the number of half-wave patterns of the electrostatic field along the A and B sides of the structure. The frequency at which dimension A is one-half wavelength is termed the *cutoff frequency,* and it therefore determines the lowest frequency that may be propagated through the waveguide. Frequencies higher than this cutoff frequency are passed readily. However, if the frequency is very much higher, other modes of operation, such as TE_{11} or TM_{11}, occur. This simply means that more than one half-wave pattern occurs across the tube. In practice, the operator will find that these higher modes of operation are not used in TV relays, and the TE_{01} or TM_{01} mode is predominant. The short (B) side is one-half the dimension of the A side. A waveguide used for the 2000-MHz band is approximately $3'' \times 1\frac{1}{2}''$, and a waveguide for the 7000-MHz band is approximately $1'' \times \frac{1}{2}''$.

The TE mode normally is used with the klystron tube; the direction of polarization depends on the direction of the short (B) side. A "straight" buttonhook normally gives horizontal polarization; a "half twist" is imparted to the waveguide to give vertical polarization.

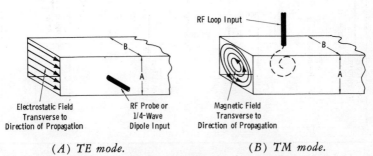

RF Loop Input

B

A

B

A

Electrostatic Field
Transverse to
Direction of Propagation

RF Probe or
1/4-Wave
Dipole Input

Magnetic Field
Transverse to
Direction of Propagation

(A) TE mode.   *(B) TM mode.*

Fig. 12-13. Modes of operation in waveguide.

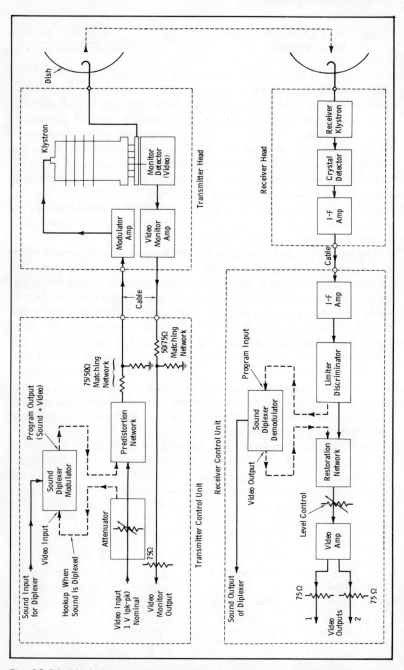

Fig. 12-14. Block diagram showing signal paths in typical microwave system.

Fig. 12-14 is a signal-path block diagram of a typical microwave system. Sound diplexing is optional, but it normally is employed when the system is used as an STL. A sound diplexer employs a frequency-modulated subcarrier (normally between 5.5 and 7 MHz) to which the picture signal is added (not mixed) in a passive combining network. A tuned sound trap is employed in the video circuitry to avoid interaction.

The cable between the control unit and the head is normally a standard camera cable that carries all voltages, control signals, and video. When the coaxial portion of the camera cable is used to carry the video signal, a 75- to 50-ohm matching network is necessary, as shown in Fig. 12-14. When the camera cable must be longer than 50 feet, video equalization is necessary; otherwise, response falls off about 0.85 dB per 100 feet at 6 MHz. With runs longer than 200 feet, it is normal practice to use an external video cable, either RG-8/U or RG-11/U. The RG-8/U cable provides a 50-ohm line. If RG-11/U cable is employed, the matching network is eliminated, and the modulator is modified to provide a 75-ohm termination rather than 50 ohms. When either of the external lines is used, equalization is required at the rate of 0.4 dB per 100 feet at 6 MHz.

Predistortion is employed when either sound or color signals (or both) must be handled. A typical circuit provides a video insertion loss of 8 dB at 60 Hz. This loss remains 8 dB through approximately 100 kHz, then tapers to 4 dB at 1 MHz, 0.9 dB at 3 MHz, and 0.25 dB at 6 MHz. This response reduces transmitter frequency deviation at low frequencies, reducing frequency excursion in the i-f section of the receiver, and permitting a reduction in differential phase shift of the 3.58-MHz color subcarrier and the higher-frequency sound subcarrier.

A restoration network with an inverse attenuation characteristic is provided in the transmitter monitor, and a similar network is used in the receiver. These networks restore the output to a flat response.

EXERCISES

Q12-1. Give the frequency bands for television-relay microwave systems.
Q12-2. Are there different frequency bands for STL service and mobile service?
Q12-3. What is the absolute minimum first-Fresnel-zone clearance recommended?
Q12-4. Define "first Fresnel zone."
Q12-5. Why should paths entirely over water be avoided when possible?
Q12-6. Is the video modulation for microwave systems normally a-m or fm?
Q12-7. What type of line couples the energy from the output stage of a microwave transmitter into the parabolic dish?
Q12-8. What determines the plane of polarization of the microwave energy?

CHAPTER 13

Television Antenna Systems
and Transmitters

A basic outline of signal propagation, transmitter circuitry, and antenna systems is presented in this chapter. Engineers concerned with new-station planning and details of transmitter location, field-intensity surveys, etc., should obtain current copies of Volumes I, II, and III of the FCC Rules and Regulations. These may be obtained from the Superintendent of Documents, U.S. Government Printing Office, Washington, D.C. 20402. You must obtain the current price from that office before ordering.

13-1. TERMINOLOGY

Field strength may be expressed in terms of dBu, or dB above 1 microvolt per meter. Power often is referred to in terms of dBk, or dB above 1 kilowatt of power. The advantages of this terminology are:

1. Transmission-line losses may be subtracted directly from transmitter power level in dB above or below 1 kilowatt.
2. Antenna gain in dB may be added directly to transmitter power level.
3. An increase of 1 dB at the transmitter results in an increase of 1 dB in received field strength.

In television broadcasting, two field-intensity contours are considered in the study of signal propagation. These are specified as:

1. Grade-A service, or a signal that should give a satisfactory noise-free picture on the "average" receiver.
2. Grade-B service, or a signal that may result in intermittent noise in the picture on the "average" receiver.

The required field strength in dBu for each of the foregoing classes of service depends on the channel of operation (frequency); values are shown

Table 13-1. Grades of Television Service

Channel	Frequency	Grade A	Grade B	Local Community
2-6 (Low VHF)	54-88 MHz	68 dBu	47 d3u	74 dBu
7-13 (High VHF)	174-216 MHz	71 dBu	56 dBu	77 dBu
14-69 (UHF)	470-806 MHz	74 dBu	64 dBu	80 dBu

Table 13-2. Hedge Chart

| Channels | Estimated Field Intensity | | |
	Probably Unsatisfactory	Questionable	Probably Satisfactory
2-6	Less Than 40 dBu	40-47 dBu	Over 47 dBu
7-13	Less Than 50 dBu	50-56 dBu	Over 56 dBu
14-69	Less Than 60 dBu	60-65 dBu	Over 65 dBu

in Table 13-1. Note also that a minimum field strength for the local community is specified; the transmitting facilities must be capable of meeting this requirement. The table lists the minimum required median field strength in dBu for the specified contour.

Under actual conditions, the extreme variations of terrain over any particular path, the variable noise levels, and the inherent differences in receiver and antenna performance preclude the possibility of having any definite basis for establishing a "satisfactory" or "unsatisfactory" signal. Therefore, it is necessary to "hedge" a little in such estimates. Table 13-2, which is based on present experience, may be used for this purpose. In many instances, modern receivers give grade-A pictures with grade-B signals. Also, both receivers and antennas have been improved (on the average) since the FCC curves were established. The "hedge chart" is actually conservative, and high-gain receiving antennas were not considered in arriving at the signal-intensity figures listed.

13-2. USE OF CHARTS AND GRAPHS

For convenience, four reference graphs are included in this chapter: Fig. 13-1 is a graph relating dBu and actual microvolts per meter (μV/meter). Fig. 13-2 is a graph relating dBk and actual watts of power. Fig. 13-3 relates the theoretical induced voltage for a 300-ohm dipole to a given field intensity (dBu) at a given frequency. This graph emphasizes that actual field strength at a given distance is meaningless unless effective antenna length is considered at the same time. (This is analyzed in Section 13-6.) Fig. 13-4 is a graph relating dB attenuation and percent efficiency. This graph is particularly useful for computations involving transmission lines.

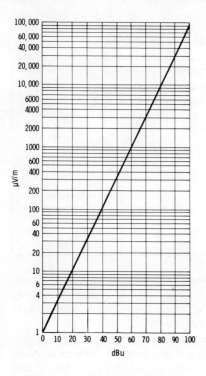

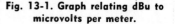
Fig. 13-1. Graph relating dBu to microvolts per meter.

Example 1. From Fig. 13-1:

$$0 \ \text{dBu} = 1 \ \ \mu\text{V/meter}$$
$$20 \ \text{dBu} = 10 \ \ \mu\text{V/meter}$$
$$40 \ \text{dBu} = 100 \ \mu\text{V/meter}$$

Example 2. From Fig. 13-2:

$$0 \ \text{dBk} = 1 \ \text{kw, or 1000 watts}$$
$$-10 \ \text{dBk} = 0.1 \ \text{kW, or 100 watts}$$
$$+10 \ \text{dBk} = 10 \ \text{kW, or 10,000 watts}$$

Example 3. Considering the necessary field strength for FCC Grade-A service (Table 13-1): The requirement for channels 2-6 is 68 dBu, or (from Fig. 13-1) 2500 μV/meter. For channels 7-13, the requirement is 71 dBu, or (from Fig. 13-1) 3600 μV/meter. For channels 14-69, the requirement is 74 dBu, or (from Fig. 13-1) 5000 μV/meter.

13-3. EFFECTIVE RADIATED POWER

Effective radiated power (erp) is a measure of the actual signal level radiated by the station. It depends on transmitter power output, transmission-line loss, and antenna gain. For the visual transmitter, erp is given in terms of peak carrier. For example, when the visual output is rated at 20,000 watts, peak power is implied. The aural carrier is rated in terms of rms values and is limited by FCC regulations to 10 to 20 percent of the visual power. With existing receivers, an aural carrier at 20 percent of the visual erp has somewhat greater range than the visual carrier; therefore, greater power for the aural carrier is not economical. Also, in color transmission, greater aural power worsens the problem of sound-carrier beats with the color subcarrier in the average receiver.

The meaning of effective radiated power should be made apparent by the following example, which shows how it is calculated. The given information is:

Transmitter output: 20,000 watts
Tower height: 1000 feet
Overall length of transmission line: 1100 feet
Type of transmission line: 3⅛-inch air-dielectric line
Antenna gain: Power gain of 6
Channel of operation: Channel 6
Efficiency of 1100 feet of above transmission line: 78 percent on channel 6

The computation involves the following steps:

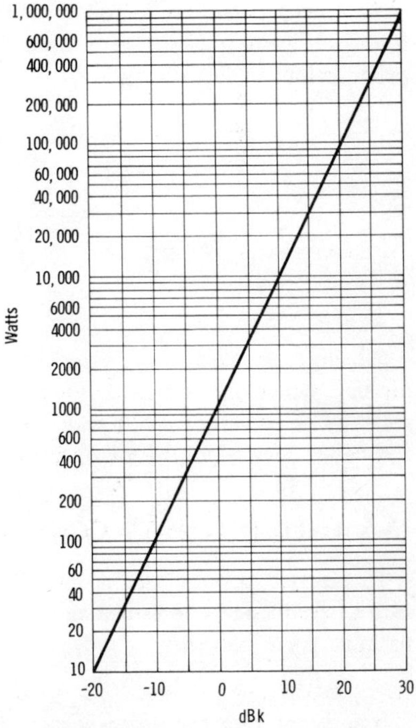

Fig. 13-2. Graph relating dBk to power in watts.

Transmitter output: 20,000 watts
Power to antenna: 20,000 watts times transmission-line efficiency, or
                20,000 × 0.78 = 15,600 watts
Effect of antenna gain: 15,600 × 6 = 93,600 watts erp

In this example, if a diplexer is used between the transmitter and antenna, any losses in the diplexer would be added to the transmission-line loss.

13-4. FREE-SPACE FIELD INTENSITY

Although free-space field intensity is only of academic interest to the practicing engineer, such information is often needed to complete FCC application forms. To find the free-space field intensity, the following formula should be used:

$$e = \frac{7\sqrt{P}}{d} \qquad \text{(Eq. 13-1.)}$$

where,

e is the free-space field intensity in volts per meter,
P is the effective radiated power in watts,
d is the distance in meters (1 mile = 1610 meters).

Assume a half-wave dipole is radiating 1 kW of power. To find the free-space field intensity at 1 mile (1610 meters), substitute the appropriate values in Equation 13-1:

$$e = \frac{7\sqrt{1000}}{1610}$$

$$= \frac{(7)\,(31.6)}{1610}$$

$$= 0.137 \text{ volt/meter}$$

$$= 137 \text{ mV/meter}$$

All antenna gains (or losses) are related to this "standard half-wave dipole," which gives a field intensity of 137 millivolts per meter at one mile for a power of 1000 watts. In free space, e is independent of the frequency (wavelength) of operation.

13-5. FIELD INTENSITY CONSIDERING GROUND EFFECTS

In practice, ground effects must be considered, and e becomes a function of frequency (or wavelength). When allowance for this is made, the formula for e becomes:

$$e = \frac{3.2ah\sqrt{P}}{d^2\lambda} \qquad \text{(Eq. 13-2.)}$$

where,

    e is the field intensity in microvolts per meter,
    a is the height of the transmitter antenna in feet,
    h is the height of the receiver antenna in feet,
    P is the effective radiated power in watts,
    d is the distance in miles,
    λ is the wavelength in meters.

Note that since λ is the denominator, a shorter wavelength results in a greater field strength. However, this increase is directly offset by a simultaneous decrease in the effective antenna length (Section 13-6).

13-6. EFFECTIVE ANTENNA LENGTH

Equation 13-2 indicates that, at a given distance and for a given power and antenna height, the field strength is inversely proportional to wavelength; stated another way, the field strength increases directly with frequency. For example, the field strength for a 600-MHz signal is 10 times the field strength of a 60-MHz signal, assuming all other factors are equal. However, field strength in itself is not significant. Receiving antennas, regardless of type and gain, are based on the dipole principle for maximum efficiency. The open-circuit voltage (E) induced in the antenna depends on the *effective length*, which, for a half-wave dipole, varies inversely with frequency. Considering only the open-circuit voltage (disregarding the effect of antenna termination), the effective length of a half-wave dipole is:

$$L = \frac{\lambda}{\pi} \qquad \text{(Eq. 13-3.)}$$

where,

    L is the effective length in meters,
    λ is the wavelength in meters.

At 60 MHz (5 meters):

$$L = \frac{5}{3.1416} = 1.6 \text{ meters (approx)}$$

At 600 MHz (0.5 meter):

$$L = \frac{0.5}{3.1416} = 0.16 \text{ meter (approx)}$$

The open-circuit induced voltage is found by using the equation:

$$E = Le \qquad \text{(Eq. 13-4.)}$$

where,

    E is the open-circuit induced voltage,
    L is the effective antenna length (in meters) parallel to the wavefront,
    e is the field intensity in volts per meter.

Example 1. Find E (open circuit) for a 60-MHz receiving antenna lying parallel to the wavefront at a point where the field intensity is 28.8 millivolts per meter.

$$E = Le$$
$$= 1.6 \times 0.0288$$
$$= 0.046 \text{ volt, or } 46 \text{ millivolts}$$

Since the effective length is actually 0.6 meter greater than one meter, the induced voltage is greater than the stress across one meter, which is 28.8 millivolts.

Example 2. Find E across a 600-MHz receiving antenna in a field of 288 millivolts per meter. (This is the field intensity for the same conditions, except frequency, as in the previous example).

$$E = Le$$
$$= 0.16 \times 0.288$$
$$= 0.046 \text{ volt or } 46 \text{ millivolts}$$

Note that exactly the same voltage is induced in each antenna although the field strength is 10 times greater at 600 MHz than at 60 MHz (with equivalent powers from a standard dipole).

Previous examples have shown the theoretical open-circuit voltage induced in an antenna. Practical examples should include the effect of terminating the antenna in its characteristic impedance, such as 72 ohms or 288 ohms. (The impedance of a folded dipole is four times the impedance of a half-wave dipole, or $4 \times 72 = 288$ ohms. Ordinarily, this is called 300 ohms.) In practice the following equation may be applied when a 72-ohm lossless line is used:

$$E = 0.32 \text{ Le} \qquad \text{(Eq. 13-6.)}$$

where,

E is the induced voltage (between the receiver antenna terminals),
L is the actual length of the antenna in meters ($\lambda/2$ for a half-wave dipole),
e is the field strength in volts per meter.

When a 288-ohm lossless line is used, the formula is modified as follows:

$$E = 0.64 \text{ Le} \qquad \text{(Eq. 13-7.)}$$

The graph of Fig. 13-3 shows the theoretical induced voltage between the terminals of a 300-ohm folded-dipole receiver antenna terminated in a 300-ohm line. (The graph is based on Equation 13-7.) The following examples illustrate the use of this graph.

Example 1. For a given field strength of 1000 μV/m (60 dBu), the induced voltage at 100 MHz is 960 μV. At 1000 MHz, the induced voltage is 96 μV.

Example 2. Note in Table 13-2 the estimated values below which reception probably will be unsatisfactory. For channels 2-6 the value is 40 dBu. At 70 MHz (approximate center of low vhf band), the theoretical induced voltage is 140 μV. For channels 7-13, the value is 50 dBu. At 200 MHz, the theoretical induced voltage is 140 μV. For channels 14-69 the value

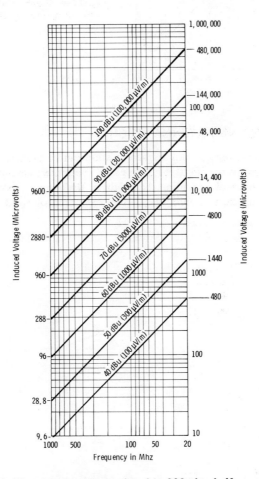

Fig. 13-3. Theoretical voltage induced in 300-ohm half-wave dipole feeding 300-ohm lossless line.

is 60 dBu. At 700 MHz, the theoretical induced voltage is 140 μV. Thus, an average spread of approximately 10 dBu is necessary between the low vhf, high vhf, and uhf bands to achieve the same induced voltage in the receiver antenna.

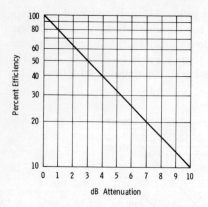

Fig. 13-4. Graph relating percent efficiency and dB attenuation.

13-7. EFFECTIVE ANTENNA HEIGHT

Before the coverage from a given location can be estimated, it is necessary to determine the effective height of the transmitter antenna. In general, the procedure involves finding the height of the antenna above the terrain that lies along each of eight radial lines extending outward from the an-

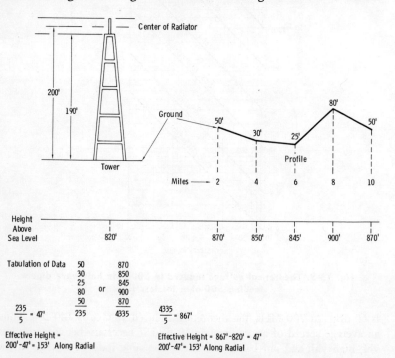

Fig. 13-5. Determination of effective antenna height.

tenna location. Only the interval between 2 and 10 miles from the antenna along each *radial* is considered.

An example of the height determination for one radial is illustrated in Fig. 13-5. This example is for a tower that extends 190 feet above ground and supports an antenna that has an overall height of 20 feet. The first step is to find the height of the *center of the radiator* above ground level at the tower location. In this example, the center of the antenna is 200 feet above ground. Plots are made on topographic maps that show elevations in feet above sea level. In cities, the heights of buildings are added to this figure. Along the radial shown in Fig. 13-5, the average elevation difference is 235 divided by the number of check points (5), or 47 feet. The antenna effective height above average terrain along this radial is then 200 − 47, or 153 feet. The greater the number of check points, the greater the accuracy will be.

In the example illustrated, the tower is located at a point lower than average terrain. If the topography were reversed (tower on a hill or knoll) and the ground sloped so that the values shown were negative, the effective height would be 200 + 47, or 247 feet.

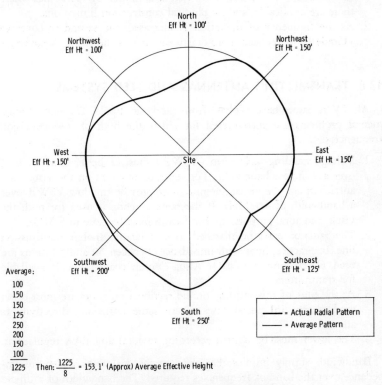

Fig. 13-6. Estimated coverage pattern.

In practice, it is found that, in general, the effective antenna height (and predicted coverage) is different along each plotted radial. Assume the conditions illustrated in Fig. 13-6. The heavy line shows the estimated coverage pattern based on the resultant effective antenna heights determined for the respective radials. This is the type of pattern that would actually be plotted on polar graph paper for FCC application data. The thin line shows a pattern based on an antenna height obtained by averaging the effective heights for the eight radials. In practice, this pattern could be obtained with the aid of a planimeter, a drafting instrument that integrates the area enclosed within a curve.

The task of estimating the coverage of a proposed television station generally is assigned to a licensed consulting engineer. However, it is interesting to note the following "rules of thumb" that apply:

1. Doubling the antenna height is approximately equivalent to increasing the power by a factor of five. Stated in another way, a power increase of approximately five times is necessary at a given effective antenna height to be equivalent to doubling the antenna height with the same erp. In practice, tower height is limited by FAA rules where these are applicable, and by physical construction limitations.
2. As the frequency of operation is increased, the secondary coverage (Grade B) is reduced more drastically than is the Grade-A coverage.

13-8. TRANSMITTING ANTENNAS AND FEED SYSTEMS

All TV radiators have evolved from the basic dipole. The same fundamental problems are encountered for either the high or low television frequencies:

1. Voltage standing-wave ratio (VSWR) must be better than 1.1 to 1 over a sufficient bandwidth. For separate antennas on the visual and aural outputs, the visual antenna must provide this low VSWR over a bandwidth of 4.5 MHz. If the same antenna is used for both the visual and aural outputs, the bandwidth must be close to 6 MHz.
2. The radiator must be matched to a standard 50-ohm transmission line. In practice, lines with impedances between 50 and 52 ohms are used (51.5 ohms nominal). Also, 75-ohm coaxial line is available for transmitters.
3. At the time of this writing, omnidirectional radiators are most often used, except in special cases where some horizontal directivity is allowed.
4. The tower must be lighted according to local and FAA regulations.

Bandwidth simply implies that the antenna maintains a definite impedance over the band of frequencies required. The bandwidth of a dipole is determined mainly by the ratio of diameter to length of the dipole arms.

Wide bandwidth of an ordinary folded dipole is obtained by using large-diameter elements.

The number of elements, or stacking of bays, affects the characteristic impedance of the radiator. Horizontal gain is achieved by the stacking of bays in the vertical direction. This cuts the power ordinarily wasted in vertical radiation and concentrates this power in the horizontal direction.

TV antennas, especially for the vhf bands, usually are built utilizing the principles of the doublet, with means provided to achieve circular radiation and power gain in the horizontal plane. Circular radiation is achieved by crossing the dipoles in a turnstile arrangement and feeding them in quadrature; that is, the currents are fed to the radiating elements 90° out of phase. This principle is illustrated in Fig. 13-7. Considering only one dipole at a time, it may be observed that the east-west dipole gives unity radiation in the north-south direction, and minimum radiation in the east-west direction (Fig. 13-7A). The north-south dipole gives unity radiation in the east-west direction, and minimum energy off the ends in the north-south direction (Fig. 13-7B). At any of the angles at which radiation is received from both dipoles, the resultant energy is the vector sum of the two fields, which have cosine distribution. Thus, toward the northeast (45° from north) the radiation from each element is 0.707, and the vector sum is unity (Fig. 13-7C). At 30° from north, the radiation from the north-south dipole is 0.866, and the radiation from the east-west dipole is 0.5. Again, the vector sum is unity (Fig. 13-7D). If the reader repeats this process around the entire circle, he will find that the vector sum at all angles is unity, and the desired circular radiation is achieved (Fig. 13-7E).

The *superturnstile,* or *bat-wing,* type of antenna popular on the vhf television bands is composed of a number of crossed dipoles that are modified in design to result in a broad-band impedance characteristic. These elements are termed *current sheets* in technical descriptions, and are fed with currents in quadrature as in the preceding illustrations. The evolution of the bat-wing antenna is illustrated in Fig. 13-8 and is described below.

In Fig. 13-8A, a metal sheet with a slot one-half wavelength long at the operating frequency is excited with rf energy at the middle of the slot. At the instant shown, the polarity of the excitation is such that current through the metal sheet is in the direction of the arrows. The slot may be considered to be two parallel conductors one-half wavelength long and shorted at the ends. Since center i⸳ed is used, the current wave is set up as shown by the dash line. Current through the metal sheet is in the direction of the arrows, and is of greatest strength through the center as indicated by the length of the arrows. The spacing of the edges of the slot is negligible at the operating frequency, and actual radiation occurs in both directions from the center of the metal sheet.

If an actual metal sheet were used, the radiation resistance would become so high at a distance slightly exceeding one-quarter wavelength from

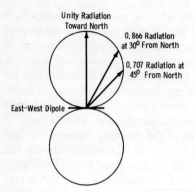

(A) Pattern from east-west dipole.

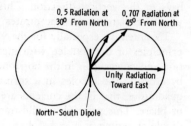

(B) Pattern from north-south dipole.

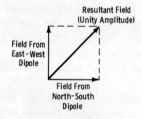

(C) Resultant field at 45°
from north.

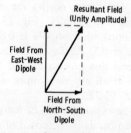

(D) Resultant field at 30°
from north.

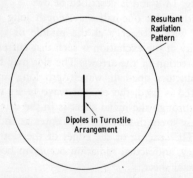

(E) Pattern from complete antenna.

Fig. 13-7. Radiation from turnstile antenna arrangement.

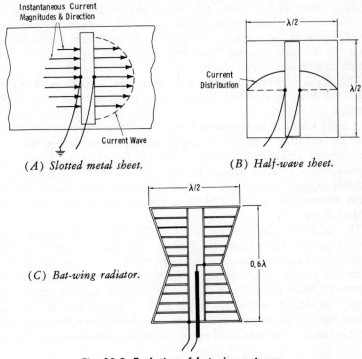

(A) Slotted metal sheet.

(B) Half-wave sheet.

(C) Bat-wing radiator.

Fig. 13-8. Evolution of bat-wing antenna.

the slot that negligible radiation would result. If, therefore, the sheet is made not more than one-half wavelength wide, as shown in Fig. 13-8B, optimum radiation is achieved. The current distribution and resulting radiation characteristics approximate those of a half-wave dipole, with maximum current and zero voltage at the center of the radiator. Such an element may be physically supported by a mount at this zero-potential point, which occurs at the center of either end of the sheet.

In practice, the "sheet" is notched at the center (Fig. 13-8C) to reduce this dimension below one-quarter wavelength and reduce the current at the midsection. The height of the current sheet is then made approximately 0.6 wavelength. This results in greater currents at the top and bottom relative to those through the center, and the vertical pattern approaches the characeristics of two horizontal dipoles spaced one-half wavelength apart vertically. This allows a gain in the horizontal plane of about 1.2 per bay. These bays may be stacked to achieve gains up to approximately 7 in the vhf band. Each bay actually consists of two "current sheets" in the turnstile, or quadrature, arrangement. The coaxial feed lines may be connected so that the outer conductor feeds one side of each element and the inner conductor feeds the other side, as in Fig. 13-8C.

As is shown in Fig. 13-8C, the current sheet is not solid, but consists of metal bars to reduce wind resistance. The spacing between the bars must be negligible at the operating wavelength. All feed lines to the stacked array are fed from common junction boxes, and are of exactly the same length to maintain correct phasing.

The physical appearance of the superturnstile antenna is shown by Fig. 13-9. More than 600 of these antennas are in operation all over the world.

A more recent superturnstile antenna from RCA is termed the "butterfly." The radiating elements are smaller and lighter in weight than in a conventional superturnstile antenna, and the bat-wings are folded back slightly toward reflector screens. The design combines features of a superturnstile antenna with the mounting and pattern-shaping flexibility of a panel antenna. The principle of operation is essentially the same as described above.

The RCA "Vee-Zee" uhf antenna and "Zee" panel uhf antenna (Fig. 13-10) are designed to meet requirements for either an omnidirectional or directional antenna that can be stacked around a tower, the top of which is used to support other antennas. (These uhf antennas also may be top mounted.) Beam tilt and null fill may be designed into the vertical patterns. Each element is complete and electrically independent in itself, and almost any desired antenna pattern can be achieved through proper relative placement of the panels and proper power division and current phasing.

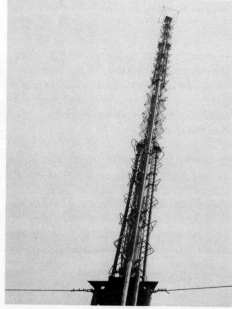

Fig. 13-9. Typical bat-wing antenna.

Courtesy RCA

Fig. 13-10. Section of uhf panel antenna.

The large aperture of each element, fed from a single feed point, strikes a balance between the mechanical complexity of many feed points and a lack of pattern flexibility as a result of too few feed points.

The "Zee" antenna comprises zig-zag radiating elements that branch two ways from a central feed point along a flat reflecting plane. The "Vee-Zee" antenna has the same configuration except that both the elements and the reflecting panel are bent along a central longitudinal line to form a forward-opening "vee" (Fig. 13-11). While both types of radiating ele-

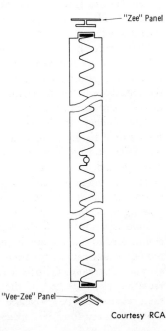

Fig. 13-11. Comparison of "Zee" and "Vee-Zee" antennas.

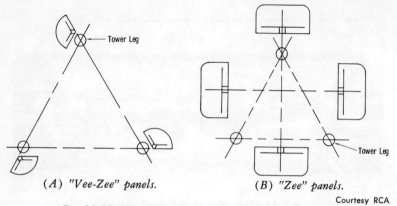

(A) "Vee-Zee" panels.   (B) "Zee" panels.

Courtesy RCA

Fig. 13-12. Use of antenna panels placed around tower.

ments are similar in electrical concept, their physical shapes affect the basic horizontal patterns, and each offers advantages for particular requirements. Where large tower structures are involved, omnidirectional or directional patterns may be obtained from the 120° pattern of the "Vee-Zee" radiator by mounting three radiators, one on each of the three tower legs, so as to fire tangentially around the tower (Fig. 13-12A). Where the antenna is to be mounted on top of the tower, "Vee-Zee" radiators (usually three in number) firing tangentially (Fig. 13-12A) may be installed, or "Zee" panels (normally four in number) with their 90° patterns oriented radially (Fig. 13-12B) can be used. The choice depends on the shape of the pattern desired.

Circularities between ±1 and ±3 dB (depending on application) are achieved by feeding equal power to all elements in a horizontal plane. Directional patterns are obtained by varying the amplitude and phase of the signals radiated, by changing relative spacings, and by choice of the directions of fire of the various elements.

The number of elements stacked vertically and the amplitudes and phases of the signals radiated by the elements determine the vertical pattern. "Sculpturing" can be done to have either zero nulls (where distant coverage and maximum gain are desired) or filled nulls (where thorough close-in coverage is necessary). Beam tilt can be achieved in all directions or only in selected directions by tilting the individual panels, by electrical phasing of successive radiators, or both.

Gain is a measure of the degree to which the vertical pattern has been compressed to concentrate the signal near the earth and the degree to which the horizontal pattern has been designed to concentrate the signal in given azimuth directions. Gain is a function of the number and orientation of radiating elements and of the phases and amplitudes of the currents in these elements. The "Vee-Zee" and "Zee" panel antennas provide flexibility of choice for each of these variables.

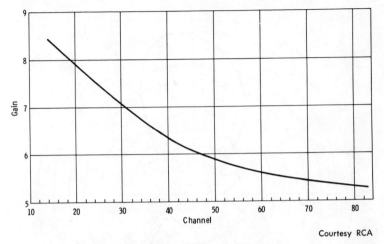

Courtesy RCA

Fig. 13-13. Gain of "Vee-Zee" antenna.

Certain relationships should be borne in mind in considering the gain to be used:

1. Effective radiated power (erp) in a given direction is the transmitter power times the efficiency of the transmission line times the antenna gain in that direction.
2. Gain in a given direction is the gain of the vertical pattern times the gain of the horizontal pattern (if the vertical pattern is the same in all directions). Thus, gain is affected by all radiators in an antenna.
3. Gain must be sacrificed (normally by from 0 to 15 percent) to obtain null fill. Thus, more vertically stacked panels may be required to obtain a desired gain in a filled pattern.
4. Approximate gains for a single layer of uhf "Vee-Zee" panels radiating omnidirectionally are shown in Fig. 13-13. Slightly higher gains are achieved by use of "Zee" panels.

A "Zee" panel directional uhf antenna is in operation at station WAND-TV (channel 17), Decatur, Illinois. This antenna is designed to be top mounted on a triangular tower approximately 1000 feet in height. A 30-kW transmitter is used to feed the antenna.

Each section of the WAND-TV antenna consists of four individual panels assembled around a square supporting frame. Four sections are fastened together and stacked to form the complete antenna assembly, which is enclosed in a removable radome for protection from the effects of the atmosphere.

Fig. 13-10 shows one section of the WAND-TV antenna. The single feed point for one panel is shown in the center of the photograph. The glass-fiber brackets located around the corners are used to mount the ra-

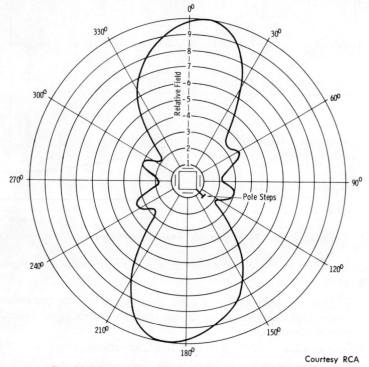

Courtesy RCA

Fig. 13-14. Horizontal pattern of WAND-TV antenna.

dome. Note the zig-zag radiating elements mounted on the reflector panel. Fig. 13-14 illustrates the horizontal pattern of the complete antenna installation.

Another panel installation is located at WCVW-TV (channel 57), Richmond, Virginia. This omnidirectional "Vee-Zee" panel antenna is mounted, ten panels high, on each of the three legs of a tower that is 577 feet high and measures 7 feet on each face. The tower also supports a top-mounted pylon antenna (Fig. 13-15.) The 30 "Vee-Zee" panels are stacked around the tower beginning approximately 98 feet below the tower top. One 30-kW transmitter feeds the channel-57 antenna, and another 30-kW transmitter feeds the top-mounted channel-23 pylon antenna. The horizontal pattern of the channel-57 antenna is illustrated in Fig. 13-16A, and the vertical pattern is shown in Fig. 13-16B.

Broadcasters are showing increased interest in tower sharing at multiple-antenna sites. This arrangement allows any TV viewer to orient his receiving antenna in one direction for all stations. Both the FCC and the FAA strongly encourage site sharing, particularly in large metropolitan areas where land is at a premium and air traffic is heavy. When costs of land,

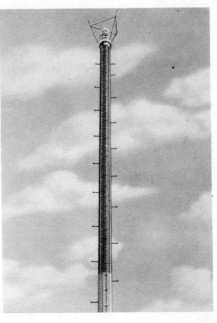

Fig. 13-15. Typical uhf pylon antenna.

Courtesy RCA

tower materials, and erection are shared by several station owners, the savings are significant compared to the cost of a number of separate antenna installations. Advantages exist in the common use of power sources, roads, and water supplies. By pooling building-maintenance manpower and equipment, one crew can service the entire installation.

The TV antenna complex on the John Hancock building (Fig. 13-17) in Chicago is one of the largest multiple antenna installations in the world. The antennas and supporting tower sections of five stations—WBBM-TV (channel 2), WFLD-TV (channel 32), WGN-TV (channel 9), WMAQ-TV (channel 5), and WSNS (channel 44)—are mounted on two 12-foot-diameter, 100-foot-long masts atop the 100-story building. The antennas for channels 5, 9, and 32 occupy the west tower, and those for channels 2 and 44 are located on the east tower (Fig. 13-18). Space is reserved on triangular sections in each tower for the addition of five more TV antennas. Construction of the complex involved erection of 140 tons of antennas and supporting structures atop the building to an overall structure height of 1450 feet.

The stacking of antenna systems always involves important mechanical and electrical considerations that greatly influence the design. In the John Hancock installation, for example, with its potential use by ten stations, the choice of antennas was affected by the need for many transmission lines down through the antennas and supporting structures, and through the

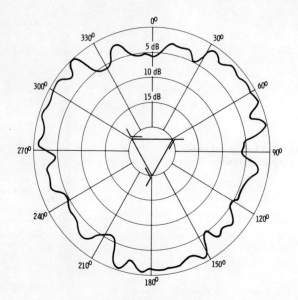

(A) Horizontal pattern.

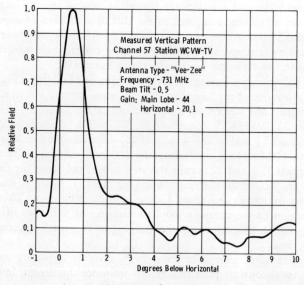

Measured Vertical Pattern
Channel 57 Station WCVW-TV

Antenna Type – "Vee-Zee"
Frequency – 731 MHz
Beam Tilt – 0.5
Gain: Main Lobe – 44
         Horizontal – 20.1

(B) Vertical pattern.

Courtesy RCA

Fig. 13-16. Radiation patterns of WCVW-TV antenna.

Fig. 13-17. Model of John Hancock Center.

Courtesy RCA

building to the transmitter rooms. The design of the antennas not only had to allow for the lines to come through, but also had to provide access to each line for tests, repair, or replacement. The strength needed in the 249-foot towers called for fairly large and rugged antennas that would provide the horizontal pattern characteristics of normally lighter and more slender antennas. Structural problems were further complicated by the requirement that the overall rigidity of each tower be such as to restrict sway of the top antenna to 0.5° in a 50-mph wind. (Holding the relatively narrow radiated beam of the top-mounted uhf antennas within these limits avoids undesirable changes of signal strength at distant points.)

Another important aspect in cases of systems where antennas are side by side is the effect that reflections and mutual couplings may have on the free-space patterns of the individual antennas. This effect must be determined before a certain configuration can be deemed acceptable. The results of tests using scale models and the experience gained in the design and installation of similar arrays were invaluable in predicting the performance of the John Hancock system. The antennas selected were the superturnstile (channel 2), "Butterfly" (channel 5), "Zee" panel (channel 9), and polygon (channels 32 and 44).

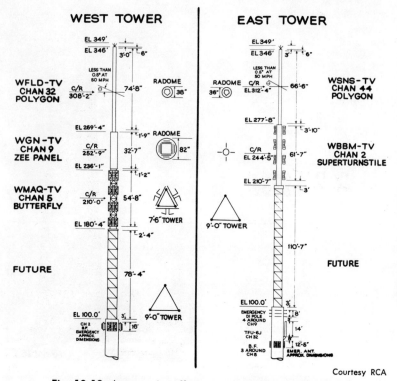

WEST TOWER   EAST TOWER

Courtesy RCA

Fig. 13-18. Antenna installation on John Hancock Center.

It was desired that the uhf antennas provide patterns that not only would cover Chicago and the west shore of Lake Michigan, but also would cover Gary, Indiana, on the south shore. The polygon antenna was selected for this application because of its five faces, each separately controllable in magnitude of radiated signal. Fig. 13-19 shows how the cardioid pattern chosen corresponds with the desired area of coverage.

Being directional, the uhf antennas could be made shorter than usual. Thus, the lengths of these antennas are only 74 and 66 feet. Omnidirectional antennas having the same gain would be 70-percent longer. Each of the uhf antennas is capable of delivering an effective radiated power of 5 megawatts—the highest allowed by the FCC for uhf stations—and each can handle an input power of 110 kilowatts.

The polygon antenna provides diversity of horizontal and vertical patterns, and it is adaptable to supporting other antenna structures in multiantenna arrays. The walls of the five-sided steel sheath can be made as thick as desired, eliminating the need for an internal tower structure. Adjacent sections of the antenna are hoisted into position and bolted together. Individual elements in each five-panel layer are excited by an external belt-

(A) Desired area of coverage.

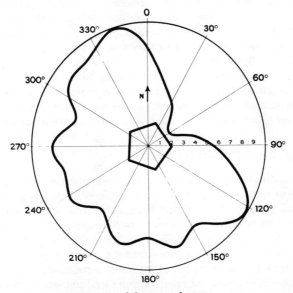

(B) Actual horizontal pattern.

Fig. 13-19. Determination of horizontal pattern for WFLD-TV and WSNS.

line feed system, and the several layers are connected and fed by a transmission line that is removable for servicing.

WMAQ-TV (channel 5) utilizes a four-section "butterfly" antenna. This antenna, being a panel type, can be arranged in triangular or square configurations and lends itself to horizontal pattern sculpturing for directional use. The design also presents a small silhouette both physically and electrically, minimizing wind load and reducing pattern scalloping that might otherwise occur because of radiation from adjacent antennas.

WGN-TV (channel 9) has a four-sided "Zee" panel antenna. As with the polygon, antenna radiation patterns of almost any shape can be achieved by varying the relative power inputs and phases to the panels. In this particular installation, however, the antenna is omnidirectional. Its long internal space permits maintenance personnel to work inside the supporting structure.

All antennas except the channel-9 "Zee" panel, the single-panel design of which makes it unfeasible to do so, are split-fed. That is, two separate transmission lines are run to each antenna; one feeds the radiators above the center of the antenna, and the other feeds the radiators below the center. This allows for emergency operation of either half of the antenna independently of the other half in case of failure or during repairs. The shift in mode of operation is made by coaxial switching in the transmitter room.

In addition to the emergency features described above, separate auxiliary antennas for channels 5 and 9 are installed at the top of the 100-foot supporting mast in the east tower. Provision also is made for a similar channel-2 auxiliary antenna on the west tower.

The supporting cylinders in the west and east towers, besides providing mounting space for the emergency antennas, house all the transmission lines for the antennas, plus power feeds for beacons, deicers, and working lights and outlets within the towers. A communication system connecting any antenna location with any transmitter room also is installed.

13-9. TRANSMITTER VIDEO STAGES

All commercial TV transmitters have certain characteristics in common regardless of make or band of operation. Some of these characteristics are determined by FCC technical standards. Other characteristics are set by the manufacturers themselves through mutual agreement by way of their coordinating organization, the Electronic Industries Association (EIA). An examination of these common features serves as a good introduction to the study of TV transmitters. (Review also Section 1-7, Chapter 1.)

All visual transmitters are rated in terms of their *peak power* output. Thus, if a 5-kilowatt visual transmitter is specified, it is capable of putting 5 kilowatts of peak power into the transmission line. In practice, the *average* power output is first measured with the transmitter modulated by the *standard black* television signal and operating into a dummy load. The dummy load must have substantially zero reactance, and resistance equal to the surge impedance of the transmission line to be used. The average power determined in this measurement is multiplied by 1.68 to obtain the peak power output. Methods are described in Chapter 14. Some control, either manual or automatic, is provided so that the peak power output can be adjusted within definite limits over the operating day.

All visual transmitters, for both the vhf and uhf bands, are amplitude modulated by the picture and control-pulse signals. The term 100-percent modulation refers to the maximum carrier amplitude on sync peaks. The blanking level (pedestal) occurs at 75 percent of the peak amplitude, within ±1 percent for any fixed picture content (such as a test pattern) and within ±2.5 percent for a variation of picture content such as occurs during regular program transmission. The reference white-level amplitude occurs at 12.5 percent ±2.5 percent of the peak carrier amplitude. Thus, maximum white level occurs between 10 and 15 percent of peak carrier amplitude. Overshoots in white signals should be avoided, since the resulting carrier cutoff causes severe buzz and noise in intercarrier receivers.

The polarity of the transmitter input picture signal is black negative. The level is a minimum of 1 volt and a maximum of 2.5 volts peak-to-peak across the standard input impedance of 75 ohms.

The polarity of the modulated transmitter output signal is black positive (negative modulation—see Section 1-7) into a standard transmission line. The output connections for picture-signal monitoring provide a black-negative signal with an amplitude of 0.5 to 2.5 volts peak-to-peak across a resistive impedance of 75 ohms.

The frequency of the visual carrier must be maintained within ± one kilohertz of the authorized frequency, and the aural carrier must be 4.5 MHz ± 1 kHz above the actual visual carrier frequency.

The function of the *visual exciter* of a TV transmitter is to generate the carrier wave at the assigned frequency, and to amplify the power to the level necessary to drive the high-level amplifiers. For the utmost in stability, a crystal-controlled oscillator is used. A typical arrangement for channels 2 through 6 consists of a crystal oscillator stage followed by a tripler and two doublers, resulting in a frequency multiplication of 12 $(3 \times 2 \times 2)$. For example, the visual carrier frequency for channel 6 is 83.25 MHz. Thus, the crystal frequency would be 6.9375 MHz (83.25 MHz/12). The multipliers step up the oscillator voltage in both frequency and amplitude to the level required to drive the modulated stage. Thus, the visual exciter contains conventional narrow-band rf circuits, which are tuned to the crystal frequency by adjusting the tank-circuit capacitors (or inductors when inductively tuned) for minimum plate current in the stage and maximum grid current to the following stage. All necessary frequency multiplication takes place prior to the stage in which modulation takes place.

The incoming video from the studio usually is fed to a stabilizing amplifier to minimize the effects of hum, noise, or sync compression. Incorporated in the stabilizing amplifier are sync-stretcher circuits for control of the relative sync-to-video amplitude. Also included, in some cases, are linearity controls to precorrect the transfer characteristic of the transmitter amplifiers. The transmitter transfer characteristic is the ratio of the rf output voltage to the video input voltage; generally, it is linear within 10 percent. This is desired at the transmitter, since the gamma at the studio (signal sources) is adjusted to result in optimum picture quality as observed by the television audience; hence, the remaining portion of the overall system should be as nearly linear (gamma = 1) as possible.

The final video amplifier stage in the transmitter is the video modulator. The dc component is reinserted at the grid of this stage by clamper circuits, and to maintain this dc component the modulator is coupled directly to the rf stage being modulated.

The rf circuits that follow the modulated stage are essentially linear rf amplifiers adjusted for maximum power output consistent with a flat frequency response throughout the upper sideband. Proper adjustment of these amplifiers results in partial cancellation of the lower sideband. If low-level modulation (which may be either grid or plate modulation) is employed, a sufficient number of linear amplifiers are used to obtain the desired vestigial-sideband response. This action is aided by inserting a notching filter adjusted to a frequency 1.25 MHz below the visual carrier frequency. For high-level modulation (which must be grid modulation), a vestigial-sideband filter must be used. The standard transmission signal from this filter is then fed to either a *bridge diplexer* or a *notch diplexer,* which couples the signal into the transmission line to the antenna system.

In the visual-exciter section of Fig. 13-20, typical values are given for a transmitter operating on channel 6. Dc restoration is shown taking place at the grids of the modulator tubes. Response of the modulator stage in most commercial visual transmitters is made flat to 5 MHz to insure freedom from phase distortion. The illustration shows partial removal of the lower sideband at the output of the final modulated stage (in this example).

The operator must become acquainted with the correct interpretation of picture and waveform displays, since the vestigial-sideband characteristic results in limitations in both transmitter and monitoring devices. The details of interpretation are included later; the physical arrangement and a description of the electrical characteristics are given here.

Typical frequency response at two monitoring points is shown in Fig. 13-20. One monitoring output is obtained from a diode demodulator at the output of the modulated final amplifier, and the other is obtained from a vestigial-sideband demodulator at the output of the vestigial-sideband filter. The typical output response of the ordinary diode demodulator is reduced approximately 50 percent at 4 MHz with a gradual rolloff from about 2.75 MHz to 5 MHz. As a result of this high-frequency rolloff, the

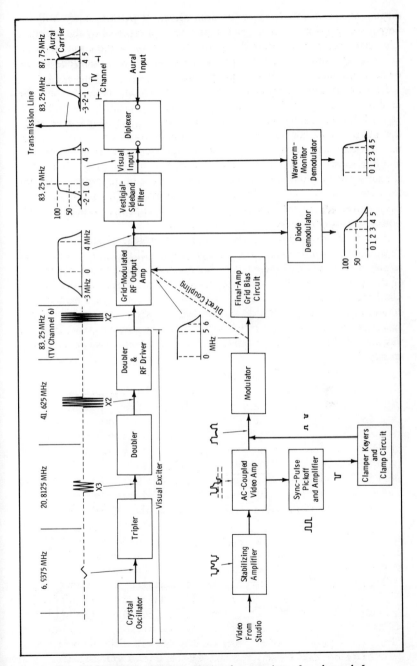

Fig. 13-20. Block diagram of visual transmitter for channel 6.

display observed on a picture monitor driven from this source will inherently lack sparkle or detail, and should be so interpreted by the operator. This diode curve results from the partial cancellation of the lower sideband in the final amplifier; addition of the upper and lower sidebands in the detector produces the typical curve shown.

Because of this characteristic, the ordinary diode detector cannot be used for picture monitoring at any point after the vestigial sideband filter, or, with low-level modulation, at any point past the modulated stage, since the sharp attenuation of the lower sideband results in a response curve that is useless for observing picture detail. Therefore, the vestigial-sideband monitor shown connected at the output of the filter in Fig. 13-20 is a special insensitive receiver circuit for picture monitoring; typical response for such a demodulator is shown in the illustration. This response curve has a longer flat top, and the sharp cutoff at the high end permits observation of any ringing effects in the picture.

For waveform monitoring, the output of this demodulator is fed to a keyer circuit, then to an oscilloscope. The purpose of the keyer circuit (also termed a *vibrator* or *chopper*) is to intermittently short-circuit the output of the detector, providing a zero-output line on the scope screen. Keep in mind the two components of the standard composite signal, namely, the dc and ac components. The dc axis must be constant; the ac video-signal axis is variable, depending on light or shade in the original scene. Periodic shorting of the demodulator produces a zero reference level representing no signal.

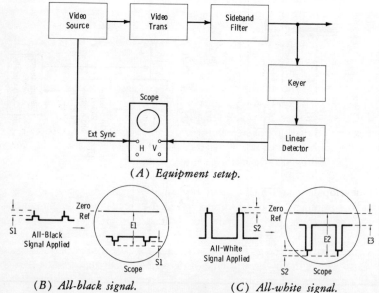

(A) Equipment setup.

(B) All-black signal.   (C) All-white signal.

Fig. 13-21. Method for measuring modulation characteristics.

The basic equivalent circuit of such waveform monitoring is shown in Fig. 13-21A; Figs. 13-21B and 13-21C illustrate its application to measuring modulation characteristics. If an all-black video signal with a sync-pulse height S1 above pedestal level is fed into the transmitter, the resultant scope pattern from such an arrangement is as shown in Fig. 13-21B. The ratio of S1 to E1 is an expression of the *modulation capability* of the transmitter for an all-black signal, with respect to the sync pulses. If the transmitter is left adjusted as before, and an all-white signal is fed to the transmitter input, the scope pattern appears as in Fig. 13-21C. The ratio of E3 to E2 is an expression of transmitter modulation capability for an all-white signal, with respect to the sync pulses. For a properly adjusted transmitter, these ratios should be practically equal. In other words, the variations of blanking and sync levels with changes in picture brightness from black to white must be held to an absolute minimum. The FCC standards limit this variation to 10 percent of the amplitude of an all-black picture. When functioning properly, modern transmitters hold well within 5 percent in going from black to white. The percent of variation under such conditions may be determined as follows:

$$\text{Blanking-level variations} = \frac{(E2 - S2) - (E1 - S1)}{E1 - S1} \times 100\%$$

$$\text{Sync-level variation} = \frac{E2 - E1}{E1} \times 100\%$$

The preceding arrangement also enables the operator to set the maximum white level of the video signal to $12\frac{1}{2}\%$ percent $\pm 2\frac{1}{2}$ percent of the peak sync amplitude.

In actual practice, the picture and waveform monitors incorporate input selector switches so that monitoring is accomplished at points other than those shown in Fig. 13-20. Usually, the switches provide for insertion of the monitors at the stabilizing-amplifier output and modulator output. This permits observation of the signal at a sufficient number of points to aid in determining stages in which trouble may have occurred. Some stations utilize an ordinary receiver monitor as an overall check, in which case precautions must be taken not to overload the receiver circuits with the strong signal prevailing at the transmitter location.

Because of FCC limitations, video transmitters can not reach the stage of development at which overall frequency response is equal to that of studio equipment. Since distortion is additive, studio equipment must be operated with as great a bandwidth as it is possible to obtain with the equipment used.

The vestgial-sideband characteristic in itself is a source of picture distortion which, in any present type of demodulation system, produces slight leading whites and trailing smears upon a transition from white to black. Such defects may be made very slight, however, in comparison to the advantage realized in gaining maximum use of the available frequency spectrum. The inherent distortion of vestigial-sideband transmission can be minimized by using phase and amplitude predistortion in portions of the transmitter circuits or in the stabilizing amplifier. Where transmitters are concerned with color signals, the transmitter and average receiver characteristics are equalized.

The overall frequency response up to the video-modulator stage in the transmitter is essentially flat to 4.2 MHz. From this point on, the response is a compromise based on economic design of circuits and the inherent nature of the standard transmission signal. The final clamping point for dc reinsertion ordinarily is found at the grid of the modulator stage. This necessitates some form of direct coupling between the modulator plate and modulated grid in order to maintain the dc component. The reader may wonder why the clamping action is not inserted at the grid of the modulated stage rather than at the modulator grid so that this direct coupling could be eliminated. This is the first point of compromise in design. The clamper *keying pulses* must have a peak-to-peak value greater than that of the actual clamping pulses and the video voltage applied to the clamped grid. If this were not the case, one of the clamper diodes might be brought into conduction during the video-signal interval rather than the blanking interval. Any advantage that might be gained by clamping the grid of the modulated stage would be offset by the larger power-handling capability required in the modulator stages, which would unnecessarily increase the initial cost. If the grid of the rf modulated stage were ac coupled, approximately 60 percent greater signal amplitude (peak-to-peak) would exist at that point than in the case of an arrangement in which dc coupling were used.

To meet the standard transmission characteristics of negative modulation, an *increase* in light content must cause a *decrease* in the amplitude of the carrier wave. This requires that the grid-modulated radio-frequency stage receive a *black positive* video signal as indicated in Fig. 13-22A. As the signal swings in the black (positive) direction, the grid bias is decreased (becomes less negative), and the plate current is increased, resulting in a greater amplitude of the rf output signal. As the video swings in the negative (white) direction, the grid bias is increased (made more negative), decreasing the plate current; the amplitude of the rf output signal is decreased.

For the signal at the grids of the modulated stage to be black positive, the signal input to the modulator grid circuit must be *black negative,* as shown in Fig. 13-22A. Fig. 13-22B illustrates a typical modulator-stage transfer characteristic. As the signal swings in the positive (white) direction, the grid voltage is made less negative, and the modulator plate current increases. Increasing the plate current causes a greater voltage drop across the modulator load, reducing the voltage at the coupled point. This results in the familiar phase reversal of 180° between plate voltage and

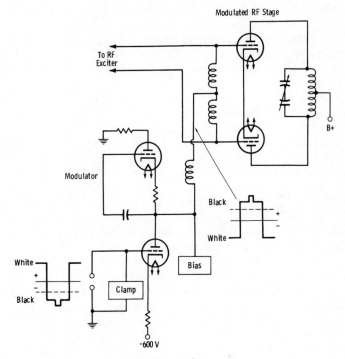

(A) *Circuit diagram.*

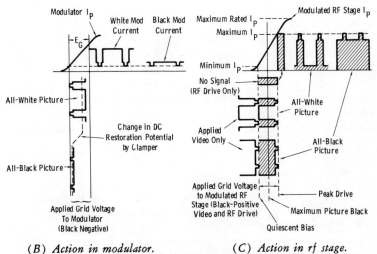

(B) *Action in modulator.*   (C) *Action in rf stage.*

Fig. 13-22. Fundamentals of video modulation.

grid voltage. As the grid voltage is increased in the positive direction, the plate voltage coupled to the modulated grids goes in the negative direction. Also notice that there is a change in dc restoration potential accomplished by the clamper or restorer stage, so that pedestal and sync levels occur at the same modulator plate current in either all-black or all-white signal conditions.

The transfer characteristic of the modulated rf stage is illustrated in Fig. 13-22C. A grid-modulated stage is operated in Class B (as are any following linear amplifiers), and therefore it is biased close to cutoff with no excitation. For negative modulation, when no video signal is received, the radio-frequency excitation from the driver stage is sufficient to drive the plate current to its maximum value. For an all-white picture signal, the bias on the grids is maximum, and the plate current is reduced (except during the blanking and sync intervals) to a point between 10 and 15 percent of the maximum level. For an all-black video signal, such as the application of pedestal and sync only, the bias on the grids is at minimum, resulting in maximum amplitude of the carrier envelope.

In practice, the quiescent grid bias is adjusted so that the video excursion about the bias point maintains the output waveform in the linear portion of the grid-plate transfer characteristic curve. Excessive bias pushes the operation down around the lower knee of the curve and results in compression of the white portions of the picture signal. Insufficient bias does not allow full advantage of the linear portion of the curve without sync compression; the resulting operation along the upper part of the curve causes the sync region to fall on the bend of the curve unless the amplitude of the applied video is held to an unreasonably low level.

From the foregoing analysis of the video-modulator action, the importance of proper clamping may be observed. This dc-restoration action at the modulator grids is fundamentally illustrated in Fig. 13-22B. Fig. 13-23 illustrates the clamper action. The average ac axis for a symmetrical waveshape is as shown by waveforms 1 and 2 in this figure; equal areas above and below the zero axis are enclosed by the waveform. Waveform 3 is a video signal equivalent to an all-white signal. It is necessary for the clamper or dc-restorer circuit to shift the ac axis in the positive direction to hold the pedestal and sync at the predetermined reference level. Compare this action to the dc restoration of Fig. 13-22B. An all-black video

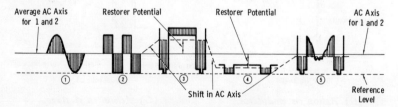

Fig. 13-23. Basic clamper operation.

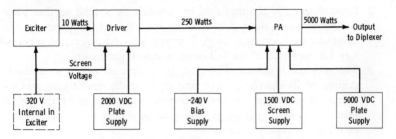

Fig. 13-24. Block diagram of fm transmitter.

signal is shown as waveform 4 in Fig. 13-23. Notice that the ac axis must be shifted in the negative direction to hold the peaks at the previously mentioned reference levels. In this case, the quiescent bias of the modulator grids is shifted in the negative direction. This shifting of the ac axis so that the reference level always occurs at the same point on the grid-voltage, plate-current transfer curve, regardless of waveshape, is equivalent to restoring the dc signal component. As shown by waveform 5 in Fig. 13-23, a video signal consisting of an exact balance between black and white would have its ac axis very nearly equal to that of a symmetrical waveshape. The slight difference occurs because of the difference between video maximum black and pedestal level, and the existence of the sync peak level.

13-10. TV AURAL (FM) STAGES

Fig. 13-24 is a block diagram of a typical TV aural transmitter. From exciter output to transmission line, there are only two radio-frequency stages. All frequency multiplication occurs in the exciter unit.

The block diagram of a typical exciter unit is shown in Fig. 13-25. The basic function of each numbered block is as follows:

1. The crystal oscillator generates a stable sine-wave signal at a relatively low-frequency submultiple of the assigned carrier frequency.

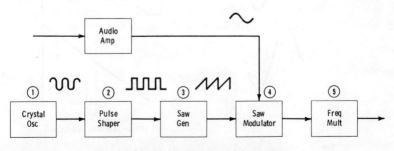

Fig. 13-25. Block diagram of fm exciter.

2. Pulse shaping is accomplished by overdriving a tube so that plate current flows in short pulses and is cut off between pulses by the automatic bias developed by grid current. This shaping action is necessary for operation of the following sawtooth generator.

3. The sawtooth is generated by charging a capacitor over a small linear portion of its changing curve, and then rapidly discharging it. Discharge occurs when the tube is driven to conduction by the pulses from the preceding stage. By this process, a sawtooth at the crystal-oscillator frequency is formed.

4. One of the tube elements, such as the control grid, receives the sawtooth from the generator, while another element, such as the cathode, receives the amplified program audio. This modulator stage is usually operated with low plate-to-cathode voltage. The audio voltage, being superimposed on the static cathode-bias voltage, shifts the point on the sawtooth slope at which the tube begins to conduct. Therefore, the phase of the conduction point is advanced or retarded at the audio rate (Fig. 13-26). The actual phase shift, which is quite low for good linearity, is about 2 radians (roughly 120°).

5. A string of frequency multipliers increases the crystal frequency to the assigned carrier frequency. As the frequency is multiplied, so is the phase shift; thus, the full 100-percent modulation (±25 kHz for TV broadcasting) can be achieved.

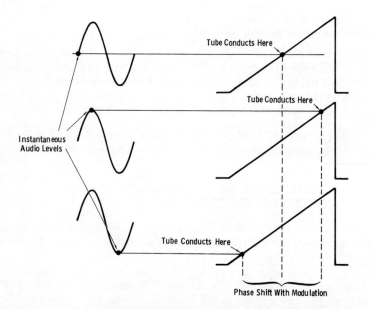

Fig. 13-26. Principle of serrasoid modulation.

The method of modulation described is termed *phase-shift modulation employing pulse techniques,* or sometimes *serrasoid modulation.* With this method, the output frequency changes only during the time the phase is being changed by modulation. Since the oscillator and modulator stages are fully isolated, the modulation process cannot affect the average carrier frequency, which is tightly controlled by the crystal.

The audio signal is pre-emphasized at the transmitter before modulation. A complementary de-emphasis circuit is employed in the receiver and in transmitter monitoring equipment. This arrangement improves the signal-to-noise ratio so that extended dynamic range is a practical reality. (Also, in a transmitter of this kind, the response of the audio circuits driving the modulator must be inversely proportional to frequency. This is necessary to make phase modulation equivalent to frequency modulation.)

NOTE: Recent solid-state modulators employ a variable-capacitance diode[1] in the oscillator circuit to produce direct fm, which is stabilized by an afc circuit. These devices overcome the drift problems inherent in earlier multitube afc systems.

13-11. REDUNDANCY IN TV TRANSMITTERS

Some TV transmitters employ *redundant* circuitry to avoid loss of air time or the necessity of purchasing a complete standby transmitter installation. Redundancy takes the form either of actual parallel operation of two transmitters, or of a special means of diplexing that allows part of the power amplifier to be used (at reduced output) while the other part (feeding a dummy load) can be repaired.

In either case, some form of combining network is employed as illustrated basically in Fig. 13-27. If the output of each transmitter is 5 kW, the normal power output with the switches in the positions shown is 10 kW. A fault in either transmitter automatically changes the position of S3, and the dummy load is connected. If the fault is in transmitter 2, S2 remains in the position shown, now feeding the dummy load, while S1 changes position so that transmitter 1 feeds the antenna directly.

13-12. SPECIAL LOWER-SIDEBAND REQUIREMENT IN COLOR OPERATION

The required lower-sideband characteristic of the transmitter for color signals is shown in Fig. 13-28. The portion of the lower sideband resulting from the color subcarrier must be attenuated by at least 42 dB. When high-level video modulation is employed, a complete vestigial-sideband filter

---

[1]See Harold E. Ennes, *Workshop in Solid State* (Indianapolis: Howard W. Sams & Co., Inc., 1970), pp. 41-44.

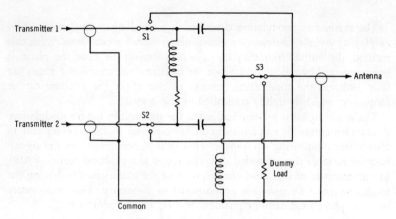

Fig. 13-27. Basic rf combining network.

(Fig. 1-19, Chapter 1) is used, and filtering of the lower sideband—including adequate attenuation of the portion corresponding to the color subcarrier—may be achieved in one unit.

When low-level video modulation is used, a large amount of the lower sideband attenuation is accomplished in the tuning of the transmitter circuits, rather than by the use of a complete vestigial-sideband filter. In this case, a separate color-subcarrier trap on the output transmission line normally is employed.

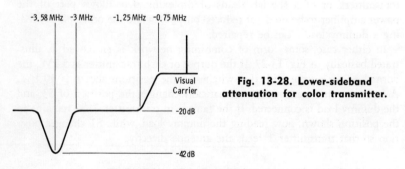

Fig. 13-28. Lower-sideband attenuation for color transmitter.

EXERCISES

Q13-1. If a station is authorized to radiate 100 kw of power, what is this rating in dBk?

Q13-2. What number of watts corresponds to (A) −10 dBk, and (B) 30 dBk?

Q13-3. If you measure a field strength of 100 μv/meter, what is this in dBu?

Q13-4. Would a field strength of 100 μv/meter give an adequate received picture?

Q13-5. What is the ratio of rms aural power to peak visual power?

Q13-6. Define erp.
Q13-7. Define effective antenna height.
Q13-8. How may directivity in the horizontal plane be obtained from a television antenna?
Q13-9. What is the maximum white level of the visual carrier wave?
Q13-10. What is the required lower-sideband attenuation at the color-subcarrier frequency?

CHAPTER 14

Operations at the Transmitter

The transmitter operator is concerned basically with the following duties:
(1) keeping circuits in proper adjustment as to tuning, power output,
modulation percentage, carrier frequency, and relative levels of blanking
to sync, (2) monitoring visual waveform, picture quality, visual modulation percentage, and visual carrier-frequency deviation, (3) monitoring
aural modulation percentage, audio quality, and aural-carrier center frequency, (4) keeping proper logs as required by the FCC, (5) performing preventive maintenance, and (6) performing emergency maintenance
to *keep the station on the air.*

14-1. TRANSMITTER MONITORING

Monitoring duties are outlined as follows:

1. Visual-transmitter amplitude-frequency characteristic (generally
   monitored, by use of a device such as a *sideband analyzer,* during
   adjustment procedures before sign-on or after sign-off).
2. Visual-modulation depth measurement in percent. Continuously
   monitored during the operating day.
3. Picture quality as observed on picture monitor.
4. Visual-carrier frequency deviation. Monitored continuously during
   the operating day.
5. Visual-transmitter power output. Monitored continuously.
6. Aural-carrier center-frequency deviation. Monitored continuously
   during the operating day.
7. Aural modulation percentage. Monitored continuously during any
   aural transmission.
8. Aural program-signal quality as monitored on a good-quality speaker.
9. Aural-transmitter power output. Monitored continuously.

A functional diagram of the RCA sideband response analyzer is shown
in Fig. 14-1. The wobbulator section consists of the conventional arrange-

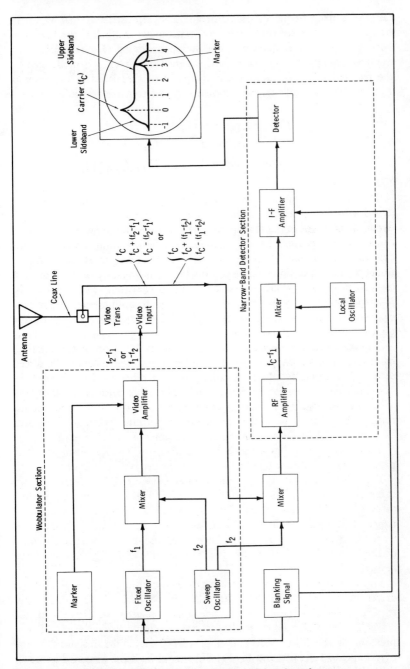

Fig. 14-1. Block diagram of RCA sideband analyzer.

ment of a fixed-frequency oscillator and a sweep oscillator that varies in frequency by approximately equal amounts above and below the fixed frequency. When the frequency of the sweep oscillator (f_2) is higher than the fixed frequency (f_1), the radio-frequency carrier (f_C) is modulated by the frequency $f_2 - f_1$. During this interval, the signal fed to the antenna (and to the sideband analyzer) contains three frequencies, as follows:

$$\text{Carrier frequency} = f_C$$
$$\text{Upper sideband} = f_C + (f_2 - f_1)$$
$$\text{Lower sideband} = f_C - (f_2 - f_1)$$

The transmitter-output signal is fed to the mixer stage, where it is heterodyned with the signal (f_2) from the sweep generator. The resulting heterodyned signal is fed to the rf amplifier, which is the first stage of the narrowband detector portion of the analyzer. As shown in the diagram, this detector accepts only the frequency $f_C - f_1$. The output is then proportional to the upper-sideband response when f_2 is greater than f_1. Similarly, the output is proportional to the lower sideband when f_2 is lower than f_1. If a sufficiently high sweep rate is used, the frequency-versus-amplitude characteristic of the transmitter is displayed on the CRO as shown for the typical curve in the illustration.

The net effect may be seen to be separation of the upper and lower sideband responses for the purpose of simultaneous presentation on the screen of an oscilloscope. In this application, the method is used to observe the vestigial-sideband response of the transmitter. This permits checking and adjusting the broadband overcoupled rf circuits used in most visual transmitters. The display with markers obviously permits optimum adjustment of stages to obtain the proper standard transmission characteristics.

This technique is known as *broad-banding* the visual transmitter. The proper bandwidth must be obtained at the proper peak power output.

Fig. 14-2 illustrates typical traces obtained with the analyzer for two common misadjustments of transmitter tuning. Fig. 14-2A shows the effect of cathode-lead resonance, and Fig. 14-2B shows the trace after this resonance is damped out. In Fig. 14-2C, improper neutralization is indicated by the inequality of the upper and lower sideband in the immediate vicinity of the carrier. Fig. 14-2D shows the trace obtained upon correction of the neutralization.

Fig. 14-3 illustrates how the trace proves useful in proper tuning of the driver stage for vestigial-sideband transmission. Fig. 14-3A indicates incorrect driver alignment (double-sideband response). In Fig. 14-3B, the driver has been aligned correctly for carrier offset to achieve partial suppression of the lower sideband. Fig. 14-3C shows the response of the overcoupled rf circuits, and Fig. 14-3D shows the curve obtained ahead of the sideband filter in most modern transmitters. Fig. 14-4 is a photograph

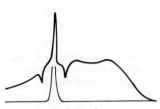

(A) Cathode-lead resonance.

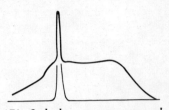

(B) Cathode resonance removed.

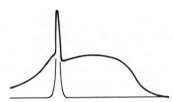

(C) Improper neutralization.

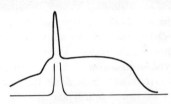

(D) Neutralization corrected.

Fig. 14-2. Transmitter adjustments by means of sideband analyzer.

of a typical trace monitored following the vestigial-sideband filter in a properly adjusted transmitter.

When a sideband-response analyzer is available, the video-sweep observation normally is made daily before the start of regular program operation. During the regular operating day, the video-modulation indicator is an oscilloscope used in conjunction with a keyer and a linear detector. (Review Section 13-9, Chapter 13.)

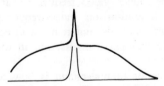

(A) Incorrect driver alignment.

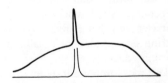

(B) Driver correctly aligned.

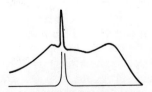

(C) Overcoupled rf circuits.

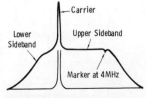

(D) Typical transmitter curve.

Fig. 14-3. Transmitter tuning with sideband analyzer.

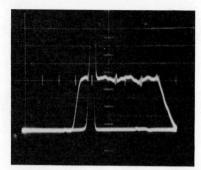

**Fig. 14-4. Typical sideband response
after adjustment within
FCC specifications.**

In Fig. 14-5 is a representation of a CRO display set to indicate depth of carrier modulation with the chopper establishing the zero-carrier line. Note that the actual video-to-sync ratio transmitted is identical to that received from the studio. The 25-percent sync refers *only* to 25 percent of the carrier, *not* to the percentage of the transmitted composite signal. (Review Section 3-6, Chapter 3.)

In the process of amplitude modulation by the grid-bias method, a small amount of incidental phase modulation may be introduced. Extra precautions are taken in the initial design of transmitters to minimize this effect, and the operator also must maintain precise adjustment so that the allowable phase modulation in the picture signal is not exceeded. The net effect of incidental phase modulation is greater in vestigial-sideband transmission than would be the case in double-sideband transmission. This is so because the extra sidebands that are produced cancel out in detection of a double-sideband a-m signal, whereas they add directly to the desired sideband in detection of a single-sideband signal. Observation of the vestigial-sideband characteristic reveals that lower video frequencies up to 0.75 MHz are actually transmitted double-sideband, whereas higher video frequencies are transmitted single-sideband. This is a limitation on video-frequency characteristics that is fixed by transmission standards, and it has a direct bearing on the amount of allowable incidental phase modulation in the transmitted signal.

The most important point for the operator to understand is how to determine in a practical manner the allowable amount of incidental phase

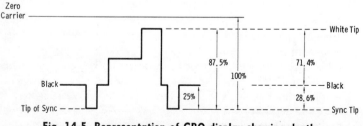

**Fig. 14-5. Representation of CRO display showing depth
of carrier modulation.**

modulation. An intercarrier-type receiver in good working order provides a most reliable basis for judgment. In this type of receiver, an intermediate frequency of 4.5 MHz is obtained as a beat between the visual and aural carriers. If the visual signal contains incidental phase modulation, buzz and noise will result from picture modulation in the sound portion of the receiver. This test must assume that other factors that would also produce noise in an intercarrier receiver, such as overmodulation in the white direction, are not present. In general, it may be assumed that picture distortion caused by phase modulation is negligible if sound distortion from this cause is negligible in an intercarrier-type receiver.

Phase modulation of the visual signal might be caused by any condition that results in an undue amount of rf feedback. This would occur in stages that are improperly neutralized.

Video currents in the grid, screen, and plate circuits of amplifiers produce corresponding voltage drops across the internal impedances of the associated power supplies. These drops obviously affect the dc potentials applied to the electrodes, and will result in picture distortion. Thus, power supplies for video circuits are designed with extremely low internal impedances, and many tubes are paralleled in regulator circuits not only to take advantage of their current-handling capabilities but also to decrease the impedance to an absolute minimum. Also, the screen voltages of rf stages usually are regulated by electronic means.

14-2. GENERAL TUNING PROCEDURES

The tuning of television transmitters is not unduly complex, but the procedures are necessarily more involved and interdependent than is the case for the conventional radio transmitter. It is necessary for the radio-station operator to gear his thinking to the circuit operation required by the nature of the standard television signal.

Fig. 13-22C, reproduced here as Fig. 14-6, should be observed during the following discussion of the basic adjustment of a grid-modulated rf stage. This stage operates essentially as a Class-B amplifier with fixed bias such that the tubes are operated near plate-current cutoff. The general procedure is as follows:

1. The grid bias of the modulated rf amplifier is adjusted without rf drive or video signal to a point such that only a small plate current is present; in other words, the bias is adjusted to very near the cutoff value. This fixes the lowest point of operation along the most linear portion of the transfer curve, shown as minimum plate current (minimum I_p) in Fig. 14-6. This is the quiescent, or static, bias of the stage and depends on the tube and circuit conditions of the particular transmitter. Normal plate voltage and loading must be used on the stage during this adjustment.

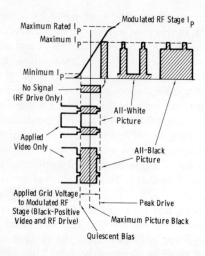

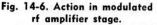

**Fig. 14-6. Action in modulated
rf amplifier stage.**

2. The rf drive (no video) is increased sufficiently to drive the plate current to the upper knee of the transfer curve. Note that the indicated minimum and maximum values are the operating values of plate current. In practice, the maximum operating value is approximately one-half the maximum *rated* plate current of the particular tubes used in the modulated rf stage. In general, therefore, the rf drive is increased to a value about one-half that required to drive the tubes to the maximum rated plate current.

3. A maximum-white video signal of approximately 30 percent sync and 70 percent video is applied to the modulators. The video gain is advanced until the modulation envelope shows the white level at between 10 and 15 percent of maximum carrier. The monitoring device may be either a diode pickup and a scope (with chopper reference line) or a special rf waveform analyzer equipped with a calibrated screen. Note from Fig. 14-6 that application of a white signal produces negative modulation, since maximum carrier amplitude occurs under no-signal conditions. This is to say that with proper rf drive applied and *no* video signal, the peak output level of the transmitter prevails. Also notice from Fig. 14-6 that when video is applied, the adjustment is such that the sync peaks of the applied video signal fall at the quiescent bias level (minimum operating plate current) of the tube. Now considering sync tips only, the carrier amplitude is the same as with rf drive only, or maximum value. At the blanking (pedestal) level, the bias is increased, and the plate current decreases a corresponding amount. The large negative swing of the picture voltage then increases the bias still further, and the carrier amplitude accordingly decreases to the minimum value. The same reasoning is applied to the all-black signal (sync and blanking levels only), and

the plate current (hence carrier amplitude) is reduced to the pedestal level over the 92 percent of the line interval between sync pulses. The sync tips represent 100-percent modulation, the blanking level 75-percent modulation, and the maximum white level 10 to 15-percent modulation (12.5 percent nominal).

The preceding discussion has considered adjustment of the modulated stage, and it has been assumed that the exciter supplying the drive to this stage has been properly tuned by conventional methods. The remaining tuning procedures concern the output circuit of the modulated stage and any following linear rf amplifiers (where this method is used). If the preceding modulation adjustment is carried out before adjustment of subsequent linear amplifiers, the monitor pickup must be from the modulated rf stage.

The TV operator is concerned with stages in which both the plate and grid circuit are tuned (double-tuned circuits in an overcoupled condition) to achieve adequate power output with satisfactory bandwidth. Since the tuning procedures for such circuits are unconventional, a brief review of fundamental theory is in order.

Fig. 14-7A illustrates a double-tuned circuit, and Fig. 14-7B shows response curves corresponding to three different conditions. The *shape factor* of the response curves is given by the relationship:

$$S = k\sqrt{Q_1 Q_2}$$

where,

S is the shape factor,
k is the coefficient of coupling,
Q_1 is the effective loaded Q of the primary,
Q_2 is the effective loaded Q of the secondary.

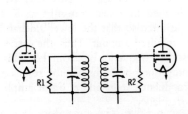

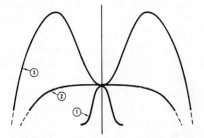

Curve 1: S < 1
Curve 2 S > 1   (Low-Q Circuits)
Curve 3: S > 1   (High-Q Circuits)

(A) Circuit diagram.   *(B) Response curves.*

Fig. 14-7. Double-tuned circuit.

When the coefficient of coupling is small, the secondary response is small for a constant-current ac in the primary, and it has the typical shape of a single-peak resonance curve. As the coupling is increased, the secondary response rises in amplitude and is broadened. If this process is continued until the resistance that the secondary couples back into the primary is just equal to the primary resistance at resonance, the point of *critical coupling* is reached. At this point, the single-hump secondary response curve attains its maximum possible amplitude. The shape factor of this curve is still less than 1, even with high-Q circuits, as indicated for curve 1 in Fig. 14-7B. (The primary or secondary Q is the ratio of the energy stored in the given circuit to the energy dissipated per cycle [X/R].) The transmitter operator tunes the primary and secondary for maximum secondary response, indicating resonance and optimum loading simultaneously.

From this point on, conditions differ from conventional a-m circuit action. When the coupling is tightened beyond the critical value, the secondary response begins to show double humps. When this occurs, the shape factor becomes greater than 1, even with low-Q circuits (curve 2, Fig. 14-7B). For a given circuit, the peaks of the humps become greater in amplitude and farther apart as the coupling is increased. Thus, the peaks may become quite pronounced with a decided valley between them, as shown by curve 3 in Fig. 14-7B. This is typical of a tightly overcoupled circuit with high circuit Q and a resulting shape factor much greater than 1.

In the visual transmitter, a curve similar to curve 2 in Fig. 14-7B is desirable. The shape factor depends on the coupling and the circuit Q. The Q in itself depends on the loading of the circuit. The operator has no control over the "designed Q." Therefore, he has only two variables, coupling and loading.

In practice, circuit constants have been designed so that the shape of the response curve will be correct when the circuit is adjusted for optimum bandpass characteristics. For a given value of coupling in a given over-coupled circuit, increasing the secondary load (by *decreasing* the effective value of R2 in Fig. 14-7A) decreases the amplitude of the humps and, at the same time, provides a more flat-topped response, as shown by curve 2 in Fig. 14-7B.

Both the primary and secondary of the overcoupled circuit must be tuned *on resonance*. This resonant frequency need not be the actual carrier frequency; indeed, this practice seldom is used in visual transmitters. The reason is that the tuned rf circuits are adjusted so that the lower sideband of the visual signal is attenuated by the required amount. Thus, the carrier frequency actually is lower than the resonant frequency by about 1.5 MHz, as illustrated in Fig. 14-8. This is accomplished by adjusting the resonant frequency of the tuned circuits in the modulated stage and rf linear amplifier to a value higher than the carrier frequency.

Since both the primary and secondary of the double-tuned, overcoupled circuit must be tuned to resonance, the operator cannot follow the con-

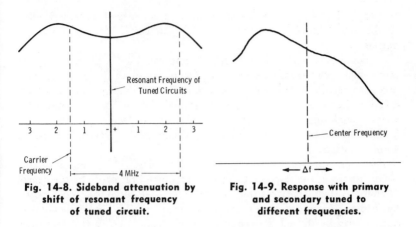

Fig. 14-8. Sideband attenuation by shift of resonant frequency of tuned circuit.

Fig. 14-9. Response with primary and secondary tuned to different frequencies.

ventional practice of tuning for maximum power output as in a-m circuits or broad-band single-tuned circuits. What actually happens when this is attempted is that the primary and secondary are tuned to different frequencies in order to find a load impedance favorable to maximum power output. The result is shown in Fig. 14-9. When the primary and secondary are tuned to the same frequency, resulting in a symmetrical bandpass characteristic, the input impedance at the center of the band (which determines the maximum power that can be developed by the tube) is at a minimum value.

If he is tuning strictly by the meter method without the aid of an oscilloscope, the operator performs adjustments with the above characteristic in mind; he tries to obtain minimum load impedance for a given value of coupling. When minimum load impedance on the driver is obtained, a minimum peak in grid current of the driven stage will occur upon "rocking" the primary capacitor back and forth through resonance. For initial adjustment, the plate voltage on the driven stage is lowered, and the primary tuning adjustment is rocked through resonance as indicated by the grid-current meter in the driven stage. If this stage uses a tetrode tube, a more accurate indication may be observed by watching the screen-current meter for the peak screen current. When primary resonance is found by this procedure, the secondary is adjusted so that a minimum peak in grid or screen current occurs as the primary is varied back and forth through resonance. This ensures that both the primary and secondary are tuned to the same frequency.

In stages using variable coupling, it may occur that the driver load impedance obtained from the foregoing procedure is too low. This will be revealed by excessive plate current compared to the effective power output, indicating high internal anode power dissipation. Under these conditions, the bandwidth usually is greater than required. The situation is remedied by using reduced coupling and repeating the above procedure. This increases the load impedance, which results in a decrease of the tube-loading condition and hence in a reduction of the plate current for a given power output.

It should be noted here that, under some conditions, increased driving power may result from *reducing* rather than increasing the coupling, as is necessary in conventional a-m transmitters. This is a characteristic of double-tuned, overcoupled rf transformers. It should also be borne in mind that when a single-tuned broad-band circuit is used between stages, the circuit is tuned in the conventional way for maximum grid current in the driven amplifier stage.

For a properly tuned overcoupled circuit, bandwidth (separation between humps in the response curve) is determined primarily by the degree of coupling. Flatness across the top of the response curve is affected mostly by loading. After the circuit has been tuned properly, the resonant frequency may be shifted the required amount above the carrier frequency by using a sweep generator, markers, and a scope.

Any type of transmission line may be used as a resonant circuit, or to present inductive or capacitive reactance (Fig. 14-10). Fig. 14-11 illustrates the use of transmission lines as tuned circuits, which is common in uhf transmitters. The lines normally are ¾ wavelength long, depending on the frequency of operation.

Coupling adjustments on ordinary link-coupled circuits are obvious; moving the links farther apart decreases the coupling, and vice versa. Adjustment of circuits using resonant lines is not as obvious, however. In general, moving the connection on the resonant line toward the open end results in increased loading. For example, consider the common case of a driver stage that is coupled to the following grounded-grid stage by tapping onto the cathode resonant line. The driver stage would be

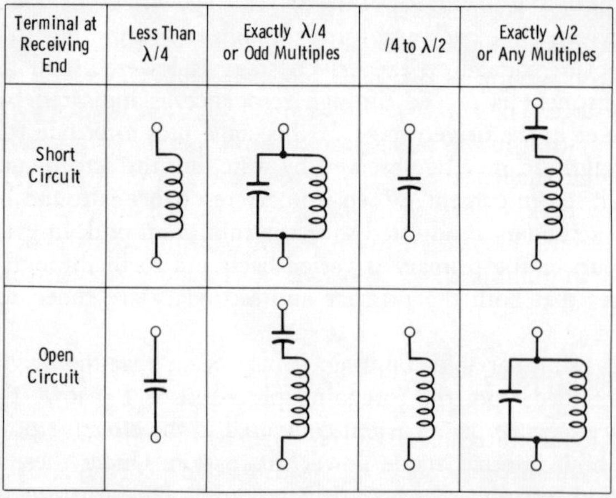

Terminal at Receiving End	Less Than $\lambda/4$	Exactly $\lambda/4$ or Odd Multiples	$/4$ to $\lambda/2$	Exactly $\lambda/2$ or Any Multiples
Short Circuit				
Open Circuit				

Fig. 14-10. Transmission lines as tuned circuits or reactive elements.

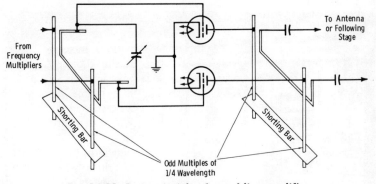

Fig. 14-11. Basic principle of tuned-line amplifier.

loaded more heavily by adjusting the point of connection toward the open end of the line, and the loading would be *decreased* by moving this connection toward the cathode terminal of the driven amplifier. In the instance of loading a final stage to the transmission line, the final amplifier is loaded more heavily by moving the transmission-line connections toward the open end of the final-plate resonant-line output circuit.

The visual and aural power amplifiers of uhf transmitters such as that pictured in Fig. 14-12A are most generally multicavity klystrons (Fig. 14-12B). The resonant circuits are integral parts of each tube and are adjustable to permit tuning to the desired channel and bandwidth. Cooling is accomplished by either water or vapor for the collector, and forced air for the gun assembly. (The klystron was described basically in Chapter 12).

Klystrons have gains in the region of 40 dB. Thus, very low driving power is required. For example, a 10-watt exciter can be used for a power amplifier that has an output in the range of 10 to 50 kW.

The klystron serving as the visual power amplifier operates as a broadband linear stage. The aural power-amplifier klystron normally is the same type as the klystron in the visual amplifier, but is tuned for a narrow bandwidth and high efficiency. Since the aural power is only about 20 percent (rms) of the peak visual power, the beam power is reduced by reducing the beam current. The instructions of the transmitter manufacturer must be followed in tuning these stages for satisfactory performance of the klystron.

14-3. TRANSMITTER POWER OUTPUT

The visual transmitter is never required to develop an average power output greater than the average of the combined pedestal and sync levels. The rating of a visual transmitter is given in peak power capability, and this peak output power must be determined under conditions of maximum

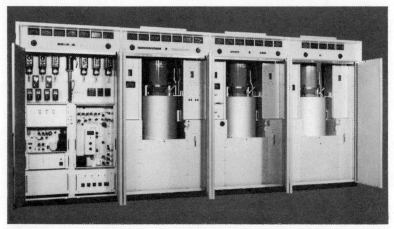

(A) Front view of transmitter cabinets.

(B) Klystron in carriage for handling.

Fig. 14-12. RCA Model TTU-110A uhf TV transmitter.

average power. Thus, in practice, the average power of a standard black signal is found, and the peak value is computed from this measurement.

A dummy load and rf wattmeter consist of a resistor element for terminating the transmission line in its characteristic impedance, and a meter for determining the amount of power dissipated. The power-dissipating section (dummy load) consists of a resistor unit immersed in a liquid that is cooled by air in low-power units, by tap water in medium-power units, and by forced water in high-power units. In order to prevent excessive use of tap water during the time the rf power is at a low level, a water saver is used in most instances. This consists of a thermostatically controlled solenoid valve that allows the water to flow only when needed. More recent dummy loads have sealed-in coolant to eliminate the necessity for external water systems.

The power-measuring section consists of a short length of transmission line, a meter, and a wattmeter element. A socket is provided on the side of the transmission-line coupling section to accommodate a calibrated wattmeter element that, when coupled to the transmission line, develops a dc current approximately proportional to the forward-wave voltage across the load resistor. This current is applied to a remote meter that is calibrated to indicate directly the power dissipated in the load.

The wattmeter element is a reflectometer that consists of a coupling loop, a crystal detector, and a filter network. The wattmeter element may be rotated 180° in the transmission-line housing. This permits it to indicate the incident power to the load or the reflected power from the load.

The transmitter is operated into the dummy load for about fifteen minutes to obtain equilibrium of temperature. The wattmeter indicates the *average* power output with the transmitter modulated by a standard black signal. This means that the modulation consists only of pedestal and sync voltages, with the pedestal level carefully adjusted to 75 percent of the peak amplitude. To obtain the *peak* power output, the measured average power level is multiplied by the factor 1.68. This measurement is made at the output of the vestigial-sideband filter, when used. For transmitters in which low-level modulation is used, the measurement is made at the output of the final linear amplifier.

The reader should understand how the multiplying factor, 1.68, is obtained. A standard black signal consists of sync pulses and a blanking (pedestal) voltage that is 75 percent of the peak sync voltage. In this standard signal, the sync pulse occupies 8 percent of the line interval, or 0.08H. The pedestal level then occupies the remaining 92 percent of the line interval, or 0.92H. Th entire line interval (100 percent, or 1H) is from the leading edge of one horizontal-sync pulse to the leading edge of the next horizontal-sync pulse. These relationships are shown in Fig. 14-13.

We can find the ratio of peak to average power if we realize that the total energy dissipated during the horizontal-sync pulse is proportional to the duration of the pulse times the *peak* power, and the total energy dissi-

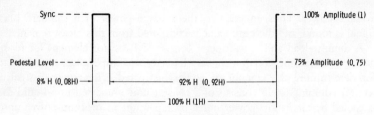

Fig. 14-13. One line of black signal.

pated during the entire line interval is proportional to the duration of one line times the *average* power. Since the sync pulse is at the peak level, for convenience the sync *voltage* level may be called unity, or 1. Then the blanking (pedestal) *voltage* level is 0.75. Since power is proportional to the square of voltage or current, the corresponding relative power levels are $(1)^2$, or 1, for sync level and $(0.75)^2$, or 0.56, for blanking level. The duration of the sync pulse is 0.08H, and the duration of the blanking level is 0.92H; therefore, the relative energy dissipated during the sync pulse is 0.08×1, or 0.08, and the relative energy dissipated during the blanking-level interval is 0.92×0.56, or 0.515. The total energy is $0.08 + 0.515$, or 0.595. The average power is proportional to the total energy divided by the total time, or 0.595/H. The peak power is proportional to the energy during the sync pulse divided by the duration of the sync pulse, or 0.08/0.08H. Therefore, the ratio of peak power to average power is:

$$\frac{\dfrac{0.08}{0.08H}}{\dfrac{0.595}{H}} = \frac{1}{0.595} = 1.68$$

Also, it can be shown that the ratio of the rms voltage or current of the carrier wave during the horizontal-sync interval to the rms voltage or current of the carrier wave during the entire line interval is equal to 1.295. The square of this factor is 1.68.

To meet the requirements of the FCC Rules and Regulations, all installations must include some indicating device that shows peak power output of the visual transmitter during operation. This meter is calibrated initially by transmitting a known power as determined, for example, by the method described above. The indicator then must be checked at periodic intervals by repeating the dummy-load power determination and comparing the indicator reading with the computed power output.

The peak-power indicator usually takes the form of a *reflectometer*. This is a combination of a directional coupling device and a peak-reading diode-detector circuit. This indicator provides a constant check on power output as well as showing the condition of the transmission line and antenna system as they affect standing waves on the line.

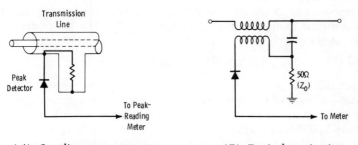

(A) Coupling arrangement.   *(B) Equivalent circuit.*

Fig. 14-14. Principle of directional coupler.

A diagram of a directional coupler is shown in Fig. 14-14. The loop may have both magnetic and capacitive coupling to the transmission line. The capacitive coupling is small, with a large reactance at the carrier frequency. Therefore, the current through the resistor is in quadrature ($90°$) with the line current. The loop and resistor voltage drops are in series. For a wave traveling in one direction, the voltage across the transmission line and the current in the line are in phase at a magnitude set by the characteristic impedance of the line. Since both the coupled voltage and the resistor voltage are in quadrature with the line current, the loop voltage is in phase with the resistor voltage, and the sum represents the directional-coupler output voltage.

In the event of a standing wave (wave on transmission line from opposite direction), the loop induced voltage is out of phase with the resistor voltage drop. Now, if the loop is adjusted so that these voltage drops are made equal, the coupler output voltage will be zero. In this way, the directional coupler can distinguish between waves traveling in opposite directions. It can be calibrated to measure power output, or, by comparing voltages of opposite wave direction, it can measure load mismatch or voltage standing-wave ratio.

It is necessary to calibrate the reflectometer and then ensure maintenance of correct calibration in terms of peak power output. Calibration is done with the dummy load and rf wattmeter as described before.

The operating power of the aural transmitter is determined by the *indirect method.* This means that the power output of the final stage is found by use of the following formula:

$$P_o = E_p \times I_p \times F$$

where,

P_o is the power output (in watts) to the transmission line,
E_p is the plate voltage (in volts),
I_p is the plate current (in amperes),
F is the efficiency factor.

For example, assume a plate voltage of 3000 volts, a plate current of 0.5 ampere, and an efficiency factor (as supplied by the manufacturer for the particular transmitter and frequency) of 0.70. The power output is then calculated to be:

$$3000 \times 0.5 \times 0.7 = 1050 \text{ watts}$$

In practice, this power—as well as the peak power of the picture transmitter—is "stepped up" by the power gain of the antenna system. The output power from the antenna is termed the effective radiated power (see Section 13-3, Chapter 13).

A more practical application of the indirect method of measuring power, from the operator's point of view, is determining the plate current required for a given plate voltage and power output. Thus, he is able to determine how much he must "load up" the output stage by adjustment of the coupling to the transmission line. For example, if a power of 1000 watts is to be fed into the transmission line, and the plate voltage is 3000 volts, the required plate current for an efficiency of 0.7 is:

$$I_p = \frac{P_o}{E_p F} = \frac{1000}{3000 \times 0.7} = \frac{1000}{2100} = 0.476 \text{ ampere}$$

All TV aural transmitters are frequency modulated, with 100-percent modulation being defined as a swing of ±25 kHz. Standard 75-microsecond pre-emphasis is employed in the audio circuits ahead of the modulator, as in regular fm broadcasting. The term *frequency swing* refers to the frequency variation back and forth through the center frequency (assigned frequency of the aural carrier) as a result of modulation. The term *frequency deviation* refers to the drift of the center frequency from the assigned value. (The FCC requires that the aural center frequency be 4.5 MHz above the actual visual carrier frequency, within ±1 kHz. The visual carrier frequency must be maintained within ±1 kHz of its assigned value.)

14-4. TYPICAL TRANSMITTER OPERATIONS

Transmitter operations may be divided roughly as follows: pre-sign-on; regular program transmission; shut-down period after sign-off; and preventive maintenance, which also takes place after sign-off.

The sign-on man generally arrives at the transmitter an hour or so before the first test-pattern signal is to be put on the air. In systems using water-cooled tubes in the high-level stages, the water pumps usually are started before any other operation takes place. In air-cooled systems, the blowers ordinarily are actuated upon application of filament voltages. Any adjustable autotransformer must be set on the tap that gives the proper primary voltage for the particular installation.

Modern transmitter circuits employ an orderly control-circuit system for two purposes: (1) to prevent improper transmitter functioning (overloaded circuits or inadequate time delays for application of high voltages), and (2) to protect the operating personnel from contacting high-voltage terminals. The first function also serves to prevent application of potentials to certain elements if they are not receiving a normal flow of cooling medium, such as water or forced air. The operator must be thoroughly familiar with the system operation and sequence of control at any particular installation.

A typical sequence of operation is as follows. When the START button is pressed, this action applies filament voltages to the tubes and initiates action in a number of other relay circuits. Blower motors are started, and, until the air stream is of sufficient strength to actuate mercury switches on the air vanes, full filament voltage is not applied. At the same instant, time-delay relay motors are started. These do not close the high-voltage circuits until a specified time, such as 30 seconds or one minute, has elapsed. Some transmitters employ switches that automatically apply high voltages upon timing out of the delay relay. When the transmitter is first placed on the air, however, these switches normally are set to the manual position so that other adjustments may be made or checked before the high voltages are applied. Filament voltages should be checked (as indicated by their respective meters), and any adjustments necessary to bring the potentials to the normal value should be made. Door-interlock switches prevent application of high voltage if any door is open in a cubicle containing high voltage. An open door also usually actuates grounding switches that short the high-voltage supplies to ground so that large capacitors cannot discharge to ground through an operator's body should he accidentally come in contact with a high-voltage conductor.

After all filament meters have been checked and normal operation is obtained, the low voltages usually are applied to the video amplifier stages and rf exciter stages. This permits checking the operation of these stages before application of high voltages to the final stage or series of high-level linear amplifiers. Any necessary adjustments such as touch-up of tuning controls are made, and the grid current of any high-level stage is observed to ascertain that normal driving power is available before application of high voltage to the high-level stage.

Fig. 14-15 illustrates the TV-fm transmitter-control console at WBAL, Baltimore. In this particular installation, the transmitter room contains a 50-kW TV transmitter, a 2-kW auxiliary TV transmitter, and a 5-kW fm transmitter.

One of the reasons for locating the studios at the transmitter site of this station was to make it possible to combine some of the functions of master control and transmitter control. It was desired that the transmitter operator become part of a team with the master-control man. To this end, a custom transmitter-control center was designed. The unit is located just outside the transmitter room in the quiet atmosphere of master control. The transmitter operator can monitor audio most efficiently, and still fulfill require-

Courtesy WBAL

Fig. 14-15. Transmitter control panels at WBAL.

ments for meter reading at proper intervals by making short trips into the transmitter room.

The top panels enclose the meters of the reflectometers and the various frequency and modulation indicators for the main and auxiliary sound and picture transmitters. Gain controls for two stabilizing amplifiers and a switch to bypass all of the terminal equipment also are incorporated here, along with an overmodulation indicator. On the extreme right side is a meter panel for the fm transmitter. Immediately below the fm-transmitter meter panel are the control panels for all three transmitters. The operator can start and stop the transmitters, reset the overload devices, and change the main and auxiliary TV transmitters from this point. A small audio control panel for transmitter monitoring and for program-level control is provided.

To the left of the transmitter-control panel is an emergency switcher. In the center of this switcher are three buttons. Normally, the button on the right is depressed. This means that the emergency switcher is bypassed completely, and everything is operating normally. If an audio failure occurs, the button on the left is punched first. By this action, audio from any of the sources with an audio counterpart is placed directly on the air by punching the proper source button. In the event there is only a video failure, the video button is punched. Thereafter, any time a source button is punched, this video signal is put directly on the air.

Finally, below the emergency switcher is a video-control panel. Here, various portions of both transmitters are monitored. The panel also provides control over each of four operating centers. Thus, much of the master control panel can be operated from this position should the need for such an arrangement arise.

High-Voltage Rectifier Housing   Dummy Load   Portion of Vestigial-Sideband Filter

Courtesy WISH-TV

Fig. 14-16. Dummy load and associated station equipment.

When the operator is assured that the transmitter is functioning properly into the antenna system, he generally removes the high voltage and connects the dummy load instead of the antenna to the final stage (Fig. 14-16). He is then in a position to check performance with test signals either from equipment at the transmitter or from the studio line. These tests include test-pattern signals, checking of waveform as received from the studio and after passing through the transmitter, adjustment of video and audio levels, etc.

Because of the nature of the fm aural transmission, some operators are at first confused about obtaining proper interpretation of the aural modulation-monitor meter readings. Recall that audio pre-emphasis in accordance with FCC standards is used in the aural transmitter. The monitor circuit that provides audio voltage to drive the sound-monitoring amplifiers then includes a de-emphasis circuit to restore the audio amplitude-frequency curve to its normal shape. It must be remembered, however, that some modulation meters show percent of modulation in terms of this pre-emphasis curve, which means that about 17 dB less audio (pure tone) is required at 15,000 Hz to modulate the transmitter 100 percent than at any frequency below approximately 400 Hz. This action in terms of average program material should be taken into consideration by the operator. A program that has a number of "highs" in the signal content should modulate the transmitter 100 percent. However, films or recordings often contain sound that is definitely lacking in highs, and 100-percent modulation will cause noticeable distortion of "lows," as a result of the overloaded condition in the audio amplifiers at the station and especially in the receiver circuits following de-emphasis. Voice transmission seldom should exceed 30- to 40-percent modulaton. The studio operator may be peaking his meter close to the 100-percent indication, but the transmitter modulation indication cannot be expected to follow the studio meter under all program conditions.

While the transmitter is on the air, a competent operator will be continually alert to picture quality, waveform level, amount of setup and proper sync-to-blanking ratio, meter readings, temperatures, and even his sense of smell. As he observes the meters and makes adjustments of the line-voltage autotransformer, he should be sensitive to the characteristic operating odors of resistors, relay solenoids, capacitors, and transformers. This practice often indicates the general location of impending trouble, even if visual observation is impossible because of the location of the component. (Meeting emergencies is discussed more fully in Section 14-5.) Meter readings must be recorded in the transmitter log at 30-minute intervals (or as required by current FCC Rules).

After sign-off, the high voltage is removed, and the rear doors are opened for visual observation of all components. The operator should be thoroughly familiar with the *feel* of the connections on high-level tubes, important capacitors, and other components for which temperature indication as revealed by this feeling process is important in case of trouble or impending failure. *CAUTION: The operator must ALWAYS ascertain that interlock switches and high-voltage shorting relays are functioning properly before he touches any component.*

After the low voltages and filament potentials are removed, blower motors or water-cooling systems sometimes continue to operate (under control of keep-alive relays) for a specified time (generally 4 to 7 minutes) to cool the high-level tubes. This marks the end of the operating day and the start of preventive-maintenance activities.

14-5. EMERGENCY PROCEDURES

The variety of corrective measures that might be called for in getting a transmitter back on the air or in clearing defective pictures is so great that a natural limitation is immediately placed on the thoroughness of presentation in this section. The treatment is therefore very general in nature.

The most important phase of meeting emergencies is in training the entire staff to be *mentally prepared* for corrective procedures. Because of the complexity and unusual expense, complete standby transmitters for emergency use are a rarity in TV broadcasting, particularly in smaller stations. It is the duty of each individual operator to study thoroughly the technical aspects of his particular installation and to prepare himself to analyze malfunctioning equipment with a certain coolness and deliberateness. Such a psychological preparedness actually minimizes the time necessary to clear transmitter faults. The chief engineer or other supervisory personnel of every station should conduct off-hours classes in which most likely and typical faults are simulated for observation of the effect on meter readings, waveform and picture content, etc. Transmitters that have been on the air for a year or more inevitably have certain peculiarities that are revealed in case histories, and these should be pointed out to all operators.

Within a few seconds after trouble has occurred, the operator should be able to analyze the fault as being in one of the following general areas: control circuits or power supplies, video amplifiers used as incoming-line amplifiers from studio or network, video modulator section, or radio-frequency section. If trouble is in the sound, the operator immediately will identify the possible source as the line amplifier from the studio, the modulator section of the aural transmitter, or the frequency-multiplier and final rf stages.

It is most helpful to observe the respective frequency and modulation monitors upon the first instant of trouble in picture or sound (or both). This is most important for the following reasons:

1. A picture, for example, may disappear from the picture monitor screen, yet still be transmitted over the air. In this instance, the picture monitor itself is obviously at fault. If this should be the case, the monitor showing the modulated rf envelope would be indicating as usual, and a spare picture monitor could be substituted for the defective one.

2. Assume that the picture disappears from the picture monitor. A quick glance at the modulation-monitor CRO shows no modulation taking place. At this time, it is possible to obtain a preliminary idea as to the possible source of trouble by observing the rf-input indication on the frequency monitor. For example, in some monitors, any deviation from the normal rf input is indicated by a lamp. If this lamp indicates a fault in the rf input level, the operator suspects that the trouble is in the rf portion of the transmitter. If the rf indication to the monitor is normal, the operator should suspect a fault in the video section (which includes the stabilizing amplifier on the incoming line and the video-amplifier and modulator stages in the transmitter) or a lack of incoming signal from the studio. These areas may be checked quickly in most installations by means of a switch that places a monitor across the incoming line, the output of the stabilizing amplifier, or the output of the modulator stage in the transmitter. Thus, the signal may be traced in this manner to isolate the faulty stage quickly.

Obviously, if the transmitter goes off the air because of tripping of the overload relays or failure of a power supply, the previous procedure is unnecessary. Visual observation of the transmitter rectifier tubes, overload-relay indicators, and meters is the initial step the operator should take. He then mentally analyzes the evidence and decides what is necessary to place the unit back in operation. Overload relays or thermal switches may have to be reset. If this results in another quick shut-down of the transmitter, the fault must be located and corrected before the high voltage is re-applied and the transmitter returned to the air.

Sometimes, visible or audible arcing occurs to give an indication of the stage being overloaded. If the arcing is not visible, aural perception usually is sufficient to tell the operator which rear doors to open so that he can look for signs such as blackened spots on the frame next to a capacitor or high-voltage terminal. Insulators must be observed for cracks or signs of breakdown, and high-voltage leads to tubes or components should be examined for bad insulation.

If the arcing cannot be located by either of the above methods, it will be necessary to carry out an emergency procedure *that must be exercised with the utmost caution and preferably with another operator standing by*. This procedure consists of opening the rear doors of the suspected unit, strapping closed the interlock circuits for that cubicle, and applying high voltage while the operator watches for the point of arcing. To do this, it is simply necessary to jumper the proper terminals associated with the particular door lock so that voltage may be applied with the door open. *This is an emergency procedure only; it is never done except when absolutely necessary.* The need for caution in this regard cannot be overemphasized; operators have been killed as a result of carelessness when working around high voltage.

The operator should be so familiar with the control-circuit diagram that he can locate terminal numbers with a minimum of delay. For example, it would be entirely possible for the contacts of a time-delay relay to open, either from improper adjustment, dirt or corrosion between contacts, or a faulty relay itself. This would be indicated by a light, usually on the control panel, designated, for example, as READY. The purpose of this light is to show that the time-delay interval after the transmitter is turned on has expired, and the high-voltage circuit is capable of being energized when the operator is ready. If this light should go out, the time-delay relay would be one possible cause, and the operator should be able to locate the proper terminal-board points to jumper for this emergency. This example, of course, is only one of many possibilities.

Overloads that trip the ac overload devices usually are caused by arcing back of the mercury-vapor rectifiers. Many transmitters employ an arc-back indicator on each rectifier tube so that the affected tube may be identified and replaced. If such indicators are not used, it is wise to replace the entire complement of mercury-vapor tubes with rectifiers known to be good, preheated, and air-tested. This occurrence usually can be avoided through proper preventive maintenance.

Some transmitters (even some older models) have built-in emergency provisions for occurrences such as improper clamping of signals in the video modulator section. This is the case in the GE TT-10-A transmitter. Several different types of trouble may cause the modulator to stop functioning in its normal, clamped manner. One instance is an input signal (from the studio) that is defective in certain ways. Another instance is failure of some of the tubes or components in the sync-operating and pulse-forming part of the modulator. As long as the failure does not involve the video-amplifier stages in the transmitter, operation may be resumed under emergency conditions. The emergency procedure is outlined by General Electric as follows:

If the input sync voltage is too low, there will not be enough sync to be separated properly, and keying pulses will be formed in an erratic manner. Note that lack of sufficient input sync voltage results either from too low a total peak-to-peak composite signal input (considerably below one volt) or from too low a sync percentage, even though the peak-to-peak value of the composite signal is one volt or greater. (GE specifications call for at least 10 percent sync at any input voltage over one volt.)

If the incoming signal has back porches narrower than standard, or split pulses resulting in narrow slots going down to black in the sync, normal operation will not be obtained. Notice that certain other types of defective signals, such as those with hum or poor low-frequency response, are improved by the clamp operation of the modulator.

In the GE TT-10-A transmitter, if such a failure occurs, the modulator may be switched to emergency operation. In this change, simple diode dc insertion on sync peaks is substituted for the back-porch-clamp type of normal operation. The switch from normal to emergency operation is accomplished by changing two 6AL5 diodes from their normal to their emergency sockets, as indicated by front-panel markings on the modulator. The switch is left in the clamp position. Readjustment of the RF GAIN, SYNC, and VISUAL PGM controls on the control panel will then be required. In emergency operation, the two stages in which the diodes are used for dc insertion are operated far outside the normal grid-resistance ratings of the tubes. Therefore, it is recommended that the emergency condition be used no longer than is absolutely necessary. It is necessary to repair whatever caused the need for the change and to return to clamp operation as soon as possible.

Troubles in rf stages of transmitters generally may be isolated by observing the meter readings of the individual stages. For example, the first stage (stage nearest the oscillator) showing lack of proper grid current may be receiving insufficient drive from the preceding stage, or there may be a defective tube or component in the observed circuit. Tubes are always the first component to be suspected in a malfunctioning stage.

A word of caution is in order at this point. When a transmitter is placed on the air at the start of operations, and during preliminary overall checks with the sideband analyzer or waveform indicator, an indication of complete detuning might result from a defective sweep generator or indicating device. Should the rf waveform be defective, yet all meters are indicating "on the nose," the operator should check his equipment first before suspecting the transmitter. This is done by checking the output trace of the sweep generator on a scope that is known to be good. If the traces are normal at the terminations and inside the sweep generator, and the rf waveform is defective as displayed on the substitute scope, the trouble may be assumed to be in the rf stages of the transmitter. Most generally, defects in the rf stages result in abnormal meter readings in these stages.

The same technique should be followed if, at any time during the operating day, the frequency meter for either the visual or aural transmitter indicates a frequency outside the authorized tolerance. Remember that the station monitor is a secondary standard, and before suspecting the transmitter, be sure the frequency monitor itself is in proper working order. This may be done by checking with a commercial frequency-measuring service authorized by the FCC. When the station monitor has been calibrated in this way, as it should be whenever its operation is suspected, then the transmitter may be adjusted accordingly. The monitor usually is checked with such a frequency-measuring service once a month as part of routine maintenance procedures.

Tubes in push-pull rf linear amplifier stages should have their plate or cathode currents balanced within 10 percent. When an unbalance greater than this amount is revealed by meter readings, it may be caused either by tubes or by components in the stage. The nature of the cause can be determined quickly by temporarily removing the rf drive. If the unbalance remains, the tubes should be suspected and replaced with a balanced pair. If the currents are balanced after removal of rf drive, the *circuits* should be suspected and examined for the cause. For example, many rf stages use cathode or filament bypass capacitors. Should one of these capacitors be defective, currents would obviously be unbalanced with rf drive. If one is shorted, the currents would be unbalanced with or without rf drive, and the cathode-current meter readings would be abnormally low. If one is open, excessive tilt across the tops of the horizontal sync pulses would be observed at the output, whereas the waveform at the modulator would be normal. Obviously unbalanced currents also could be caused by such things as defective screen or plate bypass capacitors, bad connections, misaligned link couplers, or other circuit defects.

Since any emergency changing of tubes in the rf stages during the operating day will affect to some degree the tuning of the stage, it is wise to try all the spare tubes in these stages during regular preventive-maintenance periods. Dial settings for proper tuning with these tubes should be posted. This procedure saves the alignment time that otherwise would be required.

The transmitter then may be aligned exactly with the proper test equipment after the end of the regular operating day.

EXERCISES

Q14-1. You must have 5 kW of power into the transmission line from the aural transmitter. The efficiency factor is 0.68, and the plate voltage of the final stage is 5000 volts. How much plate current must you "load" into the final power amplifier?

Q14-2. One-hundred-percent modulation of the TV aural transmitter corresponds to what frequency swing?

Q14-3. What is the maximum visual-carrier frequency deviation permitted by the FCC?

Q14-4. What is the maximum aural center-frequency deviation permitted by the FCC?

Q14-5. What is the proper sync-to-video ratio for (A) carrier levels, and (B) demodulated signal?

Q14-6. Give the two basic causes at the TV transmitter for buzz in the sound portion of a TV receiver.

Q14-7. How are the linear rf power amplifiers in a visual transmitter usually adjusted?

Q14-8. What is "broad-banding" the visual transmitter?

Q14-9. During the operating day, what device is used to indicate proper power output of the visual transmitter?

Q14-10. What is the highest video frequency for which both sidebands are transmitted?

Reference Tables

Table A-1. Television Channels

Channel No.	Frequency Limits (MHz)	Freq. of Visual Carrier (MHz)	Freq. of Aural Carrier (MHz)	Channel No.	Frequency Limits (MHz)	Freq. of Visual Carrier (MHz)	Freq. of Aural Carrier (MHz)
2	54-60	55.25	59.75	43	644-650	645.25	649.75
3	60-66	61.25	65.75	44	650-656	651.25	655.75
4	66-72	67.25	71.75	45	656-662	657.25	661.75
5	76-82	77.25	81.75	46	662-668	663.25	667.75
6	82-88	83.25	87.75	47	668-674	669.25	673.75
7	174-180	175.25	179.75	48	674-680	675.25	679.75
8	180-186	181.25	185.75	49	680-686	681.25	685.75
9	186-192	187.25	191.75	50	686-692	687.25	691.75
10	192-198	193.25	197.75	51	692-698	693.25	697.75
11	198-204	199.25	203.75	52	698-704	699.25	703.75
12	204-210	205.25	209.75	53	704-710	705.25	709.75
13	210-216	211.25	215.75	54	710-716	711.25	715.75
14	470-476	471.25	475.75	55	716-722	717.25	721.75
15	476-482	477.25	481.75	56	722-728	723.25	727.75
16	482-488	483.25	487.75	57	728-734	729.25	733.75
17	488-494	489.25	493.75	58	734-740	735.25	739.75
18	494-500	495.25	499.75	59	740-746	741.25	745.75
19	500-506	501.25	505.75	60	746-752	747.25	751.75
20	506-512	507.25	511.75	61	752-758	753.25	757.75
21	512-518	513.25	517.75	62	758-764	759.25	763.75
22	518-524	519.25	523.75	63	764-770	765.25	769.75
23	524-530	525.25	529.75	64	770-776	771.25	775.75
24	530-536	531.25	535.75	65	776-782	777.25	781.75
25	536-542	537.25	541.75	66	782-788	783.25	787.75
26	542-548	543.25	547.75	67	788-794	789.25	793.75
27	548-554	549.25	553.75	68	794-800	795.25	799.75
28	554-560	555.25	559.75	69	800-806	801.25	805.75
29	560-566	561.25	565.75	*70	806-812	807.25	811.75
30	566-572	567.25	571.75	*71	812-818	813.25	817.75
31	572-578	573.25	577.75	*72	818-824	819.25	823.75
32	578-584	579.25	583.75	*73	824-830	825.25	829.75
33	584-590	585.25	589.75	*74	830-836	831.25	835.75
34	590-596	591.25	595.75	*75	836-842	837.25	841.75
35	596-602	597.25	601.75	*76	842-848	843.25	847.75
36	602-608	603.25	607.75	*77	848-854	849.25	853.75
37	608-614	609.25	613.75	*78	854-860	855.25	859.75
38	614-620	615.25	619.75	*79	860-866	861.25	865.75
39	620-626	621.25	625.75	*80	866-872	867.25	871.75
40	626-632	627.25	631.75	*81	872-878	873.25	877.75
41	632-638	633.25	637.75	*82	878-884	879.25	883.75
42	638-644	639.25	643.75	*83	884-890	885.25	889.75

*These frequencies have been reallocated to the land mobile services. However, for an indefinite time, the licenses of television translators operating on these frequencies will be renewed on a secondary basis. That is, translators must accept interference from land-mobile stations but must not cause interference to such stations.

Table A-2. Neutral-Density Filters

Density	% Light Transmission	Increase in Stop
0.10	80	½
0.20	63	¾
0.30	50	1
0.40	40	1¼
0.50	32	1¾
0.60	25	2
0.70	20	2¼
0.80	16	2¾
0.90	13	3
1.00	10	3¼

This table shows the correlation of density, percent of light transmission, and effective f/-stop change for neutral-density filters. When it is necessary to cut down the light without disturbing the lens f/-stop setting, the neutral-density filter that accomplishes the desired result is chosen.

Glossary

A

Aberration, Chromatic: Image defect caused by variations in bending by the lens of light rays of different colors.

Aberration, Spherical: Lens defect resulting in production of a disc of light (rather than a point of light), which has a definite minimum size not reducible by focusing. It *may* be reduced by "stopping down" the lens (using smaller iris opening).

Achromat: Lens corrected for chromatic aberration.

Angular Field: The greatest angle covered horizontally or vertically by a particular lens.

Antenna Field Gain: In radio theory, a "standard doublet" is a doublet antenna that, with one kilowatt of input power, gives an effective free-space field intensity of 137.6 millivolts per meter at a distance of one mile. The field gain of an antenna is a comparison of the field intensity it produces to the field intensity produced by the standard doublet.

Antenna Height Above Average Terrain: The average height of the antenna (center of radiating portion) above the terrain between 2 and 10 miles from the antenna.

Aspect Ratio: Numerical ratio of picture width to picture height. TV standards specify an aspect ratio of 4 to 3, or 4 units wide to 3 units high.

Asymetric-Sideband Transmission: (*See Vestigial-Sideband Transmission.*)

Average Power Output: Average power of the visual transmitter is measured with an rf wattmeter while the final power amplifier is loaded by a dummy antenna. The transmitter is modulated by a standard black signal during the measurement. The peak power of the visual transmitter is found by multiplying the average power by the factor 1.68.

B

Back Porch: That portion of the composite video signal at blanking level and between the trailing edge of the horizontal-sync pulse and the trailing edge of the horizontal-blanking pulse.

Beam Current: Measurement (usually in microamperes) of the electron flow that makes up the scanning beam in a pickup tube, picture tube, etc.

Birdseye: Type of incandescent lighting unit having a self-contained reflector throwing a floodlight.

Black Compression: Amplitude limiting of the signals corresponding to black in the picture, resulting from loss of *setup* in the picture. (*See Setup.*)

Blacker-Than-Black (*Infrablack*): That portion of the standard television signal that is of an amplitude greater than black level (blanking or pedestal level). This region is occupied by the synchronizing pulses.

Black Level: The amplitude of the modulated rf video carrier at which the beam in the picture tube is extinguished so that retrace lines (which are instigated during the blanking interval) are not visible on the picture tube.

Blanking Pulses: Pulses of such amplitude as to cut off the beam in the picture tube. This sets the black level of the signal. (*See Black Level.*)

Breezeway: The time interval between the trailing edge of horizontal sync and the start of the color burst.

Bridging: The act of connecting a high-impedance device in parallel with a given source so as to extract a portion of the signal without perceptable effect on the source circuit.

Burst: A signal consisting of 8 to 10 cycles of the 3.579545-MHz reference used to synchronize the receiver color information with that transmitted.

Burst Flag: A pulse timed from horizontal sync to gate on the color-subcarrier burst at the proper time. It is deleted during the 9-line interval at the start of vertical blanking.

Buttonhook: A waveguide attached to the unit that reflects the microwave signals at the transmitter or receiver.

C

Camera Control Unit: A control panel with internal amplifiers and control circuits that supply driving pulses and other currents to the camera, and receive the video signal from the camera preamplifier in the pickup head.

Cans: Earphones.

Cavity Resonator: A space enclosed by conducting walls of given dimensions within which resonant rf fields may be excited and extracted.

Center Frequency: The frequency of the unmodulated aural carrier wave. It also is defined as the average frequency of the emitted wave when frequency modulated by a pure sine wave.

Clamping Circuit: A circuit that maintains either the blanking level or sync-tip level of the video signal voltage at predetermined dc level. This maintains the dc component in an ac-coupled amplifier, and minimizes ac sine-wave (hum) pickup.

Composite Video Signal: The complete visual waveform composed of picture information, blanking pedestal, sync pulses, and equalizing pulses.

Compound Lens: A lens consisting of two or more elements of glass, sometimes cemented together.

Contrast: Scenic range of dark to light values, in proportion to the signal-voltage variation at the grid of the picture tube.

Cover Shot: A picture that covers the entire scene of action. As used in baseball pickups, coverage of the pitcher, batter, catcher, and umpire.

D

DC Reinsertion: The act of restoring the dc picture component ordinarily lost in ac-coupled systems.

DC Restorer: A circuit providing means of restoring the dc or low-frequency components of a video signal after these components have been lost in ac-coupled systems.

DC Video Component: The portion of the composite video signal resulting from the average steady background illumination of the televised scene.

Definition: The quality of the fine details in the image produced.

Dimmer: A control used to regulate the current supplied to lights.

Dish: A large metal microwave reflector.

Dissolve: A reduction of video gain toward black level.

Dolly (Noun): A mobile camera mount.

Dolly (Verb): "Dolly in": Move toward scene. "Dolly out": Move away from scene.

Double-Tuned Circuit: Two single-tuned circuits inductively coupled.

Driving Pulses: Signals that drive the scanning generators in the cameras at the studio.

E

Effective Radiated Power (ERP): The product of the antenna power (output power from transmitter minus transmission-line losses) times (1) the antenna power gain, or (2) the square of the antenna field gain.

Electron Multiplier: An electron tube or tube section containing a number of electrodes that provide high emission of secondary electrons. These electrodes greatly multiply the initial impinging electron stream.

Encoder: A device that transforms the original R, G, and B video signals into a luminance signal and a chrominance signal for transmission.

Equalizing Pulses: A series of pulses, six before and six after the serrated vertical-sync pulses, at twice the line frequency (31,500 Hz). They are used to insure correct timing of the vertical retrace at the receiver for proper interlace of the scanning lines.

Equivalent Focus: Same as *Focal Length.*

F

Field: One complete scanning of every other picture line, including retrace to the top of the picture.

Field Frequency: The rate at which television fields are transmitted. The FCC standard is 60 fields per second, or 60 Hz. (*Also see Frame.*)

Flare: Distribution of extraneous light.

Flyback: The shorter of the two basic intervals of a sawtooth wave; the return of the electron beam at the end of each scanning line.

f/ Number: A number related to the ratio of the focal length of a lens to the diameter of the beam of light that passes through the lens.

Focal Length (of a lens): The distance from the lens to the focal point behind the lens.

Follow Shot: (*See Travel Shot.*)

Frame: One complete picture. In the standard interlaced method of scanning, one frame consists of two fields, each field containing alternate lines.

Framing: Adjustment of the film projection and pickup system so that the edges of the picture are not visible on the TV screen.

Freeze It: Use the set as is.

Front Porch: That portion of the horizontal blanking interval between the leading edge of the pedestal pulse and the leading edge of the horizontal-sync pulse.

Frost: Frost gelatin put over lamps to diffuse light.

G

Gamma: The slope of the characteristic, plotted on log-log coordinates, expressing the output amplitude as a function of the input amplitude.

Garbage Can: Slang term for a microwave-relay transmitter.

Ghost: A second, undesired image displaced from the desired image on a picture-tube screen.

Gobo: A sudden decrease of light in the picture (opposite of *Womp*).

Gradient: A progressive change in tones or shades along a monochromatic scale.

H

Head Room: An instruction given by the director when he wants more space above the person or object.

High Hat: A high camera position on a table top or other waist-high platform or mounting.

Horizontal Blanking: The act of cutting off the electron beam at the end of each scanned line for the retrace interval.

Horizontal-Blanking Pulse: A pulse of proper amplitude, duration, and timing to produce horizontal blanking.

Horizontal Retrace: (See *Flyback*.)

Horizontal-Sync Pulse: A rectangular-shaped pulse above pedestal level, the purpose of which is to time the horizontal scanning in the receiver with that in the camera at the studio.

I

Image Orthicon: A camera pickup tube employing a low-velocity beam to scan a target. An electrical image appears on the target as a result of optical focusing of an image on a separate photocathode; electrons emitted by the photocathode impinge upon the scanned target. This is the image section of the tube, which exhibits an image multiplier action. Electron multiplication also takes place in the tube by dynode action on the modulated return beam within the tube.

Infra Black: (See *Blacker-Than-Black*.)

Interlaced Scanning: A method of scanning in which one-half the total lines (262.5 lines) are scanned in one field (even-numbered lines), and the other half of the total number of lines are scanned in the next field (odd-numbered lines). Thus, adjacent lines of a complete picture (one frame) belong to successive fields.

J

Jitter: Vertical instability of the picture on a kinescope screen (for example, the effect caused by "fluttering" film in the film gate of a projector).

K

Keying Circuit: A circuit, under control of one signal, used to pass ("key on") or block ("key off") another signal.

Kinescope: The picture tube in a monitor or home receiver.

L

Linearity: The degree of straightness of the rising portion of the horizontal or vertical scanning wave. If this portion of the waveform is a straight, inclined line (indicating the same rate of rise throughout the curve), the picture elements are uniformly distributed over the image area.

Link: A channel or circuit connected in tandem with other channels or circuits. The term also may refer to an rf studio-to-transmitter link employed to relay TV programs without use of a coaxial cable.

Loosen Shot: A command given by the director when he wants more space around the object, person, or scene.

Lose One (or *Two, Three,* etc.): The director's cue to the switcher to remove a superimposition. The number one, two, etc., refers to the camera number.

M

Medium Shot: A picture that gives coverage of the specified person from the waist up. This term also may define a "medium" area of any given scene (coverage between a close-up and a general view).

Mercury-Vapor Light: A type of light in which a tiny pool of mercury is "fired" in a small quartz tube. The tube is surrounded by a larger glass tube (called capillary), through which about a gallon of water flows per minute.

Modulation Capability: In the aural transmitter, the maximum modulation that can be obtained without exceeding a given percentage of distortion. In the visual transmitter, modulation capability is defined in terms of the change in pedestal level when a change is made from an all-black to an all-white signal.

Monoscope: A camera employing a cathode-ray tube with a fixed internal test pattern. It is used for testing and adjusting purposes.

Multiplexer: A device to permit use of one TV film camera with more than one TV film projector.

N

Negative Transmission: A transmission that results from negative modulation of the visual transmitter (an increase in light content of the picture results in a decrease in carrier amplitude).

Nemo: A program picked up away from the studio; a remote pickup.

O

Odd-Line Interlace: The term used to describe the standard method of interlace. Since the total number of lines per frame (525 lines) is an odd number, each field contains a number of complete lines plus a half line (262½ lines).

P

Pairing: Failure of proper interlace. The lines of one field do not fall precisely between the lines of the previous field, resulting in separated picture lines and loss of vertical resolution.

Panning: Slowly sweeping the camera to one side or the other, or from side to side. The term also may include up-and-down movement.

Peaking Coil: An inductor used in a video amplifier to compensate for high-frequency losses.

Peak Power Output: The output power of the visual transmitter at maximum sync level. (*See Average Power Output.*)

Pedestal: The level of the video signal at which blanking of the picture-tube scanning beam occurs.

Percentage Modulation: For the visual transmitter, the modulation factor expressed as a percentage of the peak (sync) value. In the aural transmitter, the percentage is in relation to a ± 25-KHz frequency swing, which is defined as 100-percent modulation.

Pickup Head: The camera lens, pickup tube, associated preamplifiers, sweep circuits, etc., contained in the camera (complete camera *without* viewfinder).

Pickup Tube: A camera tube, such as an image orthicon, Vidicon, or *Plumbicon.*

Picture Element: The smallest part of a picture that can be defined by the scanning system. Its size cannot exceed the nominal line width.

Pix: Slang term for picture.

Polarity of Video Signal: The direction of a voltage change that represents a change to black, relative to the direction of a voltage change that represents a change to white. Polarity is given in terms of black, such as "black negative" or "black positive."

Q

Quasi-Single-Sideband Transmission: (*See Vestigial-Sideband Transmission.*)

R

Raster: The area illuminated by the modulated or unmodulated scanning beam in the picture tube.

Reference Black Level: That level corresponding to a specified maximum signal excursion that represents black. With proper *setup,* the maximum black signal level is held about 10 percent under the pedestal (blanking) level.

Reference White Level: That level corresponding to a specified maximum signal excursion that represents white. In practice, this level is between 10 and 15 percent of the sync level.

Resolution: A measurement of the smallest element that can be distinguished. (*See Picture Element* and *Resolving Power.*)

Resolving Power: The ability of the lens or the TV system to reproduce closely spaced lines without overlap or blending.

Return Trace (Retrace): The return of the electron beam after horizontal or vertical scan to begin a new line or field.

S

Sawtooth Wave: A waveform that rises linearly with time between two amplitude values, and has a quicker linear return to the minimum value.

Scanning: The process of analyzing, according to a predetermined sequence, the light values of picture elements constituting the total picture area.

Scanning Generator: The electronic circuit that generates the sawtooth wave used for scanning.

Scanning Line: A single, continuous, narrow strip across the picture area; it results from scanning.

Scanning Spot: The cross section of the electron beam used to scan the picture area. Also termed *scanning aperture.*

Series Peaking: A method of using a series coil for high-frequency amplifier compensation. (*See Peaking Coil.*)

Setup: The difference between the signal amplitude for maximum picture black and the pedestal (blanking) level. This value should be adjusted so that maximum picture black is about 10 percent lower in amplitude than the pedestal.

Shunt Peaking: A method of using a shunt coil for high-frequency amplifier compensation. (*See Peaking Coil.*)

Single Shot: A term meaning the director intends for the camera shot to include only one person.

Sky Hook: The waveguide "hook" on microwave transmitter and receiver parabolic reflectors. In general, the term refers to the entire assembly of the parabolic reflector and the exciter waveguide.

Snow: Flecks on the monitor or receiver kinescope caused by a weak signal, noisy amplifier, etc. The presence of snow indicates a high noise-signal ratio.

Spurious Signal: Any undesired signal present with the picture or sound.

Stabilizing Amplifier: A video amplifier containing corrective circuits to compensate for hum or stray noise pickup, poor sync, or generally faulty signals. Usually, it is used on incoming network or remote signals, at the studio output, and at the transmitter input.

Subcarrier Standard: A device that generates the color-subcarrier frequency of 3.579545 MHz and counts down to control the local sync generator.

Superimposition (Super): The mixing of two separate scenes so that one appears to be placed over the other on the picture-tube screen.

Sweep: The action by which the electron beam moves back and forth and up and down over the area to be scanned.

Sync (Synchronization): The maintenance of one operation in step with another. In the TV system, the sync generator supplies horizontal- and vertical-sync pulses that cause the picture-tube scanning beam to start its sweep, be blanked, and start its retrace at the same time that these events occur in the camera.

Sync Clipper: An electronic circuit that removes all or a portion of the sync signals from a composite signal.

Sync Compression: Reduction of the sync amplitude in ratio to the amplitude of the composite signal.

Sync Generator: The unit that generates properly timed control pulses and shapes them into the standard sync waveform.

Sync Level: The maximum level of the composite signal; the level of the sync-pulse tips.

Sync Pulses: The pulses (generated, shaped, and timed by the sync generator) that keep the horizontal and vertical scanning at the monitors and receivers in step with those at the camera.

T

Tearing: Horizontal disturbances in the picture caused by interference or improper adjustments of controls.

Telephoto Lens: A lens system containing a converging group of lenses followed by a diverging group. It is used to obtain close-ups of distant objects.

Test Pattern: A chart, placed before a camera or contained in a monoscope camera, having geometric lines and patterns used in making studio, transmitter, monitor, and receiver adjustments.

Tighten Shot: A command given by the director when he wants less space around the sides of the object or person. The term "tighten" also may mean to speed up dialogue.

Tilt Shot: A shot in which the camera is tilted up or down for scenic effects.

Tongue In (or *Out*): A command given to the crane operator by the director when he wants the camera boom to move right or left from its present position.

Transcriber Kinescope: A picture tube especially designed for the recording of a TV program on film.

Travel Shot: A shot in which the action is followed by dollying the camera (as distinguished from panning with the camera on a stationary mount such as a fixed tripod).

Triggering: The act of starting an action in a circuit. The circuit then continues to function for a predetermined time under control of its own time constants.

Turret: A plate, mounted on the front of the TV camera, that contains up to four different lenses. The turret is rotatable from the rear of the pickup head.

Two Shot: A camera shot that the director intends to include two persons.

V

Vestigial-Sideband Transmission: A system of transmission in which one of the generated sidebands is partially suppressed at the transmitter.

Video Signal: A signal containing picture information.

Viewfinder: A picture monitor located atop the TV camera so that the operator can see the field covered by the lens system of the pickup head.

Volume Units (VU): A measure of audio signal level as indicated by a VU meter.

VU Meter: A meter, of specified characteristics, used on audio control boards to monitor audio signal level.

W

Wedge: A fan-shaped pattern of black and white lines that converge to a minimum separation. It is contained on TV test patterns for purposes of judging the resolving power (picture resolution) of the system.

White Compression: Compression occuring in the part of the signal that corresponds to white in the picture.

White Peak Limiter: A circuit employed at some stations to avoid over-modulation of the visual transmitter in the white (minimum-carrier) direction.

Womp: A sudden flare of brightness (opposite of *Gobo*).

Woof: A word used in checking audio levels. It also means "OK, goodbye." It can mean "at this instant" in checking time signals, as "It will be 9 o'clock in 5 seconds, 4, 3, 2, 1,—woof!"

Y

"You Got Fire": An expression used to mean, "The equipment is turned on, no more ad libbing."

Z

Zoom: The act of rapidly approaching a televised subject or scene by physical or optical means.

Zoom Lens: A special camera lens that accomplishes zooming action optically.

APPENDIX **C**

Answers to Exercises

CHAPTER 1

A1-1. The pickup tube is an optical-video transducer that converts the reflected light rays passing through the lens to corresponding electrical impulses.

A1-2. Camera driving pulses are converted to horizontal and vertical scanning currents for beam deflection, and to blanking signals to cut off the scanning beam during retrace.

A1-3. If the pickup-tube scanning beam were left on during retrace, a corresponding electrical charge would appear on the target. This charge would be discharged during the next *active* line scan, causing the retrace lines to be visible.

A1-4. So that the picture-tube retrace will occur after the beam is cut off by the blanking pulse.

A1-5. This is a process in which alternate lines of a frame (one frame is a complete picture) are scanned horizontally during one downward trace of the beam, and the remaining lines are scanned horizontally during the next downward trace.

A1-6. The ratio of field frequency to frame frequency is 2:1.

A1-7. This ratio is 15,750/30 = 525, which is equal to the total number of scanned lines of the frame, or complete picture.

A1-8. Since the frequency is 15,750 Hz, the time is:

$$\frac{1}{15,750} = 63.5 \text{ microseconds (approx)}$$

A1-9. Since the frequency is 60 Hz, the time is:

$$\frac{1}{60} = 16,667 \text{ microseconds (approx)}$$

A1-10. Since this frequency is 30 Hz, then the time is:

$$\frac{1}{30} = 0.033333 \text{ sec} = 33,333 \ \mu\text{s}.$$

CHAPTER 2

A2-1. The addition of any two primaries must not produce the third primary.

A2-2. "Mr. X," or the "standard observer," who represents the average of a large number of colorimeter tests using many observers.

A2-3. This wavelength is 5200 angstroms. It corresponds to a green color.

A2-4. The yellow-green region (around 555 nm) appears brightest to the human eye because the plot of average-eye response versus wavelength of stimulus peaks in this region.

A2-5. The primaries are red, green, and blue of specified single wavelengths. Secondaries are the complements of the above colors, namely cyan, magenta, and yellow. Cyan is the combination of green and blue, magenta is the combination of red and blue, and yellow is the combination of red and green.

A2-6. "White" will turn yellow. Since blue is not present, you have only red plus green, or yellow.

A2-7. A fully saturated color contains no white. Therefore, it has a single wavelength, uncontaminated with any other wavelength. As this color becomes mixed with any amount of light with some wavelength other than its own, the result can *still* be a fully saturated color if the added single wavelength is itself pure. For example, a fully saturated red and a fully saturated green would produce a fully saturated yellow. But the next addition of any color will *desaturate* the resultant color, since it now contains some amount of "white" (white simply means a certain amount of all colors). The greater the amounts of additional colors become, the greater the "desaturation" of original color becomes.

A2-8. Since R + G + B must equal unity, or 1:

$$B = 1 - (R + G)$$
$$= 1 - (0.4 + 0.3)$$
$$= 1 - 0.7$$
$$= 0.3$$

A2-9. These colors have the same hue, but vary in degree of saturation. The closer the color is to W, the less saturated it is.

A2-10. You can see that x is about 0.32 and y is about 0.68; the total already is unity, so the value of z is negligible. You can corroborate this result by observing Fig. 2-3.

A2-11. The hue is green. As you approach illuminant C (white), the green becomes less saturated and turns slightly bluish-green before becoming white. This is because illuminant-C white is slightly toward the blue part of the spectrum compared to the "equal-energy white" previously discussed and shown by Fig. 2-6.

A2-12. Hue, saturation, and brightness.

A2-13. Hue would be analogous to frequency, since it is related to the dominant wavelength of the color. Saturation is comparable to the signal-to-noise ratio of the radiated wave, since it defines purity of the signal. Brightness would be analogous to the amplitude of the radiated wave.

A2-14. Red. This is evident from Fig. 2-13 and Table 2-1. Note that the x and y chromaticity coordinates of red (610 nm) total unity.

A2-15. This question illustrates the importance of luminance-to-chrominance proportions in color broadcasting. Brown results from a low-luminance red or pink. Remember that the white axis is representative of all (monochrome) luminances from black through shades of gray to maximum white amplitude.

A2-16. The outputs from all pickup tubes will have equal amplitudes.

A2-17. The area of fully saturated blue will produce the darkest value of gray on the picture tube.

A2-18. Red: Both I and Q are positive.

Green: Both I and Q are negative.

Blue: I is negative; Q is positive.

A2-19. Approximately 1.3 MHz for the I channel, approximately 0.5 MHz for the Q channel. Actual FCC bandwidth limitations are:

I-Channel Bandwidth:

At 1.3 MHz, less than 2 dB down

At 3.6 MHz, at least 20 dB down

Q-Channel Bandwidth:

At 400 kHz, less than 2 dB down

At 500 kHz, less than 6 dB down

At 600 kHz, at least 6 dB down

The Y-channel bandwidth (response within 2 dB) is 4.2 MHz.

A2-20. Instead of sending $R - Y$ and $B - Y$ as such over the subcarrier, each of these signals is mixed with the other so that blue minus brightness contains some red, and red minus brightness contains some blue. This mixing occurs in the I and Q signals. The receiver can then recover the green minus brightness $(G - Y)$ information according to a mathematical relationship between the other two, as shown in Fig. 2-14D.

A2-21. First, you know that the color information should be high in the video band to help avoid interference in the region where most of the monochrome picture energy lies. Energy above about 2 MHz is low in the average picture, although it can be considerable in a pattern of high-frequency content such as a test pattern. Second, the subcarrier frequency must be an odd multiple of one-half the line frequency for proper "interleaving." This technique results in minimum interference with the brightness signal. Third, the beat frequency between the subcarrier fundamental frequency and the sound carrier frequency must be an odd multiple of half the line-scanning rate.

A2-22. This is a condition in which variations in brightness by the chrominance signal cancel each other on the picture-tube face.

A2-23. Yellow and cyan (maximum saturated).

A2-24. Red and blue (maximum saturated).

A2-25. As shown by Fig. 2-18A and explained in the text, the reference level is the video setup level (picture black), not blanking level. Therefore, the 0.11 volt is this value above the reference level, and the excursion in the negative voltage direction is a minus 0.447 plus 0.11, or −0.337.

A2-26. Line OA is a rotating vector. Line AB represents the instantaneous voltage value along the plus-y axis. It is equal to the maximum voltage times the sine of angle ωt (the angular velocity times the time from reference point).

A2-27. Line OA represents the maximum voltage.

A2-28. This quantity really indicates the angle from the starting (reference) point.

A2-29. E_{max} is represented by line OA and may be calculated by finding the square root of the sum of the squares of the other two sides of triangle OAB.

$$\sqrt{(7.07)^2 + (7.07)^2} = \sqrt{100} = 10 \text{ volts}$$

To obtain the angle with the x axis: The tangent of θ = the opposite side divided by the adjacent side = $7.07/7.07 = 1$. Therefore, $\theta = 45°$ (from trigonometric tables).

A2-30. Since $R - Y$ is 0.74 for the I vector and 0.48 for the Q vector:

$$\text{Sum} = \sqrt{(0.74)^2 + (0.48)^2}$$
$$= \sqrt{0.78} = 0.88$$

A2-31. For a red signal (fully saturated), the chroma amplitudes from the I and Q matrix are:

$$R - Y = 0.614$$
$$B - Y = -0.148$$

So:

$$\text{Red} = \sqrt{(0.614)^2 + (-0.148)^2}$$
$$= \sqrt{0.377 + 0.0219}$$
$$= \sqrt{0.399}$$
$$= 0.632$$
$$\text{Tan } \theta = 0.614/-0.148 = -4.15$$

So, from trigonometric tables, $\theta = 76°$ (lagging the $-x$ axis, or burst reference). You can see that this angle is $90° + 14° = 104°$ with respect to the $+x$ axis ($B - Y$ axis). So the complete specification is: $0.632\underline{/104°}$.

A2-32. Since $I = 0.28$ and $Q = 0.525$ for magenta:

$$\text{Magenta} = \sqrt{(0.28)^2 + (0.525)^2}$$
$$= \sqrt{0.353}$$
$$= 0.593$$

Cot θ = adj side/opp side = $0.525/0.28 = 1.88$

$\theta = 28°$ with Q axis

so $\theta = 28° + 33° = 61°$ with $B - Y$ axis (reference, or zero, axis)

The complete specification is $0.593 \underline{/61°}$

A2-33. You have developed the initial unreduced amplitudes of R — Y and B — Y. Now suppose, for example, you consider only the pure, fully saturated blue signal. The R — Y component is −0.11 and the B — Y component is 0.89, so the blue chroma is:

$$B = \sqrt{(-0.11)^2 + (0.89)^2} = \sqrt{0.804} = 0.9$$

Now see Fig. 2-18A. Suppose you add the 0.9 chroma signal to the 0.11 blue luminance level. This would cause more than a 70-percent overshoot in the sync region, which far exceeds the allowable 33-percent overshoot. Yellow would overshoot about the same amount in the whiter-than-white region. (Careful tests have shown that a 33-percent overshoot on maximum saturated colors can be tolerated, but no more than this.) Therefore, the reduction factor is applied.

NOTE: It has been mentioned already that fully saturated colors as produced by a color-bar generator are not encountered in practice. You will find later in this book that even the color-bar signal is reduced to 75-percent saturation so that chroma amplitudes are held to the normal video excursion range. For testing purposes, the color-bar generator is switchable between 100-percent and 75-percent bars.

A2-34. Again review Fig. 2-14 and the associated text. Now you need to employ a reduction factor, which we will designate "a" for R — Y and "b" for B — Y. Put this down as follows:

$$R - Y = a(0.70R - 0.59G - 0.11B)$$
$$B - Y = b(-0.30R - 0.59G + 0.89B)$$

Consider now that red and blue are the two variables of the chroma (subcarrier) signal. The *red* chroma has a magnitude of 0.70a for the R — Y component and a magnitude of −0.30b for the B — Y component. The *blue* chroma has a magnitude of −0.11a for the R — Y component and a magnitude of 0.89b for the B — Y component. Now set up your two matrix equations (amplitudes without regard for polarity) as:

$$R = \sqrt{(0.70a)^2 + (0.30b)^2} = 0.63$$
$$B = \sqrt{(0.11a)^2 + (0.89b)^2} = 0.44$$

(Note that in squaring a number you can ignore negative signs, since the square of any number is a positive number.)

Now the values of a and b can be found by solving the above a and b expressions as a set of two simultaneous equations. This gives the result that a = 0.877 and b = 0.493. If you need this review, see the discussion below.

To get sufficiently exact values of the reduction factors (0.877 and 0.493), you must use more precise values of the constants in the above equations. First solve for a^2 in this expression:

$$\sqrt{(0.701a)^2 + (0.299b)^2} = 0.299 + 0.333$$

Squaring both sides gives:

$$(0.701a)^2 + (0.299b)^2 = 0.632^2$$
$$0.4914a^2 + 0.0894b^2 = 0.3994$$

Transposing $0.0894b^2$ and dividing both sides by 0.4914, you get:

$$a^2 = 0.8127 - 0.1819b^2$$

Now solve for a^2 in the following:

$$\sqrt{(0.115a)^2 + (0.885b)^2} = 0.115 + 0.333$$

Squaring both sides, you get:

$$(0.115a)^2 + (0.885b)^2 = (0.448)^2$$
$$0.0132a^2 + 0.7832b^2 = 0.2007$$

Transposing $0.7832b^2$ and dividing both sides by 0.0132, you get:

$$a^2 = 15.2045 - 59.3333b^2$$

You now have two expressions for a^2, as required. Therefore:

$$0.8127 - 0.1819b^2 = 15.2045 - 59.333b^2$$

Transposing:

$$59.1514b^2 = 14.3918$$

Dividing both sides by 59.1514, you get:

$$b^2 = 0.2433$$
$$b = 0.493 \ (\text{Reduction factor for } B - Y)$$

Substituting the value for b^2 in the first expression for a^2, you get:

$$a^2 = 0.8127 - 0.1819 (0.2433)$$
$$a^2 = 0.7684$$
$$a = 0.877 \ (\text{Reduction factor for } R - Y)$$

A2-35. The output sidebands also will change in phase by $180°$.

A2-36. At any given instant, the resulting signal has an amplitude equal to the instantaneous vector sum of the two output signals. The resultant phase lies within the $90°$ separation of the two output signals, favoring the phase of the signal having the greater amplitude.

A2-37. Saturation.

A2-38. No.

A2-39. No.

A2-40. The basic function of the burst-key generator is to generate a "keying-on" pulse at the proper time to position the reference burst at the required place on the horizontal back porch. It must also "key out" the burst during the 9-line interval of vertical sync. Therefore, the keying pulse is eliminated during this interval.

A2-41. The burst-delay-multivibrator adjustment.

A2-42. The burst-width-multivibrator adjustment.

A2-43. By using a series-resonant circuit as in Fig. 2-35.

A2-44. In the method described in this chapter, by temporarily "unbalancing" the modulators along the plus-I and minus-Q axes, initiated by properly adjusting the *ratio* of burst-key-pulse amplitudes applied to the two paths.

A2-45. Delay compensation is necessary because of the different bandwidths of the three main transmission signals, Y, I, and Q. Since Y is wideband, the Y signal is passed with less delay than I and Q. Since Q has the narrowest bandwidth, it suffers the most delay. Therefore, the Y signal must be delayed to provide time coincidence with the chroma signals at the color picture tube.

A2-46. The bandpass amplifier passes a region between 2.3 and 4.2 MHz above carrier frequency. This region is amplified and fed to the demodulators for chroma information.

A2-47. This signal consists largely of chrominance signals, since sync and blanking (and therefore color burst) are eliminated.

A2-48. The tuner should have a flat response over 6 MHz on each channel. The video i-f section should have good response from 0.75 MHz below to 4.2 MHz above the i-f picture carrier.

A2-49. This attenuation is necessary to minimize the 920-kHz beat between the sound carrier and the color subcarrier.

A2-50. Since this demodulation of color-difference signals is narrow-band, only the chroma frequencies below 500 kHz are used, and the same delay is inherent in each color channel. Therefore, only the Y signal need be delayed to achieve time coincidence.

A2-51.

Primary	Complementary
Red	Cyan
Green	Magenta
Blue	Yellow

A2-52. The chroma amplitude for the complementary color has the same amplitude as for the primary color, but differs in phase by 180°.

A2-53. The colors have the same dominant hue, but vary in degree of saturation. The closer the color is to the spectrum locus, the more saturated (pure color) it will be. The closer the color is to "C," the less saturated it will be.

A2-54. To synchronize the color-receiver chroma oscillator with the subcarrier-frequency oscillator at the sending end.

A2-55. The color subcarrier frequency was chosen to be an odd multiple of one-half the line-scanning frequency; therefore, the color-subcarrier sidebands lie between clusters of monochrome information in the video spectrum.

A2-56. The two main considerations are: (1) the difference frequency between video carrier and subcarrier, and (2) the difference frequency between subcarrier and sound carrier. To achieve compatibility, the 4.5-MHz separation between video and sound carriers must be retained. At the same time, you know it is desirable to have both difference frequencies mentioned above at some odd multiple of one-half the line rate. When you add two odd harmonics of one-half the line rate, you will get an even multiple of the line rate. The video-to-sound carrier separation will be an even multiple of the new line scanning rate. After considerable experimentation, it was found that the 286th harmonic of the desired line-scan rate would be optimum. To put this in terms of the necessary 4.5 MHz video-to-sound separation:

$$f_L = \frac{4.5\,(10^6)}{286} = 15{,}734.264 \text{ Hz}$$

A2-57. You know that the frame must consist of 525 lines, so the new field frequency is:

$$f_F = \frac{f_L}{525} \times 2 = \frac{15{,}734.26}{525} \times 2 = (29.97)(2) = 59.94 \text{ Hz}$$

The horizontal scanning frequency is required to be 2/455 of the subcarrier frequency, so the subcarrier frequency must be:

$$f_S = \frac{455}{2} \times f_L = 3.579545 \text{ MHz}$$

A2-58. Because the master oscillator in the sync generator operates at twice the line-scanning frequency.

A2-59. To gate off the chrominance section of the receiver during a monochrome transmission.

A2-60. (A) Negative. (B) Positive. This is simply the correlation of the Y channel with standard monochrome transmission. The standard composite signal sent to the video transmitter has negative-black polarity. Sync tips at the antenna are at maximum carrier value. When the video swings positive, carrier is reduced, constituting standard "negative modulation" of the transmitter.

A2-61. Two signals with a 90° phase relationship are developed.

A2-62. The synchronous demodulator detects the amplitude variations of a single phase of a multiphase modulated carrier.

A2-63. The basic difference is a 33° phase shift relative to color burst. The cw reference for I leads the one for R — Y by 33°, and the reference for Q leads the one for B — Y by 33°.

A2-64. You know that when the system is normal, white or gray areas appear during the signal interval of zero subcarrier. (I and Q are cancelled out.) If either one of the modulators (I or Q) becomes unbalanced during the active line interval, a white or gray area will become colored because of the presence of the subcarrier during this interval. Also, colored intervals of the scan may have their subcarrier cancelled by the unbalanced carrier and become desaturated or white. The overall result is a white-to-color and color-to-white error that changes with picture content and is quite objectionable.

A2-65. Yes. You know that the vector representing the unwanted carrier is of constant amplitude that will add vectorially to every color vector. A positive unbalance in the I modulator shifts all hues toward orange; a negative unbalance shifts them toward cyan. A positive unbalance in the Q modulator shifts all hues toward yellow-green; a negative unbalance shifts them toward purple. (Be sure you understand the reason for this.)

A2-66. You know that a double-balanced modulator means that both the carrier and modulating video are suppressed, leaving only the modulation sidebands. When video suppression is incomplete (video unbalance), the unwanted video will add to the luminance signal. For example, if the Q modulator exhibits a video unbalance in the positive direction, yellow-green luminance levels are lowered (darkened) while red and blue luminance levels are raised (brightened). A negative video unbalance in the Q modulator will produce the opposite effect. A positive video unbalance in the I modulator will darken reds and blues, and brighten greens.

A2-67. The receiver matrix combines the Y, I, and Q signals so that the proper amplitudes of the red, green, and blue signals result.

A2-68. Maximum for white, maximum for yellow, maximum for cyan, maximum for green, zero for purple, zero for red, and zero for blue.

A2-69. You certainly *should not!* These chapters are to color telecast equipment what Ohm's law is to the entire field of electronics. But you should be better equipped than many to continue your color learning program, provided you have thoroughly assimilated the high points covered.

CHAPTER 3

A3-1. $BW = 100/10 = 10$ MHz.

A3-2. Gaussian curve. 9 dB/octave.

A3-3. One octave represents a 2:1 frequency range. For example, the range from 100 Hz to 200 Hz is one octave.

A3-4. No. The shape of rolloff affects mainly the transient response to leading and trailing edges.

A3-5. $BW = 1/2RT = 1/0.2 = 5$ MHz.

A3-6. $BW = 0.35/RT = 0.35/0.1 = 3.5$ MHz (with gaussian rolloff shape).

A3-7. $BW = 1/2(0.05) = 1/0.1 = 10$ MHz.

A3-8. At 80 TV lines/MHz, $4 \times 80 = 320$ lines.

A3-9. 28.6 percent sync to 71.4 percent video. (Review Section 3-6.)

A3-10. About 1.4 volt (peak-to-peak). (Review Section 3-7.)

CHAPTER 4

A4-1. The red reflecting mirror also reflects some blue (Fig. 4-10). This would destroy the polarization effectiveness of quarter-wavelength plates, so blue is reflected first to remove all blue from the red surface. Where quarter-wavelength polarization plates are not used, the effective color separation in red would be destroyed if the blue mirror were not placed first. (NOTE: Many of the older color cameras did not use this principle of "blue-first" reflection.)

A4-2. The range of angles is 5:1, hence the range of field of view also is 5:1.

A4-3. At 40 mm (1.6 in.):

$$W = \frac{(1.28)(20)}{1.6} = \frac{25.6}{1.6}$$

$$= 16 \text{ ft width of field}$$

$$\text{Height of field} = (0.75)(16) = 12 \text{ ft}$$

Hence field of view = 16 ft by 12 ft.

At 200 mm (8 in):

$$W = \frac{25.6}{8} = 3.2 \text{ ft width of field}$$

Height of field = (0.75)(3.2) = 2.4 ft

Hence field of view = 3.2 ft by 2.4 ft.

Thus, the solution is that the field of view can be varied from 3.2 by 2.4 ft to 16 by 12 ft; this is a 5:1 range. (These are the fields of view at 20 ft from the camera.)

A4-4. At 18 mm (0.71 in):

$$W = \frac{(0.63)(20)}{0.71} = \frac{12.6}{0.71}$$

$$= 17.7 \text{ ft width of field}$$

Height of field = (0.75)(17.7) = 13.3 ft

Hence field of view = 17.7 ft by 13.3 ft.

At 180 mm (7.1 in):

$$W = \frac{12.6}{7.1} = 1.77 \text{ ft width of field}$$

Height of field = (0.75)(1.77) = 1.33 ft

Hence field of view = 1.77 ft by 1.33 ft.

Thus, the solution is that the field of view can be varied from 17.7 by 13.3 ft to 1.77 by 1.33 ft; this is a 10:1 range. (These are the fields of view at 20 ft from the *Plumbicon* camera.)

A4-5. Assume the picture height in the *Plumbicon* is 0.5 inch. Then: 1 scanned line = 0.5/490 = 0.001 inch (approx). This is one-half that allowable for the I.O. tube.

A4-6. To compare properly relative depth of field, you must first consider the lens that will give about the same field of view for each camera. You have already computed the I.O. depth of field for a 3½-inch lens at *f*/8 focused at 10 ft (Section 4-1). The equivalent lens for a *Plumbicon* has one-half this focal length, or 1.75 inch. For this 1.75-inch lens at *f*/8 and focused at 10 ft:

$$\text{Hyperfocal Distance} = \frac{(1.75)^2 (1000)}{(8)(12)} = \frac{3062}{96}$$

$$= 32 \text{ ft (approx)}$$

$$\text{Nearest Limit} \quad = \frac{(32)(10)}{32 + 10} = \frac{320}{42}$$

$$= 7.6 \text{ ft (approx)}$$

$$\text{Farthest Limit} \quad = \frac{320}{22} = 14.5 \text{ ft (approx)}$$

Therefore, the depth of field for the *Plumbicon* under the specified conditions is from 7.6 ft to 14.5 ft, compared to the I.O. depth of field of 8.7 to 11.8 ft.

A4-7. "Effective sensitivity" is assumed to be equal. That is, the *Plumbicon* must have the same sensitivity as the I.O. for a given signal/noise ratio. In practice, the *Plumbicon* camera actually is about one $f/$ stop more sensitive than a three-tube I.O. color camera. Theoretically, this would suggest that the *Plumbicon* would require one-half as much scene lighting for a given depth of field, but this is not true. The I.O. has essentially flat-channel noise, whereas the *Plumbicon* has peaked-channel noise characteristics. Thus, in practice, about the same amount of light is required for either type of camera.

A4-8. To equalize pickup-tube sensitivities so that proper tracking will occur from low lights to high lights.

A4-9. A range converter normally is installed on the front panel of the camera. It reduces the focal lengths of the normal variable-focal-length lens by a factor of about one-half.

A4-10. The two "normal" ranges are selected by a switch on the rear of the camera. When this selector is switched from the 4-20 inch position to the 8-40 inch position, the maximum iris opening automatically becomes $f/8$ instead of $f/4$. Since this opening is a function of the basic variable-focal-length lens assembly, this condition holds true whether the range converter is in position or not in position.

A4-11. The I.O. is a photoemissive device. The vidicon and *Plumbicon* are photoconductive devices.

A4-12. To "stiffen" the electrotatic field between the decelerator electrode and the target for better "beam landing." It helps to eliminate "blackout" of the entire picture because of excessive high light, and it minimizes artificial outline of objects.

A4-13. Very important for best performance.

A4-14. The main problem is "target lag," which results in smearing of objects during rapid motion of the object or the camera. When the vidicon is used with film projectors, ample light reaches the target area to prevent lag.

A4-15. The yoke contains the horizontal- and vertical-deflection coils that control the scanning beam in the pickup tube. The tube is mounted directly in the center of the yoke.

A4-16. By passing direct current through the deflection coils to provide a fixed reference for the deflection currents.

A4-17. In the blue region.

A4-18. It helps by increasing the response toward the red end of the spectrum.

A4-19. In parallel.

A4-20. In series.

A4-21. Skew is a condition in which the raster of one channel has a different shape than the reference raster (see Fig. 4-38). It results from slight differences in deflection yokes. It is corrected by introducing a small amount of vertical sawtooth (of the proper amplitude and polarity) into the horizontal sawtooth.

A4-22. 60-percent reflectance.

A4-23. They adjust the ratio of resistance to inductance in the horizontal-deflection circuits to achieve similar sweep waveforms for all channels.

A4-24. Yes, including an extra equalizer for the viewfinder feed.

A4-25. It is used only in the luminance channel in systems fixed for operation with one luminance channel and three chroma channels. However, some film systems are capable of having voltage removed from the luminance tube for operation as a three-channel system in which luminance is derived from the three channels by matrixing. In the latter case, aperture correction is available for all channels.

CHAPTER 5

A5-1. A slide-projector lamp illuminates the pickup tube 100 percent of the frame time. The motion-picture projector illuminates the tube only about 60 percent of the frame time to allow for film pulldown and shutter operation.

A5-2. To get (as closely as possible) the same effective $f/$ number (optical-system speed)from all sources, to avoid the necessity of pickup-tube target-voltage readjustment for each source. Neutral-density light-control systems (remotely operated) are normally used to compensate for any required additional control of illumination levels.

A5-3. A virtual image is one that appears to be produced at an imaginary focal point. Virtual images appear to be behind reflecting surfaces, as, for example, a mirror; or they result from viewing an object that is enlarged directly through a lens.

   A real image is produced when the light rays from an object converge on a surface after passing through a lens system. Only real images can be resolved on a screen.

A5-4. Only within confined limits. The lower limit is the point at which the color temperature starts toward the red end of the spectrum.

A5-5. Neither. It must be focused in the center of the field lens. (See Section 5-4 for alignment procedure.)

A5-6. To get the *most shallow* depth of field for critical focus. Then when you go to practical operations, you can stop the projector lens down one or two steps for a greater depth of field to provide a "safety margin."

A5-7. Stop the projection lens down to maximum $f/$ stop (small dot of light), and do the same for the pickup-tube lens. Sometimes, an "alignment tool" is used in place of the pickup-tube lens; this is a circular disc with a small center dot or hole. The dots of light from each projector of the multiplexed system must fall onto this center dot.

A5-8. To assure that each channel has the same gain (white level) and the same setup (black level) that corresponds to a maximum current output of each pickup tube. With this assurance, you are ready to adjust pickup-tube operating electrical parameters for equal output levels on a gray scale. This results in carrier cancellation for a gray scale, which is the criterion for "color balance." (This is detailed in Chapter 10.)

A5-9.   Derived luminance (covered in Chapter 2) consists of 0.3 red, 0.59 green, and 0.11 blue for fully saturated primaries. It is obtained by matrixing at the encoder. True luminance means the brightness signal from the separate luminance-pickup channel of a four-camera color chain. When properly set up, either method fully meets NTSC and FCC color standards.

A5-10.   No. All cameras (film and live) must be properly color balanced as revealed by a common color monitor. For film, you have the additional factor of varying film-base color temperature and color-processing variations. More details on this subject are given in Chapter 10.

A5-11.   Yes. (Review Section 5-5.)

A5-12.   The main advantage is in plant-distribution or network lines to distinguish between reference-white loss and chroma (high-frequency) loss. The reference white (100-percent) pulse always exists in the split-field pattern. Also, a comparative indicator is present in the black region where the special I and Q test signals should have the same peak-to-peak amplitude as the burst and sync pulse.

CHAPTER 6

A6-1.   The driving pulse triggers the "retrace" (blanking) interval. The interval between adjacent driving pulses then forms the "trace" (active line scan) interval.

A6-2.   Yes.

A6-3.   No. The single blanking output is composite (both horizontal- and vertical-blanking pulses). See waveforms D and H of Fig. 6-1.

A6-4.   No. The single sync output is composite (both horizontal- and vertical-sync pulses). See waveforms E, N, and P of Fig. 6-1.

A6-5.   The value indicated on detail C-C of Fig. 6-10B is the *minimum allowable* front porch. Now study Note 4 of Fig. 6-10A. Put down the stated values of x, y, and z from Fig. 6-10B:

$$x = 0.02H = \text{Minimum front porch}$$
$$y = 0.145H = \text{Minimum without front porch}$$
$$z = 0.180H = \text{Maximum with front porch}$$
$$x + y = 0.02H + 0.145H$$
$$= 0.165H = \text{Minimum with front porch}$$

The maximum-minimum range is $0.180H - 0.165H = 0.015H$. If 0.02H is minimum, then $0.02H + 0.015H = 0.035H$ maximum front porch, *not* accounting for rise and fall times of blanking. But allowance for these times must be made. The maximum specified rise and fall time is 0.004H. So $2 \times 0.004H = 0.008H$ can be taken up by this time. Then $0.035H - 0.008H = 0.027H$ is the maximum front-porch width. To convert to microseconds:

$$\text{Minimum} = 0.02H = (0.02)(63.5) = 1.27 \ \mu s$$
$$\text{Maximum} = 0.027H = (0.027)(63.5) = 1.71 \ \mu s$$
$$\text{Standard} = 0.025H = (0.025)(63.5) = 1.59 \ \mu s$$

The Tektronix Type 524 oscilloscope provides a 0.025H marker for the purpose of setting the front-porch width.

A6-6. The color-sync burst normally is maintained at the same amplitude (peak-to-peak) as the sync pulse. If the sync is adjusted to be 0.3 volt above blanking level, then the burst is 0.3 volt peak-to-peak. The FCC requirement is a peak-to-peak burst amplitude between 0.9 and 1.1 times the sync amplitude. (Refer to Detail C-C of Fig. 6-10B and Fig. 6-11.)

A6-7. The color-sync burst should have 8 to 10 complete cycles.

A6-8. No. It is eliminated during the 9H interval of equalizing and vertical-sync pulses.

A6-9. All monochrome chains must have blanking, sync, and horizontal drive delayed an amount equal to the total delay of the color system up to the common switching point.

A6-10. No. Review Section 6-6.

CHAPTER 7

A7-1. 75 ohms.

A7-2. A video or pulse-distribution cable that goes into and out of a number of amplifiers (bridging), and is then terminated in 75 ohms at the final amplifier, is said to be "looped through."

A7-3. Relays (usually of the magnetic type), or solid-state "crosspoints." The crosspoint is turned on or off by control pulses from the switcher panel; the switching action usually is timed to occur within the vertical-blanking interval.

A7-4. Noncomposite signals do not have inserted sync. Video is switched, and sync is added after the switching point. Composite signals have the sync inserted in the video from each source prior to switching.

A7-5. With coaxial cables of the proper lengths to provide the desired delays.

A7-6. An amplifier separates a selected saturated color from all other colors to form a "key-out" signal. Another video source is then "keyed in" during the "key-out" interval.

A7-7. Unlike the fader handles on a switching system, the control handles on a special-effects panel permit control of the size and shape of the inserted picture area.

A7-8. When the "AF" (audio-follow) button is depressed, pulses from the video-switching system actuate the desired audio-signal crosspoint (or relay) so that audio automatically appears on the program bus. This eliminates the need for simultaneous operation by an audio operator. This feature normally is used only for quick breaks, as between network and film, local announcer, VTR, etc. It normally is not used in studio productions.

A7-9. A stabilizing amplifier corrects for signal faults such as hum, noise modulation in the sync region, and incorrect video-to-sync amplitude ratios. It also provides stripped sync to feed to the local sync generator for genlocking purposes.

A processing amplifier provides a number of additional features, such as a completely independent sync generator in case the input signal fails, automatic video and color-burst gain controls, etc. (Review Section 7-6.)

A7-10. The major difference is the much greater number of signal sources required in TV than for normal radio broadcasting.

CHAPTER 8

A8-1.   Kinescope recording, electron-beam recording (EBR), video tape recording (VTR), and magnetic disc recording.

A8-2.   The quadruplex (four-head) system, the helical-scan system, and the magnetic-disc system.

A8-3.   The single system and the double system.

A8-4.   Negative video polarity.

A8-5.   Film transport, electron optical system, and electronics.

A8-6.   (A) 5 MHz, (B) 4.28 MHz, (C) 6.8 MHz.

A8-7.   (A) 5.79 MHz, (B) 5.5 MHz, (C) 6.5 MHz.

A8-8.   (A) 7.9 MHz, (B) 7.06 MHz, (C) 10 MHz.

A8-9.   No. The heads receive a frequency-modulated carrier.

A8-10. A given field is scanned more than once, with an electronically switched half-line delay on alternate scans. If a field is repeated three times, the playback appears to have one-third normal speed; if the field is repeated six times, the playback appears to have one-sixth normal speed; etc. The same field is repeated over and over for stop action.

CHAPTER 9

A9-1.   No. The relationship depends a great deal on lamp and luminaire construction and efficiency.

A9-2.   3000 to 3200 K.

A9-3.   To illuminate cycloramas.

A9-4.   The holding of a given K characteristic with age.

A9-5.   With a foot-candle meter held on the set and facing the camera lens.

A9-6.   With a foot-lambert meter pointed toward the set area or detail in question.

A9-7.   See Fig. 9-10.

A9-8.   (A) About 150 foot-candles. (B) About 350 foot-candles.

A9-9.   Yes, very effectively.

A9-10. (A) ½ to 1 times base light. (B) A maximum of 1½ times base light.

CHAPTER 10

A10-1. No more than 20 to 1 for ideal control. The ratio can be measured with a spot brightness meter. Also, the "whitest" material should give the same amplitude on the CRO as reference white on the chip chart gives under the same light. The "blackest" material should match step 9 on the chart (this is 3-percent reflectance). The ratio of 60 percent (white) to 3 percent (black) is 20 to 1.

A10-2. 15 IEEE units.

A10-3. Optical and electrical focusing, amount of aperture correction used (determined by signal/noise ratio), scene lighting and contrast ratio, luminance-to-chrominance ratio, and camera registration. (You should be able to see horizontal resolution of 600 lines at the center of a test pattern.)

A10-4. On three-channel systems not employing the "cancellation" technique, adjust final registration for 500 lines *minimum* horizontal resolution at the center of the test pattern. The "cancellation" technique reverses the polarity of all video signals except green. Thus, when you combine a negative picture from any one channel with the positive picture from the green channel, adjust the channel being compared to green so that *complete* cancellation takes place at least through the large center circle of the registration chart.

A10-5. The appearance of its monochrome picture; does it have "snappy" contrast?

A10-6. Yes, it is. On the small area covered by the tight shot, the color temperature of the light could be different, especially with old incandescent lamps. Also remember the effect of backgrounds on skin tones; a tight shot can eliminate this effect because the background is practically eliminated. There is a way you can rebalance the camera on such shots (if necessary) during operations, if there are areas of reference white and black in the scene. Have the camera monitor on NAM monitoring, and carefully adjust for NAM balance on black and white (black and white balance controls). This takes some experience and is not recommended unless the balance is noticeably improper.

A10-7. Degree of saturation (purity) of the background and size of the background relative to the face. For a facial close-up, the background has minimum effect; if the background is comparatively large (skin area small relative to background), its effect is maximum.

A10-8. ±100 K, or ±10 volts.

A10-9. Yes. In fact, more dimmers sometimes are necessary for color than for monochrome. This is because areas where skin tones exist (where performers face the camera) should not be dimmed. Lighting for other areas can be manipulated by means of dimmers for special effects and mood scenes.

A10-10. (A) One-half to one times base light.

(B) One to one and one-half times base light.

A10-11. 0.286 volt peak to peak, or 40 IEEE units when the scope is calibrated for 1 volt = 140 IEEE units. (Same peak-to-peak amplitude as sync.)

A10-12. No. It is deleted during the 9H vertical interval of six leading equalizing pulses, six vertical-sync pulses, and six trailing equalizing pulses. It starts after the first horizontal-sync pulse following this interval in each field.

A10-13. Eight to ten complete cycles.

A10-14. Yes. I and Q are made up of portions of the color-difference signals. The values are so chosen that the resultant phase for any hue is constant whether it is formed from I and Q or the color-difference signals. The difference is in the bandwidth of the I signal.

A10-15. The plus I axis. However, this does not mean that the Q amplitude is not important to good flesh tones, since the skin reflectance in this region must balance out properly.

A10-16. No, not in the vector mode of operation. But it does show the effect of luminance on the amplitudes and phases of the color signals, which can be checked by turning Y (luminance) off and on to note any change in phases or vector amplitudes.

A10-17. Not if you switch directly from one source to another. The receiver simply uses the burst phase that accompanies the picture for synchronous demodulation. But if you *mix* two separate sources, they *must* have the same system phase.

A10-18. The result is the "funny-paper effect," or color misregistration that is not the result of camera misregistration or picture-tube misconvergence.

CHAPTER 11

A11-1. A mobile TV camera-recorder combination (as in Figs. 11-1 and 11-2, for example).

A11-2. Yes, with special packaging of the rack equipment and control panel.

A11-3. Arrangements for adequate power. Arrangements with local telephone company for any order wires necessary. Arrangements for microwave points (when involved). Construction of any operating structures required; to be cleared with local authorities as to permits and inspection. Arrangements for any extra lighting required.

A11-4. No. The "standard" *Plumbicon* has a useful scanned-area diagonal of 0.8 inch. The 1-inch *Plumbicon* (as does the 1-inch vidicon) has a diagonal of 0.62 inch. Even smaller *Plumbicons* are under development for portable camera applications.

A11-5. (A) 17 inches. (B) 8½ inches. (C) 6 inches.

A11-6. (A) 1 inch. (B) 1 inch.

A11-7. No. The useful scanned area is the same for both tubes.

A11-8. Yes. For example, see Fig. 11-2.

CHAPTER 12

A12-1. 2 GHz (1990-2110 MHz), 7 GHz (6875-7125 MHz), and 13 GHz (12,700-13,250 MHz).

A12-2. No.

A12-3. 0.6 of the first-Fresnel-zone radius.

A12-4. The volume containing the most concentrated energy. At its outer limit, an additional half wavelength in the path length causes phase cancellation.

A12-5. Because of strong reflections from a smooth surface.

A12-6. Fm.

A12-7. Waveguide.

A12-8. Whether the direction of the "short side" of the waveguide is oriented in a horizontal or vertical plane.

CHAPTER 13

A13-1. 20 dBk.

A13-2. (A) 100 watts; (B) 1,000,000 watts (1 megawatt).

A13-3. 40 dBu.

A13-4. The adequacy of the picture depends on the channel, the receiver, the receiving antenna, and the noise level at the receiver location.

A13-5. 10 to 20 percent.

A13-6. Erp, or effective radiated power, is the product of the transmitter output power (less transmission-line losses) times the antenna power gain. (See Section 13-3.)

A13-7. The effective height, from the ground to the center of the radiating elements, as corrected for the terrain and other obstructions on the selected path (Section 13-7).

A13-8. By physical placement of antenna panels relative to other panels, and by choice of the relative power levels and phases of the rf signals fed to individual panels.

A13-9. 10 percent carrier relative to sync level (100 percent).

A13-10. At least 42 dB.

CHAPTER 14

A14-1. $I_p = \dfrac{P_o}{E_p F} = \dfrac{5000}{(5000)(0.68)} = \dfrac{5000}{3400} = 1.47$ amperes (approx).

A14-2. ± 25 kHz.

A14-3. 1-kHz deviation above or below the assigned frequency. (Always check current FCC regulations.)

A14-4. The aural center frequency must be 4.5 MHz ± 1 kHz above the visual carrier frequency. (Always check current FCC regulations.)

A14-5. (A) 25 percent to 75 percent. (B) 28.6 percent to 71.4 percent. (See Fig. 14-5.)

A14-6. (1) Overmodulation in the white signal direction (carrier cutoff), and (2) incidental phase modulation.

A14-7. By use of a sweep generator and oscilloscope.

A14-8. Obtaining the proper bandwidth at the proper peak carrier power output.

A14-9. A *properly calibrated* reflectometer.

A14-10. 0.75 MHz.

Index

648